Theory of Adiabatic Potential and Atomic Properties of Simple Metals

Theory of Adiabatic Potential and Atomic Properties of Simple Metals

V. G. Baryakhtar
Institute of Metal Physics, The Ukraine

E. V. Zarochentsev
Donetsk Physico-Technical Institute, The Ukraine

and

E. P. Troitskaya
Donetsk Physico-Technical Institute, The Ukraine

GORDON AND BREACH SCIENCE PUBLISHERS
Australia • Canada • China • France • Germany • India • Japan
Luxembourg • Malaysia • The Netherlands • Russia • Singapore
Switzerland

Amsteldijk 166
1st Floor
1079 LH Amsterdam
The Netherlands

British Library Cataloguing in Publication Data

A catalogue record for this book is available from the British Library

ISBN: 90-5699-088-8

Contents

Acknowledgements

We would like to thank Professor I.V. Abarenkov for fruitful discussions and notations. We would also like to acknowledge the help from the collaborators of the Condensed State Physics Laboratory in the preparation of the manuscript.

This work was supported, in part, by a Soros Foundation Grant awarded by the American Physical Society.

Introduction

One of the fundamental problems of solid state physics is the evaluation of properties of particular materials from the "first principles" theory. It is essential to have a theory which allows regular calculation "up to the number" of the wide range of different crystal properties in agreement with experiments within the framework of a single ideological scheme.

Nowadays, this problem can be solved with the use of the pseudopotential method which, from the very beginning, was aimed at solving the practical problems. In terms of scattering properties of the atomic core, which remains almost unchanged in any situation, the pseudopotential could be introduced through exclusion of all contributions due to the core states from the total potential. Within the one-electron approximation, this is an exact procedure, and the resulting pseudopotential is considerably weaker than the initial crystal potential. Within the pseudopotential approach, the valence electron density distribution, which defines the crystal energy, is the result of multiple scattering of originally free electrons on Bragg's planes. Hence, it is clear that in the pseudopotential theory there are no distinctions, in principle, between metals, semiconductors, and dielectrics. The distinction is only a quantitative one and is due to the fact that the pseudopotential in metals, contrary to dielectrics, is so weak that it possesses no bound states and, therefore, in the theory of metals, a small parameter appears: the ratio of the pseudopotential form-factor at the reciprocal lattice site to the Fermi energy. In alloys, the theory of metals is far advanced since it is possible to describe not only the band structure with a secular equation of a small order, but also the binding energy in the lowest approximation in pseudopotential. In dielectrics, the pseudopotential can have bound states (which are filled with valence electrons). However, here the theory of lattice dynamics and the structure is also similar to that for metals. However, the density functional theory is quite applicable here which exploits the theory of metal binding but with inhomogeneous electron density.

This review presents the modern notions of the microscopic theory of atomic properties of metals. The theory is based on concepts of pseudopotential and the interacting electrons. The theory developed is used for a quantitative description of the whole complex of atomic properties of perfect simple metals: static, dynamic and thermodynamic, in wide ranges of temperatures and pressure, including melting and polymorphous transitions.

The review is written according to the following principle: each chapter should be as independent as possible. References are also given in each chapter independently. We tried to maintain the consonance of theoretical description of metal properties with computer experiment. It means that the review contains well-advanced theory with many new developments, as well as rather simple and well-known theoretical approaches necessary for computer calculation.

The book is organized as follows. In Chapter 1, the background of the theory of atomic properties of crystals is given systematically without emphasis on metals. Many results presented are not proven (only sources are quoted) though some of them, mainly new results, are discussed in full detail in order to illustrate some peculiarities of the theory. The chapter is concluded with a detailed presentation of the adiabatic perturbation theory. The material covered in these sections (3 and 4) is almost unknown to the general reader.

The theory of the pseudopotential form-factor in metals, together with the linear screening theory, is given in Chapter 2. Here, the basic principles of the construction of the adiabatic potential models of metals are presented, and the relationship of elements of these models with the density functional theory is discussed.

Chapters 3 and 4 constitute the calculation part of the review. We shall not enumerate all the results obtained here, because there are too many of them. Let us only note that in discussion of our calculations, we will pay special attention to properties of metals under pressure and to temperature anharmonic effects. High precision, which is available with modern techniques, is demonstrated with results of phase transition calculations where the difference between energies of compared phases is 10^{-2}–10^{-4} of the metal binding energy. A review of ionic core polarization in metals is presented in Section 7 of Chapter 3 in a systematic way. The main part of this theory was published in 1989–90.

We have not attempted to present a critical analysis of all calculations of metal properties, but tried to show the progress which has been made in the field on the basis of the results of two research groups: led by Professor V. G. Vaks from the Moscow Atomic Energy Institute, and ours as well. We hope that the book will be useful for specialists in the theory and computer simulation of properties of solids.

1 Free Energy of Ideal Crystals

This chapter is an introduction to the following text. Before we discuss concrete models of any crystal and the results of the calculations, it is necessary to introduce some basic concepts and to describe well-known models of the properties of elementary excitation in some solid structures. In the present review, we are predominantly interested in elastic properties and the lattice dynamics of non-transition (simple) metals. This chapter introduces a representation of the simplest models of metals and their lattice properties. As we can see below, in many cases these simplest (and as commonly phenomenological) models are insufficient to reach a quantitative agreement of the theory with experiment, which is our final aim. In this case we introduce more complicated models, but this is made on the basis of the present chapter.

Although all properties under investigation in the present review relate to finite temperatures and may be expressed in terms of derivatives of the free energy, in many cases it is more convenient to begin the analysis with calculations at zero temperature and use the internal energy.

In 1928, F. Bloch was the first to solve the problem of an electron in a crystal. The theorems proven by Bloch are the basis of all works in the quantum theory of solids. Now it is absolutely obvious that it is impossible to understand the properties of any metal, semiconductor or dielectric under some external conditions without detailed knowledge of their electronic structure. The theory of electron band structure (and general energy spectrum) is so significantly increased, that it is impossible to describe it in any book of realistic measure. For this reason, we give no detailed description of it, but provide a short description of basic ideas with references to textbooks (see, e.g., [1–5]) and original publications. However, we must first provide some remarks about crystalline potential acting on electrons in solids with some type of chemical binding.

Besides the fact that the potential must satisfy the periodicity condition

$$V(r + \mathbf{R}_l) = V(r) \tag{1.1}$$

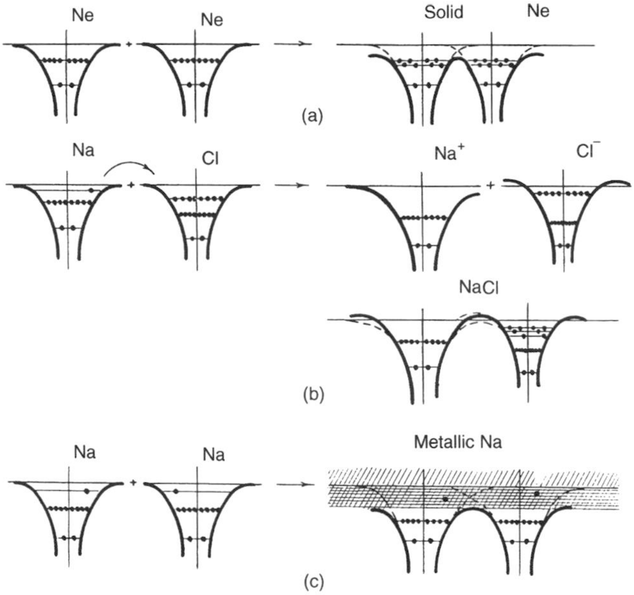

FIGURE 1.1 Formation of the crystal potential from potentials of neutral atoms: (a) rare gas, (b) ion crystal, (c) metal.

(where $\mathbf{R}_l$ is a vector of the lattice), the size and the shape of the atoms building the solid and the electron density distribution provide some restrictions for the form of $V(\mathbf{r})$.

Let us consider the crystal lattice as the result of the approach of initially non-interacting atoms so that finally they occupy positions corresponding to the crystal under investigation. The wave functions of outer electrons can either be a small perturbation from the atomic ones or be totally reconstructed. Interaction of electrons absolutely excludes the possibility of the exact equality of the electron potential of a free atom and the potential used in the band problem. But the consequences of this interaction are determined by the type of solid. For rare gases, the outer shells may be slightly deformed, but there is no principle change — each electron is tightly bound by its own atom. As a consequence of very small splitting of the atomic levels, the bands in neon are narrow with wide gaps between them. The division into valent and core electrons in the case of a neon crystal is rather conditional. The one-electron potential, for example for upper p-states, is very large and provides bound states (see Figure 1.1.a). On the other hand, the conducting bands are similar to bands of delocalized electrons which is confirmed by calculations for neon in [6].

Similar situations take place in ion crystals because valent electrons transfer from metal atoms to non-metal ones. It relates to the array of ions with closed shells (Figure 1.1.b). Ions are nearly spherical and can bind all their own electrons. The electron in the conducting zone feels the potential of charged ions. These ions are so displaced, that the actions of their charges are practically cancelled. As a result, we simply have distortion of the potential profile in the interatomic space without appreciably lowering the barrier between ionic states. It demonstrates the possibility of obtaining more narrow conducting bands in ionic compounds.

In metals and semiconductors it is possible to distinguish valent electrons from core electrons: valent electrons are delocalized, whereas at normal conditions, the wave functions of core electrons are only slightly changed relating to their state in a free atom. It means that it is possible to solve the Bloch problem — calculation of the wave functions and energies for all valence electrons. Commonly, the approximation of the same potential for all valence electrons is used (we are not considering rare earth and transition metals). Since the resulting Bloch functions of valence electrons must be orthogonal to wave functions of core electrons in the core region, the potential describing valence electrons must be very weak to eliminate the bound states. Naturally, this approach forced us to introduce the concept of the (weak) pseudopotential (see, e.g., [7–10]) which is analyzed in detail in Chapter 2.

In Chapter 3 we will use the expressions for the electronic contribution to the specific heat and the free energy to provide a comparison of the theory with experiment. Therefore, here we note that this specific heat is a linear function of the temperature [1]

$$C_e = \lambda_e T \tag{1.2}$$

and at low temperatures the electron contribution to the specific heat is much higher than the lattice contribution ($\sim T^3$). Measurements of λ_e provide important information about the density of states at the Fermi level $\nu(E_F)$, since $\lambda_e \sim \nu(E_F)$ [11].

The calculation of the temperature contribution F_e^* to the free energy is rather similar to the calculation of C_e. Omitting the details of calculation, the final result for the energy (per unit volume) is

$$F_e^* = \frac{\lambda_e T^2}{2} \tag{1.3}$$

The expressions for the phonon contribution to F will be presented in section 1, and the analysis of the electron-phonon interaction is placed in section 3.

1.1 Phonon Spectra. Lattice Anharmonicity

1.1.1 The internal and free energies

In order to investigate the lattice properties of the crystal, it is necessary to have a constructive analytical expression for the thermodynamic potential. Here we will not prove the relevant equations, but suppose that they exist and allow us to investigate the consequences following from their rather general expressions.

It is necessary to note that here and in section 4, by introducing the thermodynamic potentials of electrons and phonons and their interaction, we use implicitly the adiabatic principle which makes this division possible (see section 3).

Let us consider the rather general expression of the energy of the crystal (for simplicity with one atom per unit cell), which does not include explicitly any non-pair interaction [12]

$$E = v(\Omega) + \frac{1}{2N}\sum_{l,m} \varphi(R_{lm};\Omega). \tag{1.4}$$

Here $v(\Omega)$ and the pair potential $\varphi(R_{lm};\Omega)$ are some functions of the cell's volume Ω, $R_{lm} = |\mathbf{R}_l - \mathbf{R}_m|$ is the distance between l-th and m-th atoms, n is the number of atoms (cells) in the crystal.

Often it is more simple (in application for metals) to operate in momentum space. Using

$$\varphi(R;\Omega) = \frac{2}{N}\sum_{\mathbf{k}} \varphi(\mathbf{k};\Omega)\exp(i\mathbf{kR}), \tag{1.5}$$

where

$$\varphi(\mathbf{k};\Omega) = \frac{1}{2\Omega}\int d^3R\varphi(R;\Omega)\exp(\mathbf{kR}) \equiv \frac{\chi(\mathbf{k};\Omega)}{\Omega}, \tag{1.6}$$

we obtain the internal energy (per atom) at $T = 0$:

$$E = w(\Omega) + \sum_{\mathbf{k}} \varphi(\mathbf{k};\Omega)\, |S(\mathbf{k})|^2, \tag{1.7}$$

where

$$S(\mathbf{k}) = \frac{1}{N}\sum_{l} \exp(i\mathbf{kR}_l); \quad w(\Omega) = v(\Omega) - \frac{1}{N}\sum_{\mathbf{k}} \varphi(\mathbf{k};\Omega). \tag{1.8}$$

In this notation the dependence of E on atomic coordinates is concentrated in the structure factor $S(\mathbf{k})$. The dependence of E on the distortion tensor u_{il} is determined by the dependence of $w(\Omega)$ and $\varphi(\mathbf{k};\Omega)$ on u_{il}. The structure factor does not depend on u_{il} because both the basic vectors of the crystal lattice

$$a'_\alpha = (\delta_{\alpha\beta} + u_{\alpha\beta})a_\beta \tag{1.9}$$

and the reciprocal lattice

$$b'_\alpha(\delta_{\alpha\beta} + u_{\alpha\beta}) = b_\beta \tag{1.10}$$

change by the deformation so that the scalar product $\mathbf{kR}$ is an invariant. The change of the summation region over $\mathbf{k}$ in equation (1.7) does not take place at any deformation because

$$\sum_{\mathbf{k}}(\ldots) = \frac{V}{a^3}\int(\ldots)\, d^3\boldsymbol{\kappa},$$

where $\boldsymbol{\kappa}$ is a dimensionless vector and a is the lattice constant. It is obvious that summation over $\mathbf{k}$ factually means summation over all non-equivalent states in the reciprocal lattice.

Note the dual dependence of the Fourier ampiltude of the pair potential $\varphi(\mathbf{k};\Omega)$ on Ω. First through the dependence of $\chi(\mathbf{k};\Omega)$ in equation (1.6) on Ω which is a characteristic feature of the model, and also by the factor $1/\Omega$ which normalizes the Fourier transform.

Generally, the free energy (per atom) of the crystal in equilibrium is

$$F(\Omega,T) = E(\Omega) + E^{zp}(\Omega) + F^{*}(\Omega,T) + F_{an}, \tag{1.11}$$

where

$$E(\Omega) = w(\Omega) + \frac{1}{\Omega}\sum_{\tau} \chi(\boldsymbol{\tau};\Omega), \tag{1.12}$$

$$E^{zp}(\Omega) = \frac{1}{2N}\sum_{\lambda,\mathbf{q}} \hbar\omega_{\lambda\mathbf{q}}, \tag{1.13}$$

$$F^{*}(\Omega,T) = \frac{k_B T}{N}\sum_{\lambda\mathbf{q}} \ln\left[1 - \exp\frac{\hbar\omega_{\lambda\mathbf{q}}}{k_B T}\right]. \tag{1.14}$$

The cohesive energy of the crystal follows from (1.7) with the help of (1.8) at $S(\mathbf{k}) = \sum_{\tau}\delta_{\mathbf{k},\boldsymbol{\tau}}$, E^{zp} is the energy of zero oscillations, F^{*} is the energy of phonons with (the phonon frequency $\omega_{\lambda\mathbf{q}}$ is a function of the wave vector $\mathbf{q}$ and the polarization λ). The last terms E^{zp} and F^{*} come out by taking into account the thermal distortion of atoms $\mathbf{u}_l$ from the equilibrium positions $\mathbf{R}_l^0$, i.e., by taking into account that $\mathbf{R}_l - \mathbf{R}_l^0 + \mathbf{u}_l$ (see Subsection 2). The first three terms (1.12)–(1.14) in (1.11) are enough to calculate the first order (in the anharmonicity parameter $\varepsilon = \langle x^2\rangle/a^2$, where $\langle x^2\rangle$ is the mean square of atomic displacement) corrections to the elastic constants, because they come out by at least double derivation of $F(\Omega,T)$ over coordinates. To obtain the anharmonic corrections to the specific heat, the shift of the phonon frequency and attenuation, we take into account the lowest (in the anharmonicity parameter ε) corrections F_3 and F_4 to the anharmonic contribution in the free energy F_{an}

$$F_{an}(\Omega,T) = F_3(\Omega,T) + F_4(\Omega,T). \tag{1.15}$$

These terms relate to process participation of the minimal number of phonons: three-phonon processes (with the vertex $V_{\lambda\mu\nu}^{\mathbf{k},\mathbf{q},\mathbf{k}+\mathbf{q}}$) for the term F_3

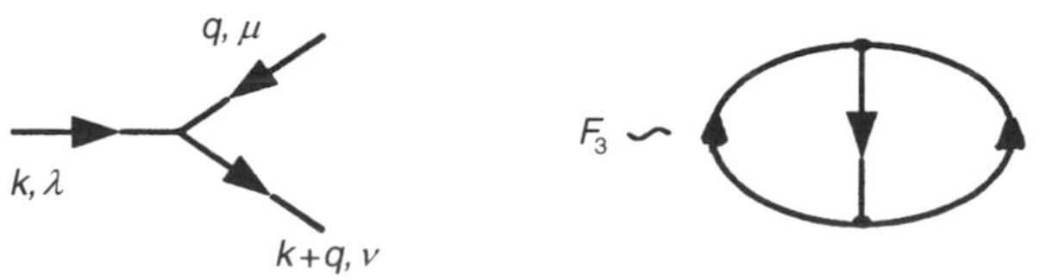

(1.16)

and four-phonon processes (with the vertex $V_{\lambda\mu\nu\eta}^{\mathbf{k},\mathbf{p},\mathbf{q},\mathbf{k}+\mathbf{p}-\mathbf{q}}$) for F_4:

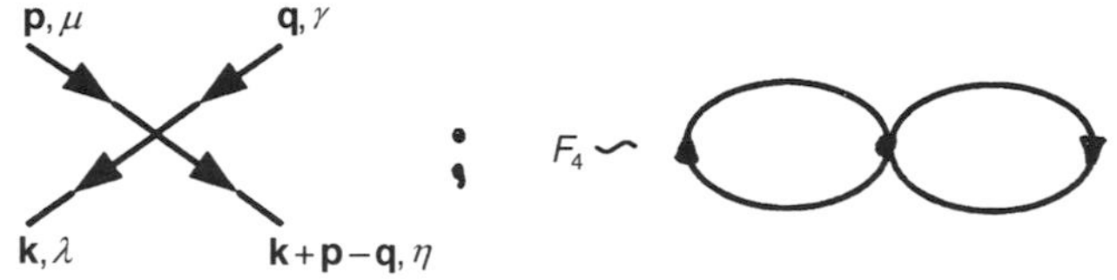

(1.17)

In diagrammatic equations (1.16), (1.17) the lines denote the phonon propagator, and the points are corresponding vertices [13].

General expressions for the anharmonic contribution to the free energy $F_{an}(\Omega, T)$ in terms of the interaction potential of phonons (vertices $V_{\lambda\mu\nu}^{\mathbf{k},\mathbf{q},\mathbf{k}+\mathbf{q}}$ and $V_{\lambda\mu\nu\eta}^{\mathbf{k},\mathbf{p},\mathbf{q},\mathbf{k}+\mathbf{p}-\mathbf{q}}$) are stated in many works, e.g., in [13]. However, the main interest in studying of anharmonic effects is the interval of non-small temperatures $T \geq \Theta_D$, because only here are these effects appreciable. For some problems, particularly for evaluation of the interval of applicability of high-temperature series, there are also interesting low-temperature values of $F_{an}(0) = F_{an}^{(0)}$ caused by zero oscillations. Following to [14], we restrict our study to the analysis of high- and low-temperature series for $F_{an}(\Omega; T)$ only. They will be denoted as F_{an}^{∞} and $F_{an}^{(0)}$, respectively. As was shown in [14], usually the variation of the term F_{an} corresponds to the high-temperature asymptotic expression at relatively low $T \sim (0.2 \div 0.3)\Theta_D$ and at lowest T the term F_{an} is extremely small. The terms F_3 and F_4 in (1.15) at high T are [13, 14][1]

$$F_3^{\infty}(\Omega, T) = -\frac{(k_B T)^2}{M^3}\frac{\Omega^2}{3 \times 2^8 \pi^6}\int d^3k \int d^3q \sum_{\lambda\mu\nu} \frac{|V_{\lambda\mu\nu}^{\mathbf{k},\mathbf{q},\mathbf{k}+\mathbf{q}}|^2}{\omega_{\lambda\mathbf{k}}^2\,\omega_{\mu\mathbf{q}}^2\,\omega_{\nu\mathbf{k}+\mathbf{q}}^2}; \tag{1.18}$$

$$F_4^{\infty}(\Omega, T) = -\frac{(k_B T)^2}{M^2}\frac{\Omega^2}{2^9 \pi^6}\int d^3k \int d^3q \sum_{\lambda\mu} \frac{V_{\lambda\lambda\mu\mu}^{\mathbf{kkqq}}}{\omega_{\lambda\mathbf{k}}^2\,\omega_{\mu\mathbf{q}}^2}. \tag{1.19}$$

Integrals over $\mathbf{k}$ and $\mathbf{q}$ are taken over the Brillouin zone (BZ), concrete expressions for the vertices of three- and four-phonon interaction $V_{\lambda\mu\nu}^{\mathbf{k},\mathbf{q},\mathbf{k}+\mathbf{q}}$ and $V_{\lambda\mu\nu\eta}^{\mathbf{k},\mathbf{p},\mathbf{q},\mathbf{k}+\mathbf{p}-\mathbf{q}}$ will be done below. The terms $F_3 = E_3^{(0)}$ and $F_4 = E_4^{(0)}$ at high $T = 0$ are given as

$$E_3^{(0)}(\Omega) = -\frac{\hbar^2}{M^3}\frac{\Omega^2}{3 \times 2^{10}\pi^6}\int d^3k \int d^3q \times \sum_{\lambda\mu\nu} \frac{|V_{\lambda\mu\nu}^{\mathbf{k},\mathbf{q},\mathbf{k}+\mathbf{q}}|^2}{\omega_{\lambda\mathbf{k}}\,\omega_{\mu\mathbf{q}}\,\omega_{\nu,\mathbf{k}+\mathbf{q}}(\omega_{\lambda\mathbf{k}} + \omega_{\mu\mathbf{q}} + \omega_{\nu,\mathbf{k}+\mathbf{q}})}, \tag{1.20}$$

[1] Here we give no proof of (1.18) and (1.19) and refer the reader to works [13–19] or books [17, 18].

$$E_4^{(0)}(\Omega) = \frac{\hbar^2}{M^2}\frac{\Omega^2}{2^{11}\pi^6}\int d^3k \int d^3q \sum_{\lambda\mu} \frac{V_{\lambda\lambda\mu\mu}^{\mathbf{kkqq}}}{\omega_{\lambda\mathbf{k}}\,\omega_{\mu\mathbf{q}}}. \tag{1.21}$$

1.1.2 Dynamics of the crystal lattice

In the framework of both adiabatic and harmonic approximations, the Hamiltonian of atomic motion may be presented as (in all formulae summation over repeated Greek indexes is implied)

$$\hat{\mathcal{H}} = \frac{1}{2}\sum_l M\dot{\mathbf{u}}_l^2 + \frac{1}{2}\sum_{l,m} V_{\alpha\beta}(l-m)u_{lm}^\alpha u_{lm}^\beta, \tag{1.22}$$

where u_l^α is α-component of the displacement of the l-th atom; $V_{\alpha\beta}$ are the force constants of the second order. Performing the Fourier transform

$$u_{lm} = \frac{1}{\sqrt{N}}\sum_{\mathbf{k}} u(\mathbf{k})\exp[\,i\mathbf{k}(\mathbf{R}_l - \mathbf{R}_m)\,] \tag{1.23}$$

we obtain

$$\mathcal{H} = \frac{1}{2}\sum_{\mathbf{k}}[\,M\dot{\mathbf{u}}(\mathbf{k})\dot{\mathbf{u}}(-\mathbf{k}) + V_{\alpha\beta}(\mathbf{k})u_\alpha(\mathbf{k})u_\beta(-\mathbf{k})\,]. \tag{1.24}$$

The force constants $V_{\alpha\beta}$ are derived as the second derivatives of the energy (1.4) with respect to R_{lm} taken at $R_{lm} = R_{lm}^0$. The Fourier components of the force constants $V_{\alpha\beta}$ form the dynamical matrix

$$D_{\alpha\beta}(\mathbf{k}) = \frac{1}{M}V_{\alpha\beta}(\mathbf{k}). \tag{1.25}$$

Its eigenvalues and eigenvectors are squares of the phonon frequency $\omega_{\lambda\mathbf{k}}^2$ (at $T = 0$) and the polarization vectors $e_{\lambda\mathbf{k}}$. In all formulae indices λ, λ', μ, ν note the phonon branch, other Greek indices relate to the Cartesian components of vectors.

From the invariance of the Hamiltonian relating arbitrary uniform displacements, it follows that $D(\mathbf{k})$ is zero at $\mathbf{k} = 0$. Therefore, it is possible to formally write

$$D_{\alpha\beta}(\mathbf{k}) = \tilde{D}_{\alpha\beta}(\mathbf{k}) - \tilde{D}_{\alpha\beta}(0). \tag{1.26}$$

The simplest way to obtain the dynamical matrix is to use the power series of the structure factor $S(\mathbf{k})$ in atomic displacements $\mathbf{u}_l$. Taking into account that $\mathbf{R}_l = \mathbf{R}_l^0 + \mathbf{u}_l$ and $\mathbf{R}_{lm} = \mathbf{R}_l - \mathbf{R}_m$ we obtain with accuracy of the order $\mathbf{u}_l^2$

$$\begin{aligned} |S(\mathbf{k})|^2 &= \frac{1}{N^2}\sum_{lm}\exp[i\mathbf{k}(\mathbf{R}_{lm}^0 + \mathbf{u}_{lm})] \\ &\approx \frac{1}{N^2}\sum_{lm}\{1 + i\mathbf{k}\mathbf{u}_{lm} - (\mathbf{k}\mathbf{u}_{lm})^2 + \cdots\}\exp\mathbf{k}\mathbf{R}_{lm}^0. \end{aligned} \tag{1.27}$$

Let us substitute (1.27) into (1.7) taking into account that

$$\frac{1}{N^2}\sum_{lm}\exp(i\mathbf{k}\mathbf{R}^0_{lm}) \equiv \sum_{\tau}\delta_{\mathbf{k}\tau}$$

where τ is the vector of the reciprocal lattice. Using the first term in (1.27) to calculate $|S(\mathbf{k})|^2$ in (1.7) we obtain the cohesive energy

$$E_{cohesiv} = w(\Omega_0) + \sum_{\boldsymbol{\tau}}\varphi(\boldsymbol{\tau};\Omega_0), \tag{1.28}$$

where Ω_0 is the equlibrium lattice volume. The contribution to the energy corresponding to the second term in the right part of (1.27) vanishes because each atom is the center of inversion and for each term $\mathbf{k}\mathbf{u}_l$ there exists a corresponding term $-\mathbf{k}\mathbf{u}_l$.

The last term in (1.27) is the dynamical matrix to be found. Let us transform this contribution to the energy using

$$u^{\alpha}_l = \frac{1}{\sqrt{N}}\sum_{\boldsymbol{\kappa}} u_{\alpha}(\boldsymbol{\kappa})\exp(i\boldsymbol{\kappa}\mathbf{R}^0_l). \tag{1.29}$$

It provides

$$\begin{aligned} u^{\alpha}_{lm}u^{\beta}_{lm} = -\frac{1}{N}\sum_{\boldsymbol{\kappa}\boldsymbol{\kappa}'} u_{\alpha}(\boldsymbol{\kappa})u_{\beta}(\boldsymbol{\kappa}')\Big[&\exp\{i(\boldsymbol{\kappa}\mathbf{R}^0_l + \boldsymbol{\kappa}'\mathbf{R}^0_m)\}- \\ &-\exp\{i(\boldsymbol{\kappa}+\boldsymbol{\kappa}')\mathbf{R}^0_l\} + l \leftrightarrow m\Big] \end{aligned} \tag{1.30}$$

(here $l \leftrightarrow m$ means all terms obtained by permutatioms of indexes l and m).

The first and the third terms in the square brackets in (1.30) provide the term $\tilde{D}_{\alpha\beta}(\mathbf{k})$ in (1.26). Calculation of this term will be considered as an example.

The substitution of both these terms into the last term in (1.27) ($\sim \mathbf{u}^2$) and usage of the results in the equation for the energy (1.7) give the energy of oscillations

$$\begin{aligned} E_{osc} &= -\frac{1}{N^2}\sum_{\mathbf{k},\alpha,\beta}\varphi(\mathbf{k};\Omega)k_{\alpha}k_{\beta}u^{\alpha}_{lm}u^{\beta}_{lm}\exp(i\mathbf{k}\mathbf{R}^0_{lm}) \\ &\to \frac{2}{N}\sum_{\mathbf{k}\boldsymbol{\tau}\boldsymbol{\tau}'\boldsymbol{\kappa}\boldsymbol{\kappa}'}\varphi(\mathbf{k};\Omega)k_{\alpha}k_{\beta}u_{\alpha}(\boldsymbol{\kappa})u_{\beta}(\kappa')\delta_{\mathbf{k}+\boldsymbol{\kappa},\boldsymbol{\tau}}\,\delta_{\boldsymbol{\kappa}'-\mathbf{k},\boldsymbol{\tau}'}. \end{aligned} \tag{1.31}$$

Let us take summation over $\mathbf{k}$ using the first Kronecker symbol (for $\mathbf{k} = \boldsymbol{\tau} - \boldsymbol{\kappa}$ and for $\boldsymbol{\kappa}'$ also) and the second one (for $\boldsymbol{\kappa}' = \tau + \tau' - \boldsymbol{\kappa}$) . It provides

$$E_{osc} \to \frac{2}{N}\sum_{\boldsymbol{\tau}\boldsymbol{\tau}'\boldsymbol{\kappa}}\varphi(\boldsymbol{\tau}-\boldsymbol{\kappa};\Omega)(\boldsymbol{\tau}-\boldsymbol{\kappa})_{\alpha}(\boldsymbol{\tau}-\boldsymbol{\kappa})_{\beta}u_{\alpha}(\boldsymbol{\kappa})u_{\beta}(\boldsymbol{\tau}+\boldsymbol{\tau}'-\boldsymbol{\kappa}), \tag{1.32}$$

where the vector $\boldsymbol{\tau}'$ appears at the argument of the last vector $\mathbf{u}(\boldsymbol{\tau}+\boldsymbol{\tau}'-\boldsymbol{\kappa})$ only. But the displacement vector $u_{\alpha}(\mathbf{k})$ is a periodic function of $\mathbf{k}$ with the vectors of the reciprocal lattice as a period. Therefore, it might be worthwhile to take its argument in the first Brillouin zone only. This means that the vector $\boldsymbol{\tau}+\boldsymbol{\tau}'$ must be taken as zero and it eliminates summation over $\boldsymbol{\tau}'$. Finally, replacing $\boldsymbol{\kappa}\to -\mathbf{k}$ we obtain

$$E_{osc}\to\frac{2}{N}\sum_{\tau\kappa}\varphi(\mathbf{p};\Omega)p_{\alpha\beta}u_{\alpha}(\mathbf{k})u_{\beta}(-\mathbf{k})=\frac{1}{N}\sum_{\mathbf{k}}\tilde{D}_{\alpha\beta}(\mathbf{k})u_{\alpha}(\mathbf{k})u_{\beta}(-\mathbf{k}), \tag{1.33}$$

where

$$\tilde{D}_{\alpha\beta}(\mathbf{k})=\sum_{\tau}\varphi(\mathbf{p};\,\Omega)p_{\alpha\beta},\quad p_{\alpha\beta}=p_{\alpha}p_{\beta},\quad p_{\alpha}=(\boldsymbol{\tau}+\boldsymbol{\kappa})_{\alpha} \tag{1.34}$$

The phonon frequencies $\omega_{\lambda\mathbf{k}}$ and the polarization vectors $\mathbf{e}_{\lambda\mathbf{k}}$ satisfy

$$\omega^2_{\lambda\mathbf{k}}\delta_{\lambda\lambda'}=e^{\alpha}_{\lambda\mathbf{k}}D_{\alpha\beta}(\mathbf{k})e^{\beta}_{\lambda'\mathbf{k}}\equiv\left[D(\mathbf{k})\right]_{\lambda\lambda'}. \tag{1.35}$$

To calculate thermodynamic characteristics of crystals, we need derivatives of the phonon frequencies with respect to some parameters $u_1,\ u_2,\ldots$ (e.g., the distortion tensor or the volume). From equation (1.35) it follows

$$\frac{\partial\omega^2_{\lambda\mathbf{k}}}{\partial u_i}\delta_{\lambda\lambda'}=\left[D^i(\mathbf{k})\right]_{\lambda\lambda'}+e^{\alpha,i}_{\lambda\mathbf{k}}D_{\alpha\beta}(\mathbf{k})e^{\beta}_{\lambda'\mathbf{k}}+e^{\alpha}_{\lambda\mathbf{k}}D_{\alpha\beta}(\mathbf{k})e^{\beta,i}_{\lambda'\mathbf{k}}. \tag{1.36}$$

Here

$$\left[D^i(\mathbf{k})\right]_{\lambda\lambda'}=e^{\alpha}_{\lambda\mathbf{k}}\frac{\partial D_{\alpha\beta}(\mathbf{k})}{\partial u_i}e^{\beta}_{\lambda'\mathbf{k}};\quad e^{\alpha,i}_{\lambda\mathbf{k}}=\frac{\partial e^{\alpha}_{\lambda\mathbf{k}}}{\partial u_i}. \tag{1.37}$$

From the orthonormality conditions for the polarization vectors

$$e^{\alpha}_{\lambda\mathbf{k}}e^{\alpha}_{\lambda'\mathbf{k}}=\delta_{\lambda\lambda'} \tag{1.38}$$

it follows

$$e^{\alpha,i}_{\lambda\mathbf{k}}e^{\alpha}_{\lambda\mathbf{k}}=-e^{\alpha}_{\lambda\mathbf{k}}e^{\alpha,i}_{\lambda\mathbf{k}}.$$

Using the last equation and the dynamic equation (1.35) we obtain

$$\frac{\partial\omega^2_{\lambda\mathbf{k}}}{\partial u_i}\delta_{\lambda\lambda'}=\left[D^i(\mathbf{k})\right]_{\lambda\lambda'}+\Delta\omega_{\lambda\lambda'}(\mathbf{k})(e^{\alpha}_{\lambda\mathbf{k}}\,e^{\alpha,i}_{\lambda'\mathbf{k}}), \tag{1.39}$$

where

$$\Delta\omega_{\lambda\lambda'}(\mathbf{k})=\omega^2_{\lambda\mathbf{k}}-\omega^2_{\lambda'\mathbf{k}}.$$

Similarly

$$\begin{aligned}\frac{\partial^2\omega^2_{\lambda\mathbf{k}}}{\partial u_i\,\partial u_j}\delta_{\lambda\lambda'}&=\left[D^{ij}(\mathbf{k})\right]_{\lambda\lambda'}+\Delta\omega^2_{\lambda\lambda'}(\mathbf{k})\left[\mathbf{e}_{\lambda\mathbf{k}}\,\mathbf{e}^{ij}_{\lambda'\mathbf{k}}-\mathbf{e}_{\lambda'\mathbf{k}}\,\mathbf{e}^{ij}_{\lambda\mathbf{k}}\right]\\&+\mathbf{e}_{\lambda\mathbf{k}}\,\mathbf{e}^{j}_{\lambda'\mathbf{k}}\frac{\partial}{\partial u_i}\Delta\omega^2_{\lambda\lambda'}(\mathbf{k})+\mathbf{e}_{\lambda\mathbf{k}}\mathbf{e}^{i}_{\lambda'\mathbf{k}}\frac{\partial}{\partial u_j}\Delta\omega^2_{\lambda\lambda'}(\mathbf{k})\\&+\frac{1}{2}\left[(\omega^2_{\lambda\mathbf{k}}+\omega^2_{\lambda'\mathbf{k}})\delta_{\mu\nu}-2D_{\mu\nu}\right]\mathbf{e}^{\mu,i}_{\lambda\mathbf{k}}\,\mathbf{e}^{\nu,j}_{\lambda'\mathbf{k}}+\mathbf{e}^{\mu,j}_{\lambda'\mathbf{k}}\,\mathbf{e}^{\nu,i}_{\lambda\mathbf{k}}.\end{aligned} \tag{1.40}$$

Here summation over repeated Greek indices is implied (excluding λ and λ' that denote the phonon branches). Note that the diagonal matrix element $[D^{ij}(\mathbf{k})]_{\lambda\lambda}$ is equal to the respective derivative of $\omega^2_{\lambda\mathbf{k}}$, however the non-diagonal element $[D^{ij}(\mathbf{k})]_{\lambda\lambda'}$ is not equal to $\partial^2\omega^2_{\lambda\mathbf{k}}/\partial u_i \partial u_j$ except in the case of deformations that do not change the shape of the unit cell.

Usually it is convenient to suppose that $D_{\mu\nu}$ depends on parameters u_i through the cell volume, vectors $\mathbf{p}$ (or $\boldsymbol{\tau}$) and the equilibrium electron density n_0 (in the case of metals). Then

$$\frac{\partial}{\partial u_i} = \frac{\partial \Omega}{\partial u_i}\left[\frac{\partial}{\partial \Omega}\right]_{n_0 \mathbf{p}} + \frac{\partial p_\beta}{\partial u_i}\left[\frac{\partial}{\partial p_\beta}\right]_{\Omega, n_0} + \frac{\partial n_o}{\partial u_i}\left[\frac{\partial}{\partial n_0}\right]_{\Omega, \mathbf{p}}. \tag{1.41}$$

Let us determine derivatives of the dynamical matrix $D_{\mu\nu}(\mathbf{k})$ with respect to the components of the distortion matrix. First, it is convenient to express $\varphi(\mathbf{p};\Omega)$ through the function $\psi(\mathbf{p}, n_0; \Omega) = \varphi(\mathbf{p}; n_0)/\Omega$. Let us denote $u_i = u_{\alpha\beta}$. Using the fact that $\psi(\mathbf{p}, n_0; \Omega)$ does not depend explicitly on Ω in the sense of (1.41), we obtain expressions for $[D^i]_{\mu\nu}$ and $[D^{ij}]_{\mu\nu}$ for metals

$$\left[\tilde{D}^{\alpha\beta}(\mathbf{k})\right]_{\mu\nu} = -\sum_{\boldsymbol{\tau}}\left[\frac{\partial}{\partial p_\beta} p_{\alpha\mu\nu} + \delta_{\alpha\beta}\, p_{\mu\nu}\, n_0 \frac{\partial}{\partial n_0}\right]\psi(\mathbf{p}, n_0), \tag{1.42}$$

$$\begin{aligned}\left[\tilde{D}^{\alpha\beta,\gamma\delta}(\mathbf{k})\right]_{\mu\nu} = \sum_{\boldsymbol{\tau}}\Bigg[&\frac{\partial^2}{\partial p_\beta\, \partial p_\delta} p_{\alpha\gamma\mu\nu} + (\delta_{\alpha\beta}\,\delta_{\gamma\delta} + \delta_{\alpha\delta}\,\delta_{\beta\gamma}) p_{\mu\nu} n_0 \frac{\partial}{\partial n_0} \\ &+ \delta_{\alpha\beta}\frac{\partial}{\partial p_\delta} p_{\gamma\mu\nu} + \delta_{\gamma\delta}\frac{\partial}{\partial p_\beta} p_{\alpha\mu\nu}) n_0 \frac{\partial}{\partial n_0} + \delta_{\alpha\beta}\,\delta_{\gamma\delta}\, p_{\mu\nu}\, n_0^2 \frac{\partial^2}{\partial n_0^2}\Bigg]\psi(\mathbf{p}, n_0),\end{aligned} \tag{1.43}$$

$$p_{\alpha\mu\nu} = p_\alpha p_\mu p_\nu; \quad \mathbf{p} = \mathbf{k} + \boldsymbol{\tau}.$$

In the case of nonmetals, corresponding expressions may be obtained from (1.42) and (1.43) by taking into account that $n_0 = z/\Omega$ (z is the valence) and by using the substitutions

$$n_0 \frac{\partial}{\partial n_0} = -\Omega \frac{\partial}{\partial \Omega}; \quad n_0^2 \frac{\partial^2}{\partial n_0^2} = \Omega^2 \frac{\partial^2}{\partial \Omega^2} + 2\Omega \frac{\partial}{\partial \Omega}.$$

In the general case, the microscopic Grüneisen parameters are defined as

$$\gamma_{\lambda\mathbf{k}} = -\frac{\partial \ln \omega_{\lambda\mathbf{k}}}{\partial \ln \Omega} = -\frac{D^\Omega_{\lambda\lambda}(\mathbf{k})}{2\omega^2_{\lambda\mathbf{k}}}, \tag{1.44}$$

where

$$D^\Omega_{\lambda\lambda}(\mathbf{k}) = e^\alpha_{\lambda\mathbf{k}} \frac{\partial D_{\alpha\beta}(\mathbf{k})}{\partial \ln \Omega} e^\beta_{\lambda\mathbf{k}}. \tag{1.45}$$

The formulae, which are obtained above, relate to the procedure of calculation of phonon frequencies in the framework of the harmonic approximation. Now we will consider the construction of the general theory of the anharmonic shift of the phonon frequency (and attenuation) in the lowest order in the anharmonicity parameter $\varepsilon = \langle x^2 \rangle / a^2$ [14, 15] i.e., by taking into account all three- and four-phonon contributions to F_{an} in (1.15).

General expressions of the anharmonic theory providing the frequency shift $\Delta_{\lambda\mathbf{k}}$ and attenuation $\Gamma_{\lambda\mathbf{k}}$ in terms of the interatomic potential $\varphi(\mathbf{r}; \Omega)$ are stated in many works, e.g., in [13]. To use these formulae in operating with the form factor $\varphi(\mathbf{k}; \Omega)$, we must first reconstruct the potential $\varphi(\mathbf{r}; \Omega)$ and its derivatives using $\varphi(\mathbf{k}; \Omega)$. However, it is more convenient (especially in the theory of metals) to operate in the momentum space rather than in coordinate space (see e.g., [9, 19]). To follow this scheme we express $\Delta_{\lambda\mathbf{k}}$ and $\Gamma_{\lambda\mathbf{k}}$ through $\varphi(\mathbf{k}; \Omega)$. To avoid difficulties relating to slow convergence of the Coulomb contribution to the energy of interaction, we use the standard Ewald method [1, 9, 20].

The anharmonic shift for the frequency $\Delta_{\lambda\mathbf{k}}$ of the phonon from the branch λ with wave vector $\mathbf{k}$ in the lowest order in anharmonicity (terms F_3 and F_4 in (1.15)) have the form

$$\Delta_{\lambda\mathbf{k}} = \omega_{\lambda\mathbf{k}}(\Omega, T) - \omega^0_{\lambda\mathbf{k}}(\Omega) = \Delta^{(3)}_{\lambda\mathbf{k}}(T, \Omega) + \Delta^{(4)}_{\lambda\mathbf{k}}(T, \Omega). \tag{1.46}$$

Here $\omega^0_{\lambda\mathbf{k}}$ is the frequency calculated within the harmonic approximation and both shifts $\Delta^{(3)}_{\lambda\mathbf{k}}$ and $\Delta^{(4)}_{\lambda\mathbf{k}}$ relating to three- and four-phonon interaction are expressed in terms of third and fourth derivatives of the interaction energy E with respect to ion coordinates.

Sometimes it is more convenient to keep some contributions to the crystal energy as sums in the coordinate space [15], i.e., to use instead of (1.7) the next expression

$$E = \sum_{\mathbf{k}\neq 0} u(\mathbf{k};\, \Omega)\, |\mathrm{S}(\mathrm{k})|^2 + \frac{1}{2N} \sum_{l,m} f(\mathbf{R}_{lm}; \Omega). \tag{1.47}$$

Now we can obtain the frequency shift $\Delta_{\lambda\mathbf{k}}$ by using both (1.47) and general formulae of anharmonic perturbation theory [13]. Since the proof of these formulae is extremely cumbersome, we give the final expression only as

$$\Delta^{(3)}_{\nu\mathbf{k}} = -\frac{\hbar}{16M^3 N\omega_{\nu\mathbf{k}}} \sum_{\mathbf{q},\lambda,\mu} \frac{L^{\mathbf{k},\mathbf{q},\mathbf{k}+\mathbf{q}}_{\nu\lambda\mu} \, |V^{\mathbf{k},\mathbf{q},\mathbf{k}+\mathbf{q}}_{\nu\lambda\mu}|^2}{\omega_{\lambda\mathbf{q}}\, \omega_{\mu,\mathbf{k}+\mathbf{q}}}, \tag{1.48}$$

$$\Delta^{(4)}_{\nu\mathbf{k}} = -\frac{\hbar}{4M^2 N\omega_{\nu\mathbf{k}}} \sum_{\mathbf{q},\lambda} \frac{V^{\mathbf{kkqq}}_{\nu\nu\lambda\lambda}}{\omega_{\lambda\mathbf{q}}} \left(N_{\lambda\mathbf{q}} + \frac{1}{2} \right). \tag{1.49}$$

Here $N_{\lambda\mathbf{q}} = [\exp(\hbar\,\omega_{\lambda\mathbf{q}}/k_B T) - 1]^{-1}$ is the Bose function, and

$$\begin{aligned} L^{\mathbf{k},\mathbf{q},\mathbf{k}+\mathbf{q}}_{\nu\lambda\mu} &= (1 + N_{\lambda\mathbf{q}} + N_{\mu,\mathbf{k}+\mathbf{q}}) \left[\frac{1}{\omega_{\nu\mathbf{k}} + \omega_{\lambda\mathbf{q}} + \omega_{\mu,\mathbf{k}+\mathbf{q}}} \right. \\ &\quad \left. + \hat{\mathcal{P}} \frac{1}{-\omega_{\nu\mathbf{k}} + \omega_{\lambda\mathbf{q}} + \omega_{\mu,\mathbf{k}+\mathbf{q}}} \right] + (N_{\lambda\mathbf{q}} - N_{\mu,\mathbf{k}+\mathbf{q}}) \\ &\quad \times \left[\hat{\mathcal{P}} \frac{1}{-\omega_{\lambda\mathbf{q}} + \omega_{\mu,\mathbf{k}+\mathbf{q}} - \omega_{\nu\mathbf{k}}} - \hat{\mathcal{P}} \frac{1}{\omega_{\lambda\mathbf{q}} - \omega_{\mu,\mathbf{k}+\mathbf{q}} - \omega_{\nu\mathbf{k}}} \right], \end{aligned} \tag{1.50}$$

where $\hat{\mathcal{P}}$ denotes the principal part. Formulae for the attenuation $\Gamma_{\nu\mathbf{k}}$ follow from (1.48)–(1.50) by substitution

$$\frac{\hat{\mathcal{P}}}{x} \rightarrow \pi\delta(x). \tag{1.51}$$

Here $V_{\lambda\mu\nu}^{\mathbf{k},\mathbf{q},\mathbf{k}+\mathbf{q}}$ and $V_{\lambda\mu\nu\eta}^{\mathbf{k},\mathbf{p},\mathbf{q},\mathbf{k}+\mathbf{p}-\mathbf{q}}$ are the three- and four-phonon vertices:

$$\begin{aligned} V_{\nu\lambda\mu}^{\mathbf{k},\mathbf{q},\mathbf{k}+\mathbf{q}} &= \sum_{\alpha,\beta,\gamma} e_{\nu\mathbf{k}}^{\alpha}\, e_{\lambda\mathbf{q}}^{\beta}\, e_{\mu,\mathbf{k}+\mathbf{q}}^{\gamma} \Bigg\{ \sum_{\mathbf{g}} [\varphi(\mathbf{k}+\mathbf{g}) + \varphi(\mathbf{q}+\mathbf{g}) - \varphi(\mathbf{k}+\mathbf{q}+\mathbf{g})]_{\alpha\beta\gamma} \\ &+ \sum_{h}{}' \Big[h_\alpha\, h_\beta\, h_\gamma\, f_3(h) + (h\delta)_{\alpha\beta\gamma}\, f_2(h)\Big] \big[\sin[(\mathbf{k}+\mathbf{q})\mathbf{h}] - \\ &- \sin(\mathbf{kh}) - \sin(\mathbf{qh})\big] \Bigg\}, \end{aligned} \tag{1.52}$$

$$\begin{aligned} V_{\nu\nu\lambda\lambda}^{\mathbf{kkqq}} &= \sum_{\alpha\beta\gamma\delta} e_{\nu\mathbf{k}}^{\alpha}\, e_{\nu\mathbf{k}}^{\beta}\, e_{\lambda\mathbf{q}}^{\gamma} e_{\lambda\mathbf{q}}^{\delta} \times \Bigg\{ \sum_{\mathbf{q}} [\varphi(\mathbf{g}) - \varphi(\mathbf{k}+\mathbf{g}) - \varphi(\mathbf{q}+\mathbf{g}) + \varphi(\mathbf{k}+\mathbf{q}+\mathbf{g}]_{\alpha\beta\gamma\delta} \\ &+ \sum_{h}{}' \Big[h_\alpha\, h_\beta\, h_\gamma\, h_\delta\, f_4(h) + (hh\delta)_{\alpha\beta\gamma\delta}\, f_3(h) + (\delta\delta)_{\alpha\beta\gamma\delta}\, f_2(h)\Big] \\ &\times [1 - \cos(\mathbf{kh})][1 - \cos(\mathbf{qh})] \Bigg\}. \end{aligned} \tag{1.53}$$

Here $\mathbf{e}_{\lambda\mathbf{k}}$ is the polarization vector of the phonon from the branch λ and with the wave vector $\mathbf{k}$, summation is taken over vectors of the direct $\mathbf{h}$ and reciprocal $\mathbf{g}$ lattices. We use the following notations:

$$\begin{aligned} [\varphi(x)]_{\alpha\beta\gamma} &= x_\alpha\, x_\beta\, x_\gamma\, \varphi(x); \\ [\varphi(x)]_{\alpha\beta\gamma\delta} &= x_\alpha\, x_\beta\, x_\gamma\, x_\delta\, \varphi(x); \\ (h\delta)_{\alpha\beta\gamma} &= h_\alpha\, \delta_{\beta\gamma} + h_\beta\, \delta_{\alpha\gamma} + h_\gamma\, \delta_{\alpha\beta}; \\ (\delta\delta)_{\alpha\beta\gamma\delta} &= \delta_{\alpha\beta}\, \delta_{\gamma\delta} + \delta_{\alpha\gamma}\, \delta_{\beta\delta} + \delta_{\alpha\delta}\, \delta_{\beta\gamma} \\ (hh\delta)_{\alpha\beta\gamma\delta} &= h_\alpha\, h_\beta\, \delta_{\gamma\delta} + h_\alpha\, h_\gamma\, \delta_{\beta\delta} + h_\alpha\, h_\delta\, \delta_{\beta\gamma} + \\ &\quad + h_\beta\, h_\gamma\, \delta_{\alpha\beta} + h_\beta\, h_\delta\, \delta_{\alpha\gamma} + h_\gamma\, h_\delta\, \delta_{\alpha\beta}; \\ f_n(h) &= \left(\frac{1}{h}\frac{d}{dh}\right)^n f(h), \quad n = 2, 3, 4\,. \end{aligned} \tag{1.54}$$

The dynamical matrix $\tilde{D}_{\alpha\beta}(\mathbf{k})$ with eigenvalues $M(\omega_{\nu\mathbf{k}}^0)^2$ in these notations has the form

$$\tilde{D}_{\alpha\beta}(\mathbf{k}) = \sum_{g} \psi_{\alpha\beta}(\mathbf{k}+\mathbf{g}) - \sum_{h}{}' \big[h_\alpha\, h_\beta\, f_2(h) + \delta_{\alpha\beta}\, f_1(h)\big] \cos(\mathbf{kh}) \tag{1.55}$$

With no pressure p, when the change of $\Omega = \Omega(T)$ is a consequence of the thermal expansion only.[2] We consider here the total anharmonic shift $\Delta^{an}_{\nu\mathbf{k}}$ in the lowest order in anharmonicity:

$$\begin{aligned}\Delta^{an}_{\nu\mathbf{k}} &= \omega_{\nu\mathbf{k}}(T, p=0) - \omega^0_{\nu\mathbf{k}}(p=0) \\ &= \Delta^{qh}_{\nu\mathbf{k}}(\Omega_0, T) + \Delta^{(3)}_{\nu\mathbf{k}}(\Omega_0, T) + \Delta^{(4)}_{\nu\mathbf{k}}(\Omega_0, T). \end{aligned} \tag{1.56}$$

Here $\Delta^{qh}_{\nu\mathbf{k}}$ is the "quasi-harmonic" shift relating to the variation of the volume $\Delta\Omega$ under the action of the phonon (thermal) pressure p^* (in particular relating to the thermal expansion):

$$\Delta^{qh}_{\nu\mathbf{k}} = -\gamma^0_{\nu\mathbf{k}}(\Omega_0)\omega^0_{\nu\mathbf{k}}(\Omega_0)\frac{\Delta\Omega(T)}{\Omega_0}, \tag{1.57}$$

where $\Delta\Omega/\Omega_0$ is the thermal expansion, Ω_0 is the volume per lattice cell at $T = 0$, and $\gamma^0_{\nu\mathbf{k}}$ are the Grüneisen parameters calculated without anharmonic contributions. In equations (1.48)–(1.50) terms with 1/2 in factors like $N_{\lambda\mathbf{q}} + 1/2$ correspond to the contribution of zero oscillations. They determine the anharmonic shift $\Delta\omega^{zp}_{\nu\mathbf{k}}$ at $T = 0$.

The observable temperature-dependent frequency shift $\Delta_{\nu\mathbf{k}}(T)$ provide terms with $N_{\lambda\mathbf{q}}$

$$\begin{aligned}\Delta^{zp}_{\nu\mathbf{k}} = \Delta^{an}_{\nu\mathbf{k}}(T=0); \quad \Delta_{\nu\mathbf{k}}(T) &= \omega_{\nu\mathbf{k}}(T) - \omega_{\nu\mathbf{k}}(0) \\ &= \Delta^{qh*}_{\nu\mathbf{k}}(T) + \Delta^{(3)*}_{\nu\mathbf{k}}(T) + \Delta^{(4)*}_{\nu\mathbf{k}}(T), \end{aligned} \tag{1.58}$$

where $\Delta^{(i)*}_{\nu\mathbf{k}}(T) = \Delta^{(i)}_{\nu\mathbf{k}}(T) - \Delta^{(i)}_{\nu\mathbf{k}}(p=0)$. If we replace in the left side of equation (1.56) $\omega^0_{\lambda\mathbf{k}}(p=0)$ with $\omega^0_{\lambda\mathbf{k}}(\Omega_0)$ (so that it will be made in calculations in Chapter 3) then in the right side of (1.56) the term $\Delta^{qh}_{\nu\mathbf{k}}$ will be replaced by $\Delta^{qh*}_{\nu\mathbf{k}}$, i.e., total variation of $\Delta\Omega$ in (1.57) will be replaced by the thermal expansion $\Delta\Omega(T)$ corresponding to terms with $N_{\lambda\mathbf{q}}$.

1.1.3 Thermodynamic characteristics

This subsection provides definitions of crystal characteristics we use below. They include: the specific heat C_V, the thermal pressure p^*, the thermal expansion $\Delta\Omega/\Omega_0$ and the coefficients of thermal expansion $\alpha(T)$. Equations that determine the temperature dependence of C_{ik} will be presented in section 2 of the present chapter. It is convenient to take the values that characterize the variation of the unit cell u (like variation of the basis vectors or the distortion tensor or the deformation tensor) and the temperature T as independent thermodynamic variables. An appropriate thermodynamic potential will be the free energy $F(u, T)$.

Now we restrict our investigation of thermodynamic properties by the lowest-order corrections in the anharmonicity parameter $\varepsilon = \langle x^2\rangle/a^2 \sim T/E(R_0)$ ($E(R_0)$ is the potential well depth in the equilibrium state R_0). To find the elastic constants, it is enough to calculate the temperature dependent part of free energy in the harmonic

[2] The effect of the pressure on the frequency shift, attenuation and the Grüneisen parameters of alkali metals is investigated in [10]

approximation (up to the term F^*). The possibility of restricting the calculation of the free energy by the term F^* only follows from the fact that, in the case of crystals with one atom per unit cell, each differentiation of phonon frequencies $\omega_{\lambda\mathbf{k}}$ provides one additional factor ε only.

In the framework of the harmonic approximation, the energy and the specific heat (per atom) are given by well-known formulae

$$E_h = E^{zp}(\Omega) + E^*(\Omega, T) = \frac{\hbar\Omega}{a^3}\sum_{\nu=1}^{3}\int \omega_{\nu\mathbf{q}}\left(N_{\nu\mathbf{q}} + \frac{1}{2}\right) d^3q, \tag{1.59}$$

$$C_V^h = \left[\frac{\partial E_h}{\partial T}\right] = \frac{\hbar^2\Omega}{a^3}\beta^2\sum_{\nu=1}^{3}\int \omega_{\nu\mathbf{q}}^2 N_{\nu\mathbf{q}}(1 + N_{\nu\mathbf{q}})\, d^3q. \tag{1.60}$$

Here $\mathbf{q} = \mathbf{k}a/(2\pi)$ is the dimensionless wave vector, and $\mathbf{k}$ relates to the first Brillouin zone, $\omega_{\lambda\mathbf{k}}^0$ is the phonon frequency in the harmonic approximation and $\beta = (k_B T)^{-1}$. To calculate first order corrections (in ε) of the thermodynamic properties relating to the energy (e.g., the specific heat), we need to take into account the next two corrections on the atomic displacements in the Hamiltonian (1.22).

Anharmonic contributions to the specific heat C_V^{an} and the pressure p_{an} follow from F_{an} in (1.15) in accordance with general thermodynamic equations [17]

$$C_V^{an} = -T^2\frac{\partial^2 F_{an}}{\partial T^2}; \quad p_{an} = -\left[\frac{\partial F_{an}}{\partial\Omega}\right]_s. \tag{1.61}$$

Within the harmonic approximation, the phonon contribution to the pressure must be determined as

$$p_h = p_0^{zp}(\Omega) + p^*(\Omega, T) = \frac{\hbar\Omega}{a^3}\sum_{\lambda=1}^{3}\int \omega_{\lambda\mathbf{k}}\,\gamma_{\lambda\mathbf{k}}^0\left[N_{\nu\mathbf{q}} + \frac{1}{2}\right] d^3k. \tag{1.62}$$

Remember, that is the zero oscillation pressure p_0^{zp} relating to the term with 1/2 in brackets; p^* is the thermal pressure vanishing at $T = 0$; $\gamma_{\lambda\mathbf{k}}^0$ is the microscopic (or one-phonon) Grüneisen parameter (1.44).

Volume expansion coefficient α in the first order in anharmonicity is determined by following

$$\alpha_1(T) = \frac{1}{B}\frac{\partial p^*}{\partial T} = \frac{1}{B}\frac{\hbar^2}{a^3}\beta^2\sum_{\lambda=1}^{3}\int \omega_{\lambda\mathbf{k}}^2\,\gamma_{\lambda\mathbf{k}}\,N_{\lambda\mathbf{k}}[1 + N_{\lambda\mathbf{k}}]\, d^3k, \tag{1.63}$$

where $B = B(\Omega)$ is the compressibility module calculated without phonons. The thermal volume $\Delta\Omega = \Omega(T) - \Omega_0$ and linear $\Delta a = a(T) - a_0$ expansion are

$$\left[\frac{\Delta\Omega}{\Omega_0}\right]_1 \equiv u_1(T) = \int_0^T \alpha_1(T')\, dT' = \frac{p^*(\Omega_0, T)}{B(\Omega_0)};$$

$$\left[\frac{\Delta a}{a_0}\right]_1 = \frac{u_1(T)}{3}. \tag{1.64}$$

respectively.

By calculating the next order contributions (in ε) to $\alpha_1(T)$, it is possible to take into account the variation of $\Omega = \Omega(T)$ due to the thermal expansion, the anharmonic correction p_{an} to the thermal pressure and phonon contributions $B_0 + B^* = -\partial p^*/\partial u$ to the compressibility modulus B. It gives

$$\alpha = \alpha_1\left[1 - u_1\frac{\partial \ln B}{\partial u}\right] + \frac{\partial}{\partial T}\left[\frac{p_{an}}{B} - u_1\frac{B_0 + B^*}{B}\right], \tag{1.65}$$

$$\frac{\Delta\Omega}{\Omega} = u_1\left\{1 + \frac{1}{2}u_1\left[1 - \frac{\partial \ln B}{\partial u}\right]\right\} + \frac{p^*_{an}}{B} - u_1\frac{B_0 + B^*}{B}, \tag{1.66}$$

$$\frac{\Delta a}{a_0} = \frac{1}{3}\left[\frac{\Delta\Omega}{\Omega} - \frac{u_1^2}{3}\right]. \tag{1.67}$$

Formally all correction terms in these formulae have the same (second) order in the anharmonicity.

The microscopic Grüneisen parameter $\gamma(T)$ is determined by the relation

$$\gamma(\Omega, T) = \frac{\Omega\alpha(T)B_S}{C_p} = \frac{\Omega\alpha(T)B_T}{C_V} = \frac{\Omega}{C_V}\left[\frac{\partial p^*}{\partial T}\right]_V. \tag{1.68}$$

We will calculate $\gamma(T)/\Omega$ at constant $\Omega = \Omega_0$ in the lowest order in the anharmonicity

$$\gamma(\Omega_0, T) \approx \frac{\Omega_0}{C_V}\left[\frac{\partial p^*_h}{\partial T}\right]_V = \frac{\sum_{\nu=1}^{3}\int \omega^2_{\lambda\mathbf{k}}\,\gamma^0_{\lambda\mathbf{k}}\,N_{\lambda\mathbf{k}}[1 + N_{\lambda\mathbf{k}}]\,d^3k}{\sum_{\nu=1}^{3}\int \omega^2_{\lambda\mathbf{k}}\,N_{\lambda\mathbf{k}}[1 + N_{\lambda\mathbf{k}}]\,d^3k}. \tag{1.69}$$

The quantity γ/Ω from (1.69), in accordance with our calculations, depends very weakly on Ω: $\partial \ln(\gamma/\Omega)/\partial u \leq 10^{-1}$, whereas performing comparisons with experiment, we will ignore variations of this value by thermal expansion, simply allowing $\gamma(\Omega, T)/\Omega = \gamma(\Omega_0, T)/\Omega_0$.

All these formulae determine the phonon (or lattice) contribution to p^* and $\partial p^*/\partial T$. To perform a comparison with experiment we must subtract the electronic contributions p^*_e and $\partial p^*_e/\partial T$ from the measured values. In our case $T \ll E_F$, it is possible to obtain the general thermodynamic relation between p^*_e/T and the electron specific heat C_e. However, for both of these values, we can restrict the consideration to terms linear in T (like in the low-temperature expansion (1.2)). Then this relation has the simplest form[3]

$$\frac{\partial p^*_e}{\partial T} = \frac{2C_e}{3\Omega} = \frac{2\lambda_e T}{3\Omega}. \tag{1.70}$$

[3] Since the electron contribution to the macroscopic Grüneisen parameter is $\lambda_e = 2/3$ [1] then using (1.68), (1.69) and (1.2) we obtain equations (1.70).

Below we will take values of λ_e from experiment.

Therefore, $\alpha(T)$, $\Delta\Omega(T)/\Omega_0$ and $\gamma(T)$ are supposed to be the lattice contributions only. Experimental values of these quantities are obtained using the relations

$$\alpha = \alpha^{lat}(T) = \alpha_{exp}^{total} - \frac{1}{B_T}\left[\frac{\partial p_e^*}{\partial T}\right]_V = \alpha_{exp}^{total} - \frac{2}{3}\frac{\lambda_e T}{\Omega_0 B_T},$$
$$\frac{\Delta\Omega}{\Omega} = \left[\frac{\Delta\Omega}{\Omega_0}\right]_{\exp}^{total} - \frac{\lambda_e T^2}{3\Omega_0 B_T}, \quad \gamma(T) = \gamma^{lat}(\Omega, T) = \frac{\Omega\alpha^{lat} B_T}{C_V^{lat}}, \tag{1.71}$$

where $B_T(\gamma) = (B_T)_{exp}$ is the total compressibility module obtained in experiment. Naturally, corrections provided by electrons are essential at low T only.

1.2 Elastic Constants

1.2.1 Elastic moduli of Brugger and Birch

To describe the crystal characteristics mentioned in subsection 3 of the previous section (excluding C_V), it is sufficient to obtain the expressions for elastic modules of the first and second order as functions of temperature.

Let us use the components of the Lagrange deformation tensor as parameters of deformation

$$\bar{u}_{\alpha\beta} = \frac{1}{2}\left[\frac{\partial x'_\gamma}{\partial x_\alpha}\frac{\partial x'_\gamma}{\partial x_\beta} - \delta_{\alpha\beta}\right] = \frac{1}{2}(u_{\alpha\beta} + u_{\beta\alpha} + u_{\gamma\alpha}\, u_{\gamma\beta}), \tag{1.72}$$

$$u_{\alpha\beta} = \frac{\partial}{\partial x_\beta}(x'_\alpha - x_\alpha). \tag{1.73}$$

The isothermic elastic constants are defined as the coefficients of the series expansion of $F(u, T)$ in powers of the deformations tensor $\bar{u}_{\alpha\beta}$, normalized in the equilibrium crystal volume. It means that the isothermic elastic constants are

$$C_{\alpha\beta\cdots\gamma\delta}(T) = \frac{1}{N\Omega}\frac{\partial^n F(u, T)}{\partial u_{\alpha\beta}\cdots\partial u_{\gamma\delta}}\bigg|_{\bar{u}=0}. \tag{1.74}$$

These are the so-called Brugger moduli [21].

The first order modulus of the cubic crystal is simply the pressure. Hence, from (1.11), it follows that the equation of state is

$$p = p_{st}(\Omega) + p^*(\Omega, T) + p_{an}(\Omega, T), \tag{1.75}$$

where

$$p\delta_{\alpha\beta} = -\frac{1}{N\Omega}\frac{\partial F}{\partial \bar{u}_{\alpha\beta}} = -\frac{1}{\Omega N}\frac{\partial F}{\partial u_{\alpha\beta}}. \tag{1.76}$$

The pressures p_{st}, p^* and p_{an} are determined from equations (1.12)–(1.15) for $E(\Omega)$, $F^*(\Omega, T)$ and F_{an} respectively.

In the framework of our approximation, the thermal pressure $p^*(\Omega, T)$ is expressed through the triple anharmonic amplitudes $V_{\nu\lambda\mu}^{\mathbf{k},\mathbf{q},\mathbf{k}+\mathbf{q}}$ (see equation (1.16) and this term fully determines the thermal expansion of the cubic crystal.

The second order elastic constants are

$$C_{iklm}(T) = C_{iklm}(\Omega) + C^*_{iklm}(\Omega, T), \tag{1.77}$$

where $C_{iklm}(\Omega)$ are the second order elastic constants at $T = 0$. They are determined not only by the energy $E(u)$, but also by the energy of the oscillations $E^{zp}(u)$.

Taking the series expansion of C_{ijklm} in $\Delta\Omega(T)$ in the first order in the anharmonicity, we obtain

$$C_{iklm}(T) = C_{iklm}(\Omega_0) + \left[\frac{\partial C_{iklm}}{\partial \Omega}\right]_{\Omega_0} \Delta\Omega(T) + C^*_{iklm}. \tag{1.78}$$

It is more convenient to introduce the Birch moduli [22] instead of C^*_{ijklm}. If a small additional stress σ_{ik} acts on a stressed crystal (in the state $\{x\}$), it provides small deformations ε_{lm} which relate to σ_{ik} by Hooke's law

$$\sigma_{ik}(x) = \mathcal{B}_{iklm}(x)\varepsilon_{lm}. \tag{1.79}$$

Coefficient $\mathcal{B}_{iklm}$ are the Birch moduli. Taking p^* in these terms as an external pressure, we obtain

$$\mathcal{B}^*_{iklm} = C^*_{iklm} - p^*(\delta_{il}\,\delta_{km} + \delta_{im}\,\delta_{kl} - \delta_{ik}\,\delta_{lm}). \tag{1.80}$$

Taking into account that $\partial/\partial\Omega = -(B/\Omega)\cdot\partial/\partial p$ and $\Delta\Omega/\Omega_0 = p^*/B$, we introduce

$$\mathcal{B}^{qh}_{iklm} - p^*\left[(\delta_{il}\,\delta_{km} + \delta_{im}\,\delta_{kl} - \delta_{ik}\,\delta_{lm}) - \frac{\partial C_{iklm}}{\partial p}\right]. \tag{1.81}$$

Let us denote parameters $\mathcal{B}^{qh}_{\alpha\beta\gamma\delta}$ as the quasi-harmonic contribution to the change of $C_{\alpha\beta\gamma\delta}$ with temperature, and $\mathcal{B}^*_{\alpha\beta\gamma\delta}$ as the anharmonic contribution respectively. For the crystal with cubic symmetry CI [22], it provides

$$\begin{aligned} B^* &= \frac{1}{3}(\mathcal{B}^*_{11} + 2\mathcal{B}^*_{12}) = \frac{1}{3}(A^*_{11} + 2A^*_{12} + 2p^*), \\ C'^* &= \frac{1}{2}(\mathcal{B}^*_{11} - \mathcal{B}^*_{12}) = \frac{1}{2}(A^*_{11} - A^*_{12} - p^*), \\ C^*_{44} &= \mathcal{B}^*_{44} = A^*_{44}, \end{aligned} \tag{1.82}$$

where

$$N\Omega_0 A^*_{11} = \frac{\partial^2 F^*}{\partial u^2_{11}}; \quad N\Omega_0 A^*_{12} = \frac{\partial^2 F^*}{\partial u_{11}\,\partial u_{22}}; \tag{1.83}$$

$$N\Omega_0 A^*_{44} = \frac{\partial^2 F^*}{\partial u^2_{12}}; \quad \frac{\partial^2 F^*}{\partial u_{\alpha\beta}\,\partial u_{\gamma\delta}} = \frac{\partial^2 F^*}{\partial \bar{u}_{\alpha\beta}\,\partial \bar{u}_{\gamma\delta}} + \delta_{\alpha\beta}\frac{\partial F^*}{\partial \bar{u}_{\gamma\delta}}.$$

In equations (1.82) and in the text below, the Voigt indexes (see, e.g. [22]) are used in all equations with elastic constants. Now each index denotes the pair of Cartesian indexes $\alpha\beta$ in accordance with the rule

$$11 \to 1; \quad 22 \to 2; \quad 33 \to 3; \quad 23 \to 4; \quad 32 \to \bar{4};$$
$$13 \to 5; \quad 31 \to \bar{5}; \quad 12 \to 6; \quad 21 \to \bar{6}.$$

In addition, to evaluate the quasi-harmonic (qh) contributions to the elastic constants, it is convenient to introduce

$$\begin{aligned} \mathcal{B}^{qh} &= \frac{1}{3}\left(\mathcal{B}_{11}^{qh} + 2\mathcal{B}_{12}^{qh}\right) = -p^* \frac{\partial B_{11}}{\partial p}, \\ C'_{qh} &= \frac{1}{2}\left(\mathcal{B}_{11}^{qh} - \mathcal{B}_{12}^{qh}\right) = -p^* \frac{\partial B_{33}}{\partial p}, \\ C_{qh} &= \mathcal{B}_{44}^{qh} = -p^* \frac{\partial B_{44}}{\partial p}. \end{aligned} \tag{1.84}$$

Derivatives $\partial B_{11}/\partial p$, $\partial B_{33}/\partial p$, $\partial B_{44}/\partial p$ will be determined below. In experimental works these values are known (not fully correctly) as $\partial B/\partial p$, $\partial C'/\partial p$ and $\partial C_{44}/\partial p$

Equations (1.78) and (1.80)–(1.84) give

$$\begin{aligned} B(T) &= B + B^* - p^* \frac{\partial B_{11}}{\partial p}, B = \frac{1}{3}(C_{11} + 2C_{12}), \\ C'(T) &= C' + C'^* - p^* \frac{\partial B_{33}}{\partial p}, C' = \frac{1}{2}(C_{11} - C_{12}), \\ C_{44}(T) &= C_{44} + C'^* - p^* \frac{\partial B_{44}}{\partial p}. \end{aligned} \tag{1.85}$$

The elastic constants B, C' and C_{44} are more convenient for our analysis than the others because they have a simple physical interpretation as volume and shear moduli and determine the slope of some phonon branches $\omega_{\nu\mathbf{k}}$. Additionally, for alkali metals $C' \ll B$, hence C' is more sensitive to the model of the structure and details of calculations (as will be seen below) than constants C_{11} and C_{12}.

In the following text, we need equations for the slope of dispersion curves $\omega_{\lambda\mathbf{q}}$ of a cubic crystal relating for some high-symmetry directions. If at small wave vectors ($a\mathbf{q} \ll 1$) these dispersion curves can be written as

$$\omega_{\lambda q} = C_\lambda(\mathbf{n})q, n = \mathbf{q}/q, \tag{1.86}$$

then there exist the following relations between B, C', and C_{44} and the sound velocities $C_\lambda(\mathbf{n})$ for different symmetrical directions $\mathbf{q}$ and some phonon polarizations $e_{\lambda\mathbf{q}}$,

1. $\mathbf{q} \parallel [\mathbf{100}]$

$$\text{a.}\quad \mathbf{e}_{\lambda\mathbf{q}} = [100], C_\lambda(\mathbf{n}) = B + 4/3C'; \tag{1.87}$$

$$\text{b.}\quad \mathbf{e}_{\lambda\mathbf{q}} = [010], [001], C_\lambda(\mathbf{n}) = C_{44}. \tag{1.88}$$

2. $\mathbf{q} \parallel [\mathbf{110}]$

$$\text{a.}\quad \mathbf{e}_{\lambda\mathbf{q}} = [110], C_\lambda(\mathbf{n}) = B + 1/3C' + C_{44}; \tag{1.89}$$

$$\text{b.}\quad \mathbf{e}_{\lambda\mathbf{q}} = [001], C_\lambda(\mathbf{n}) = C_{44}; \tag{1.90}$$

$$\text{c.}\quad \mathbf{e}_{\lambda\mathbf{q}} = [1\bar{1}0], C_\lambda(\mathbf{n}) = C'. \tag{1.91}$$

3. $\mathbf{q} \parallel [\mathbf{111}]$

$$\text{a.}\quad \mathbf{e}_{\lambda\mathbf{q}} = [111], C_\lambda(\mathbf{n}) = B + \mathbf{1/3C}_{44}; \tag{1.92}$$

$$\text{b.}\quad \mathbf{e}_{\lambda\mathbf{q}} = [1\bar{1}0], [11\bar{2}], \quad C_\lambda(\mathbf{n}) = 2/3C' + 1/3C_{44}. \tag{1.93}$$

In the case of the arbitrary direction of the vector $\mathbf{q}$ in the Brillouin zone, the values of $C_\lambda(\mathbf{n})$ are determined as a root of some cubic equation. Limiting values of the microscopic Grüneisen parameters of any branch at $\mathbf{q} \to 0$ must be determined as

$$\gamma_{\lambda\mathbf{q}\to 0} = -\frac{1}{2}\frac{d\ln C_\lambda(\mathbf{n})}{d\ln\Omega} - \frac{1}{6}. \tag{1.94}$$

Finally, let us derive the equations for the first and second derivatives of the phonon contribution to the free energy $F_{ph} = E^{zp} + F^*$ in (1.13)–(1.14), which are necessary for calculations of the microscopic Grüneisen parameters and thermal contributions to the elastic constants. Here it is convenient to combine the energy of zero oscillations E^{zp} with the energy of phonons F^* (both depend on the phonon frequencies).

Since F_{ph} are the functions of the phonon frequencies $\omega_{\lambda\mathbf{q}}$, then the derivative $\partial F_{ph}/\partial\eta_i$ [4] is a function of the derivatives of the phonon frequencies. To calculate these values, it is possible to use the well-known theorem for the differentiation of eigenvalues of the matrix in terms of some parameters. It gives,

$$\frac{\partial F_{ph}}{\partial \eta_i} = \hbar \sum_{\lambda\mathbf{q}} (N_{\lambda\mathbf{q}} + 0.5)\frac{\partial \omega_{\lambda\mathbf{q}}}{\partial \eta_i} = \frac{\hbar}{2}\sum_{\lambda\mathbf{q}} \frac{(N_{\lambda\mathbf{q}} + 0.5)[D^i]_{\lambda\lambda}}{\omega_{\lambda\mathbf{q}}}, \tag{1.95}$$

where $N_{\lambda\mathbf{q}}$ is the Planck function of the phonon with the quasi-momentum $\mathbf{q}$ and the polarization λ values $[D^i]_{\lambda\lambda'}$ are the same as in equation (1.37). Using a well-known relation [23]

$$\hbar\frac{N_{\lambda\mathbf{q}} + 0.5}{\omega_{\lambda\mathbf{q}}} = \sum_{n=-\infty}^{\infty} \frac{k_B T}{\omega_n^2 + \omega_{\lambda\mathbf{q}}^2}; \quad \omega_n = \frac{2n\pi k_B T}{\hbar}, \tag{1.96}$$

[4] Here η_i are some parameters, describing the crystal deformation. They are, for example, components of the distortion tensor u_{ij} (1.73) or components of the deformation tensor $\bar{u}_{ij}$ (1.72)

it is possible to write the equation (1.95) as the trace (Tr) of the product of two operators, using the differentiation rule

$$\frac{\partial F_{ph}}{\partial \eta_i} = \frac{k_B T}{2} \sum_{\mathbf{q}} \sum_{n=-\infty}^{\infty} \mathrm{Tr}\left[\frac{1}{\omega_n \hat{I} + \hat{D}} \hat{D}^i\right]; \quad \hat{D}^i = \frac{\partial \hat{D}}{\partial \eta_i}, \tag{1.97}$$

where $\hat{I}$ is the unit vector, and $\hat{D}$ is the operator of the dynamical matrix. Repeating differentiation of equation (1.97) with respect to η_i gives

$$\frac{\partial^2 F_{ph}}{\partial \eta_i \, \partial \eta_i} = \frac{k_B T}{2} \sum_{\mathbf{q}} \sum_{n=-\infty}^{\infty} \mathrm{Tr}\left[\frac{1}{\omega_n^2 \hat{I} + \hat{D}} \hat{D}^{ij} - \frac{1}{\omega_n^2 \hat{I} + \hat{D}} \hat{D}^i \frac{\hat{D}^i}{\omega_n^2 \hat{I} + \hat{D}}\right]. \tag{1.98}$$

Transforming back from the operator equation to equations for matrix elements, we obtain

$$\begin{aligned} A_{ij}^{zp} + A_{ij}^{*} = \frac{1}{N\Omega_0} \frac{\partial^2 F_{ph}}{\partial \eta_i \, \partial \eta_j} = \frac{\hbar}{2} \sum_{\lambda} \int \frac{\partial^3 q}{(2\pi)^3} \Bigg[& \frac{N_{\lambda \mathbf{q}} + 0.5}{\omega_{\lambda \mathbf{q}}} [D^{ij}]_{\lambda\lambda} \\ & - \left[\frac{N_{\lambda \mathbf{q}} + 0.5}{\omega_{\lambda \mathbf{q}}^3} + \frac{\hbar N_{\lambda \mathbf{q}}(N_{\lambda \mathbf{q}} + 1)}{2 k_B T \omega_{\lambda \mathbf{q}}^2}\right] [D^i]_{\lambda\lambda} [D^j]_{\lambda\lambda} \\ & + \sum_{\lambda' \neq \lambda} \left[\frac{N_{\lambda' \mathbf{q}} + 0.5}{\omega_{\lambda' \mathbf{q}}} - \frac{N_{\lambda \mathbf{q}} + 0.5}{\omega_{\lambda \mathbf{q}}}\right] \frac{[D^i]_{\lambda\lambda'} [D^j]_{\lambda'\lambda}}{\omega_{\lambda' \mathbf{q}}^2 - \omega_{\lambda \mathbf{q}}^2}. \end{aligned} \tag{1.99}$$

The first term in square brackets on the right side of (1.99) is a contribution of the fourth order anharmonicity $A_{ij}(4)$, the rest is the contribution of the third order anharmonicity $A_{ij}(3)$. Hence, it is possible to rewrite equation (1.99) as

$$\begin{aligned} A_{ij}^{zp}(3) + A_{ij}^{*}(3) = \frac{\hbar}{N\Omega_0} \sum_{\mathbf{q}\lambda} \Bigg\{ & \left[\frac{N_{\lambda \mathbf{q}} + 0.5}{2\omega_{\lambda \mathbf{q}}^3} + \frac{\hbar N_{\lambda \mathbf{q}}(N_{\lambda \mathbf{q}} + 1)}{4 k_B T \omega_{\lambda \mathbf{q}}^2}\right] [D^i]_{\lambda\lambda} [D^j]_{\lambda\lambda} \\ & - \sum_{\lambda \neq \lambda'} \frac{1}{\Delta \omega_{\lambda \mathbf{q}}^2} \frac{N_{\lambda \mathbf{q}}}{\omega_{\lambda \mathbf{q}}} [D^i]_{\lambda\lambda'} [D^j]_{\lambda'\lambda}; \end{aligned} \tag{1.100}$$

$$A_{ij}^{zp}(4) + A_{ij}^{*}(4) = \frac{\hbar}{N\Omega_0} \sum_{\mathbf{q}\lambda} \frac{N_{\lambda \mathbf{q}} + 0.5}{2\omega_{\lambda \mathbf{q}}} [D^{ij}]_{\lambda\lambda}. \tag{1.101}$$

Equations (1.99)–(1.101) are especially suitable for concrete calculations because the dynamical matrix does not depend on the distortion tensor u_{ij} (see section 4.2).

Calculations of $\mathcal{B}$, C' and C_{44} in (1.85) in the first order in anharmonicity use the unit cell volume at $T = 0$ (i.e., Ω_0). As we will see in Chapter 3, the slopes of $|dC'/dT|_{theor}$ calculated within this approximation are larger than experimental values. There are some reasons to suppose that in the case of $C'(t)$, the next order in anharmonicity plays an appreciable role. To take (qualitatively) into account the next orders of the perturbation theory, we calculate B, C' and C_{44} using equations (1.77), (1.83), (1.99), but using the value for $\Omega(T)$ from equation (1.66).

1.2.2 Second and third order elastic constants

In the previous section, the general theory of thermodynamic properties at finite temperatures in the lowest order in the anhaharmonicity parameter ε was presented. A very small value of this parameter makes all contributions of the potential anharmonicity inconspicuous. On the other hand, the elastic modules at $T = 0$ do not contain ε, but there exists an explicit contribution of higher (higher than the second) coordinate derivatives of the cohesive energy $F(T = 0) = E + E^{zp}$. As a consequence of a rapid variation of the interatomic potential with the distance, the third order elastic constants C_{ikl} may be much larger than the second order elastic constants C_{ik}. This fact is a good test for applicability of the model.

In the presence of deformations, the free energy may be written as a power series in some deformation parameter η

$$F(\eta) - F(0) = \left[\frac{\partial F}{\partial \eta_i}\right]_0 \eta_i + \frac{1}{2}\left[\frac{\partial^2 F}{\partial \eta_i\, \partial \eta_j}\right]_0 \eta_i\, \eta_j + \frac{1}{3!}\left[\frac{\partial^3 F}{\partial \eta_i\, \partial \eta_j\, \partial \eta_k}\right]_0 \eta_i\, \eta_j\, \eta_k. \tag{1.102}$$

Generally, the free (or cohesive) energy of crystal depends on the distortion tensor $u_{\alpha\beta}$, but it may also be possible to use the components of the distortion tensor $\bar{u}_{\alpha\beta}$ as parameters η. Comparison of expansions of $F(\eta_i)$ in the power series in $\bar{u}_{\alpha\beta}$ and $u_{\alpha\beta}$ gives the following general relations between elastic modules and derivatives of F on $u_{\alpha\beta}$

$$C_{\alpha\beta} = \frac{1}{\Omega}\frac{\partial F}{\partial u_{\alpha\beta}},$$

$$C_{\alpha\beta\gamma\delta} + \delta_{\alpha\gamma}\, C_{\beta\delta} = \frac{1}{\Omega}\frac{\partial^2 F}{\partial u_{\alpha\beta}\, \partial u_{\gamma\delta}}. \tag{1.103}$$

$$C_{\alpha\beta\gamma\delta\xi\sigma} + \delta_{\gamma\xi}\, C_{\alpha\beta\delta\sigma} + \delta_{\alpha\gamma}\, C_{\xi\sigma\beta\delta} + \delta_{\alpha\xi}\, C_{\gamma\delta\alpha\sigma} = \frac{1}{\Omega}\frac{\partial^3 F}{\partial u_{\alpha\beta}\, \partial u_{\gamma\delta}\, \partial u_{\xi\sigma}}.$$

In crystals investigated by us (with one atom per unit cell) micro- and macroscopic elastic constants have the same values because there are no internal deformations [20]. In this case, the free energy of cubic crystal may be presented (by using the Voigt indexes) as

$$\begin{aligned}
\frac{1}{\Omega}[F(\bar{u}) - F(0)] = {} & C_1(\bar{u}_1 + \bar{u}_2 + \bar{u}_3) + \frac{1}{2}C_{11}\left(\bar{u}_1^2 + \bar{u}_2^2 + \bar{u}_3^2\right) \\
& + C_{12}(\bar{u}_1\bar{u}_2 + \bar{u}_1\bar{u}_3 + \bar{u}_2\bar{u}_3) + 2C_{44}\left(\bar{u}_6^2 + \bar{u}_4^2 + \bar{u}_5^2\right) \\
& + \frac{1}{6}C_{111}\left(\bar{u}_1^3 + \bar{u}_2^3 + \bar{u}_3^3\right) + \frac{1}{6}C_{112}\Big[\bar{u}_1^2(\bar{u}_2 + \bar{u}_3) + \bar{u}_2^2(\bar{u}_1 + \bar{u}_3) \\
& + \bar{u}_3^2(\bar{u}_1 + \bar{u}_2)\Big] + C_{123}\, \bar{u}_1\, \bar{u}_2\, \bar{u}_3 + 8C_{456}\, \bar{u}_6\, \bar{u}_5\, \bar{u}_4 \\
& + 2C_{144}\left[\bar{u}_1\, \bar{u}_4^2 + \bar{u}_2\, \bar{u}_5^2 + \bar{u}_3\, \bar{u}_6^2\right] + 2C_{166}\, \bar{u}_6^2(\bar{u}_1 + \bar{u}_2) \\
& + \bar{u}_5^2(\bar{u}_1 + \bar{u}_3) + \bar{u}_4^2(\bar{u}_2 + \bar{u}_3).
\end{aligned} \tag{1.104}$$

In the case of cubic (CI) crystals, there exist only one first-order module C_1, three second-order modules C_{11}, C_{12}, C_{44} and six third-order modules C_{111}, C_{112}, C_{123}, C_{144}, C_{166}, C_{456}.

At $T = 0$, the free energy is reduced to the cohesive energy E (see equation (1.4)). Let us obtain here some expressions for non-zero elastic constants of cubic crystals at $T = 0$. Taking into account the possibility of using these formulae later to calculate the third order elastic constants of metals, we suppose that E depends on the volume, on the vectors of the reciprocal lattice $\boldsymbol{\tau}$ and on the equilibrium electron density $n_0 = z/\Omega_0$ (1.41). It gives

$$C_1 = A_1 - \frac{1}{2\Omega}\sum_{\boldsymbol{\tau}\neq 0} \hat{\mathcal{Z}}_x\varphi = -p. \tag{1.105}$$

$$\begin{aligned} C_{11} &= A_1 + A_2 + \frac{1}{2\Omega_0}\sum_{\boldsymbol{\tau}}{}' \{\hat{\mathcal{Z}}_x + 2\,\hat{\mathcal{S}} + 6\}\,\hat{\mathcal{Z}}_x\varphi(\boldsymbol{\tau}),\\ C_{12} &= A_1 + A_2 + \frac{1}{2\Omega_0}\sum_{\boldsymbol{\tau}}{}' \{\hat{\mathcal{Z}}_y + 2\,\hat{\mathcal{S}} + 2\}\,\hat{\mathcal{Z}}_x\varphi(\boldsymbol{\tau}),\\ C_{44} &= -A_1 + \frac{1}{2\Omega_0}\sum_{\boldsymbol{\tau}}{}' \{\hat{\mathcal{Z}}_y + 2\}\,\hat{\mathcal{Z}}_x\varphi(\boldsymbol{\tau}), \end{aligned} \tag{1.106}$$

$$\begin{aligned} C_{111} &= 3A_1 - 3A_2 + A_3 - \frac{1}{2\Omega_0}\sum_{\boldsymbol{\tau}}{}' \Big[\hat{\mathcal{Z}}_x(\hat{\mathcal{Z}} - 3\hat{\mathcal{S}} + 9) + 3(\hat{\mathcal{S}}^2 - \hat{\mathcal{S}} + 5)\Big]\hat{\mathcal{Z}}_x\varphi(\boldsymbol{\tau}),\\ C_{112} &= -A_1 + A_2 + A_3\\ &\quad - \frac{1}{2\Omega_0}\sum_{\boldsymbol{\tau}}{}' \Big[\hat{\mathcal{Z}}_x(\hat{\mathcal{S}}_y + \hat{\mathcal{S}} + 1) + 2\hat{\mathcal{Z}}_y(\hat{\mathcal{S}} + 3) + 3(\hat{\mathcal{S}}^2 + 5\hat{\mathcal{S}} + 3)\Big]\hat{\mathcal{Z}}_x\varphi(\boldsymbol{\tau}),\\ C_{123} &= A_1 + 3A_2 + A_3 - \frac{1}{2\Omega_0}\sum_{\boldsymbol{\tau}}{}' \Big[\hat{\mathcal{Z}}_y(\hat{\mathcal{S}}_z + 3\hat{\mathcal{S}} + 3) + 3(\hat{\mathcal{S}}^2 + 3\hat{\mathcal{S}} + 1)\Big]\hat{\mathcal{Z}}_x\varphi(\boldsymbol{\tau}),\\ C_{144} &= -A_1 - A_2 - \frac{1}{2\Omega_0}\sum_{\boldsymbol{\tau}}{}' \Big[\hat{\mathcal{Z}}_y(\hat{\mathcal{S}}_z + \hat{\mathcal{S}} + 3) + 3(\hat{\mathcal{S}} + 1)\Big]\hat{\mathcal{Z}}_x\varphi(\boldsymbol{\tau}),\\ C_{166} &= A_1 - A_2 - \frac{1}{2\Omega_0}\sum_{\boldsymbol{\tau}}{}' \Big[\hat{\mathcal{Z}}_y(\hat{\mathcal{Z}}_x + 3\hat{\mathcal{S}} + 6) + \hat{\mathcal{Z}}_x + 3(\hat{\mathcal{S}} + 3)\Big]\hat{\mathcal{Z}}_x\varphi(\boldsymbol{\tau}),\\ C_{456} &= A_1 - \frac{1}{2\Omega_0}\sum_{\boldsymbol{\tau}}{}' [\hat{\mathcal{Z}}_y(\hat{\mathcal{Z}}_z + 3) + 3]\hat{\mathcal{Z}}_x\varphi(\boldsymbol{\tau}), \end{aligned} \tag{1.107}$$

$$A_k = \Omega_0^{k-1}\left[\frac{\partial}{\partial\Omega_0}\right]^k E;\quad \hat{\mathcal{S}}^k = n_0^k\left[\frac{\partial}{\partial n_0}\right]^k_{\Omega_0,T};\quad \hat{\mathcal{Z}}^k = \frac{1}{\tau}\left[\frac{\partial}{\partial\tau}\right]^k. \tag{1.108}$$

The operator $\hat{\mathcal{Z}}_x = \hat{\mathcal{Z}}\tau_x^2$ acts only on functions depending on $|\boldsymbol{\tau}|$.

1.2.3 The Fuchs moduli

The Brugger moduli $C_{\alpha\beta\gamma\ldots}$ are commonly useful, but sometimes it is more convenient to use the deformation parameters γ_i, which have a simple physical meaning and to operate with the moduli B_{ikl} relating to γ_i.

Parameters γ_i are introduced so that they are equal to zero at zero deformation. Deformations described by γ_i are: the isotropic compression (γ_1), the uniaxial compressions (γ_2, γ_3) and shear deformations (γ_4, γ_5, γ_6). If there are few atoms per unit cell, then the parameters $\gamma_7, \gamma_8 \ldots$ describe relative displacement of these atoms by deformation. We investigate in the present chapter, however, crystals with one atom per unit cell only. From the definition of the distortion tensor, follow the relations between $u_{\alpha\beta}$ and γ_i:

$$\begin{aligned}
1 + u_{11} &= (1+\gamma_1)^{1/3}(1+\gamma_2)^{-1/3}(1+\gamma_3)^{-1/2} = f_1(\gamma_1,\gamma_2,\gamma_3);\\
1 + u_{22} &= (1+\gamma_1)^{1/3}(1+\gamma_2)^{-1/3}(1+\gamma_3)^{1/2} = f_2(\gamma_1,\gamma_2,\gamma_3);\\
1 + u_{33} &= (1+\gamma_1)^{1/3}(1+\gamma_2)^{2/3} = f_3(\gamma_1,\gamma_2);\\
u_{23} &= \gamma_4; \quad u_{13} = \gamma_5; \quad u_{12} = \gamma_6; \quad u_{32} = u_{31} = u_{21} = 0.
\end{aligned} \tag{1.109}$$

The components of the deformation tensor in these notations are

$$\begin{aligned}
\bar{u}_{11} &= \frac{1}{2}(f_1^2 - 1); \quad 2\bar{u}_{33} = \gamma_4 f_2 + \gamma_5 f_6;\\
\bar{u}_{22} &= \frac{1}{2}(f_2^2 - 1 + \gamma_6^2); \quad 2\bar{u}_{13} = \gamma_5 f_1;\\
\bar{u}_{33} &= \frac{1}{2}(f_3^2 - 1 + \gamma_4^2 + \gamma_5^2); \quad 2\bar{u}_{12} = \gamma_6 f_1.
\end{aligned} \tag{1.110}$$

Expansion $F(\gamma)$ into the power series in γ_i with $i = 1, 2, \ldots, 6$ only provides the macroscopic Fuchs moduli

$$B_{i\ldots l} = \frac{1}{\Omega}\left[\frac{\partial^n F(\gamma)}{\partial\gamma_i \ldots \partial\gamma_l}\right]_{\gamma=0} \tag{1.111}$$

In the case of cubic crystals (CI) [22] from the definition of $\bar{u}_{\alpha\beta}$ (γ_i), it follows that only the next moduli have non-zero values

$$\begin{aligned}
B_1 &= C_1 = -p;\\
B_{11} &= \frac{1}{3}(-C_1 + C_{11} + 2C_{12});\\
B_{33} &= C_1 + \frac{1}{2}(C_{11} - C_{12}) = \frac{3}{4}B_{22};\\
B_{44} &= B_{55} = B_{66} = C_1 + C_{44};\\
B_{111} &= -B_{11} + V;\\
B_{133} &= \frac{3}{4}B_{122} = B_{11} + \frac{4}{3}B_{33} + \frac{1}{3}M_1;
\end{aligned} \tag{1.112}$$

$$B_{144} = B_{155} = B_{166} = B_{11} + \frac{2}{3}B_{44} + M_2;$$
$$B_{233} = -2B_{33} - T_1;$$
$$B_{244} = B_{256} = \frac{2}{3}(2B_{33} - B_{44} + T_2); \quad B_{456} = B_{44} + T_3;$$
$$B_{222} = \frac{4}{3}(-B_{33} + T_1); \quad B_{266} = -\frac{2}{3}(B_{33} + B_{44} + 2T_2);$$
$$B_{344} = -B_{355} = B_{44} + T_2; \quad B_{333} = -3B_{33}; \quad B_{366} = B_{33} - B_{44}.$$

Here

$$B_{11} = -\frac{1}{3}C_1 + B; \quad B = \frac{1}{3}(C_{11} + 2C_{12});$$
$$B_{33} = C_1 + C'; \quad C' = \frac{1}{2}(C_{11} - C_{12}); \tag{1.113}$$

and

$$V = \frac{1}{9}(C_1 + C_{111} + C_{112} + 2C_{123}); \quad M_1 = -C_1 + \frac{1}{2}(C_{111} + C_{123});$$
$$M_2 = \frac{1}{3}(-C_1 + C_{144} + 2C_{166}); \quad T_1 = \frac{1}{6}(-8C_1 - C_{111} - 3C_{112} + 2C_{123});$$
$$T_2 = -C_1 + \frac{1}{2}(C_{166} - C_{144}); \quad T_3 = -C_1 + C_{456}. \tag{1.114}$$

We suppose the first six moduli B_{ikl} to be independent. Note that formulae (1.112)–(1.114) are valid at $p \neq 0$. In this case, C_{ik} and C_{ikl} are the Birch moduli depending on p (elastic constants). The moduli B_{ik} can be obtained as a combination of the Birch moduli $\mathcal{B}_{\alpha\beta}$ only, which are measured for CI crystals in ultra-sonic experiments. The Fuchs moduli B_{ik} for the cubic crystal relate to the Birch moduli $\mathcal{B}_{\alpha\beta}$ as

$$B_{11} = \frac{1}{3}(\mathcal{B}_{11} + 2\mathcal{B}_{12}); \quad B_{33} = \frac{1}{2}(\mathcal{B}_{11} - \mathcal{B}_{12}); \quad B_{44} = \mathcal{B}_{44}. \tag{1.115}$$

Taking into account the physical meaning of γ_i [24], it will be seen that V in (1.114) characterizes the pure volume deformation, and T_i is a characteristic of pure shear deformations; M_i is a measure of mixed deformations. Nonlinear reaction of the system on the external stress, described by C_{ikl}, is impossible to divide into pure volume and pure shear deformations.

It is possible to obtain the three moduli B_{111}, B_{133}, B_{144} from measurements of $\partial B_{11}/\partial p$, $\partial B_{11}/\partial p$ and $\partial B_{11}/\partial p$ at $p \cong 0$ using the relations

$$B_{111} = -B\left[\frac{\partial B_{11}}{\partial p} + 1\right] = -B\left[1 + \frac{\partial lnB}{\partial \ln \Omega}\right];$$
$$B_{133} = C' - B\frac{\partial B_{33}}{\partial p} = C'\left[1 + \frac{\partial lnB_{33}}{\partial \ln \Omega}\right]; \tag{1.116}$$
$$B_{144} = \frac{1}{3}C_{44} - B\frac{\partial B_{44}}{\partial p} = C_{44}\left[\frac{1}{3} + \frac{\partial lnB_{44}}{\partial \ln \Omega}\right].$$

Let us also write expressions for $\partial B_{ik}/\partial p$ at $p = 0$ through the volume V and the mixed deformations M_1 and M_2 introduced above:

$$\frac{\partial B_{11}}{\partial p} = -\frac{V}{B}; \quad \frac{\partial B_{33}}{\partial p} = -\frac{1}{3B}\left[C' + 3B + M_1\right];$$

$$\frac{\partial B_{44}}{\partial p} = -\frac{1}{B}\left[\frac{1}{3}C_{44} + B + M_2\right]. \tag{1.117}$$

1.2.4 The equation of state

One of the most important subjects of the present review is the study of solids under action of external stress and the calculation of respective crystal properties as a function of the stress, described in general case by the stress tensor t_{ij}. We will consider neither external electric and magnetic fields, no crystals with magnetic or ferroelectric order. Hence we can consider the stress tensor as to be symmetric: $t_{ij} = t_{ji}$. As the equation of state in the general case, we will mean the relation between the stress tensor t_{ij} and the distortion tensor u_{ij} at some given temperature and entropy.

Let us consider three states: $\{X\}$ – arbitrary initial state, $\{x\}$ – the stressed state to be considered, $\{x'\}$ – the state infinitely close to $\{x\}$. Transition from one state to another is characterized by some parameters illustrated by the next scheme

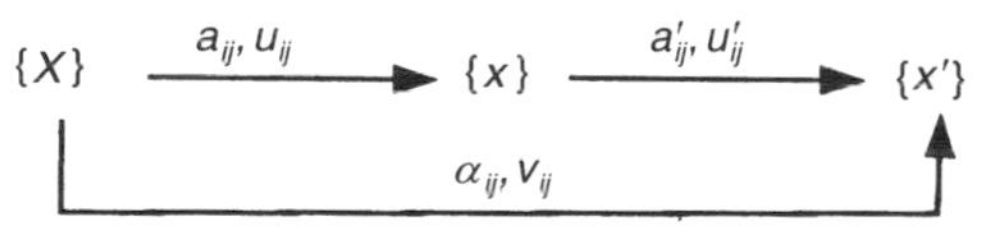

The distortion tensor is by definition

$$u_{ij} = -\delta_{ij} + \frac{\partial x_i}{\partial X_j} = -\delta_{ij} + a_{ij}, \tag{1.118}$$

tensors u'_{ij} and v_{ij} are introduced by similar equations.

In the state $\{x\}$, the stress tensor relates to the free energy per unit mass $F(T, x)$ and to the internal energy $U(S, x)$ by the following equations [22, 25]

$$t_{ij}(x) = \rho(x)\left[\frac{\partial F(T, x')}{\partial u'_{ij}}\right]_{T,u'=0}, \tag{1.119}$$

$$t_{ij}(x) = \rho(x)\left[\frac{\partial U(S, x')}{\partial u'_{ij}}\right]_{S,u'=0}, \tag{1.120}$$

where $\rho(x)$ is the density in the state $\{x\}$. Equations (1.119), (1.120) are written for an arbitrary initial state $\{x\}$. As it follows from (1.119), (1.120) the stress tensor generally has nine non-zero components $t_{ij} \neq t_{ji}$. Equations (1.119)–(1.120) may also be written in another form

$$t_{ij}(x) = \rho(x) a_{ik} \left[\frac{\partial F(T, x')}{\partial u_{ik}} \right]_T, \tag{1.121}$$

$$t_{ij}(x) = \rho(x) a_{ik} \left[\frac{\partial U(s, x)}{\partial u_{ik}} \right]_S. \tag{1.122}$$

Relations (1.119)–(1.120) or (1.121)–(1.122) are the general form of the equations of the state to be found. The free energy $F(x)$ or the internal energy $U(x)$ are functions of the distortion tensor u_{ij}

$$F(t, x) = F(T, X, u_{ij}), \tag{1.123}$$

which describes the transition from $\{x\}$ to $\{x'\}$. Equations (1.121), (1.122) describe the relation between stresses acting on the solid and deformations provided by isothermic or adiabatic processes.

The transition from (1.119), (1.120) to (1.121), (1.122) is provided as follows

$$\left(\frac{\partial}{\partial u'_{ij}} \right)_{u'=0} = \left(\frac{\partial}{\partial v_{kl}} \right)_{u'=0} = \frac{\partial v_{kl}}{\partial u'_{ij}}, \tag{1.124}$$

with

$$\frac{\partial v_{kl}}{\partial u'_{j}} = \delta_{ik} a_{ji}. \tag{1.125}$$

Substituting this equation into (1.124), we obtain

$$\left(\frac{\partial}{\partial u'_{ij}} \right)_{u'=0} = a_{jl} \left(\frac{\partial}{\partial v_{il}} \right)_{u'=0} = a_{jl} \frac{\partial}{\partial u_{il}}. \tag{1.126}$$

If the stress tensor is symmetric, then the equations of state may be written as

$$t_{ij}(x) = \frac{1}{2} \rho(x) \left[\frac{\partial F(T, x')}{\partial u'_{ji}} + \frac{\partial F(T, x')}{\partial u'_{ij}} \right]_{u'=0} \tag{1.127}$$

$$t_{ij}(x) = \frac{1}{2} \rho(x) \left[a_{il} \frac{\partial F(T, x')}{\partial u_{ji}} + a_{jl} \frac{\partial F(T, x)}{\partial u_{ij}} \right]. \tag{1.128}$$

In the general case, the free energy of any microscopic model is a function of the volume per unit cell Ω, vectors of the direct ($\mathbf{R}$) and reciprocal ($\mathbf{g}$) lattices: $F(T, x) = F(\Omega, \mathbf{R}, \mathbf{g})$. It is more convenient to use the derivatives of F over Ω and over the components of vectors $\mathbf{R}$ and $\mathbf{g}$. Let us restrict our study to the analysis of high symmetry crystals (cubic and HCP) and by stresses acting along the symmetry axis. In the coordinate system with axes oriented along the main crystal axis, the distortion tensor u_{ij} has a diagonal form. Let γ_i be its diagonal elements. The vectors **R**, **g** and the cell volume Ω may be expressed through parameters γ_i as

$$\mathbf{R} = \left\{ R_x^0(1+\gamma_1);\ R_y^0(1+\gamma_2);\ R_z^0(1+\gamma_3) \right\} \tag{1.129}$$

$$\mathbf{g} = \left\{ \frac{g_x^0}{1+\gamma};\ \frac{g_y^0}{1+\gamma};\ \frac{g_z^0}{1+\gamma} \right\}. \tag{1.130}$$

$$\Omega = \Omega_0(1+\gamma_1)(1+\gamma_2)(1+\gamma_3). \tag{1.131}$$

Variables with index 0 correspond to the non-stressed state $\{X\}$. Using (1.129)–(1.131) we obtain (no summation over repeating indexes)

$$(1+\gamma_i)\frac{\partial g_\alpha}{\partial \gamma_i} = -g_\alpha \delta_{\alpha i}, \tag{1.132}$$

$$(1+\gamma_i)\frac{\partial R_\alpha}{\partial \gamma_i} = R_\alpha \delta_{\alpha i}, \tag{1.133}$$

$$(1+\gamma_i)\frac{\partial \Omega}{\partial \gamma_i} = \Omega. \tag{1.134}$$

The equations of state (1.127) now have the form

$$\Omega\frac{\partial F}{\partial \Omega} - \sum_{\mathbf{g}} g_\alpha \frac{\partial F}{\partial g_\alpha} + \sum_{\mathbf{R}} R_\alpha \frac{\partial F}{\partial R_\alpha} = \Omega t_{\alpha\alpha}. \tag{1.135}$$

By calculations it is convenient to use the macroscopic form of the equation of the state

$$\begin{aligned} t_{ij}(x) = {} & C^T_{ijkl} u_{kl} + \left[C^T_{inkl}\delta_{jm} + C^T_{njkl}\delta_{im} + C^T_{ilmn}\delta_{jk} + C^T_{ljmn}\delta_{ik} \right. \\ & \left. + C^T_{ijln}\delta_{km} - C^T_{ijkl}\delta_{mn} - C^T_{ijmn}\delta_{kl} + C^T_{ijklmn} \right] u_{kl} u_{mn} + O(u^3). \end{aligned} \tag{1.136}$$

Here $C^T_{ijkl...} = C_{ijkl...}(X)$ are the isothermic elastic moduli in the initial non-stressed state $\{X\}$.

For high symmetry (cubic and hexagonal) crystals, which we will predominantly consider below, the equation of state must be transformed to another form. Let us take into account that, in this case, the free energy depends on two parameters of the crystal only: on the volume Ω and on the ratio c/a. In the case of the hexagonal symmetry, the equation of state is

$$\begin{aligned} p &= \frac{\partial F(T,\Omega,c/a)}{\partial \Omega}, \\ 0 &= \frac{\partial F(T,\Omega,c/a)}{\partial (c/a)}. \end{aligned} \tag{1.137}$$

In the case of cubic crystals $c/a \equiv 1$ and the second equation in (1.137) is an identity.

Finally, let us provide some notes relating to the internal deformations characterized by parameters $\gamma_7, \gamma_8, \ldots$. If the crystal contains few atoms per unit cell, then we also need to take into account internal deformations $u_i^s = \gamma_{i+6}$ [20]. They characterize distortions of atoms of s-sublattice from the equilibrium positions in the unit cell. In this case it is necessary to consider the free energy as a function of $3(n-1)$ parameters $u_i^s = \gamma_{i+6}$ (where n is number of atoms per unit cell). Besides relations like (1.135) or (1.136), the equation of state will additionally include the internal equilibrium condition

$$\frac{\partial F(T, u_{ij}, u_i^s)}{\partial u_j^s} = 0.$$

1.3 The Adiabaticity Problem and Electron–Phonon Interaction in Solids

The quantum-mechanical theory of properties of solids reflects all difficulties rising in the theory of many-particle systems. Since it is impossible to obtain an exact solution, the approximated methods play a great role. In solids, the mass of the atomic nucleus is much larger than the electron mass ($M \gg m$) and generally atoms move much slower than electrons. This fact is a basis for the so-called adiabatic approximation, which is one of the most powerful methods in the theory of molecules [20, 43] and solids [26, 27].

The adiabatic approximation was first introduced by Born and Oppenheimer in 1927. They calculated the energy of a molecule as a series in a small parameter $\kappa = (m/M)^{1/4}$ and the equation of motion for a heavy nucleus was obtained with the help of expansion in the powers of this parameter. The most important feature of this approach is a method of separation of electron and nuclear variables submitted by the authors. This division is not complete because deformation of electron states provided by motion of a nucleus determines the effective potential of nuclear motion. Therefore, within the adiabatic approximation there is a many-particle potential $\Phi_0(\{\mathbf{R}\})$ (where $\{\mathbf{R}\}$ denotes all atomic coordinates) that determines the atomic motion. This potential is the sum of the electron gas energy at fixed atomic coordinates and the electrostatic energy of interaction of ions. Note that any phenomenological theory of lattice dynamics is based on the existence of $\Phi_0(\{\mathbf{R}\})$.

The adiabatic criterion

$$\hbar\omega_D \ll |\varepsilon_m - \varepsilon_n| \tag{1.138}$$

(where ω_D is some characteristic frequency of atomic oscillations, ε_m is the energy of m-th electron state) is relatively well satisfied for dielectrics and semiconductors as a consequences of the gap in the electron spectrum.

In metals, this criterion is broken for electrons from narrow regions $|\varepsilon_\mathbf{k} - E_F| \sim \hbar\omega_D$ ($\varepsilon_\mathbf{k}$ is the electron energy, E_F is the Fermi energy). Near the Fermi surface there are electron transitions with arbitrary small energy of excitation $|\varepsilon_\mathbf{k} - \varepsilon_{\mathbf{k}+\mathbf{q}}|$, so that it is possible $\hbar\omega_\mathbf{q} \geq |\varepsilon_\mathbf{k} - \varepsilon_{\mathbf{k}+\mathbf{q}}|$ (where $\omega_\mathbf{q}$ is the frequency of the phonon with impulse $\mathbf{q}$) and hence, the adiabatic criterion (1.138) is broken. But for the main part of collective electrons $|\varepsilon_{\mathbf{k}} - \varepsilon_{\mathbf{k+q}}| \sim E_F \ll \hbar\omega_D$ and these electrons adiabatically follow oscillating nucleus.

This feature makes it possible to use the adiabatic approximation by the calculation of values determined as integrals in the electron spectrum. These problems have been analyzed in [26, 27] In [26] it has been shown that phonons may be described with great accuracy by the adiabatic approximation. Note also, that the electron–phonon interaction provides very small renormalization of the dispersion equation [19] (see also subsection 2). We will see that the negative aspect of using the adiabatic perturbation theory [27] is the complexity of practical application. Therefore, in concrete calculations in the theory of metals the Fröhlich model has been used [28, 29], which also has the form of some perturbation theory. In this model, the Hamiltonian of electron–phonon interaction is the sum of Hamiltonians of free electron and phonon fields $\hat{\mathcal{H}}_0$ and the Hamiltonian of interaction between both these fields

$$\hat{\mathcal{H}}'_F = -\sum_l (\nabla_r V(\mathbf{r} - \mathbf{R}_{0l}), \Delta\mathbf{R}_l), \tag{1.139}$$

where $V(\mathbf{r} - \mathbf{R}_{0l})$ is the potential of interaction of the l-th ion with equilibrium coordinate $\mathbf{R}_{0l}$ and the electron with a coordinate $\mathbf{r}$, $\Delta\mathbf{R}_l$ is the ion displacement from its position in equilibrium. For simplicity, it is postulated that the bare phonon frequency, i.e., the frequency of the free phonon field $\omega_0(\mathbf{q})$, is described by the acoustic dispersion law at small phonon wave vectors $\mathbf{q}$. All attempts to demonstrate equivalence of the adiabatic theory and the Fröhlich theory were unsuccessful. In [28] it is shown that it is impossible to obtain the Fröhlich model from an exact Hamiltonian for the system of electrons and phonons by any choice of the zero order Hamiltonian. The question of the limits of applicability of the Fröhlich model and the general problem of the basis of the metal theory was the main consideration in [27]. Though the Hamiltonian of the Fröhlich model is not compatible with the total Hamiltonian for the electron–ion system, some important results obtained in the frame of this approach are correct [26] (see also Subsection 2 of this section).

Recently, there have been many publications relating to the Kondo systems [30]. It is well-known that the Kondo system is described by the approach that is opposite to the adiabatic approximation, namely the effective electron mass is larger than ion mass. In nature, there are many systems where masses or the charged carriers (positive and negative) are of the same order of magnitude. Present versions of the adiabatic perturbation theory, either in canonical form or in the form of a general quantum-mechanical perturbation theory [31], are based on the use of a small parameter κ and, consequently, cannot be used in these cases ($m^* \geq M$, where m^* is the effective electron mass).

In quantum-mechanical perturbation theories (the Fröhlich model), free phonons with some dispersion law are introduced as the bare Bose-perturbations. It is well-known [26] that the electron systems could not be reduced in any self-consistent way to any system of bare electrons and phonons with some interaction between them. Including any bare phonons automatically assume the participation of electrons in the formation of bare phonons. More consistent seems to be the approach where the ion plasma waves on a fixed electronic background are used as bare Bose-perturbations [32]. As a small parameter, it is possible to use some physical parameters: relative displacements of ions and the ratio of the pseudopotential on the reciprocal lattice to the Fermi energy. The electron and phonon spectra are the solutions of a rather general system of equations and it is not necessary to classify these states as adiabatic or non-adiabatic. If the small parameter κ exists, i.e., for the adiabatic system, this approach provides the same results as the adiabatic perturbation theory.

In this section, the non-adiabatic effects are investigated in the framework of some versions of the adiabatic perturbation theory developed by Geĭlikman [27, 28, 33, 34]. As mentioned in [28], there are two versions of the canonical form of the adiabatic perturbation theory. Both these versions are the topic of subsection 1. Here we have proposed [31] to find the solution of the oscillation problem by using the perturbation theory for calculation of corrections to the energy of the adiabatic state. It allows one to calculate corrections to the energy in any order in κ^2. Subsection 2 describes the construction of the usual quantum-mechanical perturbation theory on the basis of the adiabatic expansion. We will see that this theory has a structure other than the adiabatic perturbation theory [20]. Nevertheless, both theories provide equivalent results.

This formulation of the adiabatic approach as some quantum-mechanical perturbation theory allows the introduction of the diagram technique with some distinguishable peculiarities from the standard one [23]. In subsection 3 the adiabatic theory of the electron–phonon interaction is constructed [34]. We will consider the corrections to vertices, the phonon mass-operator, and the self-energy operator for the electron. In subsection 4, we will discuss the thermodynamics of the system of interacting electrons and phonons.

1.3.1 The adiabatic perturbation theory

The Schrödinger equation for a system of electrons with coordinates $\mathbf{r}_i$ and ions with coordinates $\mathbf{R}_i$ is

$$\left[\hat{\mathcal{H}}_e + \hat{\mathcal{H}}_i + \hat{\mathcal{H}}_{ei}(\mathbf{r},\mathbf{R})\right]\Psi(\mathbf{r},\mathbf{R}) = E\,\Psi(\mathbf{r},\mathbf{R}), \tag{1.140}$$

$$\hat{\mathcal{H}}_e = \sum_i \frac{\hat{\mathbf{p}}_i^2}{2m}; \quad \hat{\mathcal{H}}_i = \sum_l \frac{\hat{\mathbf{P}}_l^2}{2M};$$

$$\hat{\mathcal{H}}_{ei}(\mathbf{r},\mathbf{R}) = V(\mathbf{r},\mathbf{R}) + V(\mathbf{r},\mathbf{r}) + V(\mathbf{R},\mathbf{R}).$$

Here $\hat{\mathcal{H}}_e$ and $\hat{\mathcal{H}}_i$ are the kinetic energies of electrons and ions, respectively; $\hat{\mathcal{H}}_{ei}$ is the total potential energy of the system.

To evaluate the order of magnitude of the oscillating energy of electrons and ions, it is possible to use the following consideration [35]. The electron energy is of the order

$$\varepsilon = \frac{p^2}{2m} \sim \frac{\hbar^2}{md^2}, \tag{1.141}$$

where d is the order of the lattice constant. The energy of ion oscillations is

$$E_{vib} = \hbar\omega = \hbar\sqrt{k/M} \tag{1.142}$$

where k is the elastic factor for the ion potential energy (within the harmonic approximation). In (1.142) it was already supposed that the ion displacements from the equilibrium were small. Since (as we can see later) the potential energy of ion oscillations is the electron energy, then

$$k = \left.\frac{\partial^2 \varepsilon}{\partial R^2}\right|_{R=R_0} \sim \frac{\varepsilon}{d^2} \tag{1.143}$$

where R_0 denotes the equilibrium. Therefore,

$$E_{vib} = \hbar\left(\frac{\varepsilon}{Md^2}\right)^{1/2} \sim \frac{\hbar^2}{d^2(mM)^{1/2}} \sim \left(\frac{m}{M}\right)^{1/2}\varepsilon. \tag{1.144}$$

the electron energy is $(M/m)^{1/2}$ times greater than the energy of ion oscillations.

It is generally accepted that the small parameter of the adiabatic theory is $(m/M)^{1/4}$. Actually, we are interested in the case of stable crystal relating to decay in the constituent parts. It means that the mean amplitude of zero oscillations a_0 is much smaller than the unit cell dimension. From the harmonic oscillator theory [36], it is well-known that $a_0^2 = \hbar/(2M\omega)$ where ω is the oscillator frequency, therefore,

$$a_0^2 \sim \frac{\hbar}{2M\omega} = \frac{\hbar^2}{2ME_{vib}}. \tag{1.145}$$

Substituting this expression into (1.44) and (1.141) and using equations (1.142) and (1.143) provide the parameter of the theory as[5]

$$\kappa = \frac{a_0}{d} \sim \left(\frac{m}{M}\right)^{1/4}. \tag{1.146}$$

However, the physical parameter characterizing small anharmonic contributions to thermodynamic values is the ratio $a_0^2/d^2 \ll 1$, which transforms at high temperatures into the well-known ratio T/ε_{at} where T/ε_{at} is the depth of the potential well. This result follows from the analysis of the series of the thermodynamic perturbation theory and is the reason to choose $a_0^2/d^2 \sim (m/M)^{1/2}$ as the adiabatic parameter. As we will see below, the final results contain the powers of κ^2 only.

In the adiabatic approximation the ion system is supposed to be frozen in some (close to equilibrium) configuration and the electron energy $E_n(\mathbf{R})$ of the n-th state is calculated. This energy provides the effective potential energy for ion oscillations and, hence, determines the phonon spectrum. It is possible to find the equilibrium

[5] The adiabatic parameter $(m/M)^{1/4}$ may be obtained in another way (see, e.g., [44])

ion configuration through the minimization of $E_n(\mathbf{R})$. Further, we can consider the ion oscillations near these equilibrium positions. On the other hand, it is possible first to solve the problem for the equilibrium configuration and calculate $E_n(\mathbf{R})$. In accordance with these possibilities, further calculations are also modified. For this reason, there are two versions of the adiabatic expansion that will be considered below.

Adiabatic expansion for arbitrary ion configuration Since the ion kinetic energy $\hat{\mathcal{H}}_i$ is κ^2 times smaller than the electron energy, it is not taken into account in the zeroth approximation and the equation is solved at some arbitrary (but close to equilibrium) (fixed) ion configuration $\mathbf{R}_l$.

$$[\hat{\mathcal{H}}_e + \hat{\mathcal{H}}_{ei}(\mathbf{r},\mathbf{R})]\psi_m(\mathbf{r},\mathbf{R}) = E_m(\mathbf{R})\psi_m(\mathbf{r},\mathbf{R}) \tag{1.147}$$

In this equation the wave function $\psi_m(\mathbf{r},\mathbf{R})$ and the electron energy $E_m(\mathbf{R})$ parametrically depends on ion coordinates.

Let us suppose we have to solve the equation (1.147) i.e., have to find the eigenvalues $E_m(\mathbf{R})$ and eigenfunctions $\psi_m(\mathbf{r},\mathbf{R})$ of electrons. In this case it is possible to search for the wave function of the general system of electrons and ions $\Psi_n(\mathbf{r},\mathbf{R})$ as a series on a complete orthogonal system of the electron wave functions $\psi_m(\mathbf{r},\mathbf{R})$:

$$\Psi_n(\mathbf{r},\mathbf{R}) = \sum_m \Phi_{nm}(\mathbf{R})\psi_m(\mathbf{r},\mathbf{R}). \tag{1.148}$$

Here index n shows that we are searching for the general Ψ-function which is close to the n-th electron state and where the main contribution corresponds to $m = n$. Let us substitute (1.148) into (1.140) then multiply it from the right by $\psi_s^*(\mathbf{r},\mathbf{R})$ and integrate it over electron coordinates. The final result for $s = n$ is

$$\left[\hat{\mathcal{H}}_i + E_n(\mathbf{R})\right]\Phi_{nn}(\mathbf{R}) + \sum_{m\neq n}\hat{C}_{nm}\Phi_{mn}(\mathbf{R}) + \hat{C}_{nn}\Phi_{nn}(\mathbf{R}) = E_n\Phi_{nn}(\mathbf{R}). \tag{1.149}$$

At $s \neq n$ we have

$$\left[\hat{\mathcal{H}}_i + E_s(\mathbf{R})\right]\Phi_{sn}(\mathbf{R}) + \sum_{m\neq n}\hat{C}_{sm}\Phi_{mn}(\mathbf{R}) + \hat{C}_{sn}\Phi_{nn}(\mathbf{R}) = E_n\Phi_{sn}(\mathbf{R}). \tag{1.150}$$

In equations (1.149) and (1.150)

$$\begin{aligned}
\hat{C}_{nm} &= \hat{A}_{nm} + \hat{B}_{nm},\\
\hat{A}_{nm} &= -\frac{\hbar^2}{M}\sum_l \int d\mathbf{r}\left(\psi_n^*(\mathbf{r},\mathbf{R})\frac{\partial}{\partial\mathbf{R}_l}\psi_m(\mathbf{r},\mathbf{R})\right)\frac{\partial}{\partial\mathbf{R}_l},\\
\hat{B}_{nm} &= -\frac{\hbar^2}{2M}\sum_l \int d\mathbf{r}\left\{\psi_n^*(\mathbf{r},\mathbf{R})\frac{\partial^2}{\partial\mathbf{R}_l^2}\psi_m(\mathbf{r},\mathbf{R})\right\}.
\end{aligned} \tag{1.151}$$

In the absence of a magnetic field, the functions ψ may be chosen to be real, i.e., $\hat{A}_{nm} = 0$ since all $\psi_n(\mathbf{r},\mathbf{R})$ are supposed to be normalized to unity for each R. The system of equations (1.149), (1.150) is exact. If we neglect the non-diagonal matrix elements then (1.149) provides the "adiabatic" equation for the energy levels ν of the oscillatory problem at some fixed n [20]

$$[\hat{\mathcal{H}}_i + E_n(\mathbf{R}) + \hat{B}_{nn}]\Phi_{nn\nu}(\mathbf{R}) = E_{n\nu}\Phi_{nn\nu}(\mathbf{R}). \tag{1.152}$$

The energy levels $E_{n\nu}$ of the oscillatory problem are determined by properties of the n-th electron state only and there are no electron transitions relating to ion motion. The equation (1.152) shows that the ion motion described by the function $\Phi_{nn\nu}(\mathbf{R})$ is determined by the effective potential energy $E_n(\mathbf{R}) + \hat{\mathcal{B}}_{nn}$, where $\hat{\mathcal{B}}_{nn} \sim \kappa^2\hat{\mathcal{H}}_i$ The main contributions to $\hat{B}_{nn}$ are given by regions located near the ions. It is generally supposed that these contributions are independent from the ion coordinates. Consequently, it is possible to set $\hat{B}_{nn} \cong$ const and to include it in the definition of the origin of the energy scale [36]. Electrons play a subordinate role. Their reaction on the ion motion is so fast, that the electron wave function depends on the instantaneous values of the ion coordinates only. Here we suppose the electron wave function $\psi_n(\mathbf{r}, \mathbf{R})$ to be a continuous function of $\mathbf{R}$. During ion oscillations, the electrons have time to follow the ion motion so that there is no change of the electron state caused by ion motion. Hence, electrons are always in the same electron state n. It means that the adiabatic approximation allows us to consider the electron states to be independent from the oscillatory states of the lattice. Each subsystem provides its own contribution to the total energy.

The interaction of these subsystems is described by non-diagonal matrix elements $\hat{C}_{nm}$ The adiabatic description, allowing electron configuration to be constant, must be modified to include non-adiabatic corrections caused by the operator $\hat{C}_{nm}$ with $m \neq n$. To do it we must solve equations (1.148) and (1.150) simultaneously. It is necessary to write non-diagonal matrix elements $\Phi_{sn}(\mathbf{R})$ in (1.150) through the diagonal elements $\Phi_{nn}(\mathbf{R})$ and to substitute the result into equation (1.149). Finally, we obtain the system of equations that determines the ion wave functions $\Phi_{nn}(\mathbf{R})$ and the energy E_n with arbitrary accuracy, if it is possible to solve the equations (1.149), (1.150) by the iterative method. To do this, it is necessary to expand all values in (1.149), (1.150) series in κ

$$E_n(\mathbf{R}) = E_n(\mathbf{R}_0) + \sum_{\lambda=1}^{\infty} U_{\lambda n},$$

$$U_{\lambda n} = \frac{1}{\lambda!}\sum_{\substack{i\ldots j\\ \alpha\ldots\beta}} \left.\frac{\partial^{(\lambda)}E_n(\mathbf{R})}{\partial R_{i\alpha}\ldots\partial R_{j\beta}}\right|_{\mathbf{R}_0} \Delta R_{i\alpha}\ldots\Delta R_{j\beta}$$

$$U_{\lambda n} \sim \kappa U_{(\lambda-1)n}, \ \Delta R_{i\alpha} = R_{i\alpha} - R_{0i\alpha};$$

$$\hat{A}_{sn}(\mathbf{R}) = \hat{A}^{(0)}_{sn} + \hat{A}^{(1)}_{sn} + \cdots = \hat{A}_{sn}(\mathbf{R}_0) + \sum_{i,\alpha}\left.\frac{\partial\hat{A}_{sn}}{\partial R_{i\alpha}}\right|_{\mathbf{R}_0}\Delta R_{i\alpha} + \cdots,$$

$$E_n = E^{(0)}_n + \kappa E^{(1)}_n + \kappa^2 E^{(2)}_n + \cdots, \tag{1.153}$$

$$\Phi_{nn}(\mathbf{R}) = \Phi^{(0)}_{nn}(\mathbf{R}) + \kappa\Phi^{(1)}_{nn}(\mathbf{R}) + \kappa^2\Phi^{(2)}_{nn}(\mathbf{R}) + \cdots.$$

Let us evaluate orders of $\hat{A}_{sn}^{(0)}$ and $\hat{B}_{sn}^{(0)}$. Remember that the electron wave function $\psi_n(\mathbf{r}, \mathbf{R})$ is periodic with period d (the lattice parameter), and the characteristic scale of the ion wave function $\Phi_{nn}(\mathbf{R})$ is of the order the amplitude of zero oscillations a_0. Hence

$$\frac{\partial \psi_n}{\partial R_{i\alpha}} \sim \frac{\psi_n}{d}, \quad \frac{\partial \Phi_{nn}}{\partial R_{i\alpha}} \sim \frac{\Phi_{nn}}{a_0}. \tag{1.154}$$

The series expansion of $\hat{A}_{sn}$ begins with a term proportional to $\kappa \hbar \omega_D \Phi_{nn}$

$$\hat{A}_{sn}^{(0)} \Phi_{nn} = -\frac{\hbar^2}{M} \sum_{i\alpha} \int d\mathbf{r}\, \psi_s^* \frac{\partial \psi_n}{\partial R_{i\alpha}} \frac{\partial \Phi_{nn}}{\partial R_{i\alpha}} \sim \frac{\hbar^2 \Phi_{nn}}{M d a_0} \sim \kappa \hbar \omega_D \Phi_{nn} \sim \kappa^3 E_F \Phi_{nn}. \tag{1.155}$$

Similar expansion for $\hat{B}_{sn}$ is

$$\hat{B}_{sn}^{(0)} \Phi_{nn} = -\frac{\hbar^2}{2M} \sum_j \int d\mathbf{r}\, \psi_s^* \Delta_{R_j} \psi_n \Phi_{nn}$$
$$\sim \kappa^2 \hbar \omega_D \Phi_{nn} \sim \kappa^4 E_F \Phi_{nn}. \tag{1.156}$$

The last evaluations in (1.155) and (1.156) were obtained by using the relation

$$\kappa^2 = \frac{a_0^2}{d^2} \sim \frac{\hbar \omega_D}{E_F}$$

Therefore, the order of $\hat{B}_{sn}$ is κ^4 and higher, whereas the expansion of $\hat{A}_{sn}$ begins from a term of the order κ^3. Taking in $\hat{A}_{sn}$ and $\hat{B}_{sn}$ the terms of the same order in κ and introducing the next notations:

$$E_n(\mathbf{R}_0) \equiv E_{s0}$$

we obtain

$$\left(E_{s0} - E_n^{(0)}\right) \Phi_{sn}^{(0)} = 0, \tag{1.157}$$

$$\left(E_{s0} - E_n^{(0)}\right) \Phi_{sn}^{(1)} - E_n^{(1)} \Phi_{sn}^{(0)} = 0, \tag{1.158}$$

$$\left(E_{s0} - E_n^{(0)}\right) \Phi_{sn}^{(2)} - E_n^{(1)} \Phi_{sn}^{(1)} - \left(\hat{\mathcal{H}}_i + U_{2s} - E_n^{(2)}\right) \Phi_{sn}^{(0)} = 0. \tag{1.159}$$

Since $E_n^{(0)} \equiv E_n(R_0) \neq E_s(R_0)$ (see below) then from (1.157) it follows that $\Phi_{sn}^{(1)} = 0$ and $\Phi_{sn}^{(2)} = 0$ The first non-zero term in $\Phi_{sn}(\mathbf{R})$ is of the order κ^3 (1.155), the next term is $\sim \kappa^4$ (1.156)

$$\Phi_{sn}^{(3)} = \frac{\hat{A}_{sn}^{(0)}}{E_{n0} - E_{s0}} \Phi_{nn}^{(0)}. \tag{1.160}$$

By repeating this procedure, it is possible to obtain Φ_{sn} with arbitrary accuracy.

The same procedure for the equation (1.149) provides

$$\left(E_{n0} - E_n^{(0)}\right) \Phi_{nn}^{(0)} = 0 \tag{1.161}$$

$$\left(E_{n0} - E_n^{(0)}\right)\Phi_{nn}^{(1)} - E_n^{(1)}\Phi_{nn}^{(0)} = 0 \tag{1.162}$$

$$\left(\hat{\mathcal{H}}_i - U_{2n} - E_{n\nu}^{(0)}\right)\Phi_{nn}^{(0)} = 0 \tag{1.163}$$

Since we are searching for $\Phi_{nn} \neq 0$, the ground state energy is $E_n(\mathbf{R}_0) = E_{n0} = E_n^{(0)}$ that was supposed at first. From (1.163) with the usage of (1.162), it follows that $E_n^{(1)} = 0$. We have introduced in (1.154) the new notation $E_{n\nu}^{(0)} \equiv E_n^{(2)}$ to emphasize that $E_{n\nu}^{(0)}$ corresponds to the energy of zero motion for ion oscillations. Index ν indicates the oscillatory quantum numbers. In the harmonic approximation, equation (1.154) is equivalent to the equation (1.152) for the effective potential energy $E_n(\mathbf{R}_0) + \hat{B}_{nn}$ that must be so [20]. Hence, $E_{n\nu}^{(0)}$ corresponds to the energy and $\Phi_{nn\nu}^{(0)}$ to the ion wave function in the harmonic approximation, which, as we can see, is a particular case of the adiabatic approximation. The total energy of a whole electron and ion system is simply the sum $E_n(\mathbf{R}_0) + E_{n\nu}^{(0)}$ and the total wave function is $\Psi_n^{(0)}(\mathbf{r}, \mathbf{R}) = \Phi_{nn\nu}^{(0)}(\mathbf{R})\psi_n(\mathbf{r}, \mathbf{R}_0)$.

Note that the theory under consideration does not calculate the energy of the ground state of the electron subsystem $E_{n0} = E_n(\mathbf{R}_0)$, as is supposed to be done. The solution of equation (1.162) confirms only the fact that it exists.

On the order of magnitude $E_{n\nu} \sim \hbar\omega_D$, therefore, correction to $E_{n\nu}^{(0)}$ will be the series in κ. It is absolutely clear that these corrections are due to anharmonic contributions (see section 4) or directly due to the operator $\hat{C}_{nm}$ in (1.151). Anharmonic contributions reach an appreciable value before contributions of the operator $\hat{C}_{nm}$ (with $m \neq n$) (see below).

The system of equations determines the corrections to the energy $E_{n\nu}^{(0)}$ up to terms of the order κ^6 inclusively having the form

$$\left(U_{3n} - E_n^{(3)}\right)\Phi_{nn\nu}^{(0)} + \left(\hat{\mathcal{H}}_i + U_{2n} - E_{n\nu}^{(0)}\right)\Phi_{nn\nu}^{(1)} = 0, \tag{1.165}$$

$$\begin{aligned}&\left(U_{4n} + \hat{B}_{nn}^{(0)} - E_n^{(4)}\right)\Phi_{nn\nu}^{(0)} + \left(U_{3n} - E_n^{(3)}\right)\Phi_{nn\nu}^{(1)}\\ &\quad + \left(\hat{\mathcal{H}}_i + U_{2n} - E_{n\nu}^{(0)}\right)\Phi_{nn\nu}^{(2)} = 0,\end{aligned} \tag{1.166}$$

$$\begin{aligned}&\left(U_{5n} + \hat{B}_{nn}^{(1)} - E_n^{(5)}\right)\Phi_{nn\nu}^{(0)} + \left(U_{4n} + \hat{B}_{nn}^{(0)} - E_n^{(4)}\right)\Phi_{nn\nu}^{(1)}\\ &\quad + \left(U_{3n} - E_n^{(3)}\right)\Phi_{nn\nu}^{(2)} + \left(\hat{\mathcal{H}}_i + U_{2n} - E_{n\nu}^{(0)}\right)\Phi_{nn\nu}^{(3)} = 0,\end{aligned} \tag{1.167}$$

$$\begin{aligned}&\left[U_{6n} + \hat{B}_{nn}^{(2)} + \sum_{m\neq n}\frac{\hat{A}_{nm}^{(0)}\hat{A}_{mn}^{(0)}}{E_{n0} - E_{m0}} - E_n^{(6)}\right]\Phi_{nn\nu}^{(0)} + \left(U_{5n} + \hat{B}_{nn}^{(1)} - E_n^{(5)}\right)\Phi_{nn\nu}^{(1)}\\ &\quad + U_{4n} + \hat{B}_{nn}^{(0)} - E_n^{(4)}\Phi_{nn\nu}^{(2)} + \left(U_{3n} - E_n^{(3)}\right)\Phi_{nn\nu}^{(3)} + \left(\hat{\mathcal{H}}_i + U_{2n} - E_{n\nu}^{(0)}\right)\Phi_{nn\nu}^{(4)} = 0.\end{aligned} \tag{1.168}$$

We can see that terms provided by the operator $\hat{C}_{nm} = \hat{A}_{nm} + \hat{B}_{nm}$ appear only starting at the sixth order in κ (see equation (1.168)), whereas the terms of order $\sim \kappa^3$ in (1.165)–(1.167) are due to the anharmonicity. Hence, with the accuracy up to sin κ^4, the ion motion within the adiabatic approximation is determined by the electron energy $E_n(\mathbf{R})$ only. Terms $\sim \kappa^5$, provided by $\hat{B}_{nm}$, give the contribution proportional to displacements, but the ion motion stays adiabatic. Starting with terms $\sim \kappa^6$, it is no longer applicable, because now the ground state n is added by other states n, as can be seen in (1.168).

In principle, solutions of the system (1.165)–(1.168) determine the zero order corrections to the ground state energy and wave functions. We do not think this method is constructive enough. The equation (1.149) has been solved in [28] by applying the Brillouin–Wigner perturbation theory, but this method is not applicable to many-particle systems [45]. The basis for this is that, in the case of N-particle system, the first term provided by this theory is proportional to N, whereas all the next terms are of the order $(N)^0$ (with no dependence on the perturbation intensity) and may be neglected in comparison with the first. The way to avoid this difficulty is to use the Rayleigh–Schrödinger perturbation theory. In [5], it is shown that accurate and consistent application of this theory provides contributions to the energy proportional to the number of atoms N in any order in perturbation, despite the appearance of non-physical terms proportional to N^2 and so on in intermediate formulae. These non-physical terms are mutually cancelled in all orders of the perturbation theory.

We will see that, in the case of the oscillating problem, it is possible to construct the Rayleigh–Schrödinger perturbation theory. The zero order Hamiltonian is taken as

$$\hat{\mathcal{H}}_0 = \hat{\mathcal{H}}_i + U_{2n}, \tag{1.169}$$

which depends on the ion coordinates ($\mathbf{R}$) only. The perturbation operator

$$\begin{aligned} \hat{\mathcal{H}}' &= U_{3n} + U_{4n} + \cdots + \hat{B}_{nn}^{(0)} + \hat{B}_{nn}^{(1)} + \cdots \\ &+ \sum_{m \neq n} \frac{\hat{A}_{nm}^{(0)} \hat{A}_{mn}^{(0)}}{E_{n0} - E_{m0}} + \sum_{m \neq n} \frac{\hat{A}_{nm}^{(1)} \hat{A}_{mn}^{(0)}}{E_{n0} - E_{m0}} + \cdots, \end{aligned} \tag{1.170}$$

is an infinite series in the parameter κ. Using the perturbation theory (1.170), it must be taken into account that each order of the perturbation theory contains infinite series over powers of κ.

The correction to the zero order energy $E_{n\nu}^{(0)}$ in the first order of the perturbation operator $\hat{\mathcal{H}}'(\mathbf{R})$ is

$$\begin{aligned} E_{n\nu}^{(1)} &= |U_{3n}|_{\nu\nu} + |U_{4n}|_{\nu\nu} + \cdots + |\hat{B}_{nn}^{(0)}|_{\nu\nu} + |\hat{B}_{nn}^{(1)}|_{\nu\nu} + \cdots \\ &+ \sum_{m \neq n} \frac{|\hat{A}_{nm}^{(0)} \hat{A}_{mn}^{(0)}|_{\nu\nu}}{E_{n0} - E_{m0}} + \cdots. \end{aligned} \tag{1.171}$$

After formal application of the perturbation theory, it is necessary to take from these series terms proportional to the power of κ. For example, the first correction to the energy of adiabatic state $E^{(0)}_{n\nu}$ is

$$\Delta E^{(1)}_{n\nu} = |U_{4n}|_{\nu\nu} + |\hat{B}^{(0)}_{nn}|_{\nu\nu} + \sum_{\mu\neq\nu} \frac{|U_{3n}|^2_{\nu\mu}}{E^{(0)}_{n\nu} - E^{(0)}_{n\mu}} \sim \kappa^2 \hbar\omega_D. \tag{1.172}$$

The next correction has the order $\kappa^4\hbar\,\omega_D$

$$\begin{aligned}\Delta E^{(2)}_{n\nu} &= |U_{6n}|_{\nu\nu} + |\hat{B}^{(2)}_{nn}|_{\nu\nu} + \sum_{m\neq n} \frac{|\hat{A}^{(0)}_{nm}\,\hat{A}_{m0}|_{\nu\nu}}{E_{n0} - E_{m0}} + \sum_{\mu}{}' \frac{|U_{4n}|^2_{\nu\mu} + |\hat{B}^{(0)}_{nn}|^2_{\nu\mu}}{E^{(0)}_{\nu\mu} - E^{(0)}_{n\mu}} \\ &+ \sum_{\mu}{}' \frac{2\Big(|U_{3n}|_{\nu\mu}\,|U_{5n} + \hat{B}^{(1)}_{nn}|_{\mu\nu} + |\hat{B}^{(0)}_{nn}|_{\nu\mu}\,|U_{4n}|_{\mu\nu}\Big)}{E^{(0)}_{n\nu} - E^{(0)}_{n\mu}} \\ &+ \sum_{\mu,\lambda}{}' \frac{3|U_{4n}|_{\nu\mu}\,|U_{3n}|_{\mu\lambda}\,|U_{3n}|_{\lambda\nu}}{\Big(E^{(0)}_{n\nu} - E^{(0)}_{n\mu}\Big)\Big(E^{(0)}_{n\nu} - E^{(0)}_{n\lambda}\Big)} \\ &\sum_{\mu,\varepsilon,\lambda}{}' \frac{|U_{3n}|_{\nu\mu}\,|U_{3n}|_{\mu\lambda}|U_{3n}|_{\lambda\varepsilon}\,|U_{3n}|_{\varepsilon\nu}}{\Big(E^{(0)}_{n\nu} - E^{(0)}_{n\mu}\Big)\Big(E^{(0)}_{n\nu} - E^{(0)}_{n\lambda}\Big)\Big(E^{(0}_{n\nu} - E^{(0)}_{n\varepsilon}\Big)}\end{aligned} \tag{1.173}$$

The prime in sums over $\lambda\varepsilon\mu$ notes that these indexes are not equal to ν.

Formally, the first correction to the energy of zero approximation must be written as

$$\Delta E_{n\nu} - |U_{3n}|_{\nu\nu},\, \Delta E_{n\nu} \sim \kappa\hbar\,\omega_D, \tag{1.174}$$

but since the mean value from the odd number of operators over non-disturbed wave functions is equal to zero [29], then[6]

$$\langle\nu|U_{3n}|\nu\rangle = 0. \tag{1.175}$$

Therefore, these corrections to the energy of the system are the series in the parameter κ^2. Each term includes phonon–phonon contributions (first terms in (1.171)–(1.173)) and the electron–phonon interaction. We can see that the first correction to the energy $\sim \kappa^2\hbar\,\omega_D$ does not contain terms describing the interaction between both subsystems: electrons and ions. This correction is predominantly a consequence of anharmonic effects. But the next correction $\sim \kappa^4\hbar\,\omega_D$ contains together with terms determined by phonon–phonon interaction, the term relating to the non-adiabatic contribution $\hat{A}_{nm}$ in equation (1.174).

As follows from the previous investigation, the non-adiabatic corrections to the energy of the electron subsystem have a dual nature: corrections determined by anharmonic effects (including all terms in (1.172) and (1.173) containing $U_{\alpha n}$ starting

---

[6] Actually the vanishing of the matrix element U_{3n} takes place, exactly speaking, only at $T = 0$. In the opposite case, if the thermal expansion of lattices takes place, this matrix element is not equal to zero. This can be simply described in terms of the Bose-condensation of phonons.

with $\alpha = 3$) and direct non-adiabatic corrections determined by the operator $\hat{C} = \hat{A} + \hat{B}$. The term $\Delta E^{(2)}_{n\nu}$ (1.173) includes cross terms like $U_{3n}\hat{B}^{(1)}$ and $U_{4n}\hat{B}^{(0)}$. In this section, we investigate only the non-adiabatic effects related to operators $\hat{A}$ and $\hat{B}$.

In [26, 28], the expressions were obtained for corrections to the adiabatic energy of a metal relating to operators $\hat{B}^{(0)}_{nn}$ and $\hat{A}^{(0)}_{nm}$. Note that the correction relating to $\hat{A}^{(0)}_{nm}$ has the order $\kappa^4\hbar\omega_D$ and the correction determined by $\hat{B}^{(0)}_{nn}$ has the order $\kappa^2\hbar\omega_D$. Let us briefly outline some expressions that are needed for calculations of these corrections to the energy [26]. First, it is necessary to calculate the matrix elements $|\hat{A}^{(0)}|_{\mu\nu}$ and $|\hat{B}^{(0)}|_{\mu\nu}$ with the help of eigenfunctions $\Phi^{(0)}_{n\nu}(\mathbf{R})$ of the adiabatic equation (1.152) at $\hat{B}^{(0)}_{nn} = 0$. Let us transform the integral in the expression for $\hat{A}^{(0)}_{mn}$ to the form (see below)

$$|\nabla_R|_{nm} = \frac{|\nabla_R \hat{\mathcal{H}}_{ei}(\mathbf{r}, \mathbf{R})|_{nm}}{E_{n0} - E_{m0}}. \tag{1.176}$$

It is possible to construct a many-particle wave function of electrons from a simple one-particle Bloch wave functions that are characterized by the wave vectors $\mathbf{k}$. Using the explicit representation of the operator $\hat{\mathcal{H}}_{ei}(\mathbf{r}, \mathbf{R})$ in (1.140), it can be seen that non-zero values have the matrix elements from (1.179) relating to one-electron transitions only. Further, let us introduce the substitution

$$\nabla_R \to \sum_j \nabla_{R_j}$$

which gives

$$|\hat{A}^{(0)}_{nm}|_{\nu\mu} = \frac{\hbar^2}{M}\sum_j \frac{|\nabla_{R_j} V(\mathbf{r}, \mathbf{R}_j)|_{nm}}{E_{n0} - E_{m0}} \int \Phi^{(0)*}_{\nu}\{\mathbf{R}\}\nabla_{R_j}\Phi^{(0)}_{\mu}\{\mathbf{R}\}\, \mathbf{d}^3 R. \tag{1.177}$$

By the derivation of this equation, it does not take into account the change of the oscillation spectrum provided by excitation of single electrons.

Let us introduce the second quantization representation for the phonons. To do this, first use the substitution $-i\hbar\nabla \to \mathbf{p}$ (where $\mathbf{p}$ is the momentum operator) [35]

$$\mathbf{p}_j = \frac{1}{i}\left(\frac{\hbar M}{2N}\right)^{1/2} \sum_{\mathbf{q}\lambda} \sqrt{\omega_{\lambda\mathbf{q}}}\, \mathbf{e}_{\lambda\mathbf{q}}[b_{\lambda\mathbf{q}} \exp(i\mathbf{q}\mathbf{R}^0_j) - b^+_{\lambda\mathbf{q}} \exp(-i\mathbf{q}\mathbf{R}^0_j)]. \tag{1.178}$$

Here, $b^+_{\lambda\mathbf{q}}$, $b_{\lambda\mathbf{q}}$ are the creation and annihilation operators for phonons with the polarization λ, the quasi-momentum $\mathbf{q}$ and the frequency $\omega_{\lambda\mathbf{q}}$. From this equation, it follows that

$$\begin{aligned}\frac{\hbar^2}{M}\int \Phi^{(0)*}_{\nu}\{\mathbf{R}\}\nabla_{R_j}\Phi^{(0)}_{\mu}\{\mathbf{R}\}\mathbf{dR} &= \left\langle \Phi^{(0)*}_{\nu}\right|\hbar\left(\frac{\hbar}{NM}\right)^{1/2}\sum_{\mathbf{q}\lambda}\sqrt{\omega_{\lambda\mathbf{q}}}\,\mathbf{e}_{\lambda\mathbf{q}} \\ &\times \left[b_{\lambda\mathbf{q}}\exp(i\mathbf{q}\mathbf{R}^0_j) - b^+_{\lambda\mathbf{q}}\exp\left(-i\mathbf{q}\mathbf{R}^0_j\right)\right]\left|\Phi^{(0)}_{\mu}\right\rangle \\ &\to \pm\,\omega_{\lambda\mathbf{q}}\hbar\left(\frac{\hbar}{2NM}\right)^{1/2}\mathbf{e}_{\lambda\mathbf{q}}\left(N_{\lambda\mathbf{q}} + \frac{1}{2} \mp \frac{1}{2}\right)^{1/2}\exp(\pm i\mathbf{q}\mathbf{R}^0_j)\end{aligned} \tag{1.179}$$

where $N_{\lambda\mathbf{q}}$ are the phonon occupation numbers. It can be seen that this matrix element corresponds to one-phonon transitions. Here, the upper index corresponds to the absorption of one phonon and the lower index corresponds to creation of one phonon. Let us take the summation over j in (1.177) and use standard classification of the phonon state in an ideal crystal with the help of quasi-momentum $\mathbf{k}$, i.e., let us set $n = \mathbf{k}$, $m = \mathbf{k}'$ and $E_{n0} = \varepsilon_{\mathbf{k}}$ as the electron energy. The quantum state of the phonon system is characterized by the phonon numbers $\nu = N_{\lambda\mathbf{q}}$. Taking into account that $\hat{A}^{(0)}$ contains the contributions from one phonon process only, the equation (1.177) may be transformed into

$$\left|\hat{A}^{(0)}_{\mathbf{kk}'}\right|_{N_{\lambda\mathbf{q}}\mp 1,\, N_{\lambda\mathbf{q}}} = \mp\frac{\hbar\,\omega_{\lambda\mathbf{q}}}{\varepsilon_{\mathbf{k}} - \varepsilon_{\mathbf{k}'}}|M_{\mathbf{kq}\lambda}|\left(N_{\lambda\mathbf{q}} + \frac{1}{2} \mp \frac{1}{2}\right)^{1/2}\Delta(\mathbf{k} - \mathbf{k}' \mp \mathbf{q}). \tag{1.180}$$

Here $\Delta(\mathbf{k})$ describes the conservation law for the momentum (up to the accuracy to the vector of the reciprocal lattice).

Now let us calculate the matrix element corresponding to the operator $\hat{B}^{(0)}_{nn}$. In accordance with (1.156), it is necessary to calculate the matrix element

$$|\Delta_R|_{nn} = |\nabla^2_R|_{nn} = \sum_p \frac{|\nabla_R\hat{\mathcal{H}}_{ei}(\mathbf{r},\mathbf{R})|_{np}\,|\nabla_R\hat{\mathcal{H}}_{ei}(\mathbf{r},\mathbf{R})|_{pn}}{(E_{n0} - E_{p0})(E_{p0} - E_{n0})} \tag{1.181}$$

using the one-electron representation described below and the identity

$$\sum_\lambda c^{\alpha}_{\lambda\mathbf{q}}\,c^{\beta}_{\lambda\mathbf{q}} = \delta_{\alpha\beta}, \tag{1.182}$$

after simple transformations, we obtain

$$\left|\hat{B}^{(0)}_{\mathbf{kk}}\right|_{N_{\lambda\mathbf{q}},\, N_{\lambda\mathbf{q}}} = \sum_{\lambda\mathbf{q}}\frac{\hbar\,\omega_{\lambda\mathbf{q}}}{\varepsilon_{\mathbf{k}} - \varepsilon_{\mathbf{k}+\mathbf{q}}}\,|M_{\mathbf{kq}\lambda}|^2. \tag{1.183}$$

Here and in (1.180) $M_{\mathbf{kq}\lambda}$ is the matrix element of the electron–phonon interaction. For the correction to the energy of the adiabatic state relating to the operator $\hat{A}$, from (1.172) and (1.180) it follows that

$$\begin{aligned}\Delta E_A = &\sum_{\lambda\mathbf{q}}\sum_{\mathbf{k}}\left(\frac{\hbar\,\omega_{\lambda\mathbf{q}}}{\varepsilon_{\mathbf{k}} - \varepsilon_{\mathbf{k}+\mathbf{q}}}\right)^2|M_{\mathbf{kq}\lambda}|^2\left\{\frac{N_{\lambda\mathbf{q}}}{\varepsilon_{\mathbf{k}} - \varepsilon_{\mathbf{k}+\mathbf{q}} + \hbar\,\omega_{\lambda\mathbf{q}}}\right. \\ &\left.+\frac{N_{\lambda\mathbf{q}} + 1}{\varepsilon_{\mathbf{k}} - \varepsilon_{\mathbf{k}+\mathbf{q}} - \hbar\,\omega_{\lambda\mathbf{q}}}\right\}n_{\mathbf{k}}(1 - n_{\mathbf{k}+\mathbf{q}}),\end{aligned} \tag{1.184}$$

where $n_{\mathbf{k}}$ are the electron occupation numbers.

The non-adiabatic correction to the energy provided by $\hat{B}^{(0)}_{nn}$ in accordance with (1.170) and (1.183) is

$$\Delta E_B = \sum_{\lambda\mathbf{q}}\sum_{\mathbf{k}}\frac{|M_{\mathbf{kq}\lambda}|^2\hbar\,\omega_{\lambda\mathbf{q}}}{(\varepsilon_{\mathbf{k}} - \varepsilon_{\mathbf{k}+\mathbf{q}})^2}n_{\mathbf{k}}(1 - n_{\mathbf{k}+\mathbf{q}}). \tag{1.185}$$

Finally, the correction to the energy is [26]

$$\Delta E = \Delta E_A + \Delta E_B = \sum_{\mathbf{k}} \sum_{\mathbf{q}\lambda} |M_{\mathbf{kq}\lambda}|^2 \frac{\hbar\,\omega_{\mathbf{q}\lambda}\, n_{\mathbf{k}}(1 - n_{\mathbf{k}})}{(\varepsilon_{\mathbf{k}} - \varepsilon_{\mathbf{k+q}})^2} \times \left\{ \left[\frac{N_{\mathbf{q}\lambda}}{\varepsilon_{\mathbf{k}} - \varepsilon_{\mathbf{k+q}} + \hbar\,\omega_{\mathbf{q}\lambda}} + \frac{N_{\mathbf{q}\lambda} + 1}{\varepsilon_{\mathbf{k}} - \varepsilon_{\mathbf{k+q}} - \hbar\,\omega_{\mathbf{q}\lambda}} \right] \hbar\,\omega_{\mathbf{q}\lambda} + 1 \right\}. \tag{1.186}$$

Here $M_{\mathbf{kq}\lambda}$ is the Bloch matrix element [37]

$$M_{\mathbf{kq}\lambda} = \left(\frac{\hbar N}{2M\omega_{\mathbf{q}\lambda}} \right)^{1/2} \int_{\mathbf{r_j=r-R_{0j}}} d\mathbf{r}_j\, \psi^*_{\mathbf{k}}(\mathbf{r}_j) [\mathbf{e}_{\mathbf{q}\lambda} \nabla_{\mathbf{r}} V(\mathbf{r}_j)] \psi_{\mathbf{k}'}(\mathbf{r}_j), \tag{1.187}$$

To evaluate the shift of the phonon frequency $\Delta\omega_{\lambda\mathbf{q}}$ provided by the electron–phonon interaction, it is necessary to find the variation over the phonon occupation numbers $N_{\mathbf{q}\lambda}$. It gives

$$\hbar\Delta\omega_{\mathbf{q}\lambda} = \sum_{\mathbf{k}} |M_{\mathbf{kq}\lambda}|^2 \left[\frac{n_{\mathbf{k}} - n_{\mathbf{k+q}}}{\varepsilon_{\mathbf{k}} - \varepsilon_{\mathbf{k+q}} - \hbar\,\omega_{\mathbf{q}\lambda}} - \frac{n_{\mathbf{k}} - n_{\mathbf{k+q}}}{\varepsilon_{\mathbf{k}} - \varepsilon_{\mathbf{k+q}}} \right]. \tag{1.188}$$

The second term in this equation describes the adiabatic contribution of electrons to the phonon frequency, i.e., the contribution relating to the electron energy $E_n(\mathbf{R_0}) = E_{n0}$ in equation (1.149) for the diagonal ion wave function. This term is of the order of the phonon frequency $\omega_{\lambda\mathbf{q}}$. The first term describes the nonadiabatic contribution to the phonon frequency and is also of the order $\omega_{\lambda\mathbf{q}}$. But the difference of these two terms for the main part of the phase space $|\varepsilon_{\mathbf{k+q}} - \varepsilon_{\mathbf{k}}| \sim E_F$ gives for (1.188)

$$\Delta\,\omega_{\mathbf{q}\lambda} \sim \kappa^4 \omega_{\mathbf{q}\lambda}. \tag{1.189}$$

The attenuation $\Gamma_{\lambda\mathbf{q}}$ is determined by the first term in (1.189) and in accordance with standard estimation of $\Gamma_{\lambda\mathbf{q}}$

$$\Gamma_{\mathbf{q}\lambda} \sim \kappa^2 \omega_{\mathbf{q}\lambda}. \tag{1.190}$$

In the narrow region of momentum (where $|\mathbf{q} - 2\mathbf{k}_F|/k_F \sim \hbar\,\omega_D/E_F$), the renormalization of the phonon frequency is stronger

$$\Delta\,\omega_{\mathbf{q}\lambda} \sim \kappa^2 \omega_{\mathbf{q}\lambda}. \tag{1.191}$$

If we use the Fröhlich model (1.139), the renormalization of the frequency of longitudinal phonons $\omega_{l\mathbf{q}}$ at small $\mathbf{q}$ is

$$\omega_{l\mathbf{q}} = \omega_0(\mathbf{q})\sqrt{1 - 2\zeta_0} \tag{1.192}$$

where $\omega_0(\mathbf{q})$ is the bare acoustic phonon frequency, ζ_0 is the so-called dimensionless Fröhlich parameter that characterizes the electron–phonon interaction $\zeta_0 < 1$

$$\zeta_0 \equiv |M^s_{\mathbf{kq}l}|^2 \left. \frac{\Omega\, m\, \mathbf{k}}{2\,\pi^2\,\hbar^3\,\omega_0(\mathbf{q})} \right|_{k=k_F,\ \mathbf{q}\to 0} \tag{1.193}$$

Here $\hbar \mathbf{k}_F$ is the Fermi momentum, $M^s_{\mathbf{kq}l}$ is the screened matrix element of the interaction of electrons with longitudinal phonons.

From (1.192) it follows that the phonon frequency is strongly renormalized (the renormalization is $\sim \omega_0(\mathbf{q})$) as a consequence of the electron–phonon interaction. On the other side, at $\zeta_0 \geq 1/2$ the lattice is unstable since the phonon frequency has imaginary values. It is the consequence of the fact that, in the Fröhlich model, the electron ion interactions, briefly speaking, are taken into account twice: first, implicitly by the choice of bare phonons with acoustic dispersion law, and secondly, by calculation of the renormalization using the first term in (1.188). However, as we have seen before in the framework of the adiabatic theory, the renormalization of the phonon frequency by taking into account non-adiabatic and anharmonic contributions in the value is of the order κ^2. Therefore, the conclusion relating to strong renormalization of the phonon frequency by electron-phonon interaction and about the instability of the lattice at $\zeta_0 \geq 1/2$ is not true [26]. The Fröhlich model gives an incorrect description in this case.

Below, we provide a detailed comparison of the Fröhlich model and the adiabatic theory presented in the form of the usual quantum-mechanical perturbation theory. This comparison allows us to explain the foundations of the Fröhlich model and the limits of its applicability.

Let us consider now the influence of electron–phonon interactions in the electron spectrum. To obtain the renormalization of the electron spectrum $\Delta\varepsilon_{\mathbf{k}}$, it is necessary to vary the electron occupation numbers $n_{\mathbf{k}}$ in equation (1.186). It gives

$$\Delta\varepsilon_{\mathbf{k}} = \sum_{\mathbf{q}\lambda} |M_{\mathbf{kq}\lambda}|^2 \times \left[(N_{\mathbf{q}\lambda} + 0.5) \frac{2(\varepsilon_{\mathbf{k}} - \varepsilon_{\mathbf{k+q}})}{(\varepsilon_{\mathbf{k}} - \varepsilon_{\mathbf{k+q}})^2 - (\hbar\omega_{\mathbf{q}\lambda})^2} \left(\frac{\hbar\omega_{\mathbf{q}\lambda}}{\varepsilon_{\mathbf{k}} - \varepsilon_{\mathbf{k+q}}} \right)^2 + \frac{\hbar\omega_{\mathbf{q}\lambda}(1 - 2n_{\mathbf{k+q}})}{(\varepsilon_{\mathbf{k}} - \varepsilon_{\mathbf{k+q}})^2 - (\hbar\omega_{\mathbf{q}\lambda})^2} \right]. \tag{1.194}$$

Far from the Fermi surface

$$\Delta\varepsilon_{\mathbf{k}} \sim \kappa^2 \hbar\omega_D, \tag{1.195}$$

whereas, in the narrow band $|\varepsilon_{\mathbf{k}} - \varepsilon_F| \sim \hbar\omega_D$, the renormalization of the electron velocity is appreciable and does not contain the small parameter $\hbar\omega_D/E_F$. The last result was first obtained by Migdal [38].

The analysis of non-adiabatic contributions to the energy and the phonon frequency provided above allows us to formulate general conclusions relating to the usage of the adiabatic approximations in metals and the role of the electron–phonon interaction [19].

1. In the frame of the adiabatic approximation, the phonons are determined with high accuracy: electron-phonon interactions provide very small (of the order of κ^2) renormalization of the phonon dispersion law.

2. Macroscopic values obtained from the microscopic analysis are calculated with accuracy no smaller than κ^2.

3. To determine the properties relating to electrons at the Fermi surface in zero order in the parameter κ^2, it is necessary, after the calculation of the adiabatic phonon spectrum and the electron spectrum in the static lattice, to take into account the renormalization of the electron energy near the Fermi surface provided by electron–phonon interaction.

As we can see from equation (1.185), the vertex of electron–phonon interactions is determined by the value $M_{\mathbf{kq}\lambda}[\varepsilon_{\mathbf{k}} - \varepsilon_{\mathbf{k+q}}]^{-1}$ and not by the matrix element $M_{\mathbf{kq}\lambda}$, as in the case of the Fröhlich model. It is very important for the virtual processes [26]. For real processes with $\varepsilon_{\mathbf{k}} - \varepsilon_{\mathbf{k+q}} = \pm\hbar\omega_{\mathbf{q}}$, the vertex of electron–phonon interaction is equal to the matrix element $M_{\mathbf{kq}\lambda}$ in the same manner as in the Fröhlich model. In this case, resuls of calculations of all values used in the theory of metal conductivity (the attenuation, the cross section of electron–phonon scattering in the kinetic coefficients, the value of the matrix element for electron transitions with creation and annihilation of phonons, etc.) are the same as when using the Fröhlich Hamiltonian.

As an example, let us consider the conductivity of normal metals. Using the adiabatic theory, let us calculate the matrix element of the transition of electrons from the state $\mathbf{k}$ to the state $\mathbf{k}'$ with creation or annihilation of one phonon. It has the form (1.180) or

$$|A^{ad}_{\mathbf{kk}'}|_{N_{\lambda\mathbf{q}}\mp 1, N_{\lambda\mathbf{q}}} = \frac{\hbar\omega_{\lambda\mathbf{q}}}{\varepsilon_{\mathbf{k}} - \varepsilon_{\mathbf{k}'}} U^B(\mathbf{k}, N_{\lambda\mathbf{q}} \mp 1; \mathbf{k}', N_{\lambda\mathbf{q}}), \tag{1.196}$$

where U^B is the matrix element of the Bloch model (see, for example). The definition of U^B can be seen clearly in the comparison of (1.196) with (1.180). Since for the real process $\varepsilon_{\mathbf{k}} - \varepsilon_{\mathbf{k}'} = \hbar\omega_{\lambda\mathbf{q}}$ then $A^{ad} = U^B$. Hence, the theory of conductivity based on the vague assumptions of the Fröhlich model is correct.

Now we have investigated the first version of adiabatic expansion for which the initial ion configurations are arbitrary. Let us now consider the second version of the adiabatic theory. It differs from the first in that the electron problem now is solved by equilibrium positions of ions.

Adiabatic expansion for equilibrium ion configuration Usually it is considered that the equilibrium structure of metal is known. The electron problem is solved for the equilibrium ion configuration $\mathbf{R}_{0n}$ [28]

$$\left[\hat{\mathcal{H}}_e + \hat{\mathcal{H}}_{ei}(\mathbf{r}, \mathbf{R}_{0n})\right]\psi_m(\mathbf{r}, \mathbf{R}_{0n}) = E_m(\mathbf{R}_{0n})\psi_m(\mathbf{r}, \mathbf{R}_{0n}). \tag{1.197}$$

Let us consider the total Hamiltonian of the system as a power series over the ion displacements from the equilibrium positions

$$\hat{\mathcal{H}} = H_e + \hat{\mathcal{H}}_{ei}(\mathbf{r}, \mathbf{R}_{0n}) + \Delta\hat{\mathcal{H}}, \tag{1.198}$$

where

$$\Delta\hat{\mathcal{H}} = H_i + \sum_{\lambda=1}^{\infty} \hat{\mathcal{H}}_\lambda; \quad \hat{\mathcal{H}}_\lambda = \frac{1}{\lambda!}\sum_{\substack{i \dots j \\ \alpha \dots \beta}} \left.\frac{\partial^{(\lambda)}\hat{\mathcal{H}}_{ei}(\mathbf{r}, \mathbf{R})}{\partial R_{i\alpha} \dots \partial R_{j\beta}}\right|_{\mathbf{R}_{0n}} \Delta R_{i\alpha} \dots \Delta R_{j\beta}.$$

In the same manner as in the previous version of the adiabatic theory, the wave function of the crystal $\Psi_n(\mathbf{r}, \mathbf{R})$ must be found in the basis of the whole system of electron wave function in equation (1.197)

$$\Psi_n(\mathbf{r}, \mathbf{R}) = \sum_m \Phi_{nm}(\mathbf{R})\psi_m(\mathbf{r}, \mathbf{R}_{0n}). \tag{1.199}$$

Performing calculations similar to those used by the derivation of equations (1.149) and (1.150), we obtain for $s = n$

$$\begin{aligned}&\left[\hat{\mathcal{H}}_i + E_n(\mathbf{R}_0)\right]\Phi_{nn}(\mathbf{R}) + \sum_{m\neq n}\sum_{\lambda=1}^{\infty} |\hat{\mathcal{H}}_\lambda|_{nm}\Phi_{mn}(\mathbf{R}) \\ &\quad + \sum_{\lambda=2}^{\infty} |\hat{\mathcal{H}}_\lambda|_{nn}\Phi_{nn}(\mathbf{R}) = \mathbf{E}_n\Phi_{nn}(\mathbf{R}),\end{aligned} \tag{1.200}$$

At $s \neq n$, we obtain

$$\left[\hat{\mathcal{H}}_i + E_s(\mathbf{R}_0)\right]\Phi_{ns}(\mathbf{R}) + \sum_{m\neq n}\sum_{\lambda=1}^{\infty} |\hat{\mathcal{H}}_\lambda|_{sm}\Phi_{mn}(\mathbf{R}) + \sum_{\lambda=1} |\hat{\mathcal{H}}_\lambda|_{sn}\Phi_{nn}(\mathbf{R}) = \mathbf{E}_n\Phi_{ns}(\mathbf{R}). \tag{1.201}$$

The third term in the left part of equation (1.200) contains summation starting with $\lambda = 2$ because

$$\left|\frac{\partial\hat{\mathcal{H}}_{ei}}{\partial R_{i\alpha}}\right|_{\substack{nn\\ \mathbf{R}=\mathbf{R}_0}} = \left.\frac{\partial E_n}{\partial R_{i\alpha}}\right|_{\mathbf{R}=\mathbf{R}_0} = 0$$

as a consequence of the equilibrium condition. Let us also prove that, in general,

$$\frac{\partial E_n(\mathbf{R})}{\partial R_{i\alpha}} = \left|\frac{\partial\hat{\mathcal{H}}_{ei}(\mathbf{r}, \mathbf{R})}{\partial R_{i\alpha}}\right|_{nn} \tag{1.202}$$

To do this, first act on the equation (1.147) by the operator $\partial/\partial R_{i\alpha}$, then multiply it from the left by the electron wave function $\psi_s^*(\mathbf{r}, \mathbf{R})$ and, finally, integrate it over electron coordinates. We obtain

$$\begin{aligned}&\int \psi_s^* \frac{\partial}{\partial R_{i\alpha}}\left[\hat{\mathcal{H}}_e\psi_n\right]d\mathbf{r} + \int \psi_s^*\left[\frac{\partial}{\partial R_{i\alpha}}\mathcal{H}_{ei}\psi_n\right]d\mathbf{r} \\ &\quad = \int \psi_s^* \frac{\partial E_n}{\partial R_{i\alpha}}\psi_n\, d\mathbf{r} + \int \psi_s^* E_n \frac{\partial\psi_n}{\partial R_{i\alpha}}\, d\mathbf{r}.\end{aligned} \tag{1.203}$$

Since $\hat{\mathcal{H}}_e$ does not depend on ion coordinates, then

$$\frac{\partial}{\partial R_{i\alpha}}\left[\hat{\mathcal{H}}_e\psi_n\right] = \hat{\mathcal{H}}_e\frac{\partial\psi_n}{\partial R_{i\alpha}}. \tag{1.204}$$

Let us transform the left part of (1.203) using (1.204)

$$\int \psi_s^* \hat{\mathcal{H}}_e \frac{\partial \psi_n}{\partial R_{i\alpha}} d\mathbf{r} + \int \psi_s^* \frac{\partial \hat{\mathcal{H}}_{ei}}{\partial R_{i\alpha}} \psi_n \, d\mathbf{r} + \int \psi_s^* \hat{\mathcal{H}}_{ei} \frac{\partial \psi_n}{\partial R_{i\alpha}} \, d\mathbf{r}$$

$$= \int \psi_s^* \left[\hat{\mathcal{H}}_e + \hat{\mathcal{H}}_{ei}\right] \frac{\partial \psi_n}{\partial R_{i\alpha}} d\mathbf{r} + \left| \frac{\partial \hat{\mathcal{H}}_{ei}}{\partial R_{i\alpha}} \right|_{sn}.$$

The operator $\hat{\mathcal{H}}_e + \hat{\mathcal{H}}_{ei}$ is a Hermitian one and consequently [35]

$$\begin{aligned} \int \psi_s^* \left[\hat{\mathcal{H}}_e + \hat{\mathcal{H}}_{ei}\right] \frac{\partial \psi_n}{\partial R_{i\alpha}} d\mathbf{r} &= \int \frac{\partial \psi_n}{\partial R_{i\alpha}} \left[\hat{\mathcal{H}}_e + \hat{\mathcal{H}}_{ei}\right] \psi_s^* \, d\mathbf{r} \\ &= \int \frac{\partial \psi_n}{\partial R_{i\alpha}} E_s(\mathbf{R}) \, \psi_s^* \, d\mathbf{r} \end{aligned} \tag{1.205}$$

By substituting (1.202) into (1.200), we obtain

$$\int \frac{\partial \psi_n}{\partial R_{i\alpha}} \psi_s^* \, E_s(\mathbf{R}) d\mathbf{r} + \left| \frac{\partial \hat{\mathcal{H}}_{ei}}{\partial R_{i\alpha}} \right|_{sn} = \int \psi_s^* \frac{\partial E_n(\mathbf{R})}{\partial R_{i\alpha}} \psi_n \, d\mathbf{r} + \psi_s^* \frac{\partial \psi_n}{\partial R_{i\alpha}} E_n(\mathbf{R}) \, d\mathbf{r}$$

which proves (1.202) at $s = n$.

If two electron functions $\psi_n(\mathbf{r}, \mathbf{R})$ differ by the change of the state of s electrons (from the state $\mathbf{k}_s$ to the state $\mathbf{k}_{s'}$), the new equilibrium positions $\mathbf{R}_{0n}$, and frequencies $\omega^n(\mathbf{q})$ will differ from the value of the order s/N. Therefore, for the states close to the ground state, the dependence of the equilibrium position $\mathbf{R}_{0n}$ and frequencies $\omega^n(\mathbf{q})$ from n may be neglected. In accordance with this note, we have set $\mathbf{R}_{0n} = \mathbf{R}_0$ and $\omega^n(\mathbf{q}) = \omega_{\lambda\mathbf{q}}$ in equations (1.200) and (1.201).

The solution of equations (1.200) and (1.201) is similar to respective solutions of equations of the first version of adiabatic expansion. Let us obtain as an example the equation for the zero order ion wave function

$$\left[\hat{\mathcal{H}}_i + |\hat{\mathcal{H}}_2|_{nn} + \sum_{m \neq n} |\hat{\mathcal{H}}_1|_{nm} \, |\hat{\mathcal{H}}_1|_{mn} (E_{n0} - E_{m0})^{-1}\right] \Phi_{nn\nu}^{(0)} = E_{n\nu}^{(0)} \Phi_{nn\nu}^{(0)}. \tag{1.206}$$

of the order of magnitude $E_{n\nu}^{(0)} \sim \hbar \omega_D$. This equation, after respective transformation, is equivalent to equation (1.164). Hence, this version of adiabatic expansion must, in principle, provide the same results as the first version.

The equations for the corrections to the energy of the adiabatic state are similar to the system (1.165)–(1.168), but they have a more sophisticated form.

Just as in the first version of adiabatic theory, if we have the solution for the electron problem, then it is possible to solve the problem of ion motion using the standard perturbation theory. Let us take a zero order Hamiltonian as

$$\hat{\mathcal{H}}_0 = \hat{\mathcal{H}}_i + |\hat{\mathcal{H}}_2|_{nn} + \sum_{m \neq n} |\hat{\mathcal{H}}_1|_{nm} \, |\hat{\mathcal{H}}_1|_{mn} (E_{n0} - E_{m0})^{-1}, \tag{1.207}$$

It is independent from the electron coordinates. Respectively, the perturbation operator is

$$\hat{\mathcal{H}}' = |\hat{\mathcal{H}}_3|_{nn} + \sum_{m\neq n}\left(|\hat{\mathcal{H}}_1|_{nm}|\hat{\mathcal{H}}_2|_{mn} + |\hat{\mathcal{H}}_2|_{nm}|\hat{\mathcal{H}}_1|_{mn}\right)(E_{n0} - E_{m0})^{-1}$$
$$+ \sum_{\substack{m\neq n\\ l\neq n}} \frac{|\hat{\mathcal{H}}_1|_{nm}|\hat{\mathcal{H}}_1|_{ml}|\hat{\mathcal{H}}_1|_{\text{ln}}}{(E_{n0} - E_{m0})(E_{no} - E_{l0})} + \cdots. \tag{1.208}$$

Using (1.208) we can obtain the correction to the energy up to arbitrary order of κ^2.

The adiabatic perturbation theory is not a typical quantum-mechanical perturbation theory. It relates to the fact that it is impossible to extract explicitly from the total Hamiltonian of the electron and ion system considered the zero order Hamiltonian $\hat{\mathcal{H}}_0(\mathbf{r}, \mathbf{R})$, which depends simultaneously on the ion and electron coordinates. Hence, it is impossible to calculate its eigenfunction and eigenvalues. This fact does not allow the use of simple quantum-mechanical perturbation theory. To do this, we need to use a two-stage solution. First, we must solve the electron problem (1.147) and (1.197), which is a real many-particle problem. The second stage is the solution of the problem of the ion motion. Here, as we have seen, it is possible to construct the perturbation theory with operators (1.170) or (1.208) and to calculate the corrections to the energy in arbitrary order in κ^2. That is the reason why the adiabatic perturbation theory is more sophisticated and is unsuitable to applications in contrast with the general quantum-mechanical perturbation theory.

1.3.2 The perturbation theory on the basis of the adiabatic expansion

As we have seen, the main defect of the adiabatic perturbation theory is that it is impossible to extract explicitly the zero order Hamiltonian $\mathcal{H}_0$ depending simultaneously on ion and electron coordinates and to calculate its eigenfunctions as a product of $\psi_n(\mathbf{r})\,\Phi_\nu(\mathbf{R})$ and also the perturbation operator $\hat{\mathcal{H}}'(\mathbf{r}, \mathbf{R})$. However, as was first shown by Geĭlikman [28], it is possible to construct the usual quantum-mechanical perturbation theory using the adiabatic approximation. To do this, it is necessary to take the zero-order Hamiltonian so that it provides the total wave function of zero order as $\Psi^0(\mathbf{r}, \mathbf{R}) = \Phi^{(0)}_{nn\nu}(\mathbf{R})\psi_n(\mathbf{r}, \mathbf{R})$ and the total energy of zero order as $E^0 = E_n(\mathbf{R}_0) + E^{(0)}_{n\nu}$ and the phonon frequencies $\omega(\mathbf{q})$ must be equal to adiabatic frequencies of the harmonic approximation. This choice may be performed in the following manner.

The zero order Hamiltonian $\hat{\mathcal{H}}_0(\mathbf{r}, \mathbf{R})$ for the electron and ion system is taken as

$$\hat{\mathcal{H}}_0(\mathbf{r}, \mathbf{R}) = \hat{\mathcal{H}}_{0e}(\mathbf{r}) + \hat{\mathcal{H}}_{0i}(\mathbf{R}), \tag{1.209}$$

where $\mathbf{R}_0 \equiv \mathbf{R}_{0n}$

$$\hat{\mathcal{H}}_{0e}(\mathbf{r}) = \hat{\mathcal{H}}_e + \hat{\mathcal{H}}_{ei}(\mathbf{r}, \mathbf{R}_0),$$

$$\hat{\mathcal{H}}_{0i}(\mathbf{R}) = \hat{\mathcal{H}}_i + U_{2n} = \hat{\mathcal{H}}_i + \frac{1}{2}\sum_{\substack{i,j\\ \alpha,\beta}} \left.\frac{\partial^2 E_n(\mathbf{R})}{\partial R_{i\alpha}\,\partial R_{j\beta}}\right|_{\mathbf{R}_0} \Delta R_{i\alpha}\,\Delta R_{j\beta} \tag{1.210}$$

The Hamiltonian (1.209) has eigenvalues $E^0_{n\nu} = E_n(\mathbf{R}_0) + E^{(0)}_{n\nu}$ and eigenfunctions $\Psi^0_{n\nu}(\mathbf{r}, \mathbf{R}) = \Phi^{(0)}_{nn\nu}(\mathbf{R})\psi_n(\mathbf{r}, \mathbf{R})$, and the phonon frequencies are equal to frequencies of the adiabatic theory, since the potential energy of oscillations is equal to U_{2n} in accordance with the zero order of the adiabatic theory.

Let us note that, generally speaking, eigenfunctions of whole orthogonal system $\Psi^0_{m\mu}$ of the operator (1.209) have no real physical sense (excluding $\Phi^{(0)}_{nn\nu}$), because in equation

$$\hat{\mathcal{H}}_0(\mathbf{r}, \mathbf{R})\Psi^0_{m\mu}(\mathbf{r}, \mathbf{R}) = \mathbf{E}^0_{m\mu}\,\Psi^0_{m\mu}(\mathbf{r}, \mathbf{R})$$

the equilibrium positions for all quantum states m are determined by the value $\mathbf{R}_{0n}$ and not by $\mathbf{R}_{0m}$ and the frequencies are equal to $\omega^n(\mathbf{q})$ and not to $\omega^m(\mathbf{q})$. However, as we have seen above, it is possible to neglect the dependence of $\mathbf{R}_{0n}$ and $\omega^n(\mathbf{q})$ on the n for the states close to the ground state and, consequently, all eigenfunctions have a direct physical sense.

In accordance with (1.209), the perturbation operator has the form

$$\hat{\mathcal{H}}'(\mathbf{r}, \mathbf{R}) = \hat{\mathcal{H}} - H_0 = -U_{2n} + \hat{\mathcal{H}}_1 + \hat{\mathcal{H}}_2 + \cdots, \tag{1.211}$$

where

$$\hat{\mathcal{H}}_p = \frac{1}{p!}\sum_{\substack{i\ldots j\\ \alpha\ldots\beta}} \left.\frac{\partial^p \hat{\mathcal{H}}_{ei}(\mathbf{r}, \mathbf{R})}{\partial R_{i\alpha}\ldots\partial R_{j\beta}}\right|_{\mathbf{R}_0} \Delta R_{i\alpha}\ldots\Delta R_{j\beta}.$$

The expression for $\hat{\mathcal{H}}'(\mathbf{r}, \mathbf{R})$ may be transformed to the form [27]

$$\hat{\mathcal{H}}' = \hat{\mathcal{H}}_1 + \Delta\hat{\mathcal{H}}_2 - \hat{\mathcal{H}}_{2n} + \sum_{s=3}^{\infty}\hat{\mathcal{H}}_s,$$

$$\Delta\hat{\mathcal{H}}_2 = \hat{\mathcal{H}}_2 - |\hat{\mathcal{H}}_2|_{nn}; \quad \hat{\mathcal{H}}_{2n} = \sum_{m\neq n}\frac{|\hat{\mathcal{H}}_1|^2_{nm}}{E_{n0} - E_{m0}}; \quad |\hat{\mathcal{H}}_1|_{nn} = 0. \tag{1.212}$$

The terms $|\hat{\mathcal{H}}_2|_{nn}$ and $\hat{\mathcal{H}}_{2n}$ are operators in the phonon degrees of freedom, and they are classical operators in the electron degrees of freedom. The perturbation theory for $\hat{\mathcal{H}}'$ in (1.212) is totally equivalent to the adiabatic perturbation theory, despite the fact that both of these theories have different structures. It is also possible to use this theory in the case of overlapping bands, because, in the case of a metal, as is well-known, the degeneration is avoided in the frame of the pure electron problem and $E_n(\boldsymbol{R})$ is determined by electrons of all overlapping bands [39, 40].

The terms in $\hat{\mathcal{H}}'$ have the following approximate values

$$\hat{\mathcal{H}}_1 \sim \frac{\hbar\omega_D}{\kappa}, \quad \Delta\hat{\mathcal{H}}_2 \sim \hat{\mathcal{H}}_{2n} \sim |\hat{\mathcal{H}}_2|_{nn} \sim \hbar\omega_D, \quad \hat{\mathcal{H}}_s \sim \kappa\hat{\mathcal{H}}_{s-1}. \tag{1.213}$$

The energy of oscillation in the zero order Hamiltonian has the order $\hbar\omega_D$ and the first correction to the energy of the adiabatic state is $\kappa^2\hbar\omega_D$. Despite the fact that the first term in the perturbation operator formally has a large order of magnitude

$$\hat{\mathcal{H}}_1 \sim \hbar\omega_D/\kappa \quad \text{and} \quad \Delta\hat{\mathcal{H}}_2 \sim \hat{\mathcal{H}}_{2n} \sim \hbar\omega_D.$$

Using the perturbation theory with operator $\hat{\mathcal{H}}'$ in (1.212), and selecting the terms with the same order in κ, we obtain

$$E^{(1)}_{n\nu 1} = |\Delta\hat{\mathcal{H}}_2|_{n\nu,n\nu} - |\hat{\mathcal{H}}_{2n}|_{n\nu,n\nu} + \sum_{m,\mu} \frac{|\hat{\mathcal{H}}_1|^2_{n\nu,m\mu}}{E^0_{n\nu} - E^0_{m\mu}}$$
$$\equiv \sum_{m,\mu} \frac{|\mathcal{H}_1|^2_{n\nu,m\mu}}{E^0_{n\nu} - E^0_{m\mu}} - \sum_{m\neq n} \frac{||\hat{\mathcal{H}}_1|^2_{nm}|_{\nu\nu}}{E_{n0} - E_{m0}}. \qquad (1.214)$$

Here each term has the order $\hbar\omega_D$ and the difference is the order of $\kappa^2\hbar\omega_D$. Actually, from the non-diagonality of the matrix elements of $\hat{\mathcal{H}}_1$ over ν, the energetic denominators in the first term of this expression not only present the difference $E_{n0} - E_{m0}$, but also present terms $\pm\hbar\omega_{\lambda\mathbf{q}}$. Hence, the value of $E^{(1)}_{n\nu}$ is of the order

$$\frac{\hbar\omega_D|\hat{\mathcal{H}}_{2n}|_{\nu\nu}}{E_F} \sim \kappa^2\hbar\omega_D.$$

Hence, despite the fact that each term in (1.214) is of the order $\hbar\omega_D$, mutual compensation of these terms gives $E^{(1)}_{n\nu 1} \sim \hbar\omega_{\lambda\mathbf{q}}$. This compensation of $\hat{\mathcal{H}}_2$ and $|\hat{\mathcal{H}}_2|_{nn}$ and also $\hat{\mathcal{H}}_1$ and $\hat{\mathcal{H}}_{2n}$ is very important not only in the first but also in all higher orders of the perturbation theory. Namely, subtraction of U_{2n} in (1.211) is very important for the theory.

Besides (1.214), there is also a second part of the correction to the energy $\sim \kappa^2\hbar\omega_D$:

$$E^{(1)}_{n\nu 2} = |\Delta\hat{\mathcal{H}}_4|_{n\nu,n\nu} + \sum_{m,\mu}\Big\{|\Delta\hat{\mathcal{H}}_2|^2_{n\nu,m\mu} + |\hat{\mathcal{H}}_{2n}|^2_{n\nu,m\mu}$$
$$+ 2\Big(|\hat{\mathcal{H}}_1|_{n\nu,m\mu}|\hat{\mathcal{H}}_3|_{m\mu,n\nu} - |\hat{\mathcal{H}}_{2n}|_{n\nu,m\mu}|\Delta\hat{\mathcal{H}}_2|_{m\mu,n\nu}\Big)\Big\}(E^0_{n\nu} - E^0_{m\mu})^{-1}$$
$$+ \sum_{m,l,\lambda\mu}\Big\{3\Big(|\hat{\mathcal{H}}_1|_{n\nu,m\mu}|\hat{\mathcal{H}}_1|_{m\mu,l\lambda}|\Delta\hat{\mathcal{H}}_2|_{l\lambda,n\nu} - |\hat{\mathcal{H}}_1|_{n\nu,m\mu}|\hat{\mathcal{H}}_1|_{m\mu,l\lambda}|\hat{\mathcal{H}}_{2n}|_{l\lambda,n\nu}\Big)\Big\}$$
$$\times (E^0_{n\nu} - E^0_{m\mu})^{-1}(E^0_{n\nu} - E^0_{l\lambda})^{-1} + \sum_{m,l,k,\lambda,\mu,\kappa} |\hat{\mathcal{H}}_1|^4_{n\nu,m\mu}(E^0_{n\nu} - E^0_{m\mu})^{-1}$$
$$\times (E^0_{n\nu} - E^0_{l\lambda})^{-1}(E^0_{n\nu} - E^0_{k\kappa})^{-1}. \qquad (1.215)$$

Using the second quantization representation in (1.214), we obtain

$$E^{(1)}_{n\nu 1} = \sum_{\mathbf{k},\mathbf{q}} |M_{\mathbf{kq}\lambda}|^2 \frac{n_{\mathbf{k}}(1 - 2n_{\mathbf{k+q}})}{\varepsilon_{\mathbf{k+q}} - \varepsilon_{\mathbf{k}}}$$
$$\times \left\{\frac{\hbar\omega_{\mathbf{q}\lambda}}{\varepsilon_{\mathbf{k+q}} - \varepsilon_{\mathbf{k}} + \hbar\omega_{\lambda\mathbf{q}}} - \frac{2N_{\mathbf{q}\lambda}\hbar^2\omega^2_{\mathbf{q}\lambda}}{(\varepsilon_{\mathbf{k+q}} - \varepsilon_{\mathbf{k}})^2 + \hbar^2\omega^2_{\lambda\mathbf{q}}}\right\}. \qquad (1.216)$$

Here we use the same notations as before. It is very interesting to note that equation (1.216) is equivalent to equation (1.186) obtained in [26].

By applying the variational procedure described below to $E^{(1)}_{n\nu 1}$, it is possible to calculate non-adiabatic corrections to the phonon frequencies $\omega_{\lambda\mathbf{q}}$ and to the electron spectrum $\varepsilon_{\mathbf{k}}$. The results of this theory totally coincide with the results of the adiabatic perturbation theory. Note here that by taking into account the anharmonic contributions, i.e., by using the whole equation (1.215), the renormalization of the adiabatic phonon frequency is of the order $\Delta\omega_{\lambda\mathbf{q}} \sim \kappa^2\omega_D\hbar$ in a major part of the phase space.

Using the perturbation theory based on the adiabatic expansion, it is possible to compare the Fröhlich model and the adiabatic approximation [33]. It can be seen from (1.211) that the perturbation operator $\hat{\mathcal{H}}'(\mathbf{r}, \mathbf{R})$ of the adiabatic theory is essentially different from the perturbation operator (1.139) in the Fröhlich model $\hat{\mathcal{H}}'_F$ ($=\hat{\mathcal{H}}_1$ in (1.211)). One can now ask, if it is possible to find the zero order Hamiltonian $\hat{\mathcal{H}}_0$, which is a sum of Hamiltonians of non-interacting electron and phonon fields, so that the perturbation operator, which is a difference of the exact Hamiltonian (1.140) and $\hat{\mathcal{H}}_0$ in (1.209), is equal to $\hat{\mathcal{H}}_1$. From comparison of (1.140), (1.209) and (1.211) it is obvious that it is impossible. Hence, it is impossible to obtain the Fröhlich model from the exact Hamiltonian. Nevertheless, it is possible to take $\hat{\mathcal{H}}_0$ so that corrections to the system energy and to the excitations energy $\hbar\omega_D$ will be determined by the term $\hat{\mathcal{H}}_1$, in the same manner as in the Fröhlich model. The rest of the terms in $\hat{\mathcal{H}}'$ determine corrections of the order $\kappa^2\hbar\omega_D$ and higher. To show this, the Hamiltonian $\hat{\mathcal{H}}_0$ must have the form

$$\hat{\mathcal{H}}_0 = \hat{\mathcal{H}}_e + \hat{\mathcal{H}}_{ei}(\mathbf{r}, \mathbf{R}_0) + \hat{\mathcal{H}}_{0i}; \quad \hat{\mathcal{H}}_{0i} = \hat{\mathcal{H}}_i + |\hat{\mathcal{H}}_2|_{nn}. \tag{1.217}$$

This form of $\hat{\mathcal{H}}_0$ corresponds to the essentially different choice of bare phonons in contrast with (1.210). The perturbation operator here is equal to

$$\hat{\mathcal{H}}' = \hat{\mathcal{H}} - \hat{\mathcal{H}}_0 = \hat{\mathcal{H}}_1 + \hat{\mathcal{H}}_2 + \sum_{s=3}^{\infty} \hat{\mathcal{H}}_s. \tag{1.218}$$

It gives correction to the energy

$$E_{n\nu 1} = \sum_{m,\mu} |\hat{\mathcal{H}}_1|^2_{n\nu, m\mu} (E^0_{n\nu} - E^0_{m\mu})^{-1} \sim \hbar\omega_D.$$

Therefore, the main correction is actually determined by $\hat{\mathcal{H}}_1$ from $\hat{\mathcal{H}}'$ and has the order of $\hbar\omega_D$. Other terms provide corrections $\sim \kappa^2\hbar\omega_D$ and higher.

Both assumptions about the acoustic dispersion law for the bare longitudinal phonon frequencies and, simultaneously, about the form of the perturbation operator $\hat{\mathcal{H}}' = \hat{\mathcal{H}}_1$, which are the basis of the Fröhlich model, are actually based on the supposition of the possibility of independently choosing a zero order Hamiltonian and the perturbation operator. But for a given total Hamiltonian of the system, it is impossible to provide any independent choice of $\hat{\mathcal{H}}_0$ and $\hat{\mathcal{H}}'$. The choice of zero order Hamiltonian in the form of (1.217) is not good, because the perturbation operator, as in the case of the Fröhlich model, contains no small parameter κ. Hence, the correction to the phonon frequency and to the system energy are of the same order of magnitude as the energy of the lattice oscillations in the zero order Hamiltonian. These oscillations do not take place near the true equilibrium positions of ions. If, in the adiabatic theory, phonons are already renormalized in zero order, then for the Hamiltonian $\hat{\mathcal{H}}_0$, they are not bare phonons by definition.

Using the well-known equation for the frequency,

$$\omega_{\lambda\mathbf{q}}^2 = \sum_{\alpha,\beta} e_{\lambda\mathbf{q}}^{\alpha} D_{\alpha\beta}(\mathbf{q}) e_{\lambda\mathbf{q}}^{\beta}, D_{\alpha\beta}(\mathbf{q}) = \frac{1}{M}\sum_{i,j} V_{\alpha\beta}(\mathbf{R}_{0ij}) \exp\left[-i\mathbf{q}(\mathbf{R}_{0i} - \mathbf{R}_{0j})\right],$$

$$V_{\alpha\beta}(\mathbf{R}_{0ij}) = \frac{\partial^2 E_{n0}}{\partial R_{0i\alpha}\,\partial R_{0j\beta}},$$

it is possible to obtain the expression for the longitudinal frequency

$$\omega_{l\mathbf{q}} = \omega_{l0}(\mathbf{q})\sqrt{1 - \bar{\omega}^2/\omega_{l0}^2}. \tag{1.219}$$

The structure of this equation is the same as in the Fröhlich model with $2\zeta = \bar{\omega}^2/\omega_{l0}^2$. Here $\bar{\omega}^2$ is

$$\bar{\omega}^2 = \sum_{\mathbf{k}} \frac{2\omega_{\lambda\mathbf{q}}}{\hbar} |M_{\mathbf{kq}\lambda}|^2 \frac{n_{\mathbf{k}} - n_{\mathbf{k+q}}}{\varepsilon_{\mathbf{k+q}} - \varepsilon_{\mathbf{k}}}. \tag{1.220}$$

It is proportional to the square of the modulus of the matrix element of the electron-phonon interaction. The bare frequency $\omega_{l0}^2(\mathbf{q})$ at small $\mathbf{q}$ is determined by

$$\omega_{l0}^2(\mathbf{q}) = \frac{4\pi Z^2 e^2}{M} + a^2\mathbf{q}^2 \tag{1.221}$$

(the term $a^2\mathbf{q}^2$ relates to behavior of $V(\mathbf{R}_i - \mathbf{R}_j)$ at small $|\mathbf{R}_i - \mathbf{R}_j|$). From (1.221), it follows that $\omega_{l0}^2(\mathbf{q})$ is the ion plasma frequency ω_p^2 and it corresponds not to acoustical, as in the case of the Fröhlich model, but rather to optical dispersion law at $\mathbf{q} \to 0$. Hence, the subtraction of $\bar{\omega}^2$ from ω_{l0}^2, in contrast to the Fröhlich model, does not provide zero or imaginary values for $\omega_{l\mathbf{q}}^2$. Therefore, we can see that application of the Fröhlich model with the bare acoustical longitudinal frequency and with an interaction like $|M_{\mathbf{k}}^s|^2$ provides incorrect results by calculation of the phonon spectrum.

1.3.3 Adiabatic theory of electron–phonon interaction

Above, some general conclusions were formulated about the role of the electron–phonon interaction and its contribution to some metal properties.

Now let us investigate this question in more detail. The interaction of electrons with phonons was first investigated by Migdal [38] in the case of the isotropic model of a metal. The physical mechanism of electron-phonon interaction is based on the fact that oscillations of the media provides its polarization. The resulting change of the electron energy is

$$E_{ep} = -e \int n(\mathbf{r})\, K(|r - r'|)\, \mathrm{div}\, \mathbf{P}(\mathbf{r}')\, d^3r\, d^3r', \tag{1.222}$$

where $n(\mathbf{r})$ is the electron density at point $\mathbf{r}$, $\mathbf{P}$ is the polarization vector and $K(|\mathbf{r} - \mathbf{r}'|)$ is the interaction function. If $|\mathbf{r} - \mathbf{r}'|$ is smaller than the lattice constant then $K(|\mathbf{r} - \mathbf{r}'|) \sim |\mathbf{r} - \mathbf{r}'|$ [23]. At large distances $K(|\mathbf{r} - \mathbf{r}'|)$ vanishes rapidly as a consequence of the screening of the polarization charge by electrons. Hence, it is possible to substitute

$$K(|\mathbf{r} - \mathbf{r}'|) \approx d^2 \delta(|\mathbf{r} - \mathbf{r}'|). \tag{1.223}$$

The polarization vector $\mathbf{P}$ is proportional to the distortion vector $\mathbf{u}(\mathbf{r})$, i.e.,

$$\mathbf{P}(\mathbf{r}) = c\mathbf{u}(\mathbf{r}).$$

The coefficient of proportionality is $c \sim ZeN/V$ (where N/V is the number of ions per unit cell and Z is the ion charge). Since the interaction energy of electrons with lattice oscillations contains the term $\mathrm{div}\, \mathbf{P} = c\, \mathrm{div}\, \mathbf{u}$, then electrons interact only with longitudinal oscillations. In accordance with (1.222) and (1.223), the operator of interaction may be written as

$$\hat{\mathcal{H}}_{ep} = e^2 d^2 c \int \hat{\psi}^+(\mathbf{r})\, \hat{\psi}(\mathbf{r})\, \mathrm{div}\, \mathbf{u}(\mathbf{r})\, \mathbf{d}^3 r. \tag{1.224}$$

Here $\hat{\psi}(\mathbf{r})$ is the operator of the electron field [23]. If we take into account that the operators $\mathbf{u}(\mathbf{r})$ may be expressed through the phonon amplitudes $\varphi(\mathbf{r})$, then it is possible to transform equation (1.224) with the help of the operators of the phonon field to

$$\hat{\mathcal{H}}_{ep} = g \int \hat{\psi}^+(\mathbf{r})\, \hat{\psi}(\mathbf{r})\, \varphi(\mathbf{r})\, \mathbf{d}^3 r, \tag{1.225}$$

where g is the parameter of interaction (the elementary vertex of electron-phonon interactions) [23, 34]

$$g^2 = \zeta_0 \nu^{-1}; \quad \nu = \frac{m\mathbf{k}_F}{2\pi^2 \hbar^2}. \tag{1.226}$$

Migdal first supposed that, within the accuracy of the order κ^2, the self-energy part contains only contributions from the lowest order diagram. The wavy line here corresponds to the phonon propagator D and the straight line corresponds to the electron propagator G.

It seems to be important that the vertex of the electron–phonon interaction Γ differs from its zero order value g on the value $\sim \kappa^2$, i.e., [41]

$$\Gamma = g(1 + O(\kappa^2)).$$

The investigation of [41], where the Fröhlich Hamiltonian was used, provided some controversial results. They include an appreciable value for the renormalization of the phonon energy and the question of the lattice stability. These results are the consequence of the fact that, here, it is impossible to use the Fröhlich model. In the previous subsection, it was shown that the adiabatic perturbation series provides only small renormalization of the phonon frequency. Hence, it very interesting to investigate the electron–phonon interaction in the framework of the adiabatic theory.

The interaction of electrons with phonons will be considered on the basis of the perturbation theory described in the previous subsection. For this purpose, it is very useful to use the diagram technique [28] which will be described below.

The diagram technique Let us transform the perturbation operator of the adiabatic theory $\hat{\mathcal{H}}'$ in (1.212), which acts on atomic distortions to the form which includes the creation $b^+_{\lambda\mathbf{q}}$ and annihilation $b_{\lambda\mathbf{q}}$ operators in

$$\mathbf{u}_l = \sum_{\mathbf{q},\lambda} \left(\frac{\hbar}{2MN\omega_{\lambda\mathbf{q}}} \right)^{1/2} \left(b_{\lambda\mathbf{q}} \exp[i\mathbf{q}\,\mathbf{R}_l] + b^+_{\lambda\mathbf{q}} \exp[-i\mathbf{q}\,\mathbf{R}_l] \right) \mathbf{e}_{\lambda\mathbf{q}}, \tag{1.227}$$

where $\mathbf{R}_l$ is the vector of the equilibrium position of the l-th atom. The phonon $D_{\lambda\lambda'}(X, X')$ and the electron $G(X, X')$ Green's function will be determined in the same manner as in [23]

$$D_{\lambda\lambda'}(X, X') = -i\langle T(\varphi_\lambda(X)\varphi_{\lambda'}(X'))\rangle, \quad X = \mathbf{R}, \mathbf{t}, \tag{1.228}$$

where in accordance with the expansion of $\mathbf{u}_l$ in (1.227)

$$\varphi_\lambda(X) = \sum_{\mathbf{q}} \left[\frac{\hbar\omega_{\lambda\mathbf{q}}}{2V} \right]^{1/2} \left[b_{\lambda\mathbf{q}} \exp[i(\mathbf{q}\mathbf{R} - \omega t)] + h.c. \right]. \tag{1.229}$$

The definition of the electron Green's function is

$$G(X, X') = -i\langle T\hat{\psi}(X)\hat{\psi}^+(X')\rangle. \tag{1.230}$$

Here V is the volume, T is the time ordering operator [23], angle brackets $\langle \cdots \rangle$ denote the mean value over the ground state of the system.

Since it more useful to use the diagram technique in the momentum space, let us transfer from Green's functions $G(X, X')$ and $D(X, X')$ to their Fourier representations $G(k)$ and $D(k)$ ($k = (\mathbf{k}, \varepsilon)$; $q = (\mathbf{q}, \hbar\omega)$) in uniform and isotropic space

$$G(X, X') = G(X - X') = \int \frac{d^4k}{(2\pi)^4} G(k) \exp[ik(X - X')],$$

$$D(X, X') = D(X - X') = \int \frac{d^4q}{(2\pi)^4} D(q) \exp[iq(X - X')].$$

The propagators of non-interacting electrons $G^{(0)}(k)$ and phonons $D^{(0)}(k)$ have representations [23]

$$G^{(0)}(k) = G^0(\mathbf{k}, \varepsilon) = [\varepsilon - \varepsilon_{\mathbf{k}} - \mu + i\delta \operatorname{sign}(\varepsilon)]^{-1}, \tag{1.231}$$

$$D^{(0)}_{\lambda\lambda}(q) = D^{(0)}_{\lambda\lambda}(\mathbf{q}, \hbar\omega) = -\frac{\omega^2_{\lambda\mathbf{q}}}{\omega^2 + \omega^2_{\mathbf{q}\lambda}}.$$

Here $\varepsilon_{\mathbf{k}} = \hbar^2\mathbf{k}^2/(2m)$ is the electron energy, $\omega(\lambda\mathbf{q})$ is the phonon frequency.

The vertices $\Gamma_s^{\{\lambda\}}$ correspond to each term $\hat{\mathcal{H}}_s$ from operator $\hat{\mathcal{H}}'$ in (1.212).[7]

Here, index s denotes the number of the phonon ends ($s = 1, 2, \ldots$) and the index $\{\lambda\}$ denotes the phonon polarization. Vertices Γ_s^λ may either contain two electron ends or contain no electron ends (as a consequence of the term $V(\mathbf{R}, \mathbf{R})$ in $\hat{\mathcal{H}}_{ei}(\mathbf{r}, \mathbf{R})$, i.e.,

$$\Gamma_s = \Gamma_{s0} + \Gamma_{s2}.$$

The second index i in Γ_{si} is the number of the electron ends. The evaluation of (1.213) provides $\Gamma_s \sim \kappa\Gamma_{s-1}$. The next diagram contains contributions provided by Γ_{s2} only.

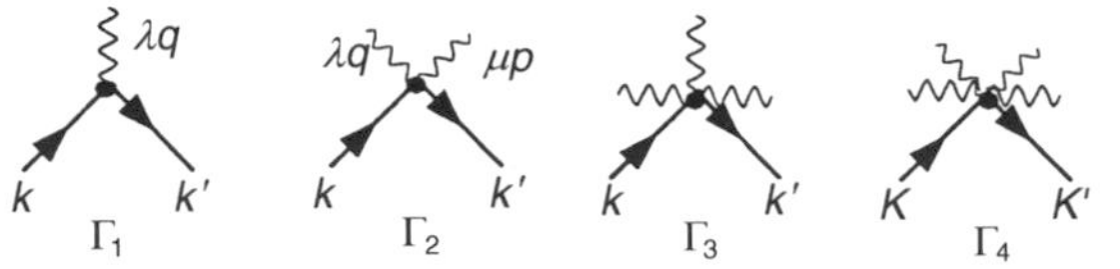

(1.232)

Here, and in the diagrams below in the present chapter, it is not necessary to correspond the Green's function to free ends of electron (straight) and phonon (wavy) lines. These ends simply show by which line this diagram is connected to the vertex. It is possible to obtain explicit expressions for Γ_s by transforming (1.211) into the second-quantization representation. In this case, we obtain, in accordance with (1.232), the expression for the vertex Γ_1^λ

$$\Gamma_1^\lambda(\mathbf{k}, \mathbf{q}) = \mathbf{M}_{\mathbf{k}\mathbf{q}\lambda}\left(\frac{2V}{\hbar\omega_{\mathbf{q}\lambda}}\right)^{1/2}. \tag{1.233}$$

Here, $M_{\mathbf{k}\mathbf{q}\lambda}$ is the matrix element of the electron–phonon interaction in (1.187)

$$M_{\mathbf{k}\mathbf{q}\lambda} = -\left(\frac{N\hbar}{2M\omega_{\mathbf{q}\lambda}}\right)^{1/2}\left(U_{\mathbf{k},\mathbf{k}+\mathbf{q}}\,\mathbf{e}_{\mathbf{q}\lambda}\right);$$

$$U_{\mathbf{k},\mathbf{k}+\mathbf{q}} = -\int \hat{\psi}^*_{\mathbf{k}}(\mathbf{r}_j)\,\nabla V(r_j)\,\hat{\psi}_{\mathbf{k}+\mathbf{q}}(\mathbf{r})\,\mathbf{d}^3 r_j.$$

The vertex $\Gamma_2^{\lambda\mu}$ is

$$\Gamma_2^{\lambda\mu}(\mathbf{k}, \mathbf{q}, \mathbf{p}) = \hbar L^{\lambda\mu}_{\mathbf{k},\mathbf{k}+\mathbf{q}}\left(\frac{4V^2}{\omega_{\lambda\mathbf{q}}\,\omega_{\mu\mathbf{p}}}\right)^{1/2}\left(\frac{N^2}{4M^2\omega_{\lambda\mathbf{q}}\,\omega_{\mathbf{p}\mu}}\right)^{1/2},$$

[7] In all equations where it is possible, we will omit the polarization indexes of phonons $\{\lambda\}$.

$$L^{\lambda\mu}_{\mathbf{k},\mathbf{k}+\mathbf{q}} = \sum_{\alpha,\beta=1}^{3} e^{\alpha}_{\lambda\mathbf{q}} e^{\beta}_{\mu\mathbf{p}} \int d^3 r_j \, \psi^*_{\mathbf{k}}(\mathbf{r}_j) \frac{\partial^2 V(\mathbf{r}_j)}{\partial r_\alpha \, \partial r_\beta} \psi_{\mathbf{k}+\mathbf{q}}(\mathbf{r}_j). \tag{1.234}$$

Remember that Cartesian indices α, β, $\gamma, \ldots$ change from one to three. The approximation of hard ions for $U_{\mathbf{kk}'}$ provides [37]

$$U_{\mathbf{k}\,\mathbf{k}'} = i(\mathbf{k}' - \mathbf{k})\tilde{\varepsilon}F(\vartheta), \tag{1.235}$$

where $\tilde{\varepsilon}$ is of the order E_F and $F(\Theta)$ is the function slowly depending on the angle between $\mathbf{k} - \mathbf{k}'$. Therefore, at small $\mathbf{k}$ and $\mathbf{k}'$, we have

$$L^{\lambda\mu}_{\mathbf{k}\,\mathbf{k}'} \simeq (\mathbf{e}_{\lambda\mathbf{q}}(\mathbf{k}' - \mathbf{k}))\,(\mathbf{e}_{\mu\mathbf{p}}(\mathbf{k}' - \mathbf{k}))\tilde{\varepsilon}.$$

Calculations in the frame of the pseudopotential approximation [1] do not destroy these momentum dependencies and evaluation of $\tilde{\varepsilon}$.

If we consider only the longitudinal branch of oscillation and acoustical dispersion law [38], then formulae for the vertices Γ_1 and Γ_2 have the simple form

$$\Gamma_1 = -i\left(\frac{\zeta_0}{\nu}\right)^{1/2}, \quad \Gamma_2 = \frac{\zeta_0}{\nu\tilde{\varepsilon}} \tag{1.236}$$

$$\zeta_0 = \frac{\nu N V \tilde{\varepsilon}}{(Mu^2)}, \quad \omega_q = uq,$$

where u is the velocity of the longitudinal sound and ν is determined by equation (1.226) and ζ_0 by (1.193). By calculating Γ_1 and Γ_2 in (1.236), we are not taking into account the umklapp process ($\mathbf{g} = 0$).

It is also necessary, in addition to (1.232), to construct the diagrammatic technique for the term U_{2n} in the perturbation operator $\hat{\mathcal{H}}'$ in (1.211). For this term, there exist the following relations (see (1.212))

$$\begin{aligned} -U_{2n} = -|\hat{\mathcal{H}}_2|_{nn} = \hat{\mathcal{H}}_{2n}, \quad -|\hat{\mathcal{H}}_2|_{nn} = \hat{\mathcal{H}}''_2 = \hat{\mathcal{H}}''_{20} + \hat{\mathcal{H}}''_{22}, \\ \hat{\mathcal{H}}_{2n} \equiv \hat{\mathcal{H}}'_{22} = \sum_m |\hat{\mathcal{H}}_1|^2_{nm}(E_{n0} - E_{m0})^{-1}. \end{aligned} \tag{1.237}$$

The vertices corresponding to $\hat{\mathcal{H}}'_{22}$, $\hat{\mathcal{H}}'_{20}{}'$, $\hat{\mathcal{H}}'_{22}{}'$ have the form

Γ''_{22} Γ''_{20} Γ''_{22} (1.238)

The diagrammatic representation of Γ_2 in (1.232) and Γ'_{22}, Γ''_{22} in (1.238) have the same form, but analytical expressions corresponding to these vertices are different as it follows from their definition.

Now we will show the analytical expression for the vertex Γ'_{22} in (1.238) only because we will use it below. Using the explicit expression of $\hat{\mathcal{H}}'_{22}$ in (1.237), we obtain

$$\Gamma_{22}^{\prime\lambda\lambda}(\mathbf{k}.\mathbf{q}) = |\Gamma_1^{\lambda}|^2\left[\left(\varepsilon_{\mathbf{k}+\mathbf{q}} - \varepsilon_{\mathbf{k}}\right)^{-1} + \left(\varepsilon_{\mathbf{k}-\mathbf{q}} - \varepsilon_{\mathbf{k}}\right)^{-1}\right]. \tag{1.239}$$

A characteristic feature of the diagram technique [28] constructed above is the existence of the vertex with many-phonon ends. It is a consequence of the very particular form of the perturbation operator (1.211). Now it is very interesting to investigate the consequences of the particular form of this diagram technique.

Corrections to the vertices The equation for the vertices Γ_1 and Γ_2 with the accuracy up to κ^2 has the diagrammatic representation

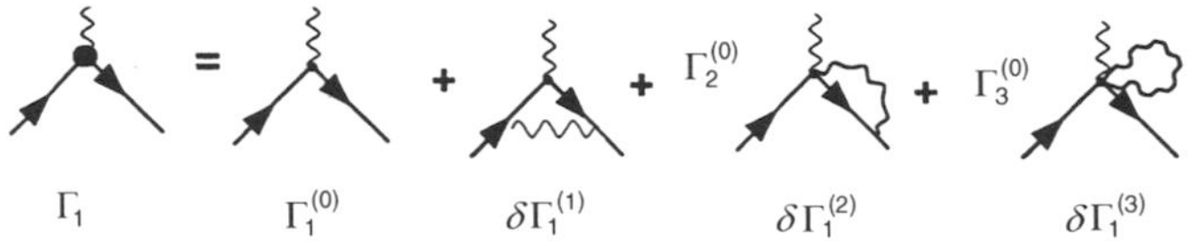

(1.240)

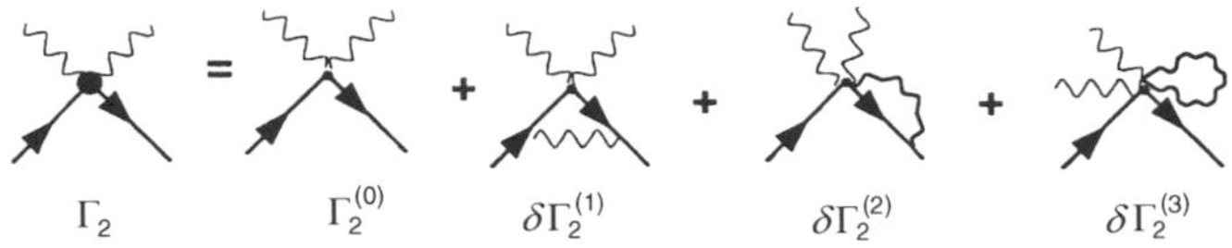

(1.241)

(Similar forms have corrections to Γ_3 and so on).

As has been shown in [38], the first correction to Γ_1 (i.e., $\delta\Gamma_1^{(1)}$) has an appreciable value at large frequencies $\omega > v_F q$ only (v_F is the phonon velocity at the Fermi surface). The first correction to Γ_2 (i.e., $\delta\Gamma_2^{(1)}$) has the same structure. At frequencies $\omega \sim \omega_D$, which provide the main contribution by calculation of the mass-operator of the phonon and the self-energy operator of electrons, these diagrams provide corrections $\sim \kappa^2$, i.e. $\delta\Gamma_1^{(1)} \sim \kappa^2\Gamma_1^{(0)}$ and $\delta\Gamma_2^{(1)} \sim \kappa^2\Gamma_2^{(0)}$.

The presence of vertices with many phonon ends provides new corrections that are absent in [38]. For the vertex Γ_1, first corrections of this type correspond to the terms $\delta\Gamma_1^{(2)}$ and $\delta\Gamma_1^{(3)}$ in (1.240). If we substitute obtained values for Γ_1 and Γ_2 into the expression for $\delta\Gamma_1^{(2)}$, then corrections are of the order $\Gamma_1^{(0)}\Sigma_2(\omega)\tilde{\varepsilon}^{-1}$, where $\Sigma_2(\omega)$ is the diagram of the self-energy part (the mass-operator) of an electron corresponding to the diagram (see below).

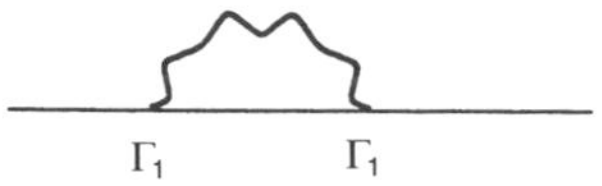

The function $\Sigma_2(\omega)$ does not depend on the momentum and may be expressed as [38]

$$\Sigma_2(\omega) = \Sigma_{02} + \mathrm{Re}\Sigma_2(\omega) + \mathrm{Im}\Sigma_2(\omega), \tag{1.242}$$

where Σ_{02} is a real constant of the order $\hbar\omega_D$, $\mathrm{Re}\Sigma_2(\omega)$ and $\mathrm{Im}\Sigma_2(\omega)$ are some complicated functions of frequency which are not larger (of the order magnitude) than $\hbar\omega_D$. Hence, at all frequencies $\delta\Gamma_1^{(2)} \sim \kappa^2\Gamma_1^{(0)}$. Since $\delta\Gamma_3 \sim \kappa^2\Gamma_1$, then the evaluation of $\delta\Gamma_1^{(3)}$ also has the form $\delta\Gamma_1^{(3)} \sim \kappa^2\Gamma_1^{(0)}$.

It is possible to generalize these results in the case of three branches of oscillation and non-acoustical dispersion laws. All calculations for corrections to the vertices are not destroyed by this generalization.

The mass-operator for the phonon It is necessary to introduce three $D_{\lambda\lambda}$-Green's functions (1.228) to consider a whole oscillation spectum of the lattice relating to each branch λ of oscillations. Hence, it is necessary to have three mass operators Π_λ. The diagrammatic representation for the phonon mass operator for a single branch up to the accuracy $\sim\kappa^2$, has the form

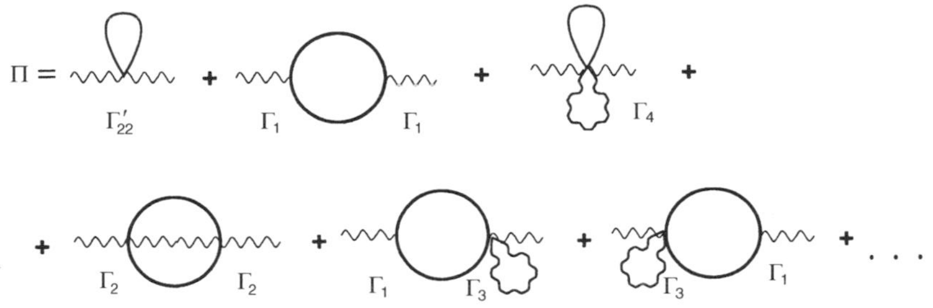

(1.243)

Taking into the account evaluations (1.213) of contributions of the perturbation operator $\hat{\mathcal{H}}'$, it can be seen that the first two diagrams have the order of κ^0 and the rest are $\sim\kappa^2$. Let us show that the sum of the first two diagrams, as a consequence of mutual cancellation, is also of the order $\sim\kappa^2$ [28].

The analytical representation of the first diagram in (1.243) is

$$\Pi_{01\lambda}(\mathbf{q}) = -2\int \frac{d^4k}{(2\pi)^4} G(k)\Gamma_{22}^{\prime\lambda\lambda}(\mathbf{k},\mathbf{q}). \tag{1.244}$$

Here $k = (\mathbf{k}, \varepsilon)$ is the four-momentum. After integration over energy (for the acoustic dispersion law), we obtain (index λ is omitted)

$$\Pi_{01}(\mathbf{q}) = 2g^2 \int \frac{d^3k}{(2\pi)^3} \frac{n_{\mathbf{k}} - n_{\mathbf{k}+\mathbf{q}}}{\varepsilon_{\mathbf{k}+\mathbf{q}} - \varepsilon_{\mathbf{k}}} = g^2\pi_0(\mathbf{q}) \tag{1.245}$$

where $\pi_0(\mathbf{q})$ is the Lindhard [42] static polarization operator. Finally, for (1.245) we obtain

$$\Pi_{01}(\mathbf{q}) = \mathbf{g}^2\nu h(q/2k_F); \quad h(x) = 1 + \frac{1-x^2}{2x}\ln\left|\frac{1+x}{1-x}\right|. \tag{1.246}$$

The second diagram in (1.243) was calculated by Migdal [38]

$$\Pi_2(\mathbf{q},\omega) = -2ig^2 \int \frac{d^4K}{(2\pi)^4}\, G(K)\, G(K+q). \tag{1.247}$$

Let us write $\Pi_2(\mathbf{q},\omega)$ as

$$\Pi_2 = \Pi_{20} + i\Pi_2'. \tag{1.248}$$

After integration over energy ε in the integral (1.247) with substitution of the Green's function $G(k)$ by the Green's function of free electron $G^{(0)}(k)$, we obtain

$$\Pi_{02}(\mathbf{q},\omega) = -2\mathbf{g}^2 \int \frac{d^3k}{(2\pi)^3} \frac{n_{\mathbf{k}} - n_{\mathbf{k+q}}}{\varepsilon_{\mathbf{k+q}} - \varepsilon_{\mathbf{k}} - \hbar\omega}. \tag{1.249}$$

If we remember the definition of g^2 following from (1.226) and (1.236),

$$g^2 = \left.\frac{V|M_{\mathbf{kq}\lambda}|^2}{\hbar\omega_{\lambda\mathbf{q}}}\right|_{k=k_F,\ \mathbf{q}\to 0}, \tag{1.250}$$

then we can see that $\Pi_{02} + \Pi_{01}$, with the accuracy up to a constant, is equal to $\hbar\Delta\omega_{\lambda\mathbf{q}}/\omega_{\lambda\mathbf{q}}$ in (1.188). For us, what is most interesting is the interval of frequencies $\omega \sim \omega_D \ll v_F q$. In this interval, it is possible to neglect ω in the denominator of Π_{02}. After integration over k, we obtain

$$\Pi_{02} \simeq -g^2\nu h(q/2k_F) = -\Pi_{01}. \tag{1.251}$$

Consequently, in accordance with (1.189) and performing some calculations we obtain

$$\Pi_{01} + \Pi_{02} \sim \kappa^4. \tag{1.252}$$

In accordance with the Dyson equation $D^{-1} = D_0^{-1} - \Pi$, the renormalization of the adiabatic phonon frequency is of the order κ^2 [28]. Terms $\sim \kappa^2$ are provided by the last four graphs of (1.253). They contain the square of ion displacements. Formal reasons why the last four graphs are $\sim \kappa^2$ are that $\Gamma_2 \sim \kappa\Gamma_1$, $\Gamma_3 \sim \kappa^2\Gamma_1$ and $\Gamma_4 \sim \kappa^3\Gamma_1$. Hence, within the accuracy of κ^2, mutual cancellation of the real and imaginary parts of the mass-operator takes place. The imaginary part Π_2 has the order κ^2 (see (1.190)). The last four graphs provide the same order contribution to Im Π also. In the Fröhlich model, Im Π = Im Π_2 and consequently, the numerical value of the coefficient at the imaginary part Π, i.e., phonon attenuation has an incorrect value. The value of Π_2' was calculated in [38]

$$\Pi_2'(\mathbf{q},\omega) = -\zeta_0 \frac{\pi m|\omega|}{\hbar k_F q}\vartheta(2k_F - q).$$

To take into account deviations from the acoustic dispersion law, we must use the substitution

$$\zeta_0 \to \zeta_0 \frac{u_\lambda^2 q^2}{\omega_{\mathbf{q}\lambda}^2}.$$

Here, as with the accuracy of κ^2, mutual cancellation in the real part of the mass-operator takes place. All corrections to the adiabatic frequency for each phonon branch are of the order $\kappa^2\omega_D$ and the phonon attenuation for each $\omega_{\lambda\mathbf{q}}$ is κ^2 times smaller than the phonon energy.

Calculations described below use the approximation of replacing the total electron Green's function G by its zero order approximation $G^{(0)}$. But really Re $G \neq$ Re $G^{(0)}$ in the narrow band of energies ε_0 is smaller than $\hbar\omega_D$. Although Im G is not equal to Im $G^{(0)}$, in this band, Im G /Re $G \ll 1$ [23]. Approximation of G by $G^{(0)}$ provides an appreciable change of the integral in (1.257) only in the region $|\varepsilon| < \hbar\omega_D$. But in the integral over ε for Π_{02}, an appreciable contribution gives the region $|\varepsilon| \sim \hbar v_F q$, which is much larger than $\hbar\omega_D$ ($v_F q \gg \omega_D$). Hence, the approximation of G by $G^{(0)}$ provides no appreciable errors for Π_{02}. For Im Π_2, the integration over ε takes place over the small region $|\varepsilon| < |\hbar\omega| \sim \hbar\omega_D$, i.e., $\Pi_2' \sim |\hbar\omega|$ [38]. Hence, by calculation of Im Π_2, substitution of G by $G^{(0)}$ really gives an appreciable error. Evaluations with the exact value of G do not change the evaluation of $\Pi_2' \sim \kappa^2$, but give wrong numerical values for the coefficients. Hence, the imaginary part of the mass-operator is determined not only by Π_2', but also by terms relating to the difference of G from $G^{(0)}$ [44].

Self-energy part for electron Diagrams for the electron mass-operator Σ to the accuracy $\kappa^2\hbar\omega_D$ inclusively are shown below (in each graph, summation over polarization $\lambda = 1.2.3$ is supposed)

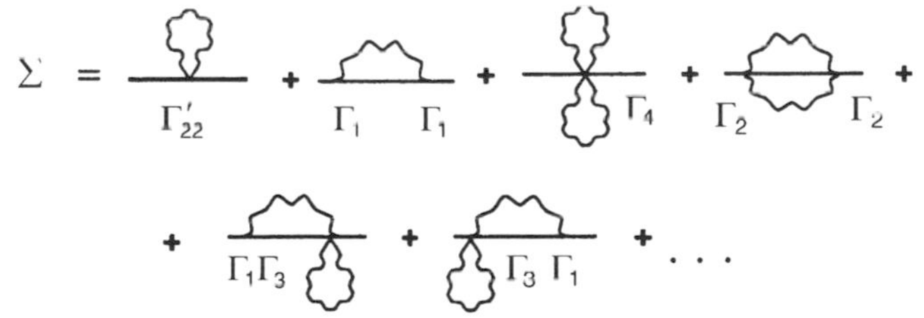

(1.253)

Let us demonstrate that, although each of the first two graphs have the order of $\hbar\omega_D$, the sum of them has the order $\kappa^2\hbar\omega_D$ or the same order as the last graphs.

Analytical expression for the first diagram in (1.253) has the form

$$\Sigma_{01}(\mathbf{k}) = \sum_\lambda \int \frac{d^4q}{(2\pi)^4} \Gamma_{22}'^{\lambda\lambda}(\mathbf{k},\mathbf{q})\, D_{\lambda\lambda}(\mathbf{q}); \quad q = (\mathbf{q}, \hbar\omega). \tag{1.254}$$

After substituting in this equation the value of $\Gamma_{22}'^{\lambda\lambda}$ from (1.239) and integrating over ω, we obtain

$$\Sigma_{01}(k) = \sum_\lambda \int \frac{d^3q}{(2\pi)^3} |\Gamma_1^\lambda(q)|^2 \left[(\varepsilon_{\mathbf{k}+\mathbf{q}} - \varepsilon_\mathbf{k})^{-1} + (\varepsilon_{\mathbf{k}-\mathbf{q}} - \varepsilon_\mathbf{k})^{-1}\right]. \tag{1.255}$$

Taking a single branch of oscillations with acoustical dispersion law $\omega_{\mathbf{q}} = uq$ for $\Gamma_1'^{\lambda}(\mathbf{q})$ and using (1.233), we obtain

$$\Sigma_{01}(k_F, 0) = -\frac{mg^2 u}{2(2\pi)^2 \hbar k_F} \int_0^{q_D} dq q^2 \ln \left| \frac{2k_F - q}{2k_F + q} \right|, \tag{1.256}$$

Here q_D is the Debye wave number [1]. This equation shows that $\Sigma_{01}(k_F, 0)$ has a real value that does not depend on frequency.

The respective equation for the second diagram is

$$\Sigma_2(\mathbf{k}, \varepsilon) = i \sum_{\lambda} \int \frac{d^4 k'}{(2\pi)^4} |\Gamma_1^{\lambda}(\mathbf{k} - \mathbf{k}')|^2 \times D_{\lambda\lambda}(\mathbf{k} - \mathbf{k}')(\varepsilon - \varepsilon' + \mu - \Sigma(\mathbf{k}', \varepsilon))^{-1}, \tag{1.257}$$

where μ is the chemical potential and Σ is the total self-energy part that is equal to the sum of all diagrams in (1.253). The last factor in (1.257) is the exact electron Green's function

$$G(k) = G(\mathbf{k}, \varepsilon) = [\varepsilon - \varepsilon_k + \mu - \Sigma(\mathbf{k}, \varepsilon)]^{-1}.$$

For the case of the single acoustic branch

$$\Sigma_2(\mathbf{k}, \varepsilon) = i\mathbf{g}^2 \int \frac{d^3 k' d\varepsilon}{(2\pi)^4} D_{\lambda\lambda}(\mathbf{k} - \mathbf{k}', \varepsilon - \varepsilon')(\varepsilon - \varepsilon' + \mu - \Sigma(k', \varepsilon'))^{-1}, \tag{1.258}$$

Let us transfer the integration over angles to the integration over the phonon momentum [23]. Now Σ_2 is the sum of two terms, namely the constant terms Σ_{02} provided by integration over $|\varepsilon'| > \delta$, $\delta \gg \hbar\omega_D$ and the integral Σ_2' corresponding to integration over $|\varepsilon'| < \delta$. By calculating Σ_{02} with the accuracy to $\kappa^2 \hbar\omega_D$, it is possible to neglect Σ in the denominator of equation (1.258). It gives

$$\Sigma_{02} = \frac{ig^2 \hbar}{(2\pi)^3 k_F} \int_{-\infty}^{\infty} d\omega \int_0^{q_D} dq q D_{ll}(\mathbf{q}, \hbar\omega) \int_{\mathbf{k_F}+\mathbf{q}}^{|\mathbf{k_F}-\mathbf{q}|} \frac{k' dk'}{\varepsilon_{k'} - \mu} \tag{1.259}$$

All evaluations for Σ_{02} are described in detail in [23], therefore, we do not reproduce them here. The final expression for Σ_{02} coincides with Σ_{01} in (1.256), but has the opposite sign. Hence, there is mutual cancellation of Σ_{01} and Σ_{02} (each of them is $\sim \hbar\omega_D$). Thus

$$\Sigma_{01} + \Sigma_{02} \sim \kappa^2 \hbar\omega_D. \tag{1.260}$$

To evaluate the total normalization of the chemical potential μ with the accuracy to $\kappa^2 \hbar\omega_D$, it is necessary to take into account the last four graphs in (1.253) and to provide calculations in (1.259) with accuracy to $\sim \kappa^2 \hbar\omega_D$.

Here only the schematic description of the adiabatic theory of electron–phonon interactions in metals is give. The main purpose is to present the simple quantum-mechanical perturbation theory (developed by Geĭlikman) of calculation of non-adiabatic corrections (electron–phonon interaction) as an alternative to using the sophisticated adiabatic perturbation theory [20]. Concrete details of calculations of electron–phonon contributions can be found in [27, 28, 33, 34].

1.3.4 Contribution of electron–phonon interaction to the thermodynamic potential

The thermodynamic potential of the electron–ion system is the sum of contributions provided by both electron (see [1] and this chapter above) and ion subsystems (see subsection 1.1.1), as well as the interaction between them Ω_{ep}. We are interested in non-adiabatic (electron–phonon) contributions to the thermodynamic potential. These contributions are determined by the next diagrammatic equation

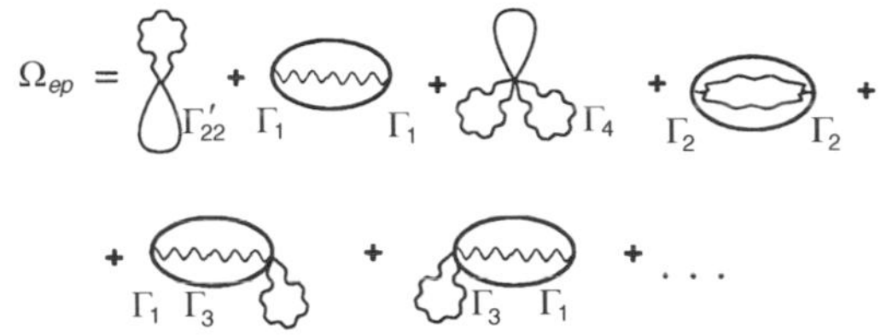

(1.261)

(It is well-known [23] that in the calculation of thermodynamic potential, it is necessary to calculate only the sum of all linked diagrams constructed from all possible electron and phonon loops). The first two diagrams have the order of $\hbar\omega_D$ and the rest are $\sim \kappa^2\hbar\omega_D$. The summation over all branches of lattice oscillations in each diagram is assumed. The contributions of the first two diagrams to the potential Ω_{ep} have the form [34]

$$\delta\Omega_1 = -2V(k_BT)^2\sum_{\lambda=1}^{3}\sum_{n,m}\int\frac{d^3kd^3q}{(2\pi)^6}\,\Gamma_{22}^{\prime\lambda\lambda}(\mathbf{k},\mathbf{q})\,G^{(0)}(\mathbf{k},\omega_n)\,\mathcal{D}_{\lambda\lambda}^{(0)}(\mathbf{q},\omega_m), \qquad (1.262)$$

$$\delta\Omega_2 = -\,2V(k_BT)^2\sum_{\lambda=1}^{3}\sum_{n,m}\int\frac{d^3kd^{3q}}{(2\pi)^6}$$
$$\times\,|\Gamma_1^{\lambda}(\mathbf{q})|^2G^{(0)}(\mathbf{k},\omega_n)G^{(0)}(\mathbf{k}+\mathbf{q},\omega_n+\omega_m)\times\mathcal{D}_{\lambda\lambda}^{(0)}(\mathbf{q},\omega_m). \qquad (1.263)$$

In these expressions, V is the system volume. The Green's functions of free electrons $G^{(0)}(\mathbf{k},\omega_n)$ and phonons $D^{(0)}(\mathbf{k},\omega_m)$are written using the Matsubara technique [27].

$$G^{(0)}(\mathbf{k},\omega_n) = (i\hbar\omega_n - \varepsilon_{\mathbf{k}} + \mu)^{-1}, \quad D_{\lambda}^{(0)} = -\frac{\omega_{\lambda\mathbf{q}}^2}{\omega_2^2 + \omega_{\lambda\mathbf{q}}^2},$$
$$\hbar\omega_n = (2n+1)\pi k_BT, \quad \hbar\omega_m = 2m\pi k_BT, \qquad (1.264)$$

where n and m are integers. Summations over frequencies in (1.262) and (1.263) are performed by using

$$\sum_{n=0}^{\infty}\frac{1}{(2n+1)^2+x^2} = \frac{1}{2x}\tanh\frac{\pi x}{2}, \quad \sum_{m=0}\frac{1}{(2m)^2+x^2} = \frac{\pi}{4x}\coth\frac{\pi x}{2}+\frac{1}{2x^2}.$$

Vertices Γ_1^{λ} (1.233) and $\Gamma_{22}^{\prime\lambda\lambda}$(1.239) were discussed in the previous subsection.

The thermodynamic potential Ω_{ep} is (see [34])

$$\Omega_{ep} = \delta\Omega_1 + \delta\Omega_2 + \Delta\Omega_{ep} = \sum_{\mathbf{kq}\lambda} |M_{\mathbf{kq}\lambda}|^2 \frac{\hbar^2 \omega_{\lambda\mathbf{q}}^2 (n_\mathbf{k} - n_{\mathbf{k+q}})}{\varepsilon_\mathbf{k} - \varepsilon_{\mathbf{k+q}}} \left[\frac{N_{\lambda\mathbf{q}} + 1}{\varepsilon_\mathbf{k} - \varepsilon_{\mathbf{k+q}} - \hbar\omega_{\lambda\mathbf{q}}} - \frac{N_{\lambda\mathbf{q}}}{\varepsilon_\mathbf{k} - \varepsilon_{\mathbf{k+q}} + \hbar\omega_{\lambda\mathbf{q}}} + \frac{2N_{(\varepsilon_{k+q}-\varepsilon_k)}(\varepsilon_\mathbf{k} - \varepsilon_{\mathbf{k+q}})}{(\varepsilon_\mathbf{k} - \varepsilon_{\mathbf{k+q}})^2 - (\hbar\omega_{\lambda\mathbf{q}})^2} \right] + \Delta\Omega_{ep}, \tag{1.265}$$

where $M_{\mathbf{kq}\lambda}$ is the matrix element of electron-phonon interaction (1.187) and $\Delta\Omega_{ep}$ is the contribution of the last four diagrams in (1.261).

In the calculation of Ω_{ep}, it is convenient to express the function $N_{(\varepsilon_{\mathbf{k+q}}-\varepsilon_\mathbf{k})}$ from the third term of (1.265) through the electron distribution functions $n_\mathbf{k}$

$$N_{(\varepsilon_{\mathbf{k+q}}-\varepsilon_\mathbf{k})} = \frac{n_{\mathbf{k+q}}[1 - n_\mathbf{k}]}{n_\mathbf{k} - n_{\mathbf{k+q}}}.$$

Following [38], we suppose that the main contribution to the thermodynamic potentials gives longitudinal phonons with the acoustic dispersion law: $\omega_{\lambda\mathbf{q}} = uq = \omega_\mathbf{q}$. Let us introduce the constant of electron–phonon interactions for the isotropic model of metal g^2 (1.250). Supposing $N_{\lambda\mathbf{q}} = N_\mathbf{q}$, we obtain for $\delta\Omega_1$ and $\delta\Omega_2$

$$\delta\Omega_1 = -Vg^2 \int \frac{d^3q\, d^3k}{(2\pi)^6} \hbar\omega_\mathbf{q}(2N_\mathbf{q} + 1) \frac{(n_\mathbf{k} - n_{\mathbf{k+q}})}{\varepsilon_\mathbf{k} - \varepsilon_{\mathbf{k+q}}}, \tag{1.266}$$

$$\delta\Omega_2 = Vg^2 \int \frac{d^3q\, d^3k}{(2\pi)^6} \hbar\omega_\mathbf{q}(n_\mathbf{k} - n_{\mathbf{k+q}}) \left\{ \frac{1 + N_\mathbf{q}}{\varepsilon_\mathbf{k} - \varepsilon_{\mathbf{k+q}} - \hbar\omega_\mathbf{q}} + \frac{N_\mathbf{q}}{\varepsilon_\mathbf{k} - \varepsilon_{\mathbf{k+q}} + \hbar\omega_\mathbf{q}} + \frac{2n_{\mathbf{k+q}}(1 - n_\mathbf{k})(\varepsilon_\mathbf{k} - \varepsilon_{\mathbf{k+q}})}{(n_\mathbf{k} - n_{\mathbf{k+q}})[(\varepsilon_\mathbf{k} - \varepsilon_{\mathbf{k+q}})^2 - \hbar^2\omega_\mathbf{q}^2]} \right\}. \tag{1.267}$$

Contributions $\delta\Omega_1$ (1.266) and $\delta\Omega_2$ (1.267) have the order of $\hbar\omega_D$ and their sum

$$\Omega_{ep1} = \delta\Omega_1 + \delta\Omega_2 = Vg^2 \int \frac{d^3q\, d^3k}{(2\pi)^6} \hbar^2\omega_\mathbf{q}^2 \frac{n_\mathbf{k} - n_{\mathbf{k+q}}}{\varepsilon_\mathbf{k} - \varepsilon_{\mathbf{k+q}}} \left\{ \frac{1 + N_\mathbf{q}}{\varepsilon_\mathbf{k} - \varepsilon_{\mathbf{k+q}} - \hbar\omega_\mathbf{q}} - \frac{N_\mathbf{q}}{\varepsilon_\mathbf{k} - \varepsilon_{\mathbf{k+q}} + \hbar\omega_\mathbf{q}} + \frac{2n_{\mathbf{k+q}}(1 - n_\mathbf{k})(\varepsilon_\mathbf{k} - \varepsilon_{\mathbf{k+q}})}{(n_\mathbf{k} - n_{\mathbf{k+q}})\left[(\varepsilon_\mathbf{k} - \varepsilon_{\mathbf{k+q}})^2 + \hbar^2\omega_\mathbf{q}^2\right]} \right\} \tag{1.268}$$

is of the order $\kappa^2\hbar\omega_D$. Hence, total non-adiabatic corrections to the thermodynamic potential of the system as a consequence of mutual cancellations are of the order $\kappa^2\hbar\omega_D$ and to calculate these, it is necessary to take the last graphs in (1.261) into account.

Note that (1.268) coincides (with g^2 as (1.250)) with the non-adiabatic correction $E_{n\nu 1}^{(1)}$ to the system energy obtained above. This equivalence of results (1.216) and

(1.268) corresponds to the well-known theorem from statistical physics [19] relating the small corrections to thermodynamic potentials. The second part of non-adiabatic correction to the energy $E^{(1)}_{n\nu 2}$ (1.215) corresponds to the last four graphs in equation (1.261).

Now we calculate Ω_{ep1} to investigate temperature dependence of the thermodynamic potential Ω_{ep}. Let us rewrite (1.268) as

$$\Omega_{ep1} = -\frac{1}{2} V g^2 \int \frac{d^3q}{(2\pi)^3} \hbar\,\omega_{\mathbf{q}} [\pi_0(\mathbf{q},\omega,T) - \pi_0(\mathbf{q},0,T)] \times \left(N_{\mathbf{q}} + \frac{1}{2}\right) - P(\mathbf{q},\omega,T). \tag{1.269}$$

Here the electron polarization operator

$$\pi_0(\mathbf{q},\omega,T) = 2\int \frac{d^3k}{(2\pi)^3} \frac{n_{\mathbf{k}} - n_{\mathbf{k+q}}}{\varepsilon_{\mathbf{k+q}} - \varepsilon_{\mathbf{k}} - \hbar\,\omega_q} \tag{1.270}$$

and also the function

$$P(\mathbf{q},\omega,T) = 4\int \frac{d^3k}{(2\pi)^3} \frac{\hbar\,\omega_{\mathbf{q}} n_{\mathbf{k+q}}[1 - n_{\mathbf{k}}]}{(\varepsilon_{\mathbf{k}} \quad \varepsilon_{\mathbf{k+q}})^2 - (\hbar\,\omega_{\mathbf{q}})^2}. \tag{1.271}$$

are introduced.

All calculations of Ω_{ep1} are provided in the frame of random phase approximations (RPA) taking the temperature corrections into account. To do this, we use equations (see for example [36, 1])

$$\pi_0(\mathbf{q},\omega,T) = \beta\left\{1 - \frac{1}{4x}\left[(\xi^2 - 1)\ln\left|\frac{\xi - 1}{\xi + 1}\right| + (1 \quad \eta^2)\ln\left|\frac{\eta - 1}{\eta + 1}\right|\right]\right\} + \frac{\pi^2}{6}(k_B T)^2 \left[\frac{\partial}{\partial\varepsilon} g(\varepsilon) + \frac{\partial}{\partial\varepsilon}\ln N(\varepsilon)\right]_{\varepsilon = E_F}, \tag{1.272}$$

where

$$g(\varepsilon) = \pi^{-2}\int_{-1}^{1} du_1 k^2 \frac{dk}{d\varepsilon} \frac{\varepsilon_{\mathbf{k+q}} - \varepsilon_{\mathbf{k}}}{(\varepsilon_{\mathbf{k}} - \varepsilon_{\mathbf{k+q}})^2 - \hbar^2\omega_{\mathbf{q}}^2},$$

$$N(\varepsilon) = \frac{(2m)^{3/2}}{2\hbar^3\pi^2}\varepsilon^{1/2},$$

$$\xi = \frac{\omega - \hbar q^2/2m}{\hbar q k_F m^{-1}} = d - x, \quad \eta = \frac{\omega + \hbar q^2/2m}{\hbar q k_F m^{-1}} = d + x,$$

$$x = q/(2k_F), \quad d = u/v_F, \quad \beta = 3n_0/(4E_F) = mk_F/(2\pi^2\hbar^2),$$

$$k_F = (3\pi^2 n_0)^{1/3}, \quad v_F = \hbar k_F/m. \tag{1.273}$$

To calculate Ω_{ep1}, it is necessary to find the difference

$$\pi_0(x,d,T) - \pi_0(x,0,T) \simeq \frac{d^2\beta}{2x}\left\{ \ln\left|\frac{x-1}{x+1}\right| - \frac{2x^2}{1-x} + \frac{4}{3}\pi^2 x_D^2 d^2 \left(\frac{T}{\theta_D}\right)^2 \left[\frac{1}{(x+1)^3} + \frac{1}{(x-1)^3}\right]\right\}. \tag{1.274}$$

Here $\Theta_D = \hbar u q_D/k_B$ is the temperature and $q_D = (6\pi^2/V_0)^{1/3}$ is the Debye radius [1, 15]. In (1.274) we have expanded the difference $\pi_0(x,d,T) - \pi_0(x,0,T)$ into a series over the ratio d of the sound velocity u to the electron velocity at the Fermi Surface (FS) v_F. Since $d \ll 1$, we restrict our calculations to the leading terms only.

The thermodynamic potential Ω_{ep1} contains two distinguished types of terms

$$\Omega_{ep1} = \delta\Omega(T=0) + \delta\Omega(T). \tag{1.275}$$

The first term is the temperature-independent contribution. Calculations for this term give

$$\delta\Omega(T=0) \simeq Ad^2\left\{ -\frac{(1-x_D^4)}{2}\ln\left|\frac{1-x_D}{1+x_D}\right| - x_D + \frac{x_D^3}{3} + 2x_D^3\left[\frac{x_D^2}{5} - \frac{3x_D}{4} + \frac{2}{3}\right]\right\} \tag{1.276}$$

where

$$A = Vg^2 k_F^6 \pi^{-4}.$$

The second term in (1.275) describes the temperature dependence of the thermodynamic potential of the electron–phonon interaction

$$\delta\Omega(T) \equiv \delta\Omega_{ep1}(T) = \delta\Omega_a(T) + \delta\Omega_b(T). \tag{1.277}$$

The main contribution to $\delta\Omega_a(T)$ and $\delta\Omega_a(T)$ has the form

$$\delta\Omega_a(T) \simeq Ad^4\left(\frac{T}{\theta_D}\right)^2 F_a(x_D), \tag{1.278}$$

$$\delta\Omega_b(T) \simeq Ad^3\begin{cases} \left(\frac{T}{\theta_D}\right)F_b(x_D), & \text{at } T \geq \theta_D, \\ \left(\frac{T}{\theta_D}\right)^4 \frac{16}{15}\pi^2 x_D^4, & \text{at } T \ll \theta_D \end{cases} \tag{1.279}$$

in these equations

$$F_a(x_D) = \frac{\pi x_D^2}{3}\left[\frac{4x_D - 1}{(x_D-1)^2} - 2\ln|x_D - 1| - 3\right] - \frac{8\pi^2}{3}\frac{x_D}{x_D^2 - 1},$$

$$F_b(x_D) = x_D\left[(x_D^2+1)\ln\left|\frac{x_D+1}{x_D-1}\right| - 2x_D\right]. \tag{1.280}$$

For metals with a near spherical Fermi-surface $d = u/v_F \sim 10^{-3}$. In the case of polyvalent metals for some regions on the Fermi-surface, the parameter d increases and can reach values $d \leq 1$. In particular, it takes place by topological transition induced by external stresses. At low temperatures $T \ll \Theta_D$, the term $\delta\Omega_a \sim d^4T^2$ is higher than $\delta\Omega_b \sim d^3T4 \ll \delta\Omega_a$, and the lattice thermodymanical potential is also small $\sim T^4$. The term $\delta\Omega_a$ provides redistribution of pure electron contribution to the total potential Ω. At high temperatures $T \geq \Theta_D$, the thermodynamic potential of electron–phonon interactions has the same temperature dependence ($\sim T$) as the thermodynamic potential of the lattice.

Therefore, we see that the contribution of non-adiabatic states is very difficult to observe because it is camouflaged at large $T \geq \Theta_D$ by the lattice contribution $\sim T$, which contains no parameter d^3. At low temperatures $T \ll \Theta_D$, the non-adiabatic contribution is also difficult to distinguish in the background of the pure electron contribution.

1.4 The Theory of Anharmonic Contributions to the Static Properties of Metals

In section 3 of the present chapter, the adiabatic approximation and the perturbation theory in the adiabatic parameter $\kappa^2 = \varepsilon = \langle x^2 \rangle / a^2$ were described. As an example of concrete application of ideas of the quantum-mechanical perturbation theory [31], which was developed for calculation of non-adiabatic corrections, the electron–phonon interaction and its contribution to the thermodynamic potentials of metals was considered. It was noted that besides electron–phonon interaction, anharmonic terms also produce non-adiabatic contributions, but these contributions correspond to higher terms of ε.

The results of section 3 demonstrate that the most convenient technique to describe anharmonic effects is the adiabatic perturbation theory with the ground state in which atoms may occupy arbitrary positions. From equations (1.711)–(1.713), it follows that the anharmonic contributions to the adiabatic potential may be simply obtained from these equations by setting $\hat{A}^{(0)} = \hat{B}^{(n)} \equiv 0$. Naturally, this procedure does not take into account the cross-contributions including both anharmonicity and electron-phonon interaction in E, but these cross-terms started with higher orders as the contributions of each interaction separately.

Now we must make some remarks about thermal expansion. In 3.1, we used the condition (1.715), i.e. supposed that the equilibrium positions of atoms at finite temperature T are known. However, it is more convenient to set the energy minimum at $T = 0$ as the equilibrium position. Here $\Delta E_{n\nu}$ in (1.714) is not equal to zero and, namely, this term determines the lowest order of thermal expansion.

The anharmonic contributions, which determine temperature dependence of the phonon frequencies and attenuation, have been briefly described in section 1. Now we will describe the theory of temperature dependence of elastic moduli of crystals (with applications to metals that we will use in Chapter 4). In this section we use a generalized definition of the elastic modules. This generalization includes elastic constants only. Our definition also includes the thermal pressure (the pressure of phonon gas) as the elastic modulus of the first order and also all temperature dependent contributions to elastic constants.

In this section, we also describe the definition of elastic moduli in dynamic and static treatment which is very suitable for some applied problems. We also prove the equivalence of both of these definitions in the pseudopotential theory by consistent definition of the order $V(\mathbf{q})$.

The examination of the equivalence of these equations by arbitrary temperature for both dynamic (calculated in the oscillating problem) and static (calculated by uniform deformation [33, 55]) elastic moduli calculated in the finite order in the pseudopotential is not trivial. Direct investigation in the case $T = 0$ of the elastic moduli obtained by static calculation from the cohesive energy E and the dynamic moduli calculated with the limit $\mathbf{q} \to 0$ from the equation for the dynamical matrix $D_{\alpha\beta}(\mathbf{q})$, for example, in the second order in the pseudopotential (see [19]), which shows that this equivalence does not take place. The outward sign of destroying this equivalence is the appearance of the derivatives $\partial E/\partial \bar{u}_{ij}$ ($\bar{u}_{ij}$ is the deformation tensor) and the derivatives over the electron density n_0 which is a characteristic feature of the theory of metals. These terms are impossible to obtain by expansion of $D_{\alpha\beta}(\mathbf{q})$ into the series over the wave vector $\mathbf{q}$.

This fact was even the reason for the opinion (see, for example, references in [19, 46]) that dynamic and static elastic moduli must not be equal. But in the series of studies (see [46]), it was shown that this equivalence takes place at $T = 0$ if a consistent procedure of taking into account all terms of some order in pseudopotential, in particular in the second order, has been used. Now, after the studies of Y. Kagan *et al.*, it is clear that the energy of the ground state of the electron subsystem is the series over the pseudopotential, but the true small parameter of the problem η is not the form factor, but is rather the ratio of its value $V(\tau)$ at the points of the reciprocal lattice τ to the Fermi energy E_F. Therefore, the elastic constants, which are formed in the static procedure, are calculated correctly since the cohesive energy contains $V(\tau)$ only. But for the dynamical problem, there are important $V(\mathbf{q})$ with arbitrary $\mathbf{q}$. At $\mathbf{q} \to 0$ the parameter $V(\mathbf{q})/E_F$ is not small and to calculate the elastic modules in some order in η, it is necessary to take into account some contributions that have, from the formal point of view, higher order in the pseudopotential (but not in the parameter η).

In the present subsection, we suggest the method for calculation of elastic moduli of any order at finite temperatures using the dynamical problem [47], which can be used instead of the long-wave technique [20]. The long-wave method is useful in calculating the second-order elastic modules only. The regular procedure of classification and calculation of all contributions in the dynamic elastic moduli of arbitrary order in parameter $\eta = V(\mathbf{q})/E_F$ is obtained. As an example, the first and second order elastic moduli are calculated in the frame of first order in anharmonicity and the equivalence of the static and dynamic moduli is demonstrated.

Below, we suggest that the potential energy E (as a function of core coordinates), in accordance with the adiabatic theory, includes the energy of the ground state of the subsystem of valence electrons of metal plus the energy of direct Coulomb interactions of cores (see section 1 of this chapter). Exactly speaking, this suggestion is valid with the accuracy to the terms of the order $\sim \varepsilon^4$ (anharmonic contributions of the fourth order) (see also section 3 and [20, 48]). Hence, using this potential energy (with neglecting the electron–phonon interaction), it is possible to consider the moduli of elastic up to the fourth order at $T = 0$ and the temperature dependence of the elastic modules of the first and second order in the first order in anharmonicity only. The calculation of other properties outside this scope, which need to take the electron–phonon interaction into account, will be considered in subsection 4 of this section (see also [49]). However, let us note that, as follows from the results of section 3, the contribution of electron transitions provide additional additive contributions to the elastic modules. Hence, investigation of the static and dynamic elastic moduli at an arbitrary temperature and establishing equality for both of these moduli for the model with the cohesive energy which does not take into account the electron-phonon interactions, is rather an independent problem. The solution of this problem provides the exact elastic moduli for the adiabatic problem of the first and second orders. This solution is also useful for the non-adiabatic problem, i.e., for calculation of the elastic moduli of higher order and their temperature dependence.

Below, we consider the isothermic elastic moduli only. In the case of finite temperatures or external stresses, it is necessary to calculate the free energy $F(X)$ describing the arbitrarily deformed state of the crystal $\{X\}$. The definition of isothermic modules for any stressed but equilibrium state $\{x\}$ is given in section 2 (equation (1.174)).

To calculate the elastic moduli in the first order in anharmonicity, it is necessary (see [50]) to restrict the investigation of the free energy by the harmonic approximation only

$$F(x, u) = E(u) + F_{ph}(T, u); \quad F_{ph} = F^{zp} + F^{*}. \tag{1.281}$$

Here, we are not considering the contributions to the elastic moduli from the energy of electron excitations F_e^* (1.3) (since first contributions are small up to the melting temperature T_m) and from the electron–phonon interactions Ω_{ep} also (see subsection 4). The contribution F_e^* will be obtained by comparison of experimental data with the theory. In (1.181), $E(u)$ is the cohesive energy of the deformed crystal and $F_{ph}(T, u)$ is the contribution of harmonic oscillations to the free energy, which includes the energy of zero oscillations F^{zp}. Below, we consider the cubic crystals with one atom per unit cell only. In section 2, it was shown that, in the first order in anharmonicity, $C_{\alpha\beta\gamma\delta}$ can be determined as

$$C_{\alpha\beta\gamma\delta}(T) = C_{\alpha\beta\gamma\delta}(0) + C^{ph}_{\alpha\beta\gamma\delta}(T) + \frac{p_{ph}(T)}{3B(0)}(C_{\alpha\beta\gamma\delta}(0) + C_{\alpha\beta\gamma\delta\varepsilon\eta}(0)). \tag{1.282}$$

Here $C_{\alpha\beta\gamma\delta}(0)$ and $C_{\alpha\beta\gamma\delta\varepsilon\eta}(0)$ are the elastic constants of the second and third order at $T = 0$; $B = (1/3)(C_{11} + 2C_{12})$. The last term in (1.182) describes the contribution of the thermal expansion to $C_{\alpha\beta\gamma\delta}(T)$ (quasi-harmonic contribution) whereas $C^{ph}_{\alpha\beta\gamma\delta}$ is the change of the elastic modules by anharmonic change of the phonon frequencies with temperature. Thermodynamic stresses $t^{ph}_{\alpha\beta}$ (the elastic moduli of the first order) and $C^{ph}_{\alpha\beta\gamma\delta}$ are determined as (see section 2)

$$t^{ph}_{\alpha\beta} = A^{ph}_{\alpha\beta};\ C^{ph}_{\alpha\beta\gamma\delta}(T) = \left(\frac{\partial^2 F_{ph}(T)}{\partial \bar{u}_{\alpha\beta}\,\partial \bar{u}_{\gamma\delta}}\right)_{\bar{u}=0} = A^{ph}_{\alpha\beta,\gamma\delta} - t^{ph}_{\beta\delta,\alpha\gamma}, \tag{1.283}$$

where

$$A^{ph}_{\alpha\beta} = \frac{1}{N\Omega}\left(\frac{\partial F_{ph}}{\partial u_{\alpha\beta}}\right)_{u=0};\quad A^{ph}_{\alpha\beta,\gamma\delta} = \frac{1}{N\Omega}\left(\frac{\partial^2 F_{ph}(T)}{\partial u_{\alpha\beta}\,\partial u_{\gamma\delta}}\right)_{u=0}. \tag{1.284}$$

Remember that concrete expressions for $A_{\alpha\beta}$ and $A_{\alpha\beta,\gamma\delta}$ are given in section 2 (equations (1.95) and (1.99)) where we have used the following notations (note, that $\eta_i \to u_{\alpha\beta}$)

$$\frac{\partial F_{ph}}{\partial \eta_i} \to A^{ph}_{\alpha\beta};\quad A^{zp}_{ij} + A^{*}_{ij} \to A^{ph}_{\alpha\beta,\gamma\delta},$$

and $p_{ph} = p^{zp} + p^{*}$ is the thermal pressure (see (1.76) with the substitution $F \to F_{ph}$) and $B(0)$ is the compressibility modulus at $T = 0$. For a cubic crystal $t^{ph}_{\alpha\beta} = -p_{ph}\delta_{ij}$. The matrix $A_{\alpha\beta,\gamma\delta}$ is obtained from (1.184) using the substitution of F_{ph} by total F. It determines the right hand side of the equation of harmonic oscillations of the crystal, i.e., is the propagation matrix [51].

Note that it is obvious from equations (1.182)–(1.184) that the proof of the equality of the elastic moduli of the first and second order in static and dynamic problems relates to the proof of the equality of the respective equations for p_{ph} and $A^{ph}_{\alpha\beta,\gamma\delta}$ in (1.184).

1.4.1 Anharmonic hamiltonian

For the crystal with one atom per unit cell, the Hamiltonian of atomic motion may be written as

$$\mathcal{H} = \frac{1}{2}\sum_{l} M\dot{\mathbf{u}}_l^2 + \frac{1}{2}\sum_{l,m} V_{\alpha\beta}(\mathbf{l}-\mathbf{m})u_{l\alpha}\,u_{m\beta} + \sum_{l,m\ldots}\sum_{n=3}^{\infty}\frac{1}{n!}V_{\alpha\beta}(\mathbf{l},\mathbf{m},\ldots)u_{l\alpha}\,u_{m\beta}\ldots, \tag{1.285}$$

where $\mathbf{u}_l$ is the displacement of the l-th atom, which corresponds to the atomic coordinate at $\mathbf{R}_l = \mathbf{R}_l^0 + \mathbf{u}_l$.

The amplitudes $V(\boldsymbol{l},\boldsymbol{m},\ldots) \equiv V(\mathbf{R}_l,\mathbf{R}_m,\ldots)$ may be expressed as the derivatives of the adiabatic potential $E \equiv E(\mathbf{R})$

$$V_{\alpha\beta\ldots}(\mathbf{l},\mathbf{m},\ldots) = \left.\frac{\partial^n E}{\partial R_{l\alpha}\,\partial R_{m\beta}\ldots}\right|_{\mathbf{R}_l=\mathbf{R}_l^0} = \left.\frac{\partial^n E}{\partial u_{l\alpha}\partial u_{m\beta}\ldots}\right|_{u=0}. \tag{1.286}$$

In equation (1.185), M is the atomic mass, $V_{\alpha\beta\ldots}$ are the force constants of n-th order, which satisfy suitable conditions of invariance (see for example [50]).

The Hamiltonian (1.185) is obtained on the basis of the adiabatic approximation. The first term is the kinetic energy of atomic motion, the next term is the potential energy for the harmonic problem (U_{2n} using the terminology of section 1.3), the last terms are contributions U_{3n}, U_{4n} and so on. Here, it is suggested that the contribution of the first order over atomic displacements is equal to zero as a consequence of the choice of the equilibrium position. But actually, for crystals with one atom per unit cell, it always vanishes as a consequence of the fact that each atom is the center of symmetry.

After the Fourier transformation

$$\mathbf{u}_l = 1/\sqrt{N}\sum_k \mathbf{u}(\mathbf{k})\exp(i\mathbf{kl}), \tag{1.287}$$

by taking in the Hamiltonian (1.185) terms of the third and fourth order in the atomic displacements $\mathbf{u}_l$, we obtain

$$\mathcal{H} = \mathcal{H}_0 + \mathcal{H}_3 + \mathcal{H}_4 \tag{1.288}$$

$$\mathcal{H}_0 = \frac{1}{2}\sum_k \left[M\dot{u}_\alpha(\mathbf{k})\,\dot{u}_\alpha(-\mathbf{k}) + V_{\alpha\beta}(\mathbf{k})\,u_\alpha(\mathbf{k})\,u_\beta(-\mathbf{k})\right] \tag{1.289}$$

$$\begin{aligned}\mathcal{H}_3 + \mathcal{H}_4 = {} & \frac{1}{3!}\sum_{1.2.3} V_{\alpha\beta\gamma}(\mathbf{1.2.3})\,u_\alpha(\mathbf{1})\,u_\beta(\mathbf{2})\,u_\gamma(\mathbf{3}) \\ & + \frac{1}{4!}\sum_{1.2.3.4} V_{\alpha\beta\gamma\delta}(\mathbf{1.2.3.4})\,u_\alpha(\mathbf{1})\,u_\beta(\mathbf{2})\,u_\gamma(\mathbf{3})\,u_\delta(\mathbf{4}),\end{aligned} \tag{1.290}$$

where $1 \equiv \mathbf{k}_1$ and so on, and $V_{\alpha\beta\ldots}(1.2,\ldots)$ are the Fourier transforms of the force constants $V_{\alpha\beta\ldots}(\boldsymbol{l},\boldsymbol{m},\ldots)$ (including the momentum conservation law). They are equal to the probability amplitudes for respective n phonon processes.

The potential energy E is taken as the adiabatic potential of the crystal which is equal to the sum of the ground state energy of the system of valence electrons E_e and the energy of the direct interaction $E_I = E_i$. We omit the energy of short-range ion repulsion E_{sr} because it does not influence the final result and may be easily added, for example, to E_i. The energy E_e is suggested to be a power series in the electron–ion interaction or in the pseudopotential. It has the form

$$E = E_e^{(0)} + \sum_{m\geq 1}(E_e^{(m)} + \delta_{m2}E_i), \tag{1.291}$$

$$E_e^{(m)} = \Omega N \sum_{\mathbf{q}_1\cdots\mathbf{q}_m} \Phi^{(m)}(\mathbf{q}_1,\ldots\mathbf{q}_m)S(\mathbf{q}_1)\ldots S(\mathbf{q}_m), \tag{1.292}$$

$$S(\mathbf{q}) = \frac{1}{N}\sum_l \exp(i\mathbf{q}\mathbf{R}_l).$$

To simplify the notations, we restrict our investigation to the approximation of local pseudopotential $V(\mathbf{q})$. It provides the factorization of $\Phi^{(m)}$

$$\Phi^{(m)}(\mathbf{q}_1,\ldots,\mathbf{q}_m) = \Gamma^{(m)}(\mathbf{q}_1,\ldots,\mathbf{q}_m)V(\mathbf{q}_1)\ldots V(\mathbf{q}_m). \tag{1.293}$$

Here $\Gamma^{(m)}(\mathbf{q}_1,\ldots,\mathbf{q}_m)$ are the universal multi-poles [19]. Remember that $\Phi^{(m)}(\mathbf{q}_1,\ldots,\mathbf{q}_m)$ are totally symmetric relating to permutation of all $\mathbf{q}_i$ and by replacing all $\mathbf{q}_i$ with $-\mathbf{q}_i$. If we restrict our consideration by terms of the order $(V(\boldsymbol{\tau})/E_F)^2$, then the concrete expression for $\Phi^{(m)}$ is not needed, excluding the value

$$\Phi^{(2)}(\mathbf{q},-\mathbf{q}) \equiv \varphi(\mathbf{q}) = \frac{1}{2}\left[\frac{4\pi(Ze)^2}{\Omega^2 q^2} - \frac{\pi(q)}{\varepsilon(q)}V^2(\mathbf{q})\right], \tag{1.294}$$

which also includes the Coulomb energy E_i (the first term in brackets). Here Z is the valency, $\pi(\mathbf{q})$ is the electron polarization operator, $\varepsilon(q) = 1 + 4\pi e^2\pi(\mathbf{q})/q^2$. Note here, that the index m is characteristic not only of the power of the pseudopotential in (1.194), but also the many-particle nature of the ion–ion interaction. For this reason, it is convenient to join the Coulomb interaction E_i with $E_e^{(2)}$. The energies $E_e^{(0)}$ and $E_e^{(1)}$ depend on the volume of the unit cell Ω only, not on the ion coordinates and consequently give no contribution to the elastic moduli $C_{\alpha\beta\gamma\delta}$ in the static problem and to amplitudes $V_{\alpha\beta\ldots}(1.2,\ldots)$ in (1.188). Using this fact, we obtain for the non-transition metal

$$\begin{aligned} V_{\alpha\beta}(\mathbf{k}) = & -\Omega\sum_{m\geq 2}\sum_{\boldsymbol{\tau}_1\ldots\boldsymbol{\tau}_m}\Delta(\boldsymbol{\tau}_1+\cdots+\boldsymbol{\tau}_m)\Big[A_m^1\Phi^{(m)}(\boldsymbol{\tau}_1,\ldots,\boldsymbol{\tau}_m)\tau_{1\alpha\beta} \\ & + A_m^2\Phi^{(m)}(\boldsymbol{\tau}_1-\mathbf{k},\boldsymbol{\tau}_2+\mathbf{k},\boldsymbol{\tau}_3,\ldots,\boldsymbol{\tau}_m)(\boldsymbol{\tau}_1-\mathbf{k})_\alpha(\boldsymbol{\tau}_2+\mathbf{k})_\beta\Big], \end{aligned} \tag{1.295}$$

$$\begin{aligned} V_{\alpha\beta\gamma}(\mathbf{k}_1,\mathbf{k}_2,\mathbf{k}_3) = & -\frac{i\Omega}{\sqrt{N}}\Delta(\mathbf{k}_1+\mathbf{k}_2+\mathbf{k}_3)\sum_{m\geq 2}\sum_{\boldsymbol{\tau}_1\ldots\boldsymbol{\tau}_m}\Delta(\boldsymbol{\tau}_1+\cdots+\boldsymbol{\tau}_m) \\ & \times\Big[(-1)^m A_m^1\Phi^{(m)}(\boldsymbol{\tau}_1,\ldots,\boldsymbol{\tau}_m)\tau_{1\alpha\beta\gamma} \\ & + A_m^2 P(3)\Phi^{(m)}(\boldsymbol{\tau}_1-\mathbf{k}_1-\mathbf{k}_2,\boldsymbol{\tau}_2-\mathbf{k}_3,\boldsymbol{\tau}_3,\ldots,\boldsymbol{\tau}_m) \\ & \times(\boldsymbol{\tau}_1-\mathbf{k}_1-\mathbf{k}_2)_{\alpha\beta}(\boldsymbol{\tau}_2-\mathbf{k}_3)_\gamma \\ & + (1-\delta_{2m})A_m^3\Phi^{(m)}(\boldsymbol{\tau}_1-\mathbf{k}_1,\boldsymbol{\tau}_2-\mathbf{k}_2,\boldsymbol{\tau}_3-\mathbf{k}_3,\boldsymbol{\tau}_4,\ldots,\boldsymbol{\tau}_m) \\ & \times(\boldsymbol{\tau}_1-\mathbf{k}_1)_\alpha(\boldsymbol{\tau}_2-\mathbf{k}_2)_\beta(\boldsymbol{\tau}_3-\mathbf{k}_3)_\gamma\Big], \end{aligned} \tag{1.296}$$

$$\begin{aligned} V_{\alpha\beta\gamma\delta}(\mathbf{k}_1,\mathbf{k}_2,\mathbf{k}_3,\mathbf{k}_4) = & \frac{\Omega}{N}\Delta(\mathbf{k}_1+\mathbf{k}_2+\mathbf{k}_3+\mathbf{k}_4)\times\sum_{m\geq 2}\sum_{\boldsymbol{\tau}_1\ldots\boldsymbol{\tau}_m}\Delta(\boldsymbol{\tau}_1+\cdots+\boldsymbol{\tau}_m) \\ & \times\Big[A_m^1\Phi^{(m)}(\tau_1,\ldots,\tau_m)\tau_{1\alpha\beta\gamma\delta} \\ & + A_m^2 P(4)\Phi^{(m)}(\boldsymbol{\tau}_1-\mathbf{k}_1-\mathbf{k}_2-\mathbf{k}_3,\boldsymbol{\tau}_2-\mathbf{k}_4,\boldsymbol{\tau}_3,\ldots\boldsymbol{\tau}_m) \\ & \times(\boldsymbol{\tau}_1-\mathbf{k}_1-\mathbf{k}_2-\mathbf{k}_3)_{\alpha\beta\gamma}(\boldsymbol{\tau}_2-\mathbf{k}_4)_\delta + A_m^2 P(3) \\ & \times\Phi^{(m)}(\boldsymbol{\tau}_1-\mathbf{k}_1-\mathbf{k}_2,\boldsymbol{\tau}_2-\mathbf{k}_3-\mathbf{k}_4,\boldsymbol{\tau}_3,\ldots\boldsymbol{\tau}_m) \\ & \times(\boldsymbol{\tau}_1-\mathbf{k}_1-\mathbf{k}_2)_{\alpha\beta}(\boldsymbol{\tau}_2-\mathbf{k}_3-\mathbf{k}_4)_{\gamma\delta} \\ & + (1-\delta_{2m})A_m^3 P(6)\Phi^{(m)}(\boldsymbol{\tau}_1-\mathbf{k}_1-\mathbf{k}_2,\boldsymbol{\tau}_2-\mathbf{k}_3,\boldsymbol{\tau}_3-\mathbf{k}_4,\boldsymbol{\tau}_4,\ldots,\boldsymbol{\tau}_m) \\ & \times(\boldsymbol{\tau}_1-\mathbf{k}_1-\mathbf{k}_2)_{\alpha\beta}(\boldsymbol{\tau}_2-\mathbf{k}_3)_\gamma(\boldsymbol{\tau}_3-\mathbf{k}_4)_\delta\Big]. \end{aligned} \tag{1.297}$$

Here A_m^ν is the arrangement number, $P(S)$ is the symmetrization operator over $(\alpha_i \mathbf{k}_i)$ providing S distinguished terms. Factors A_m^ν are the consequence of the symmetry of single terms in sums over the reciprocal lattice $\boldsymbol{\tau}_i$. Note that amplitudes formally contain all orders over $V(\mathbf{q})$.

Since the Hamiltonian is invariant relating to infinitely small translations, the amplitudes $V_{\alpha\beta\ldots}(\mathbf{k}_1, \ldots)$ vanish by the vanishing of any argument $\mathbf{k}$, i.e. it is possible to write

$$V_{\alpha\beta\ldots}(\mathbf{k}, \ldots) = \tilde{V}_{\alpha\beta\ldots}(\mathbf{k}, \ldots) - \tilde{V}_{\alpha\beta\ldots}(0, \ldots) \tag{1.298}$$

We have considered the particular form of this relation in section 1 for the dynamical matrix.

1.4.2 The elastic moduli in the static problem

The static treatment (the uniform deformations method [20]) of the elastic moduli (of arbitrary orders) corresponds to the calculation of respective derivatives of the free energy $F(x, u)$ of the crystal under uniform deformation. In the framework of this treatment, $E(u)$ determines the elastic moduli at $T = 0$ and the quasi-harmonic part (the last term in (1.284)) $F_{ph}(x, u)$ are the first anharmonic corrections to them. Using

$$F_{ph}(T, u) = k_B T \sum_{\lambda \mathbf{k}} \ln \left[2\text{sh}\left(\frac{\hbar \omega_{\lambda \mathbf{k}}(u)}{2k_B T} \right) \right]. \tag{1.299}$$

we obtain A^{ph} in (1.284). In numerical calculations, it is useful to operate with the derivatives of the dynamical matrix $D_{\alpha\beta}(\mathbf{k})$ in the distortion tensor $u_{\gamma\delta}$ [47]. The derivatives of A^{ph} are obtained from equations (1.95) and (1.99) at $\eta_i = u_{\gamma\delta}$ and from the definition of F_{ph}:

$$A^{ph} = A^{zp} + A^*.$$

Equation (1.299) is valid for a crystal with an arbitrary structure for any uniform deformed state. For the case of many (r) atoms per unit cell indices $\alpha\beta\ldots$ change from 1 to $3r$ in all equations.

The dynamical matrix of metals is fully determined by the definition of $V(\mathbf{q})$ and $\Gamma^{(m)}(\mathbf{q}_1, \ldots, \mathbf{q}_m)$. If we restrict our calculations to terms of the order $\sim \eta^2$, it is possible to use the terms with $m = 2$ only in the expression for (1.295). The amplitude $V_{\alpha\beta}(\mathbf{k})$ is the dynamical matrix. For its part $\tilde{D}_{\alpha\beta}(\mathbf{k})$, the next expression is valid

$$\tilde{V}_{\alpha\beta}(\mathbf{k}) \equiv \tilde{D}_{\alpha\beta}(\mathbf{k}) = \frac{2\Omega}{M} \sum_{\boldsymbol{\tau}} \varphi(\mathbf{q}) q_{\alpha\beta}, \tag{1.300}$$

where $\varphi(\mathbf{q})$ is determined by equation (1.294), $\mathbf{q} = \boldsymbol{\tau} - \mathbf{k}$, $q_{\alpha\beta\ldots} = q_\alpha q_\beta \ldots$ There are also terms $\sim \eta^2$ in $\tilde{V}_{\alpha\beta\ldots}(\mathbf{k}, \ldots)$ at $m = 3$ and $m = 4$ if $\mathbf{k}$ tends to zero [19]. But here, it is not necessary to take these terms into account, since by the integration over $\mathbf{k}$ contributions of these terms to A_{ph} are 'suppressed' by the weight factor $\mathbf{k}^2$.

Below, it is useful to consider the dynamical matrix to be dependent upon $u_{\gamma\delta}$ through the cell volume Ω, the equilibrium electron density n_0 and the vectors $\mathbf{q}$

(i.e. the change of the scale in reciprocal space). Hence, using the fact that the form factor $V(\mathbf{q}) \sim \Omega^{-1}$ and $D \sim \Omega^{-1}$, we obtain from (1.42) and (1.43)

$$\frac{\partial \tilde{D}_{\alpha\beta}(\mathbf{k})}{\partial u_{ij}} = -\frac{2\Omega}{M} \sum_{\tau} \left[\frac{\partial}{\partial q_j} q_{i\alpha\beta} + \delta_{ij} q_{\alpha\beta} n_0 \frac{\partial}{\partial n_0} \right] \varphi(\mathbf{q}), \tag{1.301}$$

$$\frac{\partial^2 \tilde{D}_{\alpha\beta}(\mathbf{k})}{\partial u_{ij}\, \partial u_{kl}} = \frac{2\Omega}{M} \sum_{\tau} \left[\frac{\partial^2}{\partial q_j\, \partial q} q_{ik\alpha\beta} + (\delta_{ij}\,\delta_{kl} + \delta_{il}\,\delta_{kj}) q_{\alpha\beta} n_0 \frac{\partial}{\partial n} \right.$$
$$\left. + \left(\delta_{ij} \frac{\partial}{\partial q_l} q_{k\alpha\beta} + \delta_{kl} \frac{\partial}{\partial q_j} q_{i\alpha\beta} \right) n_0 \frac{\partial}{\partial n_0} + \delta_{ij}\,\delta_{kl}\, q_{\alpha\beta}\, n_0^2 \frac{\partial^2}{\partial n_0^2} \right] \varphi(\mathbf{q}). \tag{1.302}$$

Equations (1.99), (1.299)–(1.302) are the exact solution of the problem of calculation of A^{ph} and consequently of p_{ph} and $C^{ph}_{\alpha\beta\gamma\delta}$ with the accuracy to $\sim \eta^2$ inclusively. The same procedure may be used in the calculation of the next terms in the power series expansion of $V(\mathbf{q})$ and the next anharmonic corrections to the moduli of the first $(p_{ph}\,\delta_{\alpha\beta})$ and second $(C^{ph}_{\alpha\beta\gamma\delta})$ orders. To do this, it is necessary to keep in E the terms of the higher order in the pseudopotental (precisely speaking, in parameter η) in the calculation of the dynamical matrix $\tilde{D}_{\alpha\beta}(\mathbf{k})$, and also to take into account the anharmonic amplitudes of the third $V_{\alpha\beta\gamma}$ and the fourth $V_{\alpha\beta\gamma\delta}$ orders.

1.4.3 The elastic moduli in the dynamic problem

Classical dynamic treatment of the procedure calculation of the elastic moduli at $T = 0$ (the long wave method [20]) uses the expansion of the dynamical matrix $D_{\alpha\beta}(\mathbf{k})$ as a power series over the wave vector $\mathbf{k} \ll 1$ with accuracy to k^2

$$D_{\alpha\beta}(\mathbf{k}) = [\alpha\beta, \gamma\delta] k_\gamma k_\delta, \tag{1.303}$$

with usage of the equivalence of the lattice dynamics equations and the linear elasticity theory, which defines $C_{\alpha\beta\gamma\delta}$ through the brackets $[\alpha\beta\gamma\delta]$.[8] In the framework of this method, the crystal pressure is not determined at all.

The natural generalization of this method in the case of finite temperatures is the calculation of $C_{\alpha\beta\gamma\delta}$ using the expansion of the dynamical matrix $D_{\alpha\beta} + \Delta D_{\alpha\beta}$ (here $\Delta D_{\alpha\beta}$ is the anharmonic contribution to the dynamical matrix depending on the temperature) to the power series over k [47]. But this scheme does not allow us, in principle, to calculate the elastic moduli of the first order, i.e., the thermal pressure p_{ph} and the calculation of the temperature corrections to the modules of higher than the second order is very difficult as a consequence of the necessity to analyze nonlinear equations of crystal oscillations.

In this subsection, a mainly new version of the dynamic treatment is developed. It uses the extraction of the contribution of the phonon condensate with $k = 0$ and the expansion of the free energy to the series in the powers of the tensor of thermal

---

[8] The brackets $[\alpha\beta\gamma\delta]$ are none other than the microscopic elastic moduli (see for example [20, 52]

deformations [47]. The coefficients of these series are directly expressed through the anharmonic amplitudes.

As a consequence of the relation between the uniform thermal expansion and the anharmonicity of the potential of interatomic interaction, not all Fourier components of atomic displacements $\mathbf{u}(\mathbf{k})$ have the order of $1/\sqrt{N}$. The mean values of using the of uniform atomic displacements are of the order $\sqrt{N}$, i.e. are macroscopically large. Therefore, to restore the possibility usual methods of calculations based on the fact that each degree of freedom $\mathbf{u}(\mathbf{k})$ provides statistically small contributions $\sim 1/N$ to the thermodynamic values, it is necessary to extract the terms $\langle(\mathbf{k} \to 0)\rangle$ from the Hamiltonian (1.288) (see [43]). To do this, let us present the total atomic displacement as

$$\mathbf{u}(\mathbf{k}) = \langle \mathbf{u}(\mathbf{k}) \rangle \delta_{k0} + \hat{\mathbf{u}}(\mathbf{k}), \tag{1.304}$$

where $\hat{\mathbf{u}}(\mathbf{k}) \sim 1/\sqrt{N}$ is the atomic displacement for the wave $\mathbf{k}$, the symbol $\delta_{\mathbf{k}0}$ denotes the limit $\mathbf{k} \to 0$ in all expressions below. Let us substitute $\mathbf{u}(\mathbf{k})$ into Hamiltonian (1.290). Since amplitudes $V_{\alpha\beta\ldots}(\mathbf{k}\ldots)$ are vanishing at $\mathbf{k} \to 0$ in accordance with (1.298), the final limit has the derivatives of $V_{\alpha\beta\ldots}(\mathbf{k}\ldots)$ over $\mathbf{k}$ multiplied by $\langle \mathbf{u}(\mathbf{k}) \rangle$ only, i.e.,

$$\lim_{\mathbf{k}\to\mathbf{0}} V_{\alpha\beta\gamma}(\mathbf{k},\mathbf{p},\mathbf{q})\langle u_\alpha(\mathbf{k})\rangle = \left[\frac{\partial V_{\alpha\beta\gamma}(\mathbf{k},\mathbf{p},\mathbf{q})}{i\partial k_\mu}\right]_{\mathbf{k}\to 0} \sqrt{N} u_{\alpha\mu}, \tag{1.305}$$

$$\lim_{\mathbf{k},\mathbf{q}\to\mathbf{0}} V_{\alpha\beta\gamma\delta}(\mathbf{k},\mathbf{p},\mathbf{q},\mathbf{s})\langle u_\alpha(\mathbf{k})\rangle \, \langle u_\beta(\mathbf{q})\rangle = -\left[\frac{\partial V_{\alpha\beta\gamma\delta}(\mathbf{k},\mathbf{p},\mathbf{q},\mathbf{s})}{\partial k_\mu \, \partial q_\nu}\right]_{\mathbf{k},\mathbf{q}\to\mathbf{0}} N u_{\alpha\mu} u_{\beta\nu}. \tag{1.306}$$

Here $u_{\alpha\beta}$ are components of the thermal distortion tensor. Using these facts and the momentum conservation law in amplitudes $V_{\alpha\beta\ldots}(\mathbf{k}\ldots)$ and also the symmetry, we obtain

$$\hat{\mathcal{H}}_a = \frac{1}{2}\sum_p \left[-V^{\mu}_{\alpha\gamma\delta}(\mathbf{p})u_{\alpha\mu} + \frac{1}{2} V^{\mu\nu}_{\alpha\beta\gamma\delta} u_{\alpha\mu} u_{\beta\nu}\right] \hat{u}_\gamma(\mathbf{p})\hat{u}_\delta(-\mathbf{p}), \tag{1.307}$$

where

$$V^{\mu\delta}_{\alpha\beta\gamma}(\mathbf{p}) = -\sqrt{N}\left[\frac{\partial V_{\alpha\beta\gamma}(\mathbf{k},\mathbf{p},\mathbf{q})}{i\partial k_\mu}\right]_{\substack{\mathbf{k}\to\mathbf{0}\\ \mathbf{q}\to\mathbf{p}}}, \tag{1.308}$$

$$V^{\mu\nu}_{\alpha\beta\gamma\delta}(\mathbf{p}) = -N\left[\frac{\partial V_{\alpha\beta\gamma\delta}(\mathbf{k},\mathbf{p},\mathbf{q},\mathbf{s})}{\partial k_\mu \, \partial q_\nu}\right]_{\substack{\mathbf{k},\mathbf{q}\to\mathbf{0}\\ \mathbf{s}\to\mathbf{p}}} \tag{1.309}$$

By calculating $\hat{\mathcal{H}}_a$ from $\hat{\mathcal{H}}_3 + \hat{\mathcal{H}}_4$ (1.290) the terms $\sim \hat{u}^3$ and $\sim \hat{u}^4$ are not taken into account because they describe three- and four-phonon processes which are necessary for the calculation of the shift of the phonon frequency and the attenuation only (see section 1). Note that using Hamiltonian $\hat{\mathcal{H}}_a$ is enough in calculating the first corrections to the elastic modules (not higher than the second order) only.

Let us calculate the free energy using general rules [23] treating $\hat{\mathcal{H}}_a$ as a perturbation. Corrections of the first and second orders are described by graphs (see for example [53])

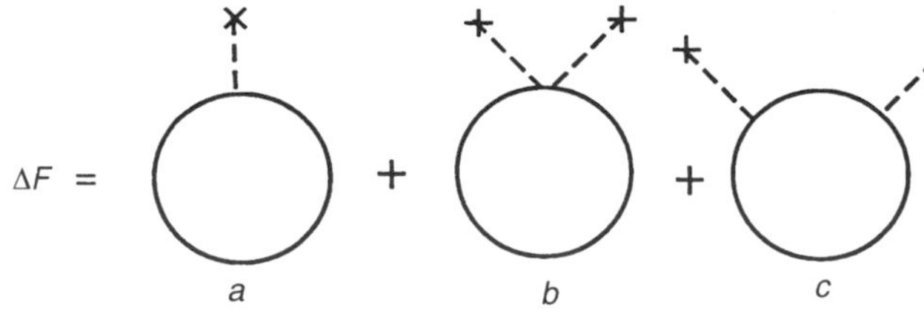

(1.310)

Here, solid lines represent the phonon Green's function $\mathcal{K}(\mathbf{k}, \omega)$ and the dashed lines with the cross denote the Bose-condensate (or the tensor $u_{\alpha\mu}$).

The phonon Green's function was already introduced with the help of classical definition (1.228) through the amplitudes of the phonon fields $\varphi_\lambda(x)$ (1.229). However, this definition is not always useful because it already uses the supposition of the possibility of introducing the phonons. More convenient for many purposes is the Green's function $\mathcal{K}_{\alpha\beta}$ [53], which is expressed through the distortion operators $\hat{u}_\alpha(l, t)$ (1.227) in the Heisenberg representation. For the crystal with one atom per unit cell

$$\mathcal{K}_{\alpha\beta}(\mathbf{l} - \mathbf{l}', t - t') = \frac{i}{\hbar}\langle \hat{T} u_{l\alpha}(t)\, u_{l'\beta}(t')\rangle, \tag{1.311}$$

$$u_{l\alpha}(t) = \exp(iHt/\hbar) u_{l\alpha} \exp(-iHt/\hbar),$$

where $\hat{T}$ is the time ordering operator. If you want to use the Matsubara representation (it will be more useful below), it is convenient to change *it* in (1.311) by the temperature T. After transition from the coordinate representation $u_{l\alpha}$ to the momentum representation $u_\alpha(\mathbf{k})$ in (1.287) and after introducing the discrete frequencies $\hbar\omega_n = 2\pi n k_B T$ $(n = 0, \pm 1, \ldots)$ instead of T, we obtain the Green's function

$$\mathcal{K}_{\alpha\beta}(\mathbf{q}, \omega_n) = \frac{1}{2\beta}\int_{-\beta}^{\beta} d\tau \exp(i\omega_n \tau)\langle \hat{T} u_\alpha(\mathbf{q}) u_\beta(-\mathbf{q})\rangle, \tag{1.312}$$

as its representation through the dynamical matrix $D_{\alpha\beta}(\mathbf{k})$. In the matrix representation

$$\hat{\mathcal{K}}(\mathbf{q}, \omega_n) = [(\omega_n^2 \hat{I} + \hat{D})^{-1}], \tag{1.313}$$

where $\hat{I}$ and $\hat{D}$ are the unity and the dynamical matrixes, respectively. In the basis of eigenfunctions of this equation, the coordinate axes are directed along the polarization vectors $e_{\lambda\mathbf{q}}$ which are eigenvectors of the matrix $\hat{\mathcal{K}}_a$ it is simple to find

$$\hat{\mathcal{K}}_{\alpha\beta}(\mathbf{q}, \omega_n) = \frac{\delta_{\alpha\beta}}{\omega_n^2 + \omega_{\lambda\mathbf{q}}^2}. \tag{1.314}$$

Let us consider contributions of diagrams (1.310) to F. The diagram "a" is determined by the amplitude of the triple process (the point with three ends) in the first order. It provides the contribution to F

$$F^{(a)} = -\frac{k_B T}{2\beta} \sum_{\mathbf{q}} V^{\mu}_{\alpha\gamma\delta}(\mathbf{q}) u_{\alpha\mu} \sum_{n=-\infty}^{\infty} \mathcal{K}_{\gamma\delta}(\mathbf{q}, \omega_n). \tag{1.315}$$

Performing the transformation to the basis of the eigenfunctions of the dynamical matrix and using the equation

$$\sum_{n=-\infty}^{\infty} \frac{k_B T}{\omega_n^2 + \omega^2} = \frac{\hbar}{2\omega} \coth\left(\frac{\hbar\omega}{2k_B T}\right) \tag{1.316}$$

we obtain that the coefficient at $u_{\alpha\beta}$ in $F^{(a)}/(N\Omega)$ is

$$-\frac{\hbar}{\Omega M N} \sum_{\lambda\mathbf{q}} \frac{N_{\lambda\mathbf{q}} + 1/2}{\omega_{\lambda\mathbf{q}}} V^{\mu\lambda\lambda}_{\alpha}(\mathbf{q}) \equiv \frac{1}{N\Omega} \frac{\partial F^{(a)}}{\partial u_{\alpha\mu}},$$

$$V^{\mu\lambda\nu}_{\alpha}(\mathbf{q}) = e^{\beta}_{\lambda\mathbf{q}} V^{\mu}_{\alpha\beta\gamma}(\mathbf{q}) e^{\gamma}_{\nu\mathbf{q}}. \tag{1.317}$$

By comparison of (1.317) with the value $A^{ph}_{\alpha\mu}$ calculated as $\partial F_{ph}/\partial u_{\alpha\mu}$ in the framework of the static treatment, we can see that for the equivalence of these equations it is necessary to be

$$M \frac{\partial D_{\beta\gamma}(\mathbf{q})}{\partial u_{\alpha\mu}} = -V^{\mu}_{\alpha\beta\gamma}(\mathbf{q}). \tag{1.318}$$

The diagram "b" is determined by the amplitude of the fourth-order process $V_{\alpha\beta\gamma\delta}$ in the first order. The value $F^{(b)}$ is obtained in (1.315) by the substitution of the factor $-V^{\mu}_{\alpha\beta\gamma\delta} u_{\alpha\mu}$ in the sum on $\mathbf{q}$ by $(1/2)V^{\mu\nu}_{\alpha\beta\gamma\delta} u_{\alpha\mu} u_{\beta\nu}$ The value of $V^{\mu\nu}_{\alpha\beta\gamma\delta}$ is determined by equation (1.309).

Now let us take into account that the coefficient of the factor $u_{\alpha\mu} u_{\beta\nu}$ in $F^{(b)}$ is a part of the derivative $\partial^2 F/(\partial u_{\alpha\mu} \partial u_{\beta\nu})$. Let us compare this factor with the first term in square brackets in the definition of the value $A^{ph}_{\alpha\mu,\gamma\nu}$ which is easy to obtain from $(A^{zp}_{ij} + A^{*}_{ij})$ in (1.99) by substitution $i \to \alpha\mu$ and $j \to \beta\nu$. This comparison gives

$$M[D^{\alpha\mu,\beta\nu}(\mathbf{q})] = e^{\gamma}_{\lambda\mathbf{q}} V^{\mu\nu}_{\alpha\beta\gamma\delta}(\mathbf{q}) e^{\delta}_{\lambda\mathbf{q}}. \tag{1.319}$$

This equation (together with the definition (1.37), generalized for the second derivative of the dynamical matrix $D_{\gamma\delta}(\mathbf{q})$ and also with the definition of $V^{\mu\nu}_{\alpha\beta\gamma\delta}$ in (1.309) and (1.297)), establishes that the equality of the terms under consideration in the framework of the static or dynamic treatment takes place if the next equation

$$M \frac{\partial^2 D_{\gamma\delta}(\mathbf{q})}{\partial u_{\alpha\mu}\, \partial u_{\beta}} = V^{\mu\nu}_{\alpha\beta\gamma\delta}(\mathbf{q}) \tag{1.320}$$

is fulfilled. The contribution of diagram "c" is

$$F^{(c)} = \frac{k_B T}{4} \sum_{\mathbf{q}} V^{\mu}_{\alpha\gamma\delta}(\mathbf{q}) V^{\nu}_{\beta\rho\sigma}(\mathbf{q}) u_{\alpha\mu} u_{\beta\gamma} \sum_{n=-\infty}^{\infty} \mathcal{K}_{\gamma\rho}(\mathbf{q}, \omega_n) \mathcal{K}_{\delta\sigma}(\mathbf{q}, \omega_n). \tag{1.321}$$

In the basis of $\{e_{\lambda\mathbf{k}}\}$, after summation over n, the coefficient at $u_{\alpha\mu}u_{\beta\nu}$ in $F^{(c)}/(N\Omega)$ takes the form

$$\frac{\hbar}{M^2 N}\sum_{\lambda\lambda'\mathbf{k}}\left[\frac{N_{\lambda\mathbf{k}}+1/2}{\omega_{\lambda\mathbf{k}}}-\frac{N_{\lambda'\mathbf{k}}+1/2}{\omega_{\lambda'\mathbf{k}}}\right]\frac{V_{\alpha}^{\mu\lambda\lambda'}V_{\alpha}^{\nu\lambda'\lambda}}{\omega_{\lambda\mathbf{k}}^2-\omega_{\lambda'\mathbf{k}}^2} \tag{1.322}$$

This expression is fully equivalent to the second term in (1.99) if the condition (1.318) is fulfilled.

Hence, equation (1.317) determines thermal stresses and the sum of the coefficients at $u_{\alpha\mu}u_{\beta\nu}$ (contributions of graphs "b" and "c") provides actual anharmonic contributions to the elastic moduli for the dynamic treatment. The equivalence condition of the long-wave method and the method of uniform deformation are written in equations (1.318) and (1.320).

If we keep in the equations for all amplitudes $V_{\alpha\beta\ldots}$ (1.295)–(1.297) for both treatments (static and dynamic) terms with the same m, i.e., with the same power of pseudopotential $V(\mathbf{q})$, then equations (1.318) (1.320) are not true because the operator $\partial/\partial\mathbf{u}_{\alpha\mu}$ in the left part contains $\partial/\partial n_0$ (1.311)–(1.312), whereas the right part does not contain this derivative. The reason for this discrepancy is that the true parameter of expansion of the cohesive energy is not $V(\mathbf{q})$, but rather $\eta = V(\tau)/E_F$. The value $V(\mathbf{q})/(\varepsilon(\mathbf{q})E_F)|_{\mathbf{q}\to 0}\sim 1$ and by calculating the cohesive energy up to terms of the order η^2 (i.e., $m=2$) to obtain the elastic moduli of some order over $V(\mathbf{q})$ in the framework of the dynamic treatment at $T=0$, it is necessary to also take some terms with $m=3$ and 4 in the dynamical matrix [19]. It will be seen below (for the simple case of static treatment with terms up to η^2) to satisfy conditions (1.318), (1.320), we need also to keep some terms with $m=3$ and 4 or of the order $V^2(\mathbf{q})$ and $V^4(\mathbf{q})$ in the expressions for $V_{\alpha\beta\gamma}$ and $V_{\alpha\beta\gamma\delta}$.

To understand the structure of terms restoring equality $A_{st}^{ph}=A_{dyn}^{ph}$, let us use the relations first obtained by Brovman and Kagan [19.46] by generalization of the sum rule for electron compressibility. In [46], the recurrent relations for the electron multi-poles $\Gamma^{(n)}$ (or $\Lambda^{(n)}$) are established. These relations, with the help of equation

$$\lim_{\mathbf{q}\to 0}\frac{\pi(q)V(\mathbf{q})}{\varepsilon(q)}=-n_0;\quad \pi(0)=\frac{dn_0}{d\mu},$$

following from the sum rules for the compressibility by taking into account the independence of $V(\mathbf{q})$ from n_0, are transformed to the form that is more useful for our purposes

$$\lim_{\mathbf{q}\to 0}\Phi^{(3)}(\mathbf{k},-\mathbf{k}-\mathbf{q},\mathbf{q})=\frac{n_0}{3}\frac{\partial\varphi(q)}{\partial n_0} \tag{1.323}$$

$$\lim_{\mathbf{k},\mathbf{p}\to 0}\Phi^{(4)}(\mathbf{k},\mathbf{p},\mathbf{q},-\mathbf{k}-\mathbf{p}-\mathbf{q})=\frac{n_0^2}{12}\frac{\partial^2\varphi(q)}{\partial n_0^2} \tag{1.324}$$

Now it is clear that by calculation of $V_{\alpha\beta\gamma}$ and $V_{\alpha\beta\gamma\delta}$, which are used in the left part of (1.318) and (1.320), it is necessary to keep terms with $\Phi^{(3)}$ and $\Phi^{(4)}$ using arguments from (1.323) and (1.324)

$$\tilde{V}_{\alpha\beta\gamma}(\mathbf{k}_1, \mathbf{k}, -\mathbf{k}-\mathbf{k}_1) = \frac{2i\Omega}{\sqrt{N}} \sum_{\boldsymbol{\tau}} \Big[\varphi(\mathbf{q}-\mathbf{k}_1)(\mathbf{q}-\mathbf{k}_1)_{\alpha\gamma\delta} - 3\Phi^{(3)}(-\mathbf{k}, -\mathbf{q}+\mathbf{k}_1, \mathbf{q}) k_{1\alpha} q_{\gamma\delta}\Big]; \tag{1.325}$$

$$\begin{aligned}\tilde{V}_{\alpha\beta\gamma\delta}(\mathbf{k}_1, \mathbf{k}_2, \mathbf{k}, -\mathbf{k}-\mathbf{k}_1-\mathbf{k}_2) = -\frac{2\Omega}{N} \sum_{\boldsymbol{\tau}} \Big[&\varphi(\mathbf{q}-\mathbf{k}_1-\mathbf{k}_2)(\mathbf{q}-\mathbf{k}_1-\mathbf{k}_2)_{\alpha\beta\gamma\delta} \\ &+ 3\Phi^{(3)}(-\mathbf{k}_1-\mathbf{k}_2, -\mathbf{q}+\mathbf{k}_1+\mathbf{k}_2, \mathbf{q})(\mathbf{k}_1+\mathbf{k}_2)_{\alpha\beta} q_{\gamma\delta} \\ &- 3\Phi^{(3)}(-\mathbf{k}_2, -\mathbf{q}+\mathbf{k}_1+\mathbf{k}_2, \mathbf{q}-\mathbf{k}_1) k_{2\alpha}(\mathbf{q}-\mathbf{k}_1)_{\beta\gamma\delta} \\ &- 3\Phi^{(3)}(-\mathbf{k}_1, \mathbf{q}+\mathbf{k}_1+\mathbf{k}_2, \mathbf{q}-\mathbf{k}_2) k_{1\alpha}(\mathbf{q}-\mathbf{k}_2)_{\beta\gamma\delta} \\ &+ 12\Phi^{(4)}(-\mathbf{k}_1-\mathbf{k}_2, -\mathbf{q}+\mathbf{k}_1+\mathbf{k}_2, \mathbf{q}) k_{1\alpha} k_{2\beta} q_{\gamma\delta}\Big],\end{aligned} \tag{1.326}$$

where $\mathbf{q} = \boldsymbol{\tau} - \mathbf{k}$; $q_{\alpha\beta\ldots} = q_\alpha q_\beta \ldots$.

Let us demonstrate for the particular case of $V_{\alpha\beta\gamma}(\mathbf{k}_1, \mathbf{k}_2, \mathbf{k}_3)$ in (1.296) the transformation of equations for anharmonic amplitudes to the form (1.325), which is more suitable for proving (1.318) [47].

Equations (1.317), (1.322) for the elastic moduli contain the first derivative of $V_{\alpha\beta\gamma}(\mathbf{k}, \mathbf{q}, \mathbf{p})$ over k_μ with subsequent transition to the limit $\mathbf{k} \to 0$, i.e., the value of $V^{\mu}_{\alpha\beta\gamma}(\mathbf{k})$. Therefore, in (1.296), it is necessary to keep terms with $\mathbf{k}_1$, which transforms later into $\mathbf{k}$ (with taking into account the conservation law $\Delta(\mathbf{k}_1+\mathbf{k}_2+\mathbf{k}_3)$), and with $m \leq 3$ only (as it is clear from equation (1.323)). Hence, equation (1.296) gives

$$\begin{aligned}V_{\alpha\beta\gamma}(\mathbf{k}, \mathbf{p}, -\mathbf{k}-\mathbf{p}) = \frac{2i\Omega}{\sqrt{N}} \Bigg\{ &\sum_{\boldsymbol{\tau}} \Big[\varphi(\boldsymbol{\tau}-\mathbf{k}-\mathbf{p})(\boldsymbol{\tau}-\mathbf{k}-\mathbf{p})_{\alpha\beta\gamma} - \varphi(\boldsymbol{\tau}-\mathbf{k})(\boldsymbol{\tau}-\mathbf{k})_{\alpha\beta\gamma}\Big] \\ &- 3 \sum_{\boldsymbol{\tau}_1, \boldsymbol{\tau}_2, \boldsymbol{\tau}_3} \Delta(\boldsymbol{\tau}_1+\boldsymbol{\tau}_2+\boldsymbol{\tau}_3) \\ &\times \Big[\Phi^{(3)}(\boldsymbol{\tau}_1-\mathbf{k}-\mathbf{p}, \boldsymbol{\tau}_2+\mathbf{k}+\mathbf{p}, \boldsymbol{\tau}_3)(\boldsymbol{\tau}_1-\mathbf{k}-\mathbf{p})_{\alpha\beta}(\boldsymbol{\tau}_2+\mathbf{k}+\mathbf{p})_\gamma \\ &+ \Phi^{(3)}(\boldsymbol{\tau}_1-\mathbf{k}, \boldsymbol{\tau}_2-\mathbf{p}, \boldsymbol{\tau}_3+\mathbf{k}+\mathbf{p})(\boldsymbol{\tau}_1-\mathbf{k})_\alpha(\boldsymbol{\tau}_2-\mathbf{p})_\beta(\boldsymbol{\tau}_3+\mathbf{k}+\mathbf{p})_\gamma \\ &+ \Phi^{(3)}(\boldsymbol{\tau}_1+\mathbf{k}, \boldsymbol{\tau}_2-\mathbf{k}, \boldsymbol{\tau}_3)(\boldsymbol{\tau}_2-\mathbf{k})_\alpha(\boldsymbol{\tau}_1+\mathbf{k})_{\beta\gamma}\Big]\Bigg\}.\end{aligned} \tag{1.327}$$

Let us discuss now which terms kept are essential for the proof of the condition (1.318), and the procedure of calculating derivatives with respect to the electron density, paticipating in (1.301) and (1.302).

The first two terms are explicitly the second order over $V(\mathbf{q})$. After applying the operator $\partial/i\partial k_\mu$, they provide (taking into account (1.298)) the first operator in the square brackets in equation (1.301). The first term with $\Phi^{(3)}$ in (1.327) provides a contribution of the order of $(V(\tau)/E_F)^2$, i.e., the derivative $\partial/\partial n_0$, but at the point $\mathbf{p} \to 0$ only (see (1.323)). However, after integration over $\mathbf{p}$ in calculating $C^{ph}_{\alpha\beta\gamma\delta}$, the contribution of this term is equal to zero as a consequence of both the slow integration volume and the statistical weight $\mathbf{p}^2$. For the second term with $\Phi^{(3)}$, the point $\tau = 0$ is essential only. Using (1.323), and performing $(\partial/\partial k_\mu)_{\mathbf{k}\to 0}$, we obtain the contribution of this term

$$2\Omega \sum_{\boldsymbol{\tau}} \delta_{\alpha\beta}(\boldsymbol{\tau} - \mathbf{k})_{\beta\gamma}\, n_0 \frac{\partial}{\partial n} \varphi(\boldsymbol{\tau} - \mathbf{k}). \tag{1.328}$$

Finally the last term in (1.327) provides the contribution to $V^{\mu}_{\alpha\beta\gamma}(\mathbf{k})$ at $\boldsymbol{\tau}_2 = 0$ only. It is

$$-2\Omega \sum_{\boldsymbol{\tau}} \delta_{\alpha\mu}\, \tau_{\beta\gamma}\, n_0 \frac{\partial}{\partial n_0} \varphi(\boldsymbol{\tau}). \tag{1.329}$$

Therefore, keeping the terms in $V_{\alpha\beta\gamma}$ (1.327), which are needed, we obtain, after trivial transformation of τ_i, equation (1.325). Providing a similar procedure for $V_{\alpha\beta\gamma\delta}$ in (1.297), we obtain equation (1.326).

We have obtained the elastic moduli in the pseudopotential theory in the framework of the static and the dynamic treatment. It was necessary to generalize the dynamic treatment in the case of finite temperatures. Simultaneously, the correctness of conditions (1.318)–(1.320), i.e., the equivalence of $A^{ph}_{\alpha\beta}$ and $A^{ph}_{\alpha\beta,\gamma\delta}$ and also of p_{ph} and $C^{ph}_{\alpha\beta\gamma\delta}$ for dynamic and static treatments were proven.

1.4.4 The contribution of electron–phonon interaction to elastic modules

Let us now consider the calculation of contributions from the thermodynamic potential of electron–phonon interaction Ω_{ep} (1.265) to elastic moduli of the crystal [49]. Let us consider the elastic moduli of the second order $C_{ijkl}(T)$ only, calculated in the first order of the perturbation theory in the anharmonic parameter (1.78)–(1.83). This approximation is enough for metals [56] although it is not difficult to calculate the elastic modules of higher orders and higher order terms of the perturbation theory. The expressions for temperature dependence of the elastic moduli of arbitrary order provided by contributions from the adiabatic phonons only are obtained and investigated in detail in previous sections (see also [47]). The elastic moduli C_{ijkl} at $T = 0$ include the contribution of the electron subsystem and the static lattice [47, 56]. The contribution of thermal excitation electrons to $C^{*e}_{ijkl}(T)$ can be simply calculated as (see (1.3))

$$C^{*e}_{ijkl} = \frac{1}{\Omega_0} \left\{ \frac{\partial^2}{\partial u_{ij} \partial u_{kl}} \left[\frac{\pi^2}{6} (k_B T)^2 \frac{\partial}{\partial \varepsilon} (\varepsilon N(\varepsilon)) \right]_{\varepsilon = E_F} \right\}, \tag{1.330}$$

(Here u_{ij} is the deformation tensor, Ω_0 is the cell volume at $T = 0$.) The tensor $C^{*e}_{ijkl}(T)$ provides appreciable contribution $\sim T^2$ to temperature dependence of elastic moduli at low temperatures $T \ll \Theta_D$ when the lattice contribution

$C^{*lat}_{ijkl}(T) \sim T^4$. The thermal pressure (the first order moduli) provided by electrons is

$$p_e^* = \frac{\partial}{\partial\Omega_0}\left[\frac{\pi^2}{6}(k_BT)^2\frac{\partial}{\partial\varepsilon}(\varepsilon N(\varepsilon))\right]_{\varepsilon=E_F}. \tag{1.331}$$

This term is also important at low temperatures.

Let us now consider the non-adiabatic corrections, i.e., the contributions of electron–phonon interactions. The corrections from $\delta\Omega(T=0)$ in (1.275) are not interesting because they are small as a consequence of the small values of the parameter $d = u/v_F$ in comparison with the large term $C_{ijkl}(0)$ provided by lattice as follows from (1.276).

Let us calculate now the temperature corrections to $C^{*ep}_{ijkl}(T)$ provided by electron–phonon interactions (terms $\delta\Omega(T)$ in (1.275)) and compare them with the elastic constants provided both by lattice $C^{*lat}_{ijkl}(T)$ [47, 56] and by electrons $C^{*e}_{ijkl}(T)$ (1.330). In the framework of the static treatment [47], which is most useful here, it is possible to differentiate Ω_{ep} in the distortion tensor $\mathbf{u}_{ij}$ taking into account equations (1.78)–(1.83) with substitution of F by Ω_{epi}. The differentiation is suitable to give, taking into account that Ω_{ep} depends on u_{ij} through the dependence of the sound velocity $u(u_{ij})$, the Fermi momentum $k_F(u_{ij})$ and the volume of the unit cell $\Omega_0(u_{ij})$ on the distortion tensor. Hence,

$$\frac{\partial}{\partial u_{ij}} = \left(\frac{\partial}{\partial u_{ij}}\right)\left(\frac{\partial}{\partial u}\right)_{k_F,\Omega_0} + \frac{\partial k_F}{\partial u_{ij}}\left(\frac{\partial}{\partial k_F}\right)_{u,V_0} + \frac{\partial\Omega_0}{\partial u_{ij}}\left(\frac{\partial}{\partial\Omega_0}\right)_{k_F,u} \tag{1.332}$$

and the derivatives of u, k_F and Ω_0 over u_{ij} are

$$\frac{\partial u}{\partial u_{ij}} = \frac{1}{2\rho u}\frac{\partial B}{\partial u_{ij}}; \quad \frac{\partial k_F}{\partial u_{ij}} = \frac{k_F}{3\Omega_0}\delta_{ij}; \quad \frac{\partial\Omega_0}{\partial u_{ij}} = \Omega_0\delta_{ij}. \tag{1.333}$$

Here, we are taking into account that, in equations (1.277)–(1.280) for $\Omega_{epi} - \delta\Omega(T)$, there exists the contribution of longitudinal phonons only for which the sound velocity is expressed through the volume compressibility module as $u = \sqrt{B/\rho}$.

In this way, it is simple to obtain the contributions $C^{*e}_{ijkl}(T)$ that are needed. However, more informative is the other treatment [49]. Let us write Ω_{ep} as

$$\Omega_{ep} = \sum_{\mathbf{q},\lambda}\hbar\delta\omega_{\lambda\mathbf{q}}\left(N_{\lambda\mathbf{q}} + \frac{1}{2}\right) + \sum_{\mathbf{k}}\delta E(\{n_{\mathbf{k}}\})n_{\mathbf{k}}. \tag{1.334}$$

From the comparison of (1.334) with (1.265) (see also [26]), it follows that

$$\delta\omega_{\lambda\mathbf{q}} = \sum_{\mathbf{k}}|M_{\mathbf{kq}\lambda}|^2\left[\frac{n_{\mathbf{k}} - n_{\mathbf{k+q}}}{\varepsilon_{\mathbf{k}} - \varepsilon_{+\mathbf{q}} - \hbar\omega_{\lambda q}} - \frac{n_{\mathbf{k}} - n_{\mathbf{k+q}}}{\varepsilon_{\mathbf{k}} - \varepsilon_{\mathbf{k+q}}}\right], \tag{1.335}$$

$$\delta E[\{n\}] = \sum_{\mathbf{q}\lambda}|M_{\mathbf{kq}\lambda}|^2\frac{\hbar\omega_{\lambda\mathbf{q}}(1 - n_{\mathbf{k+q}})}{(\varepsilon_{\mathbf{k}} - \varepsilon_{\mathbf{k+q}})^2 - (\hbar\omega_{\lambda\mathbf{q}})^2}. \tag{1.336}$$

From equation (1.334) for Ω_{ep}, it can be seen that, formally, the energy of electron–phonon interaction is divided into two terms (this remains in any order of the adiabatic theory) containing either $N_{\mathbf{q}\lambda}+1/2$ or $n_{\mathbf{k}}$. The first term relates to the lattice (phonon) part of the system energy, the second term relates to the electron subsystem. For the renormalized (in this way) lattice contribution to the elastic constants, the former relations are fulfilled (see [56] and also section 2).

$$A_{ij}^{ph}=\frac{\hbar}{2\Omega_0 N}\sum_{\mathbf{q}\lambda}\frac{N_{\lambda\mathbf{q}}+1/2}{\omega_{\lambda\mathbf{q}}}[\bar{D}^{ij}]_{\lambda\lambda}, \tag{1.337}$$

$$A_{ij,nl}^{ph}=\frac{\hbar}{2\Omega_0 N}\sum_{\mathbf{q}\lambda}\frac{N_{\lambda\mathbf{q}}+1/2}{\omega_{\lambda\mathbf{q}}}\left([\bar{D}^{ij,nl}]_{\lambda\lambda}+\sum_{\lambda'}\frac{2[\bar{D}^{ij}]_{\lambda\lambda'}[\bar{D}^{nl}]_{\lambda'\lambda}}{\omega_{\lambda\mathbf{q}}^2-\omega_{\lambda'\mathbf{q}}^2}\right). \tag{1.338}$$

Here, the dynamic matrix is renormalized:

$$\bar{D}_{\alpha\beta}(\mathbf{q},\omega_{\mathbf{q},\lambda})=D_{\alpha\beta}^{lat}(\mathbf{q})+\delta D_{\alpha\beta}^{ep}(\mathbf{q},\omega_{\mathbf{q},\lambda}). \tag{1.339}$$

In equations (1.338), (1.339) the value of $\delta\omega_{\mathbf{q}\lambda}$ is determined by (1.335) and the values $\delta\bar{D}_{\alpha\beta}^{ep}$ follow from the equation

$$[\bar{D}^{ij}]_{\lambda\lambda'}=e_{\lambda\mathbf{q}}^{\alpha}\frac{\partial\bar{D}_{\alpha\beta}}{\partial u_{ij}}e_{\lambda'\mathbf{q}}^{\beta}.$$

To obtain $\delta\bar{D}_{\alpha\beta}^{ep}(\mathbf{q},\omega_{\lambda\mathbf{q}})$, we have used the fact that the squared modulus of the matrix element of electron–phonon interaction $|M_{\mathbf{kq}\lambda}|^2$, which is proportional to the value of $\delta\omega_{\mathbf{q}\lambda}$, contains $e_{\lambda\mathbf{q}}^{\alpha}\,q_\alpha q_\beta\,e_{\lambda\mathbf{q}}^{\beta}$. Hence,

$$\delta D_{\alpha\beta}^{ep}(\mathbf{q},\omega_{\lambda\mathbf{q}})=\frac{N\hbar}{M}\sum_{\mathbf{k}}|M_{\mathbf{kq}\lambda}|^2(n_{\mathbf{k}}-n_{\mathbf{k+q}})\left[\frac{B_\alpha B_\beta}{\varepsilon_{\mathbf{k}}-\varepsilon_{\mathbf{k+q}}-\hbar\,\omega_{\lambda\mathbf{q}}}-\frac{B_\alpha B_\beta}{\varepsilon_{\mathbf{k}}-\varepsilon_{\mathbf{k+q}}}\right], \tag{1.340}$$

where

$$B_\alpha=\int d\mathbf{r}\,\psi_{\mathbf{k}}^*(\mathbf{r})\frac{\partial}{\partial r_\alpha}V(\mathbf{r})\psi_{\mathbf{k}'}(\mathbf{r}). \tag{1.341}$$

The second term in (1.334) has the same structure as the electron contributions to the energy and we can combine both these terms as

$$E_e=\sum_{\mathbf{k}}\left[E_{\mathbf{k}}+\delta E(\{n_{\mathbf{k}}\})\right]n_{\mathbf{k}} \tag{1.342}$$

and calculate using general rules [1]. It provides the contribution $\sim d^3$ to $\varepsilon N(\varepsilon)$ only, and this term does not change the temperature dependence of $C_{ijkl}^{*e}(T)$ and P_e^*.

In accordance with the above discussed contributions of electron–phonon interaction to the thermodynamic potential, we note that at low temperatures $T\ll\Theta_D$ the corrections from Ω_{ep} to C_{ijkl} have the order d^4T^2 and they are very small in the measure of small value d. Additionally, this term is very difficult to separate from pure electron contribution and to observe it in the background of very high ($\sim d^0$)

lattice contribution. However, it is posible to obtain this separation using a large difference in the temperature dependence of both terms: $C_{ijkl}^{*lat} \sim T^4$ and $C_{ijkl}^{*e+ep} \sim T^2$. Conversely, at high temperatures, the contribution of the electron–phonon interaction is impossible to distinguish from the lattice contribution $\sim T$.

1.5 Resumé

Chapter 1 plays a special role in the review. On the one hand, we discuss the main definitions and concepts well-known in the theory of electrons and phonons in crystals and in particular, in metals. On the other hand, it also contains the analysis of the complicated and less well-known adiabatic perturbation theory, which is the ideological and mathematical basis for description of common, and what seem to be trivial atomic properties of solids (sections 1 and 2).

The traditional and well-known (see, for example, the book of Ziman [1]) basic theorems relating to the electron subsystem in solids are not considered here. We call the readers attention to the following facts:

1. The investigation of the electron subsystem independent of the nuclear (ion) subsystem, in particular the introduction of the band energy, already supposes adiabatic separation of the ion and electron degrees of freedom.
2. It is necessary to consider the band electron spectrum not originally as one electron (this picture was typical until the sixties), but rather in the sense of the density functional theory [54] as a spectrum of one-particle excitations of a many-electron system, i.e., the energy determined as the pole of the whole electron propagator. The basis of the adiabatic approximation was discussed in section 3.

The conceptions of the phonon dispersion (section 1) and the elastic moduli (section 2) are very close. Since our further purpose is the calculation of atomic properties of metals, these questions need special description. For this purpose, we have described definitions of some elastic moduli to demonstrate advantages of one or another in some problems and we have tried to establish the relations between these moduli. Especially interesting from our point of view are two aspects. First, let us emphasize that the possibility of introduction of elastic constants of higher order is a consequence of going beyond the limits of the actual adiabatic approximation [20]. Further, we pay special attention to the definition of temperature dependence (the expansion of the free energy into a power series in the finite parameters of deformation) must take place in the framework of the theory of finite deformations. For this problem, there is some disadvantage of the Lagrange formalism relating to the fact that the elastic moduli, generally speaking, are increasing with the increase of their order and at finite deformations (for example the thermal deformations provided by the thermal expansion). For this reason the expansion of the free energy in the deformation parameter is divergent. In this case, it is necessary to avoid the series expansion. The calculations are possible only if the energy can be calculated exactly (with no expansion). For this case, in the end of section 2, we consider, as an example, the theory of the general equation of state $F(t_{\alpha\beta}, u_{\alpha\beta}, T) = 0$ providing relations between components of the stress tensor $t_{\alpha\beta}$, the deformation tensor $u_{\alpha\beta}$ and the temperature T.

As we have already noted, section 3 contains detailed descriptions of the adiabatic approximation and also of the theory of electron–phonon interaction as an example of elimination of the limits of adiabatic approximations while remaining in the framework of the adiabatic theory. This section, in many aspects, follows to the works by Geĭlikman. We consider in detail the adiabatic perturbation theory, which is rather difficult in applications and is unusual in form, as a consequence of the form of the perturbation operator. Let us emphasize that calculation of corrections to the adiabatic approximation takes place in two stages. First, the problem of the calculation of the energy in the static ion potential (provided by the ions occupying equilibrium, or arbitrary, but close to equilibrium, positions) is solved. The second stage is the solution of the oscillating problem for ions. The difficulties in the calculation of corrections to the Born–Oppenheimer adiabatic approximations [20] are connected with the impossibility to extract in explicit form the Hamiltonian of zero approximation, depending, in the same manner, on electron and ion coordinates and also the perturbation operator.

Further, the quantum-mechanical perturbation theory obtained by Geĭlikman on the basis of the adiabatic expansion is analyzed. The perturbation operator of this theory, $\hat{H}'(\mathbf{r}, \mathbf{R})$, has a complicated and original structure. It is a series in the powers of terms describing phonon–phonon and electron–phonon interaction. The particularity of $\hat{\mathcal{H}}'(\mathbf{r}, \mathbf{R})$ is that the main term is $\hat{\mathcal{H}}_1 \sim \hbar\,\omega_D/\kappa$. The next two terms are $\sim \hbar\,\omega_D$, whereas the energy of zero oscillations in the zero order Hamiltonian is also $\sim \hbar\,\omega_D$. However, mutual cancellation provides the first correction to the energy of the order $\kappa^2\hbar\,\omega_D$, the next $\sim \kappa^4\hbar\,\omega_D$ and so on.

The perturbation theory on the basis of the adiabatic expansion is applied to the analysis of interactions of electrons with phonons in metals. This problem is very important for the theory of superconductivity. The most important fact that we try to demonstrate is the establishment of mutual cancellations of the leading in the adiabatic parameter contributions to the energy, which, however, have no physical sign. These contributions are the consequence of the peculiarity of the perturbation operator that we have noted above. Similar cancellations take place also in the mass-operators for phonons and electrons and also for the thermodynamic potential of electron–phonon interaction.

Note that in the perturbation theory in the parameter $\kappa^2 = \langle x^2 \rangle / a^2$, which we consider here, that there are two types of contributions corresponding to two physical effects: electron–phonon contributions and anharmonic (phonon–phonon) contributions. The anharmonic effects were discussed in section 4. Here, we not only describe the procedure for calculating elastic constants as derivatives of respective free energy (the adiabatic potential at $T = 0$ or the phonon free energy at $T \neq 0$), calculated in the framework of the perturbation theory in the anharmonic Hamiltonian, but also propose some original (which are known by the authors) generalization of the long-wave method for the case of finite temperatures. The idea of this generalization is the separation of the condensate of phonons providing a unique introduction to the theory of the thermal deformation tensor (provided by thermal expansion).

This method of calculating the dynamic elastic constants provides the possibility of solving the problem of establishment of the equivalence of the dynamic and static elastic constants at finite temperatures. Note that the similar problem at $T = 0$ was solved in [55]. The authors have no possibility of eliminating the restriction to $T = 0$ because (as we have already noted) all methods of calculating dynamic elastic constants, besides those described by us, are suitable at zero temperature only.

It is well known that, if we restrict the calculation of the static and dynamic problem to the same order in the pseudopotential, the static moduli differ from the dynamic ones by the existence of derivatives with respect to the electron density. It was shown that these derivatives in the static treatment arrise as a consequence of using limiting relations for electron multi-poles of distinguished orders. These relations are the natural generalization of the well-known sum rules for the electron compressibility.

The proof of the equivalence of static and dynamic elastic constants here was done for the moduli of the first and second orders in the first order of the perturbation theory in the anharmonicity parameter ε. Only pair contributions to cohesive energy were taken into account, i.e., terms $\sim \eta^2$. However, the present status of the theory of simple metals and the experiment needs a generalization of the proof. The moduli of higher orders (in particular, third and fourth) are measured in some metals (see reference in [47]). Here the combined methods were used like measurements of the sound velocity at static stresses.

Hence, to calculate anharmonic contributions of the order ε^p to the elastic moduli of the order n with accuracy to the terms of the order η^l in the static treatment, we must take contributions to the anharmonic Hamiltonian to terms $\hat{\mathcal{H}}_{2p}$ (with amplitudes $V_{\alpha\beta\ldots}$ calculated in l-th order in V_q) or (that is the same) calculate the free energy to terms ε^{p-1}. It provides in moduli of the order n the maximum order derivatives like $\partial^n/\partial n_0^n$. In the framework of the dynamic treatment, it is necessary to use anharmonic Hamiltonian up to terms of the order V_q^{l+n}.

Finally, let us draw some conclusions about the contribution of the electron phonon interaction to the elastic constants. First note that, in the first approximation (in the anhamonicity parameter κ^2), the phonon–phonon and the electron–phonon interactions provide additive contributions and may be considered as independent. Analysis of temperature contributions provided by the process of the scattering of electrons by phonons shows that this process provides contributions to C_{ijkl}, which are impossible to select from electron contributions at low temperatures and which are impossible to select from lattice contributions at high temperatures.

2 Pseudopotentials and Screening Used in Models of the Adiabatic Potential of Metals

In the previous chapter the pseudopotential $V_{\mathbf{q}}$ was introduced formally as some external static field. The nature and the form of the pseudopotential were not very important for conclusions drawn in this chapter. The only supposition relating to the value of the pseudopotential is the possibility of using a power series expansion in the parameter $\eta = V_{\mathbf{q}}/E_F$.

The present chapter contains descriptions and foundations of the pseudopotential models used by investigations and numerical calculations of a great number of static, phonon, and thermodynamic properties of non-transition metals. Some results of these calculations will be presented in Chapters 3 and 4.

The pseudopotential concept created and further developed in the framework of the band theory provides a formalism which was the most successful in the understanding and further description of properties of solids [1–5]. The pseudopotential concept provides the model which simultaneously allows an adequate description of experimentally observed properties of solids, as well as simple and regular procedures for concrete calculations. In the present chapter, we will describe methods which were most important in developing the pseudopotential theory.

The pseudopotential concept uses three fundamental suppositions [1–4]. The first is the self-consistent-field approximation. This scheme uses some average potential in describing the electron–ion interaction. This potential depends also on the electron configuration, but in concrete calculations of the band structure, the self-consistent field $V_{\mathbf{q}}$ actually is not calculated.

The next important approximation is the possibility of dividing all electron orbitals of atoms in the crystal into two groups: core and valence states. Actually, but not necessarily, orbitals like closed shells of noble gases are considered as inner shells, and the number of conducting electrons is equal to the number of the row of the Periodic Table for the metal under consideration.

Wave functions of inner shells are supposed to be well-localized near their own nucleus. Their overlapping with wave functions of neighboring atoms under normal conditions is supposed to be small. Following this supposition, the core–core interaction is actually considered to be the Coulomb repulsion of point-like charges Z_l [5]. Additionally, the changing of the potential within the region of inner shells, provided by electron cores of neighboring atoms and by conducting electrons, is neglected. As a consequence of this approximation, it is possible to present the total periodic crystal potential as a simple superposition of atomic potentials [2], located in each lattice site $\mathbf{R}_m$.

$$V(\mathbf{r}) = \sum_m v(\mathbf{r} - \mathbf{R}_m). \tag{2.1}$$

Additionally, in the case of metals the applicability of the perturbation theory is also supposed. It allows the construction of a theory of atomic and thermodynamic properties without direct calculations of the electron band structure. However, note that the supposition about applicability of the perturbation theory needs further numerical investigations (see Chapter 3).

The main idea of the pseudopotential method is the replacing of the large potential (2.1) of the ion core by the essentially weaker pseudopotential

$$V^{ps}(\mathbf{r}) = \sum_m v^{ps}(\mathbf{r} - \mathbf{R}_m), \tag{2.2}$$

which, nevertheless, provides a true energy spectrum for valence electrons. The pseudowave functions $\Phi_{\mathbf{k}}(\mathbf{r})$ of valence electrons, coinciding with the true wave functions outer of ion cores, now have no oscillations inside cores. But since ion cores are actually small ($\sim 5\%$ from the cell volume), then $\Phi_{\mathbf{k}}(\mathbf{r})$ provides an adequate description of the electron distribution in the main part of the crystal. It needs the introducion of the non-local operator $v^{ps}(\mathbf{r}', \mathbf{r})$. Its form factor depends not only on the momentum $\mathbf{q}$ transfer to the ion by scattering, but also on the initial electron momentum $\mathbf{k}$ and also on the energy E, i.e.,

$$\langle \mathbf{k} + \mathbf{q} | v^{ps} | \mathbf{k} \rangle \equiv v^{ps}(\mathbf{k}, \mathbf{q}, E). \tag{2.3}$$

Taking into account the fact that the most important arguments of v^{ps} are, nevertheless, $\mathbf{q}$, it is possible to reduce the number of arguments in (2.3), by using the on-Fermi-sphere approximation (see [4, 6]):

$$|\mathbf{k}| = |\mathbf{k} + \mathbf{q}| = k_{F0}, \quad E = E_F, \tag{2.4}$$

where k_{F0} is the radius of the Fermi-sphere, E_F is the exact Fermi energy (chemical potential). This approximation yields a correct description of electron states near the Fermi surface (see also section 2).

Concrete calculations, in particular those listed in the present review (see Chapter 3), show that, in many cases, the local pseudopotential approximation is well motivated and works well. The pseudopotential form-factor, if local, can be written as

$$v^{ps}(\mathbf{q}) = \frac{1}{\Omega} \int v^{ps}(\mathbf{r}) \exp(i\mathbf{q}\mathbf{r})\, d^3r. \tag{2.5}$$

We must take into account that $\Omega v^{ps}(\mathbf{q})$ is some given function of $\mathbf{q}$ and does not depend on the atomic volume. It means that the core of each element composing the compound has the same scattering properties as in any other compound and in a free ion, and the interaction of valence electrons with a core does not depend on its arrangement. Nowadays, there are two main directions in developing the microscopic theory of non-transition metals: non-empirical and model calculations. Non-empirical (*ab initio*) calculations provide a good basis for the second, very powerful approach in which interaction of valence electrons with the ion system is treated by using the *ab inito* concept, but construction of the pseudopotential for each concrete element is provided in the frame of some model with fitting parameters.

From the point of view of the present-day situation in solid state physics, a good description of ideal metals (and also using this theory in the case of real crystals) is actually achieved in the frame of models with fitting parameters. The aim of this activity is to minimimize the number of fitting parameters. The calculations described in this review are predominantly related to models with fitting parameters.

The choice of the most adequate calculation scheme is provided using the next criteria:

(a) the same universal scheme must provide calculations of the maximal number of atomic properties;
(b) The calculation scheme (the form of the pseudopotential, screening, the fitting procedure for the parameters, etc.) must demonstrate a stability of the pseudopotential description, and the possibility to transfer the form-factor and the whole calculation scheme to other crystal structures.

2.1 Model Potentials

2.1.1 Model potential of an isolated ion

A lot of calculations of statical and dynamic properties of simple metals with different forms of the model pseudopotential show that the choice of the potential and its parameters appreciably effects the results of calculations. This fact explains the existence of a large number of publications dealing with construction of new ("optimized", "modified", "effective", and so on) potentials (see e.g. [7–19]). In the present section, we consider some "ab initio" model potentials mostly used in calculations, which simultaneously seem to be mostly well-grounded from the physical point of view. Using these examples, we will show some possibilities to choose the form of the potential and to determine the potential parameter, as well as to provide a localization. A very important aim of the present chapter is the demonstration of a proximity of the local form of the first-principle model Heine–Abarenkov [1, 2] and Rasolt–Taylor [12–15] potentials and also the two-parametric local Animalu–Heine–Abarenkov potential with fitting parameters [17].

The basic idea of introducing model potentials is the statement that it is possible to consider each ion as some "black box" whose properties are taken to provide an adequate description of the wave function of valence electrons outside the ion radius

R_c. In concrete calculations, this "black box" is provided by changing the deep ion (core) potential by some weak model potential. As follows from the pseudopotential theory [1, 2], it is necessary to introduce for each orbital momentum l its own potential $V_l(\mathbf{r})$. Additionally, each potential $V_l(\mathbf{r})$ depends on the electron energy E. By substitution into the Schrödinger equation, this potential must provide correct values for the energy E of valent electrons, and the wave functions obtained must coincide with the true wave functions at $r > R_c$. A very important feature of this approach is the possibility to treat this weak model potential as a perturbation.[1]

In the present subsection, we will consider three model potentials with the same general expression (see Figure 2.1):

$$\begin{aligned}\hat{V}_M(\mathbf{r}) &= \sum_l \left(V_l(\mathbf{r})\,\theta(\mathbf{R}_l - \mathbf{r}) - \frac{Z}{r}\theta(\mathbf{r} - \mathbf{R}_l)\right)\hat{P}_l \\ &\equiv \sum_l \left(A_l(E)\,\theta(\mathbf{R}_l - \mathbf{r}) - \frac{Z}{r}\theta(\mathbf{r} - \mathbf{R}_l)\right)\hat{P}_l,\end{aligned} \tag{2.6}$$

where

$$\theta(x) = \begin{cases} 1, & x \geq 0 \\ 0, & x < 0 \end{cases} \tag{2.7}$$

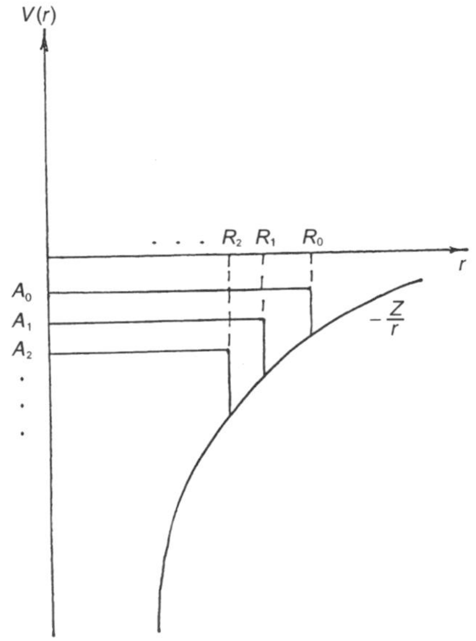

FIGURE 2.1 Schematic plots of the model potentials $V(r)$: Heine–Abarenkov, Rasolt–Taylor, and Shaw.

---

[1] This approximation results in a restriction for the applicability region of the pseudopotential method for a system with weakly non-uniform electron density $n(\mathbf{r})$ in contrast to the density functional theory (see, for example, [20]), which may be applied to arbitrary $n(\mathbf{r})$.

is the Heaviside step-function, and

$$\hat{P}_l = \sum_{m=-l}^{l} |Y_l^m >< Y_l^m| \tag{2.8}$$

is the projection operator which picks out the state Y_l with the momentum l, A_l and R_l are the parameters of the model potential, Z is the ion valency.

Introduction of this form of the potential corresponds to replacing the deep core potential by square potential walls $A_l(E)$ inside the radius R_l. At $r > R_l$ this potential is treated as the Coulomb one. As had been noted in [11], namely using square potential walls $A_l(E)$ minimizes dependence of the potential parameters upon the energy E. This conclusion follows from the comparison of different forms of $V_l(\mathbf{r})$ (2.6) (see Figure 2.2) with $R_l = R_M$ for all l.

Let us consider some concrete forms of the model potential. We will consider here the most frequently used, the Heine–Abarenkov [10, 11], Rasolt–Taylor [12–15] and the optimized Shaw [21] potentials. The difference between these potentials corresponds to relations between parameters for different l and also to physical ideas used in determining these parameters.

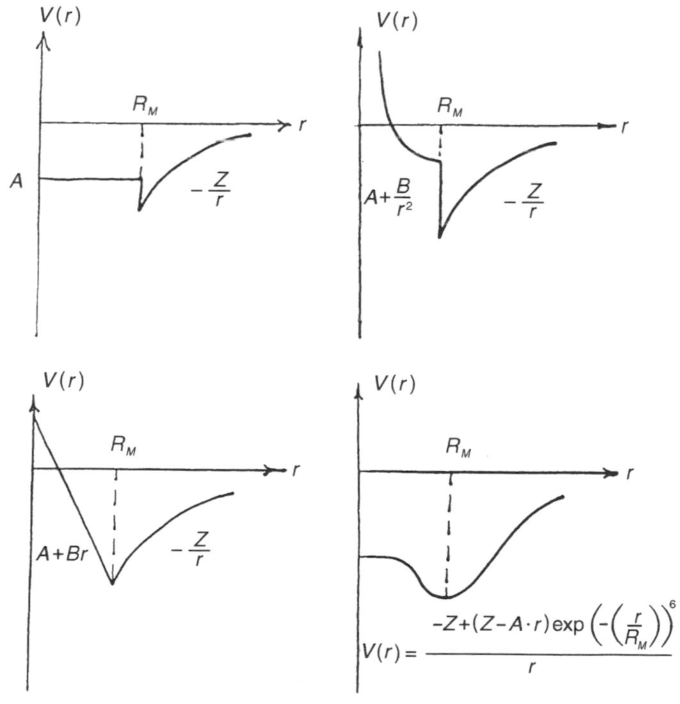

FIGURE 2.2 Different model pseudopotential considered in [11].

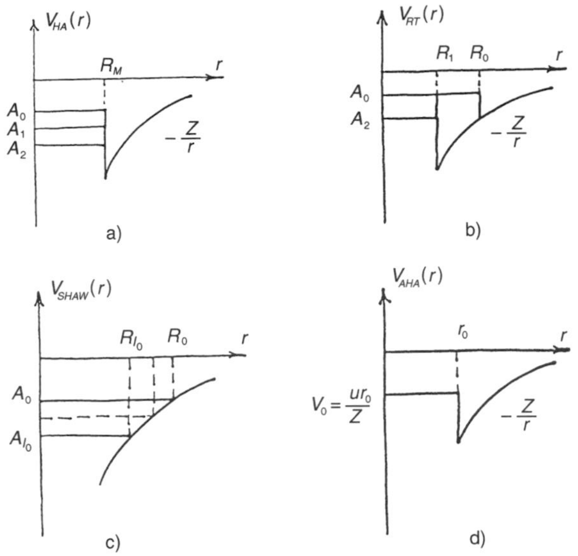

FIGURE 2.3 Different model pseudopotential (PP): (a) Heine–Abarenkov $V_{HA}(r)$ [10, 11]; (b) Rasolt–Taylor $V_{RT}(r)$ [12–15]; (c) optimized Shaw PP $V_{Shaw}(r)$ [21]; (d) fitted Animalu–Heine–Abarenkov PP V_{AHA} [17].

Heine and Abarenkov [10, 11] have supposed $R_l = R_M$ for all l and $A_l = A_2$ for all $l \geq 2$. By using this supposition, equation (2.6) has the form (see Figure 2.3a)

$$\hat{V}_{HA}(\mathbf{r}) = -\frac{Z}{r} - \theta(R_M - r)\left(\left(A_2 - \frac{Z}{r}\right) + (A_0 - A_2)\hat{P}_0 + (A_1 - A_2)\hat{P}_1\right). \quad (2.9)$$

In the plane-wave (PW) representation, the matrix elements of this potential are

$$\langle \mathbf{k}+\mathbf{q}|\hat{V}_{HA}|\mathbf{k}\rangle = \frac{4\pi Z}{\Omega_0 q^2}[M_L(\mathbf{q}) + M_{NL}(\mathbf{k}+\mathbf{q},\mathbf{k})], \quad (2.10)$$

where Ω_0 is the volume of the unit cell at $T = 0,\ p = 0$;

$$M_L(\mathbf{q}) = \frac{A_2 R_M}{Z}\frac{\sin(qR_M)}{qR_M} - \left(\frac{A_2 R_M}{Z} + 1\right)\cos(qR_M); \quad (2.11)$$

$$M_{NL}(\mathbf{k}',\mathbf{k}) = q^2\left(\frac{A_0 - A_2}{z}\right)K_0(\mathbf{k}',\mathbf{k},R_M) + 3\cos\frac{A_1 - A_2}{z}K_1(\mathbf{k}',\mathbf{k},R_M), \quad (2.12)$$

θ is the angle between $\mathbf{k}'$ and $\mathbf{k}$;

$$K_l(\mathbf{k}',\mathbf{k},R) = \int_0^R r^2 j_l(k'r) j_l(kr)\,dr; \quad (2.13)$$

$j_l(r)$ is the spherical Bessel function of the l-th order;

$$K_0(\mathbf{k}', \mathbf{k}, R) = \left(\frac{\sin kR}{k}\cos(k'R) - \frac{\sin(k'R)}{k'}\cos(kR)\right) \Big/ (k^2 - k'^2),$$

$(\mathbf{k} \neq \mathbf{k}')$

$$K_0(\mathbf{k}, \Delta\mathbf{k}, R) = \frac{2kR - \sin(2kR)}{4k^3} + \frac{\Delta K}{8k^4}[3\sin(2kR) - 2kR(2 + \cos(2kR)] + O\Big((\Delta k)^2\Big), \quad (\mathbf{k}' = \mathbf{k} + \Delta\mathbf{k}); \tag{2.14}$$

$$K_1(\mathbf{k}', \mathbf{k}, R) = \frac{R}{k^2 - k'^2}[j_1(kR)\sin(k'R) - j_1(k'R)\sin(kR)], \quad (\mathbf{k} \neq \mathbf{k}'),$$

$$K_1(\mathbf{k}, \Delta\mathbf{k}, R) = \left(-\frac{R}{k^2}j_1(kR)\sin(kR) + \frac{2kR - \sin(2kR)}{4k^3}\right)\left(1 - \frac{2\Delta k}{k}\right) + \frac{\Delta k}{8k^4}[-3\sin(2kR) + 2kR(2 + \cos(2kR)) + O((\Delta k)^2],$$
$$\mathbf{k}' - \mathbf{k} + \Delta\mathbf{k}, \tag{2.15}$$

The next example of the model potential is the Rasolt–Taylor potential [12–15]. It corresponds to the general expression (2.6) with $A_1 = A_0$, $A_l = A_2$ and $R_l = R_1$ for all $l \geq 2$. This form of the potential mostly corresponds to calculations of the phase shift [12] (see Figure 2.3b)

$$\begin{aligned}\hat{V}_{RT}(\mathbf{r}) &= \left[-\frac{Z}{r} - \left(A_0 - \frac{Z}{r}\right)\theta(R_0 - r)\right]\hat{P}_0 \quad (l = 0)\\ &= \left[-\frac{Z}{r} - \left(A_0 - \frac{Z}{r}\right)\theta(R_1 - r)\right]\hat{P}_1 \quad (l = 1)\\ &= \left[-\frac{Z}{r} - \left(A_2 - \frac{Z}{r}\right)\theta(R_0 - r)\right]\hat{P}_l \quad (l \geq 2)\end{aligned} \tag{2.16}$$

Matrix elements of this potential in the plane-wave representation correspond to (2.10) with

$$M_L(\mathbf{q}) = \frac{R_1 A_2}{Z}\frac{\sin(qR)}{qR_1} - \left(\frac{R_1 A_0}{Z} + 1\right)\cos(qR_1); \tag{2.17}$$

$$M_{NL}(\mathbf{k}', \mathbf{k}) = q^2\left[\frac{A_0}{Z}K_0(\mathbf{k}', \mathbf{k}, R_0) - \frac{A_2}{Z}K_0(\mathbf{k}', \mathbf{k}, R_1) + 3\cos\theta\,\frac{(A_0 - A_1)}{Z}K_1(\mathbf{k}', \mathbf{k}, R_1)\right] + \frac{I(|\mathbf{k} + \mathbf{k}'|, R_0, R_1) - I(|\mathbf{k} - \mathbf{k}'|, R_0, R_1)}{2kk'}; \tag{2.18}$$

$$I(k, R_0, R_1) = \int_{kR_0}^{kR_1} \cos\rho \frac{d\rho}{\rho}; \tag{2.19}$$

where K_0 and K_1 were determined in (2.14) and (2.15).

In the case of the optimized model Shaw potential [16], the core potential is replaced by A_l only for $l \leq l_0$ corresponding to the core states near valence states. For large $l > l_0$, it is necessary to use the true core potential which is, generally speaking, unknown. However, for this l as $r \to 0$, the centrifugal potential $l(l+1)/r^2$ plays the main role in the Schrödinger equation (1.29). Hence, a detailed structure of the core potential at small r is not important and it is possible to take it as $-Z/r$ (see Figure 2.3c). Thus,

$$\hat{V}_{Shaw}(\mathbf{r}) = \sum_{l=0}^{l_0} \left[A_l \theta(\mathbf{R}_l - \mathbf{r}) - \frac{Z}{r} \theta(\mathbf{r} - \mathbf{R}_l) \right] \hat{P}_l - \sum_{l=l_0+1}^{\infty} \frac{Z}{r} \hat{P}_l. \tag{2.20}$$

For $l_0 = 1$, taking into account the relations between parameters A_l and R_l obtained by Shaw, the matrix elements of this potential written as (2.10) are

$$M_L(\mathbf{q}) = -\mathbf{1}; \tag{2.21}$$

$$M_{NL}(\mathbf{k}', \mathbf{k}) = q^2 \left[\frac{A_0}{Z} K_0(\mathbf{k}', \mathbf{k}, R_0) + J(\mathbf{k}', \mathbf{k}, R_0) \right.$$

$$\left. + 3\cos\theta \left[\frac{A_1}{Z} K_1(\mathbf{k}', \mathbf{k}, R_1) + J_1(\mathbf{k}', \mathbf{k}, R_1) \right] \right]; \tag{2.22}$$

where

$$J_l(\mathbf{k}', \mathbf{k}, R) = \int_0^R r J_l(k'r) J_l(kr)\, dr; \tag{2.23}$$

$$J_0(\mathbf{k}', \mathbf{k}, R) = \frac{1}{2kk'} \left[\ln\left|\frac{k+k'}{k-k'}\right| - \int_{|k-k'|R}^{(k+k')R} \cos\frac{d\rho}{\rho} \right], \quad (\mathbf{k} \neq \mathbf{k}')$$

$$J_0(\mathbf{k}', \Delta\mathbf{k}, R) = \left[\frac{1}{2k^2} \int_0^{2kR} (1-\cos\rho) \frac{d\rho}{\rho} \right] \left(1 - \frac{\Delta k}{k} \right) + \frac{\Delta k}{k} \frac{(1-\cos 2kR)}{4k^2}$$

$$+ O\left((\Delta k)^2 \right), \quad (\mathbf{k}' = \mathbf{k} + \Delta\mathbf{k}) \tag{2.24}$$

$$J_1(\mathbf{k}', \mathbf{k}, R) = -\frac{1}{2kk'}\left[1 + kk'R^2 J_1(kR)\, J_1(k'R) - \cos(kR)\cos(k'R)\right] - \frac{(k^2 + k'^2)}{2kk'} J_0(\mathbf{k}', \mathbf{k}, R) \tag{2.25}$$

2.1.2 Choice of parameters of the model potential

In this subsection, we briefly consider general features and procedures used by authors of [10–15, 22] by fitting the potential parameters.

The procedure used by Abarenkov and Heine [10, 11] includes the fitting of the depths of the potential wells $A_l(E)$ for the model potential (2.9) to experimentally measured energy spectrum of free atoms (for alkali metals) or of ions with one valent electron (like Al^{++}) for all other metals. The depths of potential wells for orbitals with distinguished l have different values. This fact corresponds to a non-locality of the potential relating to the energy [4]. Values A_l also depend on R_M. For this reason, this procedure includes also variation of R_M within some interval, and values A_l have been fitted for each R_M. It is necessary to note that each set of the parameters A_l, obtained for any R_M from this interval, provides a correct definition of the model potential. Hence, it is possible to use this degree of freedom, for example, to find R_M corresponding to minimal differences between A_l with different l. However, for each l, there is a small dependence of the depth of potential well on the energy. Abarenkov and Heine overcame this difficulty by using the linear extrapolation of $A_l(E)$ as a function of E for $E = E_{FO}$ (where $E_{FO} = (3\pi^2 n_0)^{2/3}/2m$ is the Fermi energy of electron gas with the density n_0) corresponding to the Fermi energy of the crystal.

Rasolt and Taylor [13, 14] have used two different methods of fitting of parameters of their potential. We denote these methods as RT1 and RT2.

In the RT1 method, these parameters are fitted using the condition that the difference between values of the phase shifts for scattering by the model potential (calculated for electrons with the energy equal to the Fermi energy in the crystal) and scattering by the atomic potential must be equal to some integer number times π at $r \geq R_l$. For the atomic potential, they have used the potential calculated using the Hartree–Fock method with additional corrections provided by the exchange, screening, and so on. Thus, from the methodological point of view, the RT1 potential corresponds exactly to the main idea about using scattering amplitudes, which is the basis of the pseudopotential method. However, the accuracy of this method is essentially restricted by approximations made by calculation of the atomic potentials.

The second method, RT2, includes calculations of the electron density $n(\mathbf{r})$ near the ion in the frame of the first-order perturbation theory in the non-local weak model potential of the ion immersed in the uniform electron gas. To fit the potential parameters, Rasolt and Taylor have supposed that $n(\mathbf{r})$ must coincide with the electron density calculated by Dagens [23] in the frame of the Kohn–Hohenberg–Sham formalism [24–26]. This method uses the statement that the ground state energy of the electron system is a functional of the electron density and, consequently, the potential providing correct description of $n(\mathbf{r})$ must also provide exact values for the energy of the electron system. Since the RT2 potential describes the distribution of $n(\mathbf{r})$ in the first-order of the perturbation theory, then, strictly speaking, we must use it to calculate the adiabatic potential in the second order in the potential [6, 27].

A general feature of these methods of fitting the potential parameters is using the isolated ion approximation, but not an ion in a crystal. To improve this potential and make it closer to a typical "crystal" potential, it is necessary to add the potential provided by valence electrons in the region of the ion core (the core shift [1, 4]). Calculating this potential is a separate non-trivial problem. Actually, the core shift is approximated by some constant term, but this procedure is valid at small R_l only [12]. The accuracy of this approximation at actual values of R_l is not clear.

To determine parameters of the optimized Shaw potential [16], the variation procedure [2] was used. The idea of this procedure is the searching of the mostly smooth model wave function $\chi(\mathbf{r})$, i.e., the minimization of the integral

$$I = \int d^3r\, |\nabla\chi(\mathbf{r})|^2 \Big/ \int d^3r \chi^*(\mathbf{r})\chi(\mathbf{r}). \tag{2.26}$$

For this procedure, there exist the next relations between A_l and R_l^*:

$$A_l(R_l^*) = \frac{Z}{R_l^*},$$

$$\left.\frac{\partial A_l}{\partial R_l}\right|_{R_l^*} = 0. \tag{2.27}$$

The value R_l^* is determined by the next condition

$$\left.\frac{\partial}{\partial R_l}\frac{\langle\chi|W|\chi\rangle}{\langle\chi\,|\,\chi\rangle}\right|_{R_l^*} = 0, \tag{2.28}$$

where W is the whole model potential of the crystal, which includes not only the model potential of the electron–ion interaction but also the potential provided by conducting electrons. Values R_l^* depend on the electron energy and, actually, the values corresponding to the Fermi-level are used.

2.2 Local Pseudopotentials

2.2.1 Localization with respect to the electron density

In the rest of this chapter we will consider foundations of the schemes of calculation. But first, it is necessary to recall the problem of the non-locality of pseudopotentials, which is a characteristic feature of the pseudopotential concept. Note that quantitative calculations of atomic properties are very difficult even for models with terms up to $V_{\mathbf{q}}^2$ only. From the present-day point of view, it is very important to provide calculations including higher orders in the pseudopotential. However, in the case of a non-local pseudopotential, these calculations are impossible. For this reason, we must use some localized forms of the pseudopotential. There are many different points of view to this problem, and we consider some of them below. Note, however, that the choice of the localization method depends on properties considered and also on the model used in calculations. For this reason, there were a lot of localized versions of model potentials that were proposed. Let us now consider some basic ideas of the localization procedure.

The localization procedure actually used is the localization of the non-local potential in the framework of the on-Fermi-sphere approximation. Namely, this approximation was most frequently used in calculations of the localized form of non-local model potentials [17, 28]. This procedure yields the local potential $V_{\mathbf{q}}$ relating to the non-local potential $\langle \mathbf{k}+\mathbf{q}|\hat{V}|\mathbf{k}\rangle$ through the equation

$$V_{\mathbf{q}} = \langle \mathbf{k}+\mathbf{q}|\hat{V}|\mathbf{k}\rangle. \tag{2.29}$$

Here, in calculating $\langle \mathbf{k}+\mathbf{q}|\hat{V}|\mathbf{k}\rangle$, we must set

$$\begin{aligned} E = E_{F0}; \quad |\mathbf{k}| = |\mathbf{k}+\mathbf{q}| = k_{F0} \text{ at } q \leq 2k_{F0}, \\ |\mathbf{k}| = k_{F0}, \mathbf{k} \parallel -\mathbf{q} \text{ at } q > 2k_{F0}. \end{aligned} \tag{2.30}$$

To defend this choice, the next reasons are actually taken into account [1, 4, 6]. By calculating the energy of the electron subsystem in metals in the second order in the non-local potential using the ring diagram approximation, the following expression has been obtained (see, for example, [5] and subsection 2.2)

$$\begin{aligned} E_e^{(2)} = \frac{m}{\hbar^2}\sum_{\mathbf{q}}|S(\mathbf{q})|^2\Bigg\{2\sum_{\mathbf{k}}\frac{|\langle \mathbf{k}+\mathbf{q}|\hat{V}|\mathbf{k}\rangle|^2}{k^2-(\mathbf{k}+\mathbf{q})^2} \\ -\frac{8\pi m e^2}{\hbar^2\Omega_0 q^2\varepsilon(q)}\left|2\sum_{\mathbf{k}}\frac{\langle \mathbf{k}+\mathbf{q}|\hat{V}|\mathbf{k}\rangle}{k^2-(\mathbf{k}+\mathbf{q})^2}\right|^2\Bigg\} \end{aligned} \tag{2.31}$$

Equation (2.31) shows that the most important contribution to $E_e^{(2)}$ provides the region of **k**-space where the denominator $k^2-(\mathbf{k}+\mathbf{q})^2$ is small, i.e., $k = |\mathbf{k}+\mathbf{q}|$ at $q \leq 2k_{F0}$. At $q > 2k_{F0}$, the denominator is never equal to zero. Its minimum corresponds to an antiparallel orientation of **k** and **q** and k must be equal to k_{F0}. We see that the choice of the local potential in the framework of the on-Fermi-sphere approximation provides a rather good description of the main contributions to (2.31). This procedure includes scattering processes for the electrons located at the Fermi-sphere only (it explains the name of the approximation). For the "umklapp processes" only, the scattering back is taken into account.

As follows from the definitions (2.29)–(2.31), the on-Fermi-sphere approximation together with the second-order perturbation theory in the pseudopotential should provide satisfactory results in calculations of properties obtained by averaging over the electron spectrum. However, in using this pseudopotential in calculations, for example, the phonon frequencies and the elastic moduli of metals in the third order of the perturbation theory doesn't provide a satisfactory agreement with experiment [6, 29]. This disagreement is a consequence of very large contributions given by terms $\sim V_{\mathbf{q}}^3$. This fact is not surprising because the term $E_e^{(3)} \sim V_{\mathbf{q}}^3$ contains two denominators with minima corresponding to two different regions of the **k**-space. However, the definition (2.30) ignores this fact.

There exists also the localization method using the criterion of reproducing the electron density distribution in calculations of the ground state energy of the electron system [6]. Here, the local potential must provide the same electron density distribution calculated in the first-order perturbation theory in both local and non-local potentials. This localization method is based on the general theory of the non-uniform electron gas [20, 24–26] applied to weakly non-uniform electron distribution in simple metals. In accordance with this method, the local potential $V_{\mathbf{q}}$ relates to the non-local one $\langle \mathbf{k}+\mathbf{q} \,|\hat{V}\,|\, \mathbf{k}\rangle$ as

$$V_{\mathbf{q}} = \left[\sum_{\mathbf{k}} \frac{\langle \mathbf{k}+\mathbf{q} \,|\hat{V}\,|\, \mathbf{k}\rangle}{\varepsilon_{\mathbf{k}} - \varepsilon_{\mathbf{k}+\mathbf{q}}}\right] \Big/ \left[\sum_{\mathbf{k}} \frac{1}{\varepsilon_{\mathbf{k}} - \varepsilon_{\mathbf{k}+\mathbf{q}}}\right]. \tag{2.32}$$

Here, $\varepsilon_{\mathbf{k}}$ is the band electron energy. It is necessary to note that the "localization" (2.32) is similar to the band calculation approach in the calculation of the cohesive energy, i.e., equation (2.32) would not take into account the exchange and correlation effects. The exact expression for $V(\mathbf{q})$, conserving the mean charge density in the first order in $V_{\mathbf{q}}$, has the form

$$V(\mathbf{q}) = -\frac{t^{(1.1)}(\mathbf{q})}{\pi(\mathbf{q})}. \tag{2.33}$$

In the frame of the local approximation, the block $t^{(1.1)}(\mathbf{q})$ is determined in [95] equation (2.16); $\pi(\mathbf{q})$ is the polarization operator. The averaging in (2.32) is an exact procedure if we take into account many-electron effects.

The importance of the reproduction of the true charge density by the pseudopotential used in calculations of atomic properties was at first established independently in [13] and [14]. Later, this idea was used for the localization of the pseudopotential in [27]. In [13], the many-particle self-consistent Hohenberg–Kohn–Sham equation [24–26] was solved and the electron distribution provided by the total ion potential (of a single ion immersed into the electron gas) was obtained. This calculation is exactly equivalent to a whole summation of some series of the perturbation theory in the pseudopotential. Then, the parameters of the non-local Heine–Abarenkov pseudopotential [10, 11] were fitted to provide the most perfect reproduction of the electron density outside of the core. This method was successfully used in [14] in calculations of the phonon frequencies of some simple metals in the frame of the second-order perturbation theory in the pseudopotential. This approach allows the calculation of higher order contributions to the lattice dynamics.

The effective local pseudopotential, reproducing the density distribution from [13, 14], was constructed in [27]. A very important statement emphasized in [27] is that the localized form of the pseudopotential must correspond to the model used in calculations of the cohesive energy. If the cohesive energy is calculated while taking into account all terms up to the second order, then the local pseudopotential $V^{(1)}(\mathbf{q})$ reproducing the true electron distribution $n(\mathbf{q})$ outside the core (in [27] the density distribution calculated in [13] was used) can be obtained from the following equation

$$n(\mathbf{q}) = -\pi_0(\mathbf{q}) \frac{V^{(1)}_{\mathbf{q}}}{\tilde{\varepsilon}(\mathbf{q})} \tag{2.34}$$

Calculations in [27] were also performed including the term $E_e^{(3)}$ (2.113) into the cohesive energy. In these calculations, the local pseudopotential $V^{(2)}_{\mathbf{q}}$ was obtained from the following equation

$$n(\mathbf{q}) = -\pi_0(\mathbf{q}) \frac{V^{(2)}_{\mathbf{q}}}{\tilde{\varepsilon}(\mathbf{q})} + \sum_{\mathbf{q}_1} \frac{3\Lambda_0^{(3)}(\mathbf{q}, \mathbf{q}_1, -\mathbf{q} - \mathbf{q}_1)}{\tilde{\varepsilon}(\mathbf{q})} \frac{V^{(2)}_{\mathbf{q}_1}}{\tilde{\varepsilon}(\mathbf{q}_1)} \frac{V^{(2)}_{-\mathbf{q}-\mathbf{q}_1}}{\tilde{\varepsilon}(\mathbf{q} + \mathbf{q}_1)}. \tag{2.35}$$

Now the general rule is clear: the local pseudopotential reproducing $n(\mathbf{q})$ must be used in the equation $n(\mathbf{q}) = F(V_{\mathbf{q}})$. The last equation must include powers of $V_{\mathbf{q}}$ up to the highest power of $V_{\mathbf{q}}$ in the expression for the cohesive energy minus one.

The effective local pseudopotential determined by this procedure is sufficiently weaker (approximately in order of magnitude, see Figure 1 in Ref. 27) than the "ab initio" potentials (for the on-Fermi-sphere approximation) calculated using characteristics of isolated atoms. This is a very important statement and we will return to it later. Further, note that this potential $V_{\mathbf{q}}$ partially takes into account effects of higher order terms in the electron–ion interaction. Small values of this potential allow usage of the perturbation theory. But this fact does not mean that this pseudopotential really provides small high-order contributions (higher than are the order of perturbations taken into account by calculation of this potential). These high-order terms cannot be included in calculations at all. Hence, all calculations of these high order contributions using the potentials calculated in the framework of lower order approximation are not correct (and not instructive).

The form-factors of the local pseudopotential [27] have been calculated numerically and were used in calculations of the phonon spectrum, i.e., only a single characteristic property was considered. If we are interested in applying this model to description of a wide number of properties of metals (and under external stresses), then we must conclude that it is impossible to provide without any suppositions about the analytical form of the pseudopotential.

2.2.2 Local form-factor of the Animalu–Heine–Abarenkov potential. The Brovman–Kagan fitting condition

There exist many forms of the local pseudopotential, but we consider here only those widely used in calculations of atomic properties of non-transition metals. The most popular is the local form of the Heine–Abarenkov potential (HA) [21, 22] with the cut-off factor [17, 30], which we will denote further as the Animalu–Heine–

Abarenkov (AHA) pseudopotential. The form-factor of the AHA pseudopotential has the form

$$V_{AHA}(\mathbf{q}) = -\frac{4\pi e^2 Z}{q^2 \Omega}\left[(1+u)\cos qr_0 - u\frac{\sin qr_0}{qr_0}\right]\exp(-\xi q^4), \qquad (2.36)$$

where Z is the valency, and Ω is the volume of the elemental cell. This expression for $V_{\mathbf{q}}$ (excepting the cut-off factor $\exp(-\xi q^4)$) is the Fourier component of the HA pseudopotential which, in the coordinate space, has the form

$$V_{HA}(r) = \begin{cases} -U_0, & r \le r_0 \\ -\frac{Ze^2}{r}, & r > r_0 \end{cases} \qquad (2.37)$$

This form corresponds to the main idea of the cancellation theorem (see [1–4]): the cancellation of the electron–ion interaction in the core region r_0. From the physical point of view the parameter r_0 is close to the ion core radius. The characteristic parameter of the cancellation is the dimensionless constant

$$u = U_0 \Big/ \left(\frac{Ze^2}{r_0}\right). \qquad (2.38)$$

The cut-off factor $\exp(-\xi q^4)$ was first introduced in [30] to eliminate non-physical oscillations of $V_{\mathbf{q}}$ at large q provided by a step-like variation of $V(\mathbf{r})$ in (2.37). The value of the parameter ξ is taken here to be $0.03(rk_{F0})^{-4}$ (see [17]).

Both parameters of the pseudopotential (2.36) (the only parameters of the theory) are free parameters of the theory. Their definition seems to be a non-trivial problem because, *a priori*, it is not clear from which conditions they must be determined. Here, we, following Brovman and Kagan [5], will use the next conditions (p is the pressure, C_{44} is the shear elastic modulus, Ω_0 is the cell volume):[2]

$$p(\Omega_0; u, r_0) = 0; \quad C_{44}(\Omega_0; u, r_0) = C_{44\,exper}, \qquad (2.39)$$

These conditions must be satisfied for a given crystal structure and the value of the lattice parameter. The pseudopotential obtained by this procedure will be further denoted as $V_{pC}(\mathbf{q})$.

The first condition in (2.39) means that all calculated values must correspond to the equilibrium. The value $p(\Omega_0)$ is determined by values of the form-factor corresponding to sites of the reciprocal lattice, in particular, by the value corresponding to the center of the Brillouin zone. This last value is especially important for metals for which the contribution given by electron–ion interaction is small in comparison with the Coulomb contribution (for example, in alkali metals, see text below and Chapter 3). For this reason, the main role in (2.39) plays the value of $V(\mathbf{q})$ at $q = 0$

$$V(q=0) = \frac{b}{\Omega}. \qquad (2.40)$$

[2] General expressions for p and C_{44} have been given in Section 1, Chapter 1. Now we must use in these expressions concrete formulae for the electron energy E for different models (2.46), (2.81), (2.88).

Here

$$b = \frac{2}{3}\pi e^2 Z r_0^2 (3 + 2u), \tag{2.41}$$

is the non-Coulomb contribution to the pseudopotential (see [5]). Therefore, the first condition relates predominantly to the behavior of $V(\mathbf{q})$ at small q.

Since the dependence $p(\Omega_0)$ is a volume property which, generally speaking, weakly depends on the crystal structure [31], it can be used in determining the behavior of $V(\mathbf{q})$ at small q.

As the second condition for the fitting of $V_\mathbf{q}$, it is natural to use some structure-dependent characteristic. We can take it as the condition for the shear modulus C_{44}, which depends on values of $V_\mathbf{q}$ at sites of the reciprocal lattice with $\boldsymbol{\tau} \neq 0$ only. This structure-dependent condition is especially important because, as it is well-known [31], the volume-dependent values (the cohesive energy, pressure, compressibility) are changing at approximately 10% by changing the crystal structure.

The fitting scheme for the pseudopotential parameters (2.39) is not unique. We have also used for alkali metals the fitting of u and r_0 using experimental data for two phonon frequencies (see [32] and the text below). The potential, obtained in the frame of this procedure, we denote as $V_{ph}(\mathbf{q})$.

Similar definitions of the pseudopotential parameters (but using another analytical expressions) were performed by other authors, too. The approach described allows hope that this method corresponds to some model description of the non-local effects. But the main feature of this approach is that the pseudopotential obtained includes information about properties of an ion in the crystal.

Since expressions for $p(\Omega_0)$ and $C_{44}(\Omega_0)$ depend also on the number of terms $E_e^{(n)}$ considered and on the method of taking into account the many-electron effects, then the potential depends on the model used. This method of fitting of $V_\mathbf{q}$ is, evidently, useful in constructing qualitative accurate models. However, it could not be used in a comparison of different contributions and in considering different methods of taking into account the many-electron effects. Therefore, we, in accordance with our aim of considering contributions to $E_e^{(3)}$ of the order $\tilde{\gamma}_6$, will use the localized form-factors of the Rasolt–Taylor and Heine–Abarenkov model potentials. Nevertheless, note that the local form-factor of the non-local potential in the form (2.36) also includes some information about approximations made in describing the Fermi-liquid properties of the electron system of the metal under consideration. This information is included in the dependence of $\pi(q)$ on r_s.

2.2.3 Local form-factors by Harrison and Krasko–Gurskij

If we set $u = 0$ (or $V_0 = 0$) in (2.36), then this form-factor transforms into the well-known "empty-core" Ashcroft pseudopotential [33]. At $u = 1$ ($V_0 = Ze^2/r_0$), the form-factor (2.36) corresponds to the local form-factor of the Shaw model potential (2.20). Thus, the AHA form-factor (2.36) is rather flexible and general enough.

However, at the present time, there exist many different forms of pseudopotentials. We consider here two of them. The Harrison pseudopotential [1] comes from the point-like-ion model. This model takes into account the Coulomb potential provided by the ion charge, whereas all other terms in the non-screened form-factor are replaced by the δ-function. This (screened) form-factor turned out to be good for the calculation of electronic properties. But this form-factor does not decreases rapidly enough as $q \to \infty$ and remains finite. The last circumstance makes the model of the point-like-ion inadequate in calculating atomic properties.

To provide a correct asymptotic expression for $V_{\mathbf{q}}$, Harrison has used the fact that, from the general Austin–Heine–Sham [1–4] expression for a non-screened pseudopotential, it follows that, the decreasing of the repulsive part of the pseudopotential as $q \to \infty$ is determined by the decreasing of the Fourier transform of the atomic wave function of the core electron. The leading term in the series expansion of the Fourier amplitude in powers of $1/q$ corresponds to the s-function and has the form $\sim [1 + (qr_c)^2]^{-2}$, where r_c is some constant of the order of the Bohr radius. This fact was used by Harrison, who has supposed the following form of the potential:

$$\Omega V_H(\mathbf{q}) = \frac{4\pi Z}{q^2} + \frac{\beta}{\left[1 + (qr_c)^2\right]^2}. \tag{2.42}$$

Here β and r_c are two fitting parameters.

The derivation of (2.42) is not exactly well-founded, but the factor with r_c provides an almost correct decreasing of the behavior of the form-factor at large q. Nevertheless, the expression (2.42) does not correspond to the general Austin–Heine–Sham form [1].

The main disadvantage of the HA pseudopotential is the gap in the coordinate representation (see equation (2.37)). As a consequence, its form-factor oscillates and does not provide rapid enough (like q^{-4}) vanishing. For this reason, it is necessary to introduce some cut-off factor in the AHA form-factor. To improve this disadvantage, Krasko and Gurskij [34, 35] have proposed a new expression for the form-factor. Their suppositions (which are typical for many works) are the following. Let r_c be some radius which characterizes the core dimension. It is obvious that, at $r \gg r_c$, all non-screened model potentials must decrease like the Coulomb potential, i.e., like $-Z/r$. The most important region is the region $r \leq r_c$, because the variation of the potential here determines characteristic properties of the metal considered. It is well-known [1–4] that orthogonality properties of the wave functions of conducting electrons and electrons of inner shells in the region $r \leq r_c$ provide appreciable mutual cancellation of attractive and repulsive parts of the potential, so that as $r \to 0$, the potential must be replaced by some finite value (the cancellation theorem), but the sign of this value is *a priori* unknown. The behaviour of the potential in this region determines the behavior of the form-factor in the reciprocal space at $g \geq 1/r_c$, which, as we have already mentioned, must be proportional to q^{-4} at large q. The radial wave functions of inner shells can be expressed in terms of some products of polynomials and exponents. This fact was used by Krasko and Gurskij who have introduced their own form of the pseudopotential

$$V_{KG}(\mathbf{r}) = Z\left[\frac{\exp(-r/r_c) - 1}{r} + \frac{a}{r_c}\exp(-r/r_c)\right], \tag{2.43}$$

with the form-factor

$$\Omega V_{KG}(\mathbf{q}) = \frac{4\pi Z}{q^2} \frac{(2a-1)(qr_c)^2 - 1}{[1+(qr_c)^2]^2}. \tag{2.44}$$

This form-factor has two fitting parameters r_c and a.

In subsection 2.5, we will compare some form-factors from the point of view of the description of different atomic properties of metals.

2.2.4 *Second-order models for the adiabatic potential of metals and parameters of the pseudopotential*

A theoretical basis for construction of the adiabatic potential of non-transition metals was discussed in [1–5, 95]. Here, we introduce expressions needed and discuss approximations used in concrete calculations of atomic properties. We determine the cohesive energy as the value of the adiabatic potential when all atoms occupy their own equilibrium positions, i.e., the structure factor $S(\mathbf{q})$ for monoatomic crystals considered here is

$$S(\mathbf{q}) = \frac{1}{N}\sum_l \exp(i\mathbf{q}\mathbf{R}_l) - \sum_{\boldsymbol{\tau}} \delta_{\mathbf{q},\boldsymbol{\tau}}, \tag{2.45}$$

where $\mathbf{R}_l$ and $\boldsymbol{\tau}$ are vectors of the direct and reciprocal lattices.

In the frame of the second-order model (M2), the cohesive energy is determined as (all energies in the text below are given per one atom)

$$E = E_i + E_p + E_e^{(0)} + E_e^{(1)} + E_e^{(2)} + E^{zp}; \quad E_p = E_{sr} + E_w. \tag{2.46}$$

All terms in these equations (except for the last, see section 7, Chapter 3) were already discussed [95]. Using the Ewald method (see, for example, [1, 36]), the energy of the direct Coulomb interaction of point-like ions E_i immersed in the neutralizing background is usefully written as the rapidly converging sums over the real and reciprocal lattices

$$E_i = \frac{Ze^2}{2}\left\{\sum_{\tau}{}' \frac{4\pi}{\Omega\tau^2} e^{-\tau^2/4\sigma} - \frac{2\sqrt{\sigma}}{\sqrt{\pi}} + \sum_{l,l'}{}' \frac{1}{NR} G(\sqrt{\sigma}R) - \frac{\pi}{\sigma\Omega}\right\}, \tag{2.47}$$

where Ω is the cell volume (or the volume per atom): $R = |\mathbf{R}_l - \mathbf{R}_{l'}|$; and $\sigma = \pi/2a$ is the Ewald parameter regulating the convergence of both sums over the direct and reciprocal lattices;

$$G(x) = \frac{2}{\sqrt{\pi}} \int_x^{\infty} e^{-t^2} dt. \tag{2.48}$$

The electroneutrality condition for the whole electron–ion system provides cancellation of all divergent terms in the original Hamiltonian at $q = 0$. For this reason, E_i and $E_e^{(2)}$ do not contain terms with $\tau = 0$ (it is noted by the prime sign in sums over $\boldsymbol{\tau}$).

The core–core interaction, provided by overlapping their electron densities, is considered here using its simplest form, the parameters of which for the metals under consideration have been calculated in [38]

$$E_p \approx E_{sr} = \frac{1}{2}\sum_{\mathbf{R}} A \exp\left(-R/\rho\right). \tag{2.49}$$

The van der Waals interaction E_w [37] is taken into account by suitable choice of the parameters, see section 7, Chapter 3.

The contribution of the non-Coulomb part of the pseudopotential was obtained above in (2.40). Formula (2.41) gives the value of b for the AHA pseudopotential (2.36). Recall that the equations for $E_e^{(1,2)}$ used here were obtained within the local potential approximation when the electron multipoles were separated from the pseudopotential

$$E_e^{(1)} = Zb/\Omega \tag{2.50}$$

$$E_e^{(2)} = -\frac{\Omega}{2}\sum_{\mathbf{q}} \frac{\pi\left(\mathbf{q}\right)}{\varepsilon\left(\mathbf{q}\right)} \left|V_{\mathbf{q}} S\left(\mathbf{q}\right)\right|^2. \tag{2.51}$$

For concrete calculations, it is useful to join $E_e^{(2)}$ (2.51) with the first contribution to E_i in (2.47), and the repulsive energy of the core interaction E_{sr} with the third term in E_i, and $E_e^{(1)}$ (2.50) with the rest of E_i, because these last terms depend on the volume only.

Here, we do not use any approximations in the calculations of E_i and E_{sr} (calculating parameters of E_{sr} is a rather independent problem, and we consider these parameters to be given for metals under consideration, except Cs, see below).

To investigate the role of different approximations, we present here the results of calculations of atomic properties obtained using five different approximations for $\pi(q)$ (see the next subsection). *A priori*, it is not clear which approximation is the best. Therefore, our test of the models includes also a test of approximations for $\pi(q)$. Note in advance, that the best, from the theoretical point of view, the GT approximation [39, 40] provides also the best description of properties of alkali metals, whereas, in the case of Al, the VS approximation [44] provides approximately the same accuracy.

Finally, the last term, E^{zp}, is the energy of zero oscillations

$$E^{zp} = \frac{\hbar}{2}\sum_{\lambda\mathbf{k}} \omega_{\lambda\mathbf{k}}, \tag{2.52}$$

the effect of which is also considered ($\omega_{\lambda\mathbf{k}}$ is the phonon frequency of the phonon with the polarization λ and the momentum $\mathbf{k}$). It turns out, as a rule, that contributions E^{zp} are small and we actually neglect this term, if possible. Concrete formulae for the pressure, the elastic moduli, the dynamical matrix, and so on, are presented in Chapter 1, section 2 for the general case. In Chapter 3, sections 1, 2, 5, these formulae are given for the case of cubic metals. For each approximation of the

TABLE 2.1 Parameters of the V_{AHA} pseudopotential (2.36) for alkali metals for the second–order model (M2) for the BCC structure. $Z = 1$, u (dimensionless), r_0 (in a.u.), $E_p = E_{sr} + E_w$.

approximation for $\pi(\mathbf{q})$	parameters of $V(\mathbf{q})$	Na	K	Rb	Cs	source
		V_{pC}; $E_p = 0$; $E^{zp} = 0$				
RPA [41]	$-u$	0.313	0.490	0.604	0.654	[41, 45, 46]
	r_0	2.038	2.931	3.340	3.783	
GV [42]	$-u$	0.363	0.538	0.605	0.690	
	r_0	2.037	2.970	3.318	3.817	
TW [43]	$-u$	0.413	0.566	0.628	0.709	
	r_0	2.114	3.011	3.360	3.623	
VS [44]	$-u$	0.464	0.601	0.662	0.738	
	r_0	2.174	3.082	3.443	3.957	
GT [39]	$-u$	0.475	0.632	0.697	0.768	
	r_0	2.168	3.110	3.487	3.997	
		V_{pC}; $E_p = 0$, $E^{zp} \neq 0$				
GT [39, 40]	$-u$	0.392	0.589	0.665	0.746	[46, 47]
	r_0	2.095	3.042	3.423	3.936	
		V_{ph}; $E_p = 0$, $E^{zp} = 0$				
GT [39, 40]	$-u$	0.343	0.665	0.443	0.766	[32, 46]
	r_0	1.965	3.120	2.962	3.947	
		V_{pC}; $E_p \approx E_{sr} \neq 0$, $E^{zp} = 0$				
GT [39, 40]	$-u$	0.480	0.654	0.777	0.903	[48, 49]
	r_0	2.173	3.140	3.641	4.375	

polarization operator (RPA*[41], see also [61]), GT [42], TW [43], GT [39, 40], VS [44]) the parameters r_0 and u of the AHA pseudopotential $V_{pC}(\mathbf{q})$ were calculated using conditions (2.39), in which p and C_{44} have been calculated using formulae (2.46) for the energy. The results are listed in Table 2.1 (for Na, K, Rb, Cs) and in Table 2.2 (for Li, Al, Pb).

The second-order model with the fitting of the parameters of the pseudopotential $V_{pC}(\mathbf{q})$ in accordance with (2.39), as we will see in Chapter 3, provides a rather good description of the static properties of metals under consideration, but higher branches of the calculated phonon spectrum are located somewhat higher than experimental data. Apparently, this is a general feature of models with the local pseudopotential. Therefore, by describing pure phononic properties, like the specific heat $C_V(T)$, it seems to be useful to also consider the other version of the model, in which parameters of $V_{\mathbf{q}}$ are fitted using the experimental data relating to higher

TABLE 2.2 Parameters of the pseudopotential V_{AHA} (2.36) for lithium (BCC, $Z = 1$), aluminium (FCC, $Z = 3$), lead (FCC, $Z = 4$)

approximation for $\pi(\mathbf{q})$	parameters of $V(\mathbf{q})$	Li	Al	Pb	source
		second-order model (M2)			
		V_{pC}; $E_p = 0$, $E^{zp} = 0$			
CV [42]	$-u$	0.334			[41, 46, 50, 51]
	r_0	1.512			
TW [43]	$-u$	0.371	1.215	0.843	
	r_0	1.534	2.464	2.040	
VS [44]	$-u$	0.450	1.205		
	r_0	1.600	2.366		
GT [39, 40]	$-u$	0.408	1.215		
	r_0	1.559	2.459		
		V_{pC}; $E_p = 0$, $E^{zp} \neq 0$			
GT [39, 40]	$-u$	0.044			[47, 48, 52]
	r_0	1.369			
		fourth-order model (M4)			
		V_{pC}; $E_p = 0$, $E^{zp} = 0$			
GT [39, 40]	$-u$	0.339	1.223		[53]
	r_0	1.510	2.615		
VS [44]	$-u$		1.221		
	r_0		2.593		

frequencies of the phonon spectrum. This fitting is, probably, not well-founded, but if this fitting were successful, then the result could be considered as a very useful interpretation, which can be used for the description of results and further predictions, as well as the accuracy test for other models.

Using this consideration, we have provided the fitting of the parameters r_0 and u in (2.36) for alkali metals using experimental data for two phonon frequencies in point N of the Brillouin zone (see Chapter 3, section 2 and also [54])

$$\omega(N_1') = \omega(N_1')_{exper}; \quad \omega(N_3') = \omega(N_3')_{exper} \tag{2.53}$$

Of course, the choice of conditions (2.53) is rather free and was adopted to provide an accurate description of higher and middle phonon branches. The GT approximation [39] was used for Na, K, Rb, Cs. Experimental data for the phonon frequencies are taken from [55–58]. The values of the parameters of $V_{AHA}(\mathbf{q})$ obtained by this procedure are listed in Tables 2.1 and 2.2 (under the captions V_{ph} and V_{pC}). General data relating to metals under consideration are collected in Table 2.3.

TABLE 2.3 Data used in atomic properties calculations. Values of a, r_s, M, T_m and ω_D are taken from experiment, $\omega_p = (e^2/\pi M\Omega_0)^{1/2}$ is the plasma frequency

metal	structure	a, 10^{-10} m	r_s, a.u.	Z	A, Ry	ρ, 10^{-10} m	M, 10^{-26} kg	T_m, K	$\theta_D = \omega_D/k_B$, K	ω_p, THz
Li	BCC	3.478	3.236	1	280	0.116	1.153	453	835.4	17.41
Na	BCC	4.225	3.931	1	1631	0.168	3.817	370.7	342.8	7.143
K	BCC	5.239	4.875	1	3021	0.231	6.493	336.6	190.4	3.967
Rb	BCC	5.585	5.197	1	3837	0.258	14.191	312.5	117.0	2.438
Cs	BCC	6.045	5.625	1	4700	0.280	22.068	301.7	83.31	1.736
Ca	BCC	10.550	3.270	2						
Al	FCC	4.032	2.065	3	8452	0.118	4.480	933	830.9	15.74
Pb	FCC	4.914	2.286	4	15655	0.191	34.402	601	257.3	6.070

Tables 2.1 and 2.2 show that zero oscillations appreciably redefine the pseudopotential parameters for Li and Na only. For Li, the parameter u is decreasing in order of magnitude. This parameter characterizes the cancellation of the electron–ion interaction in the core region. The core radius varies more weakly. For heavier atoms, this re-definition of u, r_0 reduces and reaches values of about $1-2\%$ for Cs. Some consequences of this redefinition of the parameters of $V_{\mathbf{q}}$ will be discussed below (see Chapters 3 and 4). Note additionally, that the "pseudopotential depth" $V_0 = -ur_0/Ze^2$ has almost constant values for K, Rb and Cs. The core radius r_0 is increasing approximately proportionaly to the radius of the electron sphere

$$r_s = \left(\frac{3}{4\pi n_0}\right)^{1/3} \tag{2.54}$$

(from the physical point of view, it should be said vice versa: the cell radius increases almost proportionally to the core radius). This fact explains the similarity of the properties of K, Rb, Cs observed in experiments and in theoretical calculations (see Chapters 3 and 4).

Taking into account the repulsive energy, E_{sr} provides stronger values of the pseudopotential (2.36) (Table 2.1), i.e., it increases the well depths and the core radius. This effect is especially notable for Cs. For Li and Al (Table 2.2), the term E_{sr} does not provide any remarkable re-definition. Note that the parameters of E_{sr} in (2.49) were calculated in [59] for Na, K, Rb, Pb and Al. For Li and Cs, these parameters were evaluated by us using the similarity of properties of alkali metals and the supposition that $A \sim r_0^3$, $\rho \sim r_0$ [38, 48].

For all calculations presented below, we will suppose that the mass of the band electron m is equal to 1 (in atomic units). Maybe it is not a well-founded approximation, especially if exchange and correlation effects in the polarization operator and in $E_e^{(0)}$ are taken into account. As was shown in [27], it is possible to obtain a very good agreement between theoretical and experimental results for the phonon frequencies, supposing that some increase of the mass of band electrons takes place. But, apparently, it is true only for calculations with no parameter fitting. Here, we suppose that all effects like $m \neq 1$, the orthogonalization hole and so on, are automatically taken into account by the fitting of the pseudopotential parameters to experimental data for crystals.

2.2.5 Comparison of different second-order models with fitting of local form-factors

Actually the parameters $(u, r_0), (\beta, r_c)$ and (a, r_c) of the Animalu–Heine–Abarenkov (2.36), Harrison (2.42), and the Krasko–Gurskij form-factors (2.44) are determined from the fitting to some experimentally measured characteristics of metals. In *ab initio* calculations, which used ion properties in determining the pseudopotential, there appear more complicated, non-local pseudopotentials. For this reason, the number of metals and metal properties considered in the frame of this approach are much smaller. Here, we consider some local pseudopotentials of this type [60], but in calculations of a large number of properties. As a test metal, we take sodium.

Despite wide usage of the pseudopotential method, the question relating dependence of results of calculation upon the form of the potential was hardly ever discussed, although the importance of this problem is obvious. If the pseudopotential were obtained using *ab initio* methods mentioned above (for example, the method of the model Hamiltonian, the orthogonalized plane-wave method [1–4] or the method based on solution of the Kohn–Sham equation [20]), then results obtained by the exact solution of the Schrödinger equation would be independent of the form of the pseudopotential [4]. But, factually, in calculations using the model pseudopotential within the lowest order approximations in $V_{\mathbf{q}}$, and taking into account the many-particle effects, this dependence takes place. Estimations of this dependence are important, for example, evaluating the predictability of the theory. Results, which are slightly sensitive to the form of $V_{\mathbf{q}}$, can, naturally, be considered as more accurate and reliable. On the other side, large sensitivity to the form of $V_{\mathbf{q}}$ demonstrates the insufficiency of the model used for the properties considered. This fact can also emphasize a necessity of improving the model using some *ab initio* estimations. As we will see below (see also Chapter 3, section 4), in the case of alkali metals, a simple calculation scheme (the second-order model) with the AHA pseudopotential can provide a rather accurate and complete description (excluding some properties of Li). The accuracy of this model, in many cases is, not smaller than the accuracy of present-day experimental measurements. Here, we will provide comparative calculations using the pseudopotentials discussed above and compare the results with experimental data.

For any given type of $V_{\mathbf{q}}$, results of calculations can also depend on the choice of the fitting procedure for the pseudopotential parameters. The scheme used in [32, 41, 45–53] with the fitting in accordance with conditions (2.39) seems to be one of the most successful schemes in describing atomic properties. Here, we will use it for all pseudopotentials under consideration.

The dependence of results upon the approximation for the polarization operator $\pi(\mathbf{q})$ will be discussed below (see section 4). Comparison of five different approximations for $\pi(\mathbf{q})$: RPA [41], Geldart–Vosko (GV) [42], Toigo–Woodruff (TW) [43], Vashishta–Singwi (VS) [44] and Geldart–Taylor (GT) [39, 40], shows that the last approximation provides the best results for all alkali metals except lithium. For lithium, in the frame of the local scheme, more accurate results were obtained using the simple GV approximation [42] (which probably corresponds to some imitation of the non-locality effects in the pseudopotential). The results for the TW and VS approximations for all metals lie between the results for the GV and GT approximations. For this reason, in the present subsection, we will use predominantly the GV approximations for lithium and the GT approximation for other metals.

Computational formulae of the second-order model are written as the sums of expressions containing squares of $V(\boldsymbol{\tau})$ or $V(\boldsymbol{\tau}+\mathbf{k})$ over the reciprocal lattice $\boldsymbol{\tau}$. The summation for the AHA ($\xi = 0.03$) and AHA-2 ($\xi = 0.06$) pseudopotentials (2.36) was spread to $\tau_{\max} \geq 10\,k_{F0}$; for the Krasko–Gurskij (2.44) and Harrison (2.42) pseudopotentials and to $22\,k_{F0}$, and for the HA pseudopotential (2.36) with $\xi = 0$ to $30\,k_{F0}$ (the last pseudopotential corresponds to the slowest convergence of summation as a consequence of the "step-like" variation of $V(\mathbf{r})$ at $r = r_0$). These calculations provide accuracy up to five or six decimal digits for the energy, the elastic moduli B_{ik}, the phonon frequencies $\omega_{\lambda\mathbf{k}}$, and not worse than three decimal digits for the Grüneisen parameters $\gamma_{\lambda\mathbf{k}}$ and the volume derivatives $dB_{ik}/d\Omega$.

TABLE 2.4 Parameters of the Krasko–Gurskij (2.44), Harisson (H) (2.42), Heine–Abarenkov (HA) (2.37) model potentials (in a.u.), fitted to conditions (2.39): $\rho(\Omega_0) = 0$ and $C_{44}(\Omega_0) = (C_{44})_{\text{exp}}$

pseudo-potential	approx. of $\pi(\mathbf{q})$	paramets, a.u.	Li	Na	K	Rb	Cs
KG	GV	r_c	0.3536	0.4898	0.6800	0.7490	0.8440
		a	3.210	3.242	2.963	2.802	2.586
	GT	r_c	0.3612	0.5008	0.6993	0.7719	0.8647
		a	3.101	3.021	2.711	2.544	2.385
H	GV	r	0.3376	0.4633	0.6396	0.7025	0.7885
		β	11.78	22.30	39.53	45.64	54.11
	GT	r	0.3415	0.4724	0.6552	0.7208	0.7980
		β	11.72	21.92	38.70	44.66	52.10
HA	GT a_0	r_0	1.533	2.175	3.106	3.486	4.016
	(Li-GV)	U_0	0.2250	0.2100	0.1968	0.1961	0.19
AHA-2	GT	r_0		2.1691	3.067	3.422	3.909
		U_0		0.2266	0.2003	0.1961	0.1880

TABLE 2.5 Static properties of sodium at $T = 0$: the sublimation energy E_{subl}, the Fuchs moduli B_{ik} (see Section 1.2) and their derivatives, the Debye temperature θ_D (2.55), and the macroscopic Grüneisen parameter γ_0 (2.57)

approx. for $\pi(\mathbf{q})$	pseudo-potential	E_{subl}, 10^{-3}Ry/at.	B_{11}, kbar	B_{dyn} kbar	B_{33}, kbar	B_{44}, kbar	$\frac{dB_{11}}{dp}$	$\frac{d\ln B_{33}}{d\ln d\Omega}$	$\frac{d\ln B_{44}}{d\ln \Omega}$	θ_D, K	γ_0
	KG	86	73.6	111	7.82	62.2	3.7	2.3	1.6	154.5	0.90
	H	86	73.5	108	7.78	62.2	3.6	2.3	1.7	154.3	0.91
GT	HA	87	73.8	97	7.55	62.2	3.7	2.5	1.7	153.4	0.97
	AHA	88	74.4	93	7.25	62.2	3.8	2.6	1.8	152.0	1.01
	AHA–2	89	75	91	7.20	62.2	3.8	2.9	1.8	152.1	1.06
Exper.		83	78		7.74	62.2	3.74	2.7	1.5	152	
		±0.5	±4		±0.3	±1.5	±0.3	±0.2	±0.1	±2	
					7.2						
					±0.1						

TABLE 2.6 Equation of state $p(\Omega)$ (in kbar) at $T = 0$

approx. of $\pi(\mathbf{q})$	pseudo-potential	$\Delta V/V_0$		
		0.1	0.3	0.5
	KG	9.41	50.1	178
GT	H	9.38	50.2	179
	HA	9.46	51.1	185
	AHA	9.52	51.9	187

Table 2.4 shows parameters of these form-factors obtained by using (2.39). Parameters of the AHA potentials are listed in Tables 2.1 and 2.2.

Table 2.5 demonstrates the results of the calculations of chemical properties of sodium at $T = 0$: the sublimation energy E_{subl} (which is treated as the difference between the calculated cohesive energy and the experimental value of the ionization potential E_I), the elastic moduli B_{ik} and their derivatives with respect to the volume, the Debye temperature θ_D and the low temperature (lattice) macroscopic Grüneisen parameter γ_0. Values θ_D and γ_0 are determined by B_{ik} and $dB_{ik}/d\Omega$:

$$\theta_D = \frac{2\pi\hbar}{k_B a}\left[\frac{2}{9}\sum_{\lambda=1}^{3}\int v_\lambda^{-3}(\mathbf{n})dO_{\mathbf{n}}\right]^{-1/3} \tag{2.55}$$

$$\gamma_0 = \frac{1}{3}\sum_\lambda \gamma_\lambda(\mathbf{q}\to 0);\quad \gamma_\lambda(\mathbf{q}\to 0) = -\frac{1}{2}\left[1 + \frac{d\ln B_\lambda(\mathbf{n})}{d\ln\Omega}\right], \tag{2.56}$$

where squares of the sound velocities $v^2(\mathbf{n})$ are eigenvalues of the sound propagation matrix

$$A_{\alpha\gamma} = (2a^3/M)\mathcal{B}_{\alpha\beta\gamma\delta}n_\beta n_\delta;\quad \mathbf{n} = \frac{\mathbf{q}}{q}, \tag{2.57}$$

$\mathcal{B}_{\alpha\beta\gamma\delta}$ are the Birch elastic moduli (see section 2, Chapter 1), which determine the Fuchs elastic moduli B_{ik}; values $B_\lambda(\mathbf{n})$ denote combinations of values $\mathcal{B}_{\alpha\beta\gamma\delta}$, which are obtained from the sound velocities of different modes $C_\lambda(\mathbf{n})$ (1.86), if we replace values C_{ijkl} by $\mathcal{B}_{\alpha\beta\gamma\delta}$; a is the lattice constant, m is the atomic mass.

Results of calculations of the equation of state $p(\Omega)$ for all metals (except cesium) demonstrate very small sensitivity to the form $V(\mathbf{q})$ and $\pi(\mathbf{q})$. For this reason, we show in Table 2.6 results for Na only, which demonstrate this small sensitivity.

Very important characteristics of the phonon spectra $\omega_{\lambda\mathbf{q}}$ of alkali metals are the values of $\omega_{\lambda\mathbf{q}}$ in the high-symmetry points of the Brillouin zone. For this reason, in Table 2.7, we present results for these points only. The accuracy of the experimental results is given in units the last decimal digit. Table 2.8 presents results for the thermal expansion. This table gives accuracy of calculations of the Grüneisen parameter, averaged over the spectra.

Table 2.9 presents results for the differences between values of the cohesive energy ΔE and the cell volume $\Delta\Omega_0$ calculated for different structures. The value of ΔE

TABLE 2.7 Phonon frequencies for the high-symmetry points in the Brillouin zone of sodium (in units of the plasma frequency $\omega_p = (4\pi e^2/m\Omega)^{1/2}$)

approx. of $\pi(\mathbf{q})$	wave vector, $(2\pi/a)$	$\frac{1}{2},\frac{1}{2},0$			$\frac{1}{2},\frac{1}{2},\frac{1}{2}$	1,0,0
		branch				
	pseudo-potential	N_1'	N_2'	N_4'	P_4	H_{15}
	KG	0.631	0.372	0.136	0.469	0.555
	H	0.624	0.370	0.132	0.464	0.551
GT	HA	0.600	0.370	0.132	0.450	0.543
	AHA	0.590	0.369	0.125	0.442	0.537
	AHA-2	0.585	0.369	0.125	0.439	0.534
experiment		0.536	0.360	0.131	0.404	0.503
		±10	±7	±3	±6	±6

TABLE 2.8 Thermal expansion of sodium $\Delta V(T)/V_0$ (in %)

approx. for $\pi(\mathbf{q})$	T, K	$k_B T/\hbar\omega_p$	pseudopotential				experiment
			KG	H	HA	AHA	
	140	0.41	1.31	1.34	1.48	1.51	1.43
GT	270	0.79	3.41	3.49	3.84	3.91	4.01

determines stability of the structure and the temperature of possible martensitic transformations. The sign and the value of $\Delta\Omega_0$, as will be discussed below (Chapter 4), are very important for describing the phase transitions under pressure. The results from Table 2.9 will be discussed in detail in sections 1 and 2 of Chapter 4 with reference to the theory of martensitic transformations in Li and Na and polymorphic transformations in K, Rb and Cs. Here, we discuss only the main results relating to the choice of the pseudopotential. First, we see that the best approximations for calculating all properties were the GT approximations for $\pi(\mathbf{q})$ and the form-factor $V_\mathbf{q}$ in the AHA form (2.36). The same approximations also provide correct results for sequences of phases for Li and Na ($E_{HCP} < E_{BCC}$). This prediction agrees with the HCP phase observed in low-temperature experiments [62, 63] (see also Chapter 4). Only with increasing T are these metals transformed into the BCC phase. The same results also provide the GV approximation for $\pi(\mathbf{q})$. This fact confirms our statement that the GV approximation simulates the non-locality effects in Li.

In the case of heavy alkali metals K, Rb and Cs, the AHA pseudopotential with the GT approximation also provides a correct description of the experimentally

TABLE 2.9 Structure modifications: differences of energies and atomic volumes

metal	$\pi(\mathbf{q})$	$\Delta E = E - E_{BCC}$, 10^{-5} Ry/at.								$10^3(\Omega/\Omega_{BCC} - 1)$				
	PP	GV			GT					GV	GT			
		KG	H	AHA	KG	H	HA	AHA	AHA-2	H	KG	H	HA	AHA
Li	FCC	−7.1	−7.5	−8.8	−7.6	−7.9		−9.3		1.7	2.2	2.2		2.0
	HCP	−4.2	−7.4	−9.6	−5.7	−8.7		−10.9		2.0	2.6	2.8		2.7
Na	FCC	0.0	−0.4	−3.3	0.9	0.5	−0.8	−2.4	−2.9	1.8	1.9	2.0	2.2	1.9
	HCP	3.1	0.5	−3.1	3.3	1.2	−0.4	−2.5	−3.1	2.0	2.2	2.4	2.6	2.4
K	FCC	3.3	2.9	−0.2	4.8	4.5	3.2	2.3	3.7	1.8	1.8	1.9	2.2	2.6
	HCP	5.4	4.0	0.1	6.5	5.9	3.2	2.5	4.0	2.1	1.9	2.2	2.6	3.0
Rb	FCC	3.3	3.0	0.7	4.9	4.6	3.6	3.3	5.2	1.8	1.8	1.9	2.3	2.9
	HCP	5.0	4.2	1.0	6.2	6.2	4.0	3.6	5.4	2.2	1.9	2.3	2.7	3.3
Cs	FCC	3.0	2.8	1.9	1.0	2.2	3.9	4.6	6.5	1.9	−6.5	−3.1	2.9	3.6
	HCP	4.3	4.3	2.1	3.0	4.5	4.3	4.8	6.8	2.2	−3.7	−0.9	3.6	4.4

observed sequences of phases for these metals. At $T = 0$ the lowest energy corresponds to the BCC phase in accordance with experimental data. With increasing T from 0 to the melting temperature T_m, there are no phase transitions in heavy alkali metals. However, with increasing pressure in Cs and Rb, there are phase transitions from the BCC to the FCC phase [38]. In accordance with Table 2.9, the FCC phase corresponds to the lowest value of the energy after the BCC phase.

Tables 2.4–2.9 show that, generally speaking, all properties under consideration are not very sensitive to the form of the potential used. Especially close are results for static properties (Table 2.5). Naturally, this fact reflects the method of the fitting of the parameters of the potential which uses these properties. However, the sensitivity to the form of the pseudopotential increases in the next sequence: static properties — phonon spectra — Grüneisen parameters — thermal expansion — differences between values of the energy for different structures.

As we can see from these tables (it will also be discussed below), for Na (and other metals except Li), the accuracy of calculations actually increases in the next sequence of the pseudopotentials: KG–H–HA–AHA. This "accuracy hierarchy" qualitatively agrees with the hierarchy obtained in discussion of the "smoothing degree" of the electron density as the condition providing the best convergence of series of the perturbation theory in the pseudopotential [60]. At the same time, replacing the AHA with the AHA-2 approximation actually provides a worse description (with an exception, maybe, of the case of sodium). Hence, at least for K, Rb and Cs, the value $\xi = 0.03$ for the AHA pseudopotential seems to be near the optimum value from the point of view of describing atomic properties.

The results for the KG (2.44) and H (2.42) pseudopotentials actually are very close together. The difference between the results for the HA and AHA (and AHA-2) pseudopotentials also seems to be small. However, for some values, which are very sensitive to the form of the pseudopotential (not only for ΔE in Table 2.8, but also for B_{33} (1.112), $dB_{33}/d\Omega$ (1.117), and for values from Table 2.5 relating to these two values: θ_D, γ_0, and also for the value of γ in the N_4' point, and so on), this spread of results is not as small as should be expected if we take into account small values of the cut-off factor ξ in (2.36). As we have already mentioned, the values of B_{33} and $dB_{33}/d\Omega$ demonstrate a maximal sensitivity to the form of the pseudopotential in comparison with other static properties. Data shown in Table 2.5 provide a conclusion that the accuracy of calculations increases by replacing the KG with the AHA pseudopotential as well as by replacing AHA-2 with AHA. The same fitting procedure for the pseudopotential parameters was also used in calculating higher branches of the phonon spectra. We see that for all pseudopotentials, the results of the calculations are located higher than the experimental data. However, Table 2.7 shows that this difference decreases (in factor 1.5–2) in the same sequence of the pseudopotentials: from KG to HA.

According to [32, 46], we can suppose that the thermal expansion in Na can be described with high enough accuracy using the AHA pseudopotential. Table 2.8 demonstrates increasing accuracy in the sequence of the pseudopotentials from KG to AHA.

The most interesting seems to be the results relating to the difference between energies of different structures ΔE (see Table 2.9). For Na, the stability of the HCP structure observed at $T = 0$ (in the lowest approximations in anharmonicity, i.e., without contributions of zero oscillations), is reproduced by AHA and by AHA-2 pseudopotential, too. We also show that the values of ΔE can change rapidly by a relatively small variarion of the pseudopotential, for example, by transition from the HA to AHA form in the case of sodium and from the AHA to AHA-2 for K and Rb. For this reason, it is not surprising that there exists a large spread of results of calculations of ΔE and the temperatures of the martensitic transformation in Li and Na (see Chapter 4). Many of them also disagree with experimental data. Thus, for describing properties which relate to the structural phase transitions, the accuracy of the pseudopotential (and the model) must be much higher than for other properties.

In conclusion, we briefly consider here the results of calculations of properties of heavy alkali metals while taking into account the core interaction energy $E_p = E_{sr} + E_w$, which includes the Born–Mayer repulsive term E_{sr} and the van der Waals interaction E_w. These terms will be discussed in Chapter 3 (section 7). In accordance with these results, parameters of the AHA pseudopotential demonstrate no appreciable renormalization. There exists partial cancellation of contributions E_{sr} and E_w, especially in calculations of higher derivatives of the energy. Results for Cs obtained in section 3.7 are appreciably worse than results of calculations without taking into account the energy E_p [38, 48]. Apparently, this fact is a consequence of the unsuccessful approximations of parameters of E_p used in section 3.7. We will return to this question in Chapter 3, but here we note that, in [38], some more accurate approximation for E_{sr} for Cs has been proposed. Note also that parameters of E_{sr} for K and Rb used in section 3.7 were calculated in [59] without taking into account the E_w term, i.e., at $E_p = E_{sr}$.

2.2.6 *The third and fourth-order models with a priori external potential*

To describe static properties of polyvalent metals and, apparently, Li, it is necessary to escape the limit of the second-order approximation in the pseudopotential. Such attempts were made in [5, 6, 29], but in these works, calculations of the third-order contribution in the pseudopotential were performed in the framework of the ring diagram approximation. However, many-electron effects must play an important role not only in the second-order contribution to the energy $E_e^{(2)}$, but also in the next terms $E_e^{(n)}$ in the expansion of the energy in the pseudopotential (Mn models). This problem was discussed in [64], where the exchange and correlation contributions to $E_e^{(3)}$ and $E_e^{(4)}$ were analyzed and some simple, but adequate for typical metallic densities, approximations were proposed. To evaluate the number of terms needed in the series expansion of E (the order of the model), it is necessary to provide calculations of metal properties while taking into account the higher-order contributions (higher than $E_e^{(2)}$). Using these numerical estimations, we can evaluate how important many-electron effects are for each high-order contribution.[3]

[3] In [87, 88], the third-order model with asymptotic values for $\tilde{\gamma}_6(0.0.0)$ was used. This model is an alternative to the model described in the present subsection. It will be discussed in detail in the next subsection.

In Chapter 3, some static characteristics (the cohesive energy, the pressure, and the second- and third-order elastic modules) will be calculated for the sequences of simple metals Al, Pb, Li, Cd, Na with variation of r_s from 2 to 4 (see Table 2.1). These calculations use the local pseudopotentials obtained from the non-local *ab initio* pseudopotentials: Heine–Abarenkov [10, 11] – $V_{HA}(\mathbf{q})$; and Rasolt–Taylor [12–14] – $V_{RT}(\mathbf{q})$. These local form-factors have no fitting parameters. The localization was made in the framework of the on-Fermi-sphere approximation (2.30), as well as by using the condition (2.32) for the electron density. We will discuss these form-factors in detail in Subsection 8. Here, we consider some models of the adiabatic potentials which include the third- and fourth-order terms in the parameter $\eta = V(\boldsymbol{\tau})/E_F$, i.e., the terms $E_e^{(3)}$ and $E_e^{(4)}$ in addition to the adiabatic potential of the second-order model (2.46). We will note these models as the third- and fourth-order models.

In [95] the theory was constructed in which many-electron effects are taken into account as completely as possible. But from the point of view of practical calculations, these expressions for $E_e^{(3)}$ and $E_e^{(4)}$ are extremely sophisticated and concrete calculations with these formulae would be very difficult. For this reason, we consider here an approximation scheme for taking into account the many-electron effects. The many-electron effects providing the vertex $\lambda_1(\mathbf{q})$ and participating in the polarization operator $\pi(\mathbf{q})$ will be taken into account completely. For all remaining terms, we restrict our consideration to the first order in the screened-Coulomb potential $w_c(\mathbf{q})$. In the framework of this scheme, the following diagrams must be included in the thermodynamic potential (in addition to the second-order contribution E (2.46)), which are

$E_e^{(3)} + E_e^{(4)} \approx$

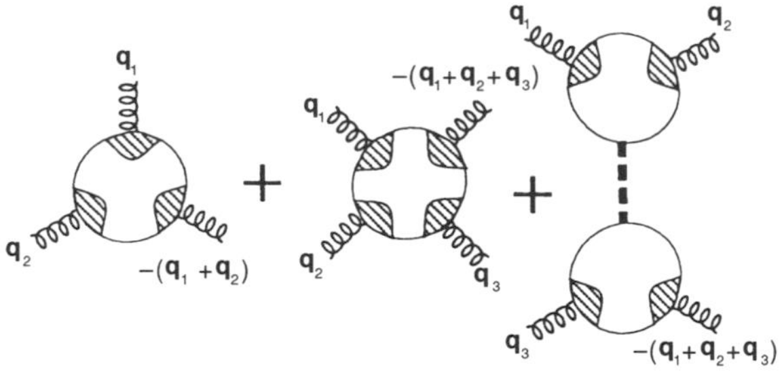

(2.58)

Definitions of lines in these diagrams were described in [95]. Here, we briefly recall that the helix denotes the linear-screened pseudopotential $W(\mathbf{q})$, the shaded segment is the local vertex $\lambda_1(q)$, the bold dashed line corresponds to $w_c(\mathbf{q})$, the full line is the electron Green's function $G_{\mathbf{k}}^{(0)}$. We emphasize that, in the first order in w_c and neglecting the exchange effects, there are no other diagrams, since the irreducible vertex functions $\tilde{\gamma}_6$ and $\tilde{\gamma}_8$ appear only starting with the third-order terms in w_c.

As was already mentioned, calculations of metallic properties, including of $E_e^{(3)}$ and $E_e^{(4)}$, are very complicated as a consequence of additional summations over the reciprocal lattice. For problems considered here, there is no reason to make the calculation scheme very complicated by including exchange and correlation effects in (2.58). As it was shown in previous sections and will be also shown in Chapters 3 and 4, these effects provide contributions $\sim 5\%$ only. These effects are important for describing very fine effects, like phase transitions, which we do not consider here. The aim of the present chapter is to obtain a model which provides an adequate description of atomic properties of single-phase systems.

The contribution of the last diagram in (2.58) is cancelled by the contribution of $E_{\mu}^{(4)}$ (2.74) as $\mathbf{q}_1 + \mathbf{q}_2 \to 0$. Here, we do not consider either of these terms at all.

Approximation (2.58) is equivalent to the next analytical representation (see [53, 65]).

$$E_{e0}^{(3)} = \Omega \sum_{\boldsymbol{\tau}_1\boldsymbol{\tau}_2\boldsymbol{\tau}_3} \Lambda_0^{(3)}(\boldsymbol{\tau}_1, \boldsymbol{\tau}_2, \boldsymbol{\tau}_3) \frac{V(\boldsymbol{\tau}_1)}{\tilde{\varepsilon}(\boldsymbol{\tau}_1)} \frac{V(\boldsymbol{\tau}_2)}{\tilde{\varepsilon}(\boldsymbol{\tau}_2)} \frac{V(\boldsymbol{\tau}_2)}{\tilde{\varepsilon}(\boldsymbol{\tau}_3)} \Delta(\boldsymbol{\tau}_1 + \boldsymbol{\tau}_2 + \boldsymbol{\tau}_3), \tag{2.59}$$

$$\begin{aligned} E_{e0}^{(4)} = \Omega \tilde{\sum_{\boldsymbol{\tau}_1 \cdots \boldsymbol{\tau}_4}} \Big[& \Lambda_0^{(4)}(\boldsymbol{\tau}_1, \boldsymbol{\tau}_2, \boldsymbol{\tau}_3, \boldsymbol{\tau}_4) + \frac{9}{2} \Lambda_0^{(3)}(\boldsymbol{\tau}_1, \boldsymbol{\tau}_2, -\boldsymbol{\tau}_1 - \boldsymbol{\tau}_2) \\ & \times w_c(\boldsymbol{\tau}_1 + \boldsymbol{\tau}_2) \Lambda_0^{(3)}(\boldsymbol{\tau}_1 + \boldsymbol{\tau}_2, \boldsymbol{\tau}_3, \boldsymbol{\tau}_4) \Big] \frac{V(\boldsymbol{\tau}_1)}{\tilde{\varepsilon}(\boldsymbol{\tau}_1)} \frac{V(\tau_2)}{\tilde{\varepsilon}(\boldsymbol{\tau}_2)} \frac{V(\boldsymbol{\tau}_2)}{\tilde{\varepsilon}(\boldsymbol{\tau}_3)} \frac{V(\boldsymbol{\tau}_4)}{\tilde{\varepsilon}(\boldsymbol{\tau}_4)} \\ & \times \Delta(\boldsymbol{\tau}_1 + \boldsymbol{\tau}_2 + \boldsymbol{\tau}_3 + \boldsymbol{\tau}_4) + E_{an}. \end{aligned} \tag{2.60}$$

Here $\tilde{\varepsilon}(\boldsymbol{\tau})$ is the effective dielectric function. The symbol "tilde" in the sum in (2.60) means that, in the second term in the square brackets, all terms with $\boldsymbol{\tau}_1 + \boldsymbol{\tau}_2 = 0$ and $\boldsymbol{\tau}_3 + \boldsymbol{\tau}_4 = 0$ must be omitted. As we have already mentioned, these terms are cancelled by $\Delta E_{\mu}^{(4)}$. In the first term in the square brackets only, the contributions with non-equal pairs of vectors $\boldsymbol{\tau}_1$, $\boldsymbol{\tau}_2$ and $\boldsymbol{\tau}_3$ are kept. The terms with pairs of equal vectors $\boldsymbol{\tau}_i$ are collected in the "anomalous" contribution E_{an}, with the next diagrammatic representation

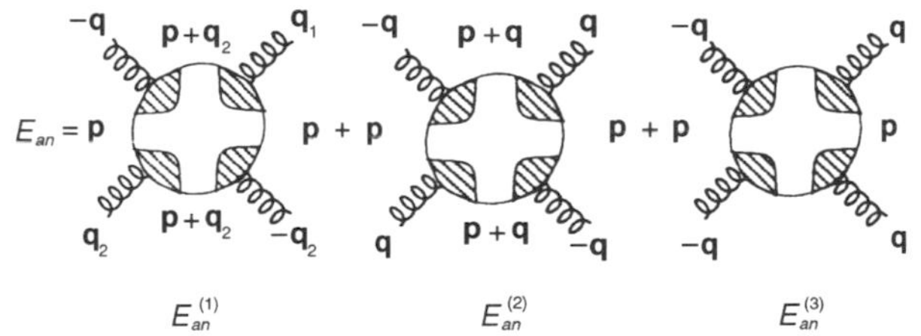

(2.61)

Diagrams (2.61) contain parts of the Green's function with coinciding arguments. To calculate the contribution of these diagrams to the energy E, it is necessary to consider these diagrams in the framework of the Matsubara technique at $T \neq 0$ and further consider the limit $T \to 0$ [5]. In calculating E_{an}(2.61), we will use the following relations [66]

$$\lim_{T\to 0} k_B T \sum_{l=-\infty}^{\infty} (i\hbar\omega_l - \varepsilon)^{-1} = n(\varepsilon) = \begin{cases} 1, & \text{if } \varepsilon \leq 0 \\ 0, & \text{if } \varepsilon > 0 \end{cases} \tag{2.62}$$

$$\lim_{T\to 0} k_B T \sum_{l=-\infty}^{\infty} (i\hbar\omega_l - \varepsilon)^{-2} = \delta(\varepsilon). \tag{2.63}$$

The term E_{an} (obtained by applying the Matsubara technique to the representation (2.61) and by using (2.62) and (2.61), see [65, 67]) consists of the following contributions:

$$E_{an}^{(1)} = \frac{1}{2}\sum_{\mathbf{p},\mathbf{q},\mathbf{k}} \left\{ \frac{-\delta(\varepsilon_{\mathbf{p}} - \mu_0)}{(\varepsilon_{\mathbf{p}} - \varepsilon_{\mathbf{p+q}})(\varepsilon_{\mathbf{p}} - \varepsilon_{\mathbf{p+k}})} + \frac{n(\varepsilon_{\mathbf{p}})}{(\varepsilon_{\mathbf{p-q}} - \varepsilon_{\mathbf{p}})(\varepsilon_{\mathbf{p}} - \varepsilon_{\mathbf{p+k}})^2} \right.$$

$$+ \frac{n(\varepsilon_{\mathbf{p}})}{(\varepsilon_{\mathbf{p+k}} - \varepsilon_{\mathbf{p}})(\varepsilon_{\mathbf{p}} - \varepsilon_{\mathbf{p+q}})^2} + \frac{n(\varepsilon_{\mathbf{p+q}})}{(\varepsilon_{\mathbf{p+q}} - \varepsilon_{\mathbf{p+k}})(\varepsilon_{\mathbf{p}} - \varepsilon_{\mathbf{p+q}})^2}$$

$$\left. + \frac{n(\varepsilon_{\mathbf{p+q}})}{(\varepsilon_{\mathbf{p+k}} - \varepsilon_{\mathbf{p+q}})(\varepsilon_{\mathbf{p}} - \varepsilon_{\mathbf{p+k}})^2} \right\} W^2(\mathbf{q}) W^2(\mathbf{k}), \tag{2.64}$$

$$E_{an}^{(2)} = \frac{1}{2}\sum_{\mathbf{p},\mathbf{k}} \left\{ -\frac{\delta(\varepsilon_{\mathbf{p}} - \mu_0) + \delta(\varepsilon_{\mathbf{p+k}} - \mu_0)}{(\varepsilon_{\mathbf{p}} - \varepsilon_{\mathbf{p+k}})^2} \right.$$

$$\left. + \frac{2[n(\varepsilon_{\mathbf{p+q}}) - n(\varepsilon_{\mathbf{p}})]}{(\varepsilon_{\mathbf{p}} - \varepsilon_{\mathbf{p+k}})^3} \right\} W^4(\mathbf{k}) \tag{2.65}$$

$$E_{an}^{(2)} = \frac{1}{2}\sum_{\mathbf{p},\mathbf{k}} - \left\{ \frac{\delta(\varepsilon_{\mathbf{p}} - \mu_0)}{(\varepsilon_{\mathbf{p}} - \varepsilon_{\mathbf{p+k}})(\varepsilon_{\mathbf{p}} - \varepsilon_{\mathbf{p-k}})} + \frac{2n(\varepsilon_{\mathbf{p}})}{(\varepsilon_{\mathbf{p+k}} - \varepsilon_{\mathbf{p}})(\varepsilon_{\mathbf{p}} - \varepsilon_{\mathbf{p-k}})^2} \right.$$

$$\left. + \frac{2n(\varepsilon_{\mathbf{p+k}})}{(\varepsilon_{\mathbf{p+k}} - \varepsilon_{\mathbf{p-k}})(\varepsilon_{\mathbf{p}} - \varepsilon_{\mathbf{p+k}})^2} \right\} W^4(\mathbf{k}). \tag{2.66}$$

Here, $n(\varepsilon_{\mathbf{p}})$ is the occupation number for the Fermi distribution of electrons with the energy $\varepsilon_{\mathbf{p}}$, μ_0 is the chemical potential of the free electron gas. The pseudopotential $W(\mathbf{q})$ is the linearly screened potential, which includes the structure factor $S(\mathbf{q})$, because the summation is performed over arbitrary values of $\mathbf{p}, \mathbf{q}, \mathbf{k}$.

The terms with $\delta(\varepsilon_{\mathbf{p}} - \mu_0)$ are given by deformations of the Fermi-sphere which, as we already mentioned, are important starting with terms $\sim V_{\mathbf{q}}^4$. This statement can be proved either by using the band structure approach described [68], or by replacing the exact Fermi-momentum $k_F(\theta, \varphi)$ by $k_{FO}[1 + \Delta(\theta)]$, [69] ($k_{FO}$ is the radius of the Fermi-sphere and the function $\Delta(\theta)$ depends on the polar angle θ only). In particular, the term $E_{an}^{(2)}$ (2.65) provides the contribution of the order of V_q^4 to the series expansion of Ω_e [70], i.e., it is a contribution to E corresponding to the coherent reconstruction of the band structure provided by a Bragg plane.

All integrals over $\mathbf{p}$ in (2.64)–(2.66) can be calculated exactly because even the most complicated of these, $E_{an}^{(1)}$, depends on two vectors $\mathbf{q}$ and $\mathbf{k}$ only. To calculate these integrals, we can use the methods described in [71]. For example,

$$\sum_{\mathbf{p}} \frac{\delta(\varepsilon_{\mathbf{p}} - \mu_0)}{(\varepsilon_{\mathbf{p}} - \varepsilon_{\mathbf{p+q}})(\varepsilon_{\mathbf{p}} - \varepsilon_{\mathbf{p+k}})} = \frac{4\Omega k_{F0} m^3}{(2\pi)^3 q^2 k^2 \hbar^6} \int \frac{dO_{\mathbf{p}}}{(1+\mathbf{r l}_1)(1+\mathbf{r l}_2)} = \frac{4\Omega k_{F0} m^3}{(2\pi\hbar^2)^3 q^2 k^2} \Phi_1, \tag{2.67}$$

where

$$\mathbf{r} = \mathbf{p}/p; \quad \mathbf{l}_1 = \frac{2k_{F0}}{q}\frac{\mathbf{q}}{q}; \quad \mathbf{l}_2 = \frac{2k_{F0}}{k}\frac{\mathbf{k}}{k};$$

$$\Phi_1 = \begin{cases} \frac{1}{2zc}\left\{ \ln\left|\frac{z+1-(a/2)}{z-(a/2)}\right| - \ln\left|\frac{z+1+(a/2)}{z+(a/2)}\right| \right\}; & d > 0 \\ \frac{1}{zc}\left\{ \arctan\left(-\frac{a}{2z}\right) - \arctan\left(\frac{2-a}{2z}\right) \right\}; & d < 0, \end{cases} \tag{2.68}$$

$$z^2 = |d|; \; d = c^{-2}\left[c - \mathbf{l}_1^2\mathbf{l}_2^2 + (\mathbf{l}_1\mathbf{l}_2)^2\right];$$
$$d = b/c; \; b = 2(\mathbf{l}_1^2 - \mathbf{l}_1\mathbf{l}_2);$$
$$c = (\mathbf{l}_1 - \mathbf{l}_2)^2.$$

Similarly, but more easily, the integrals with δ-functions participating in $E_{an}^{(2)}$ (2.65) and in $E_{an}^{(3)}$ (2.66) can be calculated.

Two remaining contributions to $E_{an}^{(2)}$ and $E_{an}^{(3)}$ provided by deformation of the Fermi-surface can be expressed using the next integral

$$\sum_{\mathbf{p}} \left[\frac{1}{\varepsilon_{\mathbf{p}} - \varepsilon_{\mathbf{p+k}}} + \frac{1}{\varepsilon_{\mathbf{p}} - \varepsilon_{\mathbf{p-k}}}\right] \frac{1}{\varepsilon_{\mathbf{p}} - \varepsilon_{\mathbf{p+k}}} - \frac{\Omega m}{2\pi^2\hbar^6 k_{F0} x^3}\left[\frac{4x}{1-x^2} - \ln\left|\frac{1+x}{1-x}\right|\right], \tag{2.69}$$

where $x = k/2k_{F0}$.

Returning to equations (2.59) and (2.60), note that the explicit expression for the three-pole $\Lambda_0^{(3)}$ in the ring diagram approximation is known (see [5])

$$\Lambda_0^{(3)}(\mathbf{q}_1, \mathbf{q}_2, \mathbf{q}_3) = \frac{2m_e^2 q_R^2}{3\pi^2\hbar^4 q_1 q_2 q_3}\left\{ \sum_m \cos\theta_m \ln\left|\frac{2k_{F0} + q_m}{2k_{F0} - q_m}\right| - \Delta\begin{bmatrix} \ln|(1-\Delta A)/(1+\Delta A)|, & k_{F0} \le q_R \\ 2\arctan(\Delta A), & k_{F0} > q_R \end{bmatrix}\right\} \tag{2.70}$$

where

$$\Delta = \sqrt{|(k_{F0}/q_R)^2 - 1|}, \ \cos\theta_i = -(\mathbf{q}_k\mathbf{q}_l)/q_k q_l,$$
$$A = q_1 q_2 q_3/\{(2k_{F0})^3[1 - (q_1^2 + q_2^2 + q_3^2)/8k_{F0}^2]\}, \tag{2.71}$$
$$\mathbf{q}_3 = -(\mathbf{q}_1 + \mathbf{q}_2), \quad q_R = q_1 q_2 q_3 \Big/ 2\sqrt{q_1^2 q_2^2 - (\mathbf{q}_1\mathbf{q}_2)^2}.$$

The indexes i, k, l in the formula for $\cos\theta_i$ (2.71) are the cyclic permutations of integers 1, 2, 3.

The four-pole $\Lambda_0^{(4)}(\mathbf{q}_1, \mathbf{q}_2, \mathbf{q}_3, \mathbf{q}_4)$, in the general case of non-equal $\mathbf{q}_1, \mathbf{q}_2, \mathbf{q}_3, \mathbf{q}_4$, is expressed through the one-dimensional integral [67], which cannot be expressed in terms of elementary functions. It provides some computational difficulties. For this reason, we, supposing that the local pseudopotential has been obtained from the non-local pseudopotential by using the localization procedure (2.32), also give averaging of $\Lambda_0^{(3)}$ and $\Lambda_0^{(4)}$ over all directions of $\mathbf{q}_i$, in the sense of the approximation (2.32).

Providing calculations of $\Lambda_0^{(4)}$ in the framework of the ring diagram approximation for arbitrary momenta $\mathbf{q}_1, \mathbf{q}_2, \mathbf{q}_3, \mathbf{q}_4$, we must consider integrals like ([5, 67])

$$\int \frac{d^3p}{(2\pi)^3} \frac{n_{\mathbf{p}}}{(\varepsilon_{\mathbf{p}} - \varepsilon_{\boldsymbol{\kappa}_1})(\varepsilon_{\mathbf{p}} - \varepsilon_{\boldsymbol{\kappa}_2})(\varepsilon_{\mathbf{p}} - \varepsilon_{\boldsymbol{\kappa}_3})} \tag{2.72}$$

where $\boldsymbol{\kappa}_n = \mathbf{p} + \mathbf{q}_1 + \cdots + \mathbf{q}_n$. This integral can be expressed through some one-dimensional integral, which cannot be calculated in terms of elementary functions. However, it is easy to see that the exact expression for (2.72) is not especially important, because $E_e^{(4)}$ includes only the triple sum of this term over vectors of the reciprocal lattice $\boldsymbol{\tau}_i$ (corresponding to values of $\mathbf{q}_i$ for the ideal lattice). Using this fact, the following approximation was substituted [65]

$$\sum_{\mathbf{p}} \prod_{j=1}^{n} (\varepsilon_{\mathbf{p}} - \varepsilon_{\boldsymbol{\kappa}_j})^{-1} \to \prod_{j=1}^{n} \sum_{\mathbf{p}} (\varepsilon_{\mathbf{p}} - \varepsilon_{\boldsymbol{\kappa}_j})^{-1}. \tag{2.73}$$

Below, we use

$$\sum_{\mathbf{p}} (\varepsilon_{\mathbf{p}} - \varepsilon_{\mathbf{p}+\mathbf{q}})^{-1} \sim \pi_0(\mathbf{q}), \tag{2.74}$$

where $\pi_0(\mathbf{q})$ is the polarization operator of the free-electron gas.

To provide an effective description of the influence of the electron–electron interaction on the ring diagrams, we replace in (2.73), (2.74) $\pi_0(\mathbf{q})$ with the whole polarization operator $\pi(\mathbf{q})$. For $\Lambda_0^{(3)}$ and $\Lambda_0^{(4)}$ in expressions (2.59) (2.60) for $E_e^{(3)}$ and $E_e^{(4)}$, this replacement corresponds to

$$\Lambda_0^{(4)}(\boldsymbol{\tau}_1, \boldsymbol{\tau}_2, \boldsymbol{\tau}_3, \boldsymbol{\tau}_4) \to \frac{1}{8n_0^2} \hat{S}\pi(\boldsymbol{\tau}_1)\pi(\boldsymbol{\tau}_2)\pi(\boldsymbol{\tau}_3), \tag{2.75}$$

$$\Lambda_0^{(3)}(\boldsymbol{\tau}_1, \boldsymbol{\tau}_2, \boldsymbol{\tau}_3) \to \frac{1}{4n_0} \hat{S}\pi(\boldsymbol{\tau}_1)\pi(\boldsymbol{\tau}_2), \tag{2.76}$$

where $\hat{S}$ is the symmetrization operator for indexes in $\boldsymbol{\tau}_i$ which reflects an exact symmetry of $\Lambda_0^{(3)}$ and $\Lambda_0^{(4)}$ with respect to permutations of arguments $\boldsymbol{\tau}_i$ [5, 67] . Here, we have taken into account that the expression for the energy includes a summation over $\boldsymbol{\tau}_i$. Sometimes in calculating $E_e^{(3)}$, we will use also the exact expression for $\Lambda_0^{(3)}$ (2.70), in which case this will be especially marked.

Finally, let us present the expression for $E_e^{(4)}$ while including (2.75) and E_{an} (2.61)–(2.69). Let us present the energy $E_e^{(4)}$ as the sum of two terms

$$E_{e0}^{(4)} = E_e^{(4F)} + E_e^{(4B)}. \tag{2.77}$$

The first term is provided by deformations of the Fermi-surface, the second one corresponds to the change of the dispersion law for electrons. Collecting all terms with $\delta(\varepsilon_{\mathbf{p}} - \mu_0)$ corresponding to the the Fermi-surface deformation, we obtain

$$E_e^{(4F)} = \frac{\Omega m^3 k_{F0}}{\pi \hbar^6} \left\{ 2 \sum_{\mathbf{p} \neq \mathbf{q}} \frac{\Phi_1}{p^2 q^2} [W(\mathbf{p}) W(\mathbf{q})]^2 \right.$$
$$\left. + \sum_{\boldsymbol{\tau}} \frac{1}{\tau^3} \left[\frac{2\tau}{(2k_{F0})^2 - \tau^2} + \frac{1}{\tau} \ln \left| \frac{2k_{F0} + \tau}{2k_{F0} - \tau} \right| \right] [W(\tau)]^4 \right\}. \tag{2.78}$$

Using approximations (2.75) and (2.76), we can calculate all other integrals in the expressions for E_{an}. It gives

$$E_c^{(4B)} = -\frac{\Omega}{16 n_0^2} \sum_{\mathbf{p},\mathbf{q},\mathbf{k}} \pi(\mathbf{p}+\mathbf{q}) [\pi(\mathbf{q}) \pi(\mathbf{p}+\mathbf{q}+\mathbf{k}) + \pi(\mathbf{q}) \pi(\mathbf{k})]$$
$$\times W(\mathbf{p}) W(\mathbf{q}) W(\mathbf{k}) W(\mathbf{p}+\mathbf{q}+\mathbf{k}) + \frac{\Omega}{32 n_0} \sum_{\mathbf{p}} w_c(\mathbf{p}) [\pi(\mathbf{p})]^2$$
$$\times \sum_{\mathbf{q} \neq \mathbf{p}} W(\mathbf{p}) W(\mathbf{p}+\mathbf{q}) [\pi(\mathbf{q}) + 2\pi(\mathbf{p}-\mathbf{q})]^2 + \frac{\Omega}{8 n_0^2} \sum_{\mathbf{p} \neq \mathbf{q}} \pi(\mathbf{p}) \pi(\mathbf{q})$$
$$\times [\pi(\mathbf{p}) - \pi(\mathbf{p}+\mathbf{q})][W(\mathbf{p}) W(\mathbf{q})]^2 + \frac{\Omega m^3}{3\pi^2 \hbar^6 k_{F0}^3} \sum_{\mathbf{k}} \Phi_4 [W(\mathbf{k})]^4; \tag{2.79}$$

$$\Phi_4 = \frac{1}{x^4} \left\{ \frac{1}{x} \ln \left| \frac{2+x}{2-x} \right| \quad \frac{1-x^2}{2x} \ln \left| \frac{1+x}{1-x} \right| \right\}. \tag{2.80}$$

In equations (2.74)–(2.80), $\mathbf{p}, \mathbf{q}, \boldsymbol{\tau}$ are vectors of the reciprocal lattice, n_0 is the mean electron density, $x = k/2k_{F0}$, and m is the electron mass.

Hence, in the framework of the fourth-order model the cohesive energy has the following form (the $M4_0$ model)

$$E = E_i + E_p + E^{zp} + E_e^{(0)} + E_e^{(1)} + E_e^{(2)} + E_e^{(3)} + E_e^{(4B)} + E_e^{(4F)}. \tag{2.81}$$

The first six terms in this formula, corresponding to the second-order models, were analyzed in Subsection 4 of this section. In Chapter 3, we present the results of calculations of static properties of Li and Al in the framework of the fourth-order model (2.81). These calculations were performed neglecting the contribution E_p, since, for these metals, the core contribution is small and has practically no influence on the metal properties at $p = 0$. The energy $E_0^{(3)}$ and its contribution to static properties were calculated using (2.59).

The parameters u and r_0 of the form-factor of the AHA pseudopotential (2.36) have been calculated for Li and Al using (2.39) for different approximations of the polarization operator (see Table 2.2).

2.2.7 *The third-order model (the pseudopotential with fitted parameters)*

As we have already mentioned, the most useful approximation within the pseudopotential approach is that based on the second order model (M2). A good qualitative description of a wide number of properties of alkali metals is achieved in the frame of this model (see Chapters 3 and 4). But attempts to apply this approximation to simple polyvalent metals (Ba, Al and so on) provides lesser success. It would be a consequence of relatively large values for $V_{\mathbf{q}}$ for polyvalent metals. The most important corrections to the second-order model in the framework of the local potential approximation include the main third-order in $V_{\mathbf{q}}$ contribution, as well as the non-analytical in $V_{\mathbf{q}}$ contributions which appear by the Fermi-surface approaching the face of the BZ. The effect of the third-order terms in $V_{\mathbf{q}}$ on the lattice properties of Na, Mg and Al was studied [5]. The influence of the third-order in $V_{\mathbf{q}}$ terms on static properties of alkali metals (from Li to Cs) and Ba and Al was considered [87] and the case of iridium was considered [88]. In this subsection, we follow the results of both these works and consider only the influence of $E_e^{(3)}$ on static properties and parameters of the pseudopotential.

The three-particle contribution to the metal energy can be written as (see [5]):

$$E^{(3)} = \Omega_0 N \sum_{\mathbf{g}_1, \mathbf{g}_2, \mathbf{g}_3} \Gamma^{(3)}(\mathbf{g}_1, \mathbf{g}_2, \mathbf{g}_3)\ V(\mathbf{g}_1)V(\mathbf{g}_2)V(\mathbf{g}_3) \times S(\mathbf{g}_1)S(\mathbf{g}_2)S(\mathbf{g}_3)\ \Delta(\mathbf{g}_1 + \mathbf{g}_2 + \mathbf{g}_3). \tag{2.82}$$

Here, $\Gamma^{(3)}$ is the three-vertex diagram which characterized the properties of the electron Fermi-liquid only, V is the non-screened potential of the electron–ion interaction, S is the structure factor, Ω is the volume per one atom in metal, and Δ denotes the delta function. The summation is spread over vectors of the reciprocal lattice $\mathbf{g}$. Formulae for the ring diagram approximation of the three-pole $\Gamma^{(3)}(\mathbf{q}_1, \mathbf{q}_2, \mathbf{q}_3)$ are given [4]. If we replace $\Gamma^{(3)}$ by $\Lambda^{(3)}$ in (2.82), then the non-screened form-factor of the pseudopotential is replaced by the screened one

$$W(\mathbf{q}) = \frac{V(\mathbf{q})}{\tilde{\varepsilon}(\mathbf{q})} S(\mathbf{q}).$$

Here

$$\tilde{\varepsilon}(\mathbf{q}) = \pi_0(\mathbf{q}) \frac{\varepsilon(\mathbf{q})}{\pi(\mathbf{q})}$$

is the effective dielectric function.

The method was proposed [89] which allows us to calculate the exchange and correlation corrections to the ring diagram approximation for $E_e^{(3)}$ (2.113), (2.114)

$$\Lambda^{(3)}(\mathbf{q}_1,\mathbf{q}_2,\mathbf{q}_3) = \lambda_1(\mathbf{q}_1)\,\lambda_2(\mathbf{q}_2)\,\lambda_3(\mathbf{q}_3)\Big[\Lambda_0^{(3)}(\mathbf{q}_1,\mathbf{q}_2,\mathbf{q}_3) + (1/3)\tilde{\gamma}_6(\mathbf{q}_1,\mathbf{q}_2,\mathbf{q}_3)\pi_0(\mathbf{q}_1)\pi_0(\mathbf{q}_2)\pi_0(\mathbf{q}_3)\Big]. \tag{2.83}$$

It is convenient to approximate the vertex function by (see [95])

$$\tilde{\gamma}_6(0,0,0) = \frac{1}{2}\frac{d}{dn}\left[\frac{1}{\pi_0(q\to 0)} - \frac{1}{\pi(q\to 0)}\right], \tag{2.84}$$

where n_0 is the electron density. In calculations of the elastic moduli B_{ik} in [87], the derivatives of the energy with respect to the deformations parameter have been obtained analytically (cumbersome expressions are given [88, 90], see also section 1, Chapter 1). The values B_{ik} were determined by applying the numerical differentiation procedure with respect to the respective shear parameters to the expression for the total energy [5] (see section 1, Chapter 1).

In a similar manner to the calculations in the beginning of the present chapter, all calculations here are made using the local Animalu–Heine–Abarenkov pseudopotential

$$V_{\mathbf{q}} = -\frac{4\pi Ze^2}{\Omega q^2}\left[(1+u)\cos(qr_0) + u\frac{\sin(qr_0)}{(qr_0)}\right]\exp\left(-\xi\frac{q^4}{16k_{F0}^4}\right). \tag{2.85}$$

Here, r_0 and u are the pseudopotential parameters. The exponential cut-off factor in $V_{\mathbf{q}}$ removes non-physical oscillations as $|q|\to\infty$ and improves convergence by summation over vectors of the reciprocal lattice. Introducing this factor corresponds to the "spreading" of the ion–core border by removing the gap at $r = r_0$.

The parameters r_0 and u are determined here by the next condition

$$p(\Omega_0) = 0; \quad \theta_D(\Omega_0) = \theta_D^{exp}(\Omega_0), \tag{2.86}$$

where Ω_0 is the equilibrium volume, p is the pressure, θ_D is the Debye temperature (2.55). In some cases, it was impossible to obtain any values which satisfy the system (2.86). In these cases, the parameters r_0 and u were determined by minimization of the function

$$f(r_0,u) = \alpha p^2(\Omega_0) + [\theta_D(\Omega_0) - \theta_D^{exp}(\Omega_0)]^2. \tag{2.87}$$

Here, the factor α determines the "relative weight" of the minimum pressure condition in comparison with the minimum of the second term. It was taken to be $\alpha = 100$ (if p is measured in units of 10^8 Pa, and θ_D in K) [87].

Results of calculations in the framework of the third-order model

$$E = E_i + E_e^{(0)} + E_e^{(1)} + E_e^{(2)} + E_e^{(3)} \tag{2.88}$$

depend rather appreciably on the value ξ in (2.85) (in actual calculations, ξ does not change). As we can see from Table 2.10 (see experimental data in Table 2.5 from

TABLE 2.10 ξ-dependence of static properties calculated in the framework of the third-order model $M3_\gamma$ with the AHA pseudopotential

ξ	0.03	0.06	0.09	0.10	0.15
lithium					
r_0 (a.u.)	1.3724	1.3671	1.0833	1.2221	1.4550
u	−0.0270	−0.0234	1.0342	0.3967	−0.2452
E(Ry)	−0.5390	−0.5382	−0.5331	−0.5358	−0.5408
$P(10^8$ Pa)	0.0	0.0	0.0	0.0	0.0
$B(10^8$ Pa)	136.1	139.4	141.9	145.3	148.3
$B_{33}(10^8$ Pa)	10.70	10.77	11.48	11.74	12.46
$B_{44}(10^8$ Pa)	113.6	119.1	133.6	130.1	122.5
$\theta_D(\theta_D^{(e)} = 342.5$ K)	324.3	328.7	342.5	342.5	342.5
sodium					
r_0 (a.u.)	1.8249	1.8185	2.0591	2.0995	2.2244
u	0.0421	0.0372	−0.3781	−0.4311	−0.5718
E(Ry)	−0.4534	−0.4536	−0.4571	−0.4578	−0.4597
$P(10^8$ Pa)	0.0	0.0	0.0	0.0	0.0
$B(10^8$ Pa)	70.51	74.83	77.12	77.26	77.42
$B_{33}(10^8$ Pa)	6.87	6.97	7.26	7.38	7.64
$B_{44}(10^8$ Pa)	61.80	65.24	62.37	61.72	59.44
$\theta_D(\theta_D^{(e)} = 152.5$ K)	149.7	152.5	152.5	152.5	152.5

Ref. 95), this dependence is large for Li and Na. The authors of [87] could not determine the parameters of (2.85) from the conditions (2.86) at $\xi \geq 0.09$ for Li and at $\xi \leq 0.03$ for Na. For these values of ξ, the pseudopotential parameters in Table 2.10 were determined by minimization of (2.87). At $\xi \geq 0.09$ for Li and $\xi \geq 0.06$ for Na, the results of calculations of elastic moduli are appreciably dependent on ξ. Similar results were obtained also for all other alkali metals and for metals like Ba and Al. The results of similar calculations for different values of ξ in the framework of the second-order model are close together [60]. In contrast to the case of the second-order model, the results, obtained in the framework of the third-order model, are very sensitive to the form of $V_{\mathbf{q}}$.

To evaluate the three-particle contribution, we set $\xi = 0.1$ for alkali metals, $\xi = 0.2$ for Ba, and $\xi = 0.3$ for Al. For this choice of ξ, the pseudopotential parameters r_0 and u (2.85) were determined from conditions (2.86). It allows us to investigate many alkali metals in the framework of the same general model. For Ba and Al, the sensitivity of the results of calculations to the variation of ξ in the vicinity of the choosen values of ξ turned out to be rather bad. The result of the calculations in the framework of the third-order model (2.88) are less sensitive to the approximation for $\pi(q)$ than to the variation of ξ. For this reason, we present here only the results for the GT approximation [39] (remember, that this approximation also provides the best results for the second-order model).

Calculations of elastic properties of alkali metals, Al, and Ba in the frame of the model discussed (the model M3γ) will be presented in section 7, Chapter 3. In this chapter we will also discuss and compare the results obtained in the framework of this third-order model (with $\tilde{\gamma}_6$).

2.2.8 Local form-factors of model potentials

The local Animalu–Heine–Abarenkov (AHA) potential with fitted parameters provides a successful description of a large number of static, dynamic and thermodynamic properties of alkali metals (see [3, 12–15, 87, 91], and Chapters 3 and 4; for alkaline-earth metals, as well as for Al and Pb these results were satisfactory). There are questions relating to the relationship between this potential and the *ab initio* non-local model pseudopotentials, of which the form and parameters were determined on the basis of the "more microscopic" models described above. To investigate these questions, we give the comparison of the form-factors of the Heine–Abarenkov (HA) (see (2.36) with $\xi = 0$) and the Rasolt–Taylor (RT) (2.16) model potentials.

The experimental data used in calculations of properties of alkali and polyvalent metals considered here are shown in Tables 2.11 and 2.12. Now we will consider static properties of these metals in the framework of the models which include those higher than the second order in the pseudopotential. An application of these models in the case of Al, Pb, Ca and Li is motivated by the fact that, in these metals, there

TABLE 2.11 Main parameters of pseudopotentials used in calculations of metallic properties (a, Ω_0, r_s in a.u.): Animalu–Heine–Abarenkov (AHA) [41, 51], Heine–Abarenkov (HA) [11] and Rasolt–Taylor (RT) [14].

metal	structure	PP	a	Ω_0	r_s
		AHA	6.5969	143.6	3.258
Li	BCC	HA	6.6175	144.9	3.258
		RT	6.5818	142.6	3.241
		AHA	7.9841	254.5	3.931
Na	BCC	HA	7.9843	254.5	3.931
		RT	8.0011	256.1	3.939
Ca	BCC	HA,RT	10.5496	293.5	3.270
		AHA	7.6190	110.6	2.065
Al	FCC	HA	7.6357	111.3	2.069
		RT	7.6212	110.7	2.065
Pb	FCC	AHA	9.2861	200.2	2.286
		HA	9.3350	203.4	2.298

TABLE 2.12 Values of parameters u and r_0 (in a.u.) of the form factor V_{pC} (2.39), (2.36) for different approximations for the screening (the second-order model in the pseudopotential)

metal	parameter set	GT [39, 40]		VS [44]		TW [43]	
		$-u$	r_0	$-u$	r_0	$-u$	r_0
Al	upper	1.2153	2.4588	1.2046	2.3652	1.2151	2.4637
	lower	0.4106	1.3034			0.4170	1.2883
Pb	upper	1.1199	2.7051			0.8431	2.0398
	lower	0.8044	2.0114				
Na	upper	0.4077	1.5586				
Li	upper	0.4754	2.1685				

exist large deformations of the Fermi-surface. Additionally, these metals have relatively large values of the pseudopotential so that, the value $(V_\tau/E_F)^3$ corresponding to the first sites of the reciprocal lattice τ is of the order of the value $(V_{\mathbf{g}}/E_F)^2$ corresponding to the second sites $\mathbf{g}$.

Note also, that by definition of parameters of $V_{pC}(\mathbf{q})$ using the conditions $p(\Omega_0) = 0$ and $C_{44}(\Omega_0) = (C_{44})_{exp}$ (2.39), it turns out that there is more than one solution (r_0, u) (in particular, two solutions for Al and Pb). For this reason, Table 2.12 contains two sets of parameters, both of which provide approximately the same accuracy in describing static properties of these metals.

Since the localization provides some local potential, then it seems to be natural to write, in the sense of [5],

$$\lim_{\mathbf{q}\to 0} V_{\mathbf{k},\mathbf{k}+\mathbf{q}} = -\frac{4\pi Ze}{\Omega_0 q^2} + \frac{b}{\Omega_0}. \tag{2.89}$$

For the Animalu–Heine–Abarenkov potential, it is easy to obtain an analytical expression for b (see (2.41)). For other pseudopotentials, the value of b was obtained by using the following procedure. Let us write $V_{k,k+\mathbf{q}}$ as

$$V_{\mathbf{k},\mathbf{k}+\mathbf{q}} = V_L(\mathbf{q}) + V_{NL}(\mathbf{q}), \tag{2.90}$$

where $V_L(\mathbf{q})$ is the local part of the pseudopotential and $V_{NL}(\mathbf{q})$ is the non-local part. It is possible to extract analytically the contribution of $V_L(\mathbf{q})$ to b. We denote it as b_L. The contribution to b from $V_{NL}(\mathbf{q} \to 0)$ was obtained numerically. As $\mathbf{q} \to 0$, the limiting value

$$\lim_{\mathbf{q}\to 0} \frac{1}{\pi(\mathbf{q})} \sum_{\mathbf{k}} \frac{\langle \mathbf{k}+\mathbf{q}\,|\mathbf{V}_{NL}|\,\mathbf{k}\rangle}{\varepsilon_{\mathbf{k}} - \varepsilon_{\mathbf{k}+\mathbf{q}}} = b_{NL}/\Omega_0 \tag{2.91}$$

tends to some value independent of $\mathbf{q}$. We denote it as b_{NL}/Ω_0. Then, in equation (2.89), the value of b is equal to

$$b = b_L + b_{NL}. \tag{2.92}$$

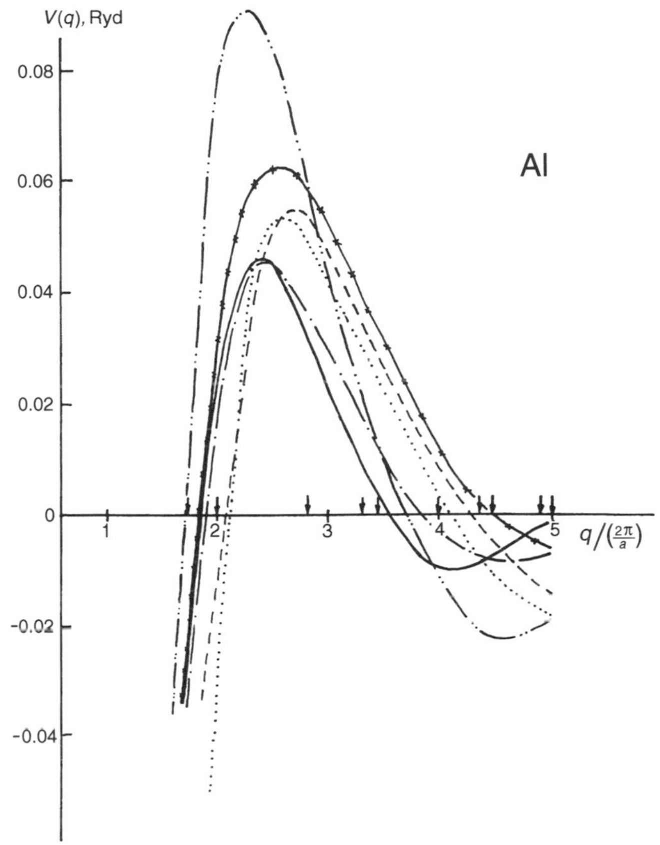

FIGURE 2.4 Local non-screened preudopotential form-factors for aluminium Notations in Figs. 2.4–2.8: — ·· —, Heine–Abarenkov PP $V_{HA}(\mathbf{q})$ within the on-Femi-sphere approximation; — · — the Heine–Abarenkov PP $V_{HA}(\mathbf{q})$ localized using (2.32); ····· Rasolt–Taylor PP $V_{RT1}(\mathbf{q})$ localized using (2.32); - - - - $V_{RT2}(\mathbf{q})$ (see text). Localized PP $V_{AHA}(\mathbf{q})$ with parameters fitted to crystal properties using (2.39): ——— upper parameter set, —×—×— lower parameter set. Arrows show sites of the reciprocal lattice.

The values of b, obtained by this procedure, are rather close to the values obtained using $V_{pC}(\mathbf{q})$ (see equation (2.41)).

Figures 2.3–2.8 show the non-screened form-factors $V(\mathbf{q})$ for simple metals: Al, Li, Ca, Na. This sequence corresponds to increasing r_s from 2 to 4 (see Table 2.11).

First, note that the value of the form-factor in the framework of the on-Fermi-sphere approximation (2.4) for all these metals (except Na) appreciably differs from results obtained by other approximations. It refers to the location and the highest of the maximum, as well as the value of q_0 — the first zero of $V(\mathbf{q})$. The localized HA, AHA and RT form-factors demonstrate quite similar behavior and all of these functions are close together. This similarity includes approximately the same values of q_0 and (it is especially important) approximately the same slope of these potentials in the first sites of the reciprocal lattice. The last fact is extremely important because,

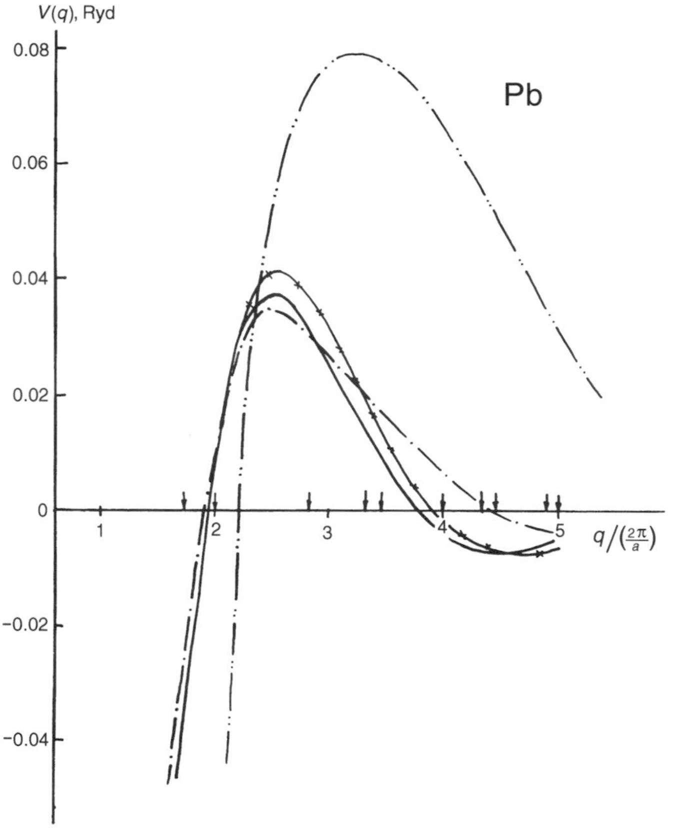

FIGURE 2.5 Form-factors of the local non-screened preudopotential for lead for different models.

namely the first derivative of the potential provides the main contribution to elastic properties of metals. The most important are contributions of these derivatives of $V(\mathbf{q})$ to the shear elastic moduli B_{33} and B_{44} (see Chapter 3).

For Al, there exist well-seperated groups of form-factors: the RT form-factors (RT1 and RT2, see section 2) and the HA and AHA form-factors. Within each group, the form-factors demonstrate rather similar behavior (see Figure 2.3).

For Al and Pb (see Figures 2.3–2.4), two plots of the AHA form-factors (2.36) are shown. They correspond to two different sets of the parameters u and r_0. For both of these sets, calculated values of static properties of metals are close together and to experimental data, and the height of the first maximum has different values. However, for static properties, this difference is not very important, since both of these sets correspond to practically the same values of q_0 and the slope of $V(\mathbf{q})$ at the first sites of the reciprocal lattice.

Let us note the following fact, very important in our opinion. As can easily be seen from Figures. (2.3–2.8), all form-factors $V(\mathbf{q})$ calculated in the framework of the on-Fermi-sphere approximation demonstrate rather different behavior for the metals

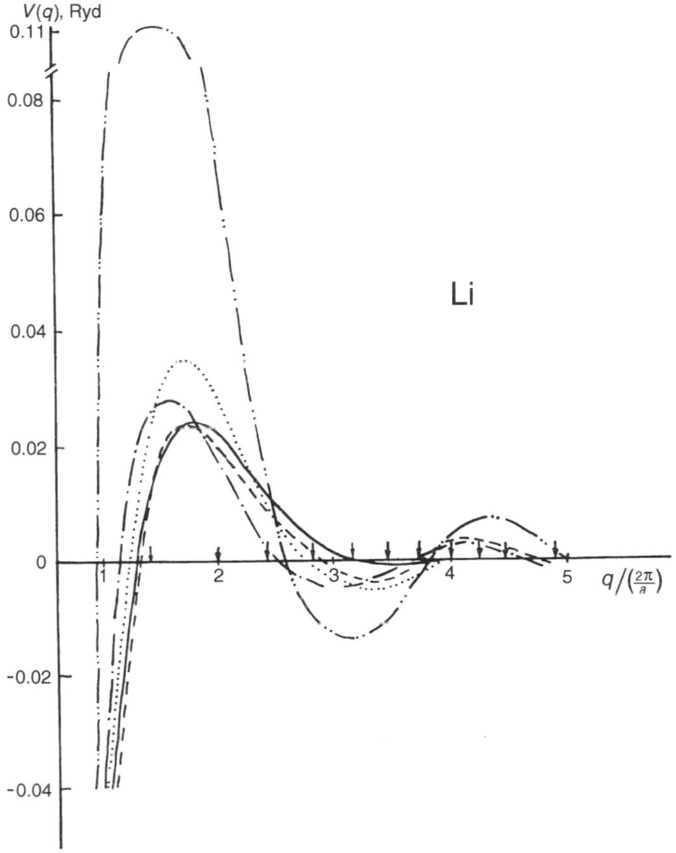

FIGURE 2.6 Form-factors of the local non-screened preudopotential for lithium for different models.

under consideration. However, after the localization using the criterion for the electron density, all these form-factors become smoother and they demonstrate rather similar behavior for different metals. Since the energy depends on the screened pseudopotential, then, besides the localization of the form-factor of the pseudopotential $V_{\mathbf{q}}$, it is necessary to localize the dielectric function also:

$$\varepsilon^{-1}(\mathbf{k},\mathbf{k}') \approx \varepsilon^{-1}(\mathbf{k}-\mathbf{k}') \tag{2.93}$$

The non-locality of ε^{-1} follows predominantly from the non-locality of the exchange and correlation interaction. Let us consider some qualitative reasons for using the local exchange and correlation potential. First, let us compare v_{xc} (2.154) in the local Slater form [92]

$$v_{xc}^{0} \approx -\frac{3\alpha}{2\pi}\left[3\pi^{2}n(\mathbf{r})\right]^{1/3} \tag{2.94}$$

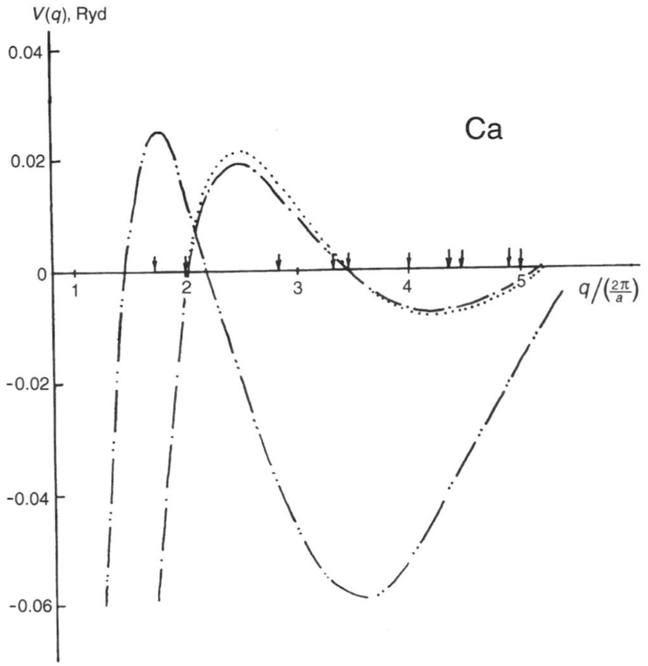

FIGURE 2.7 Form-factors of the local non-screened preudopotential for calcium for different models.

(here ($n(\mathbf{r})$ is the local electron density) with the non-local exchange Hartree–Fock (HF) potential, which is the integral operator with the kernel

$$v_{HF} = -\frac{n^{(1)}(\mathbf{r}_1, \mathbf{r}_2)}{|\mathbf{r}_1 - \mathbf{r}_2|}, \tag{2.95}$$

where $n^{(1)}(\mathbf{r}_1, \mathbf{r}_2)$ is the one-particle density matrix (see [93])

$$n^{(1)}(\mathbf{r}_1, \mathbf{r}_2) = \sum_i n_i \varphi_i^*(\mathbf{r}_1)\,\varphi_i(\mathbf{r}_2). \tag{2.96}$$

If all the occupation numbers n_i for some complete basis φ_i were equal to unity, then the next condition could be satisfied

$$n^{(1)}(\mathbf{r}_1, \mathbf{r}_2) = \delta(\mathbf{r}_1 - \mathbf{r}_2). \tag{2.97}$$

This fact forced us to suppose that, in the high-density limit, the function $n^{(1)}(\mathbf{r}_1, \mathbf{r}_2)$ must have a sharp maximum at $r_1 \sim r_2$. For this reason, we can allow $n^{(1)} \sim \delta(\mathbf{r}_1 - \mathbf{r}_2)$, and the value $|\mathbf{r}_1 - \mathbf{r}_2|$ can be replaced by the mean inter-electron distance r_s. Now the operator v_{HF} is local, and its approximate expression is

$$v_{HF} = -\frac{\gamma}{r_s}\delta(\mathbf{r}_1 - \mathbf{r}_2), \tag{2.98}$$

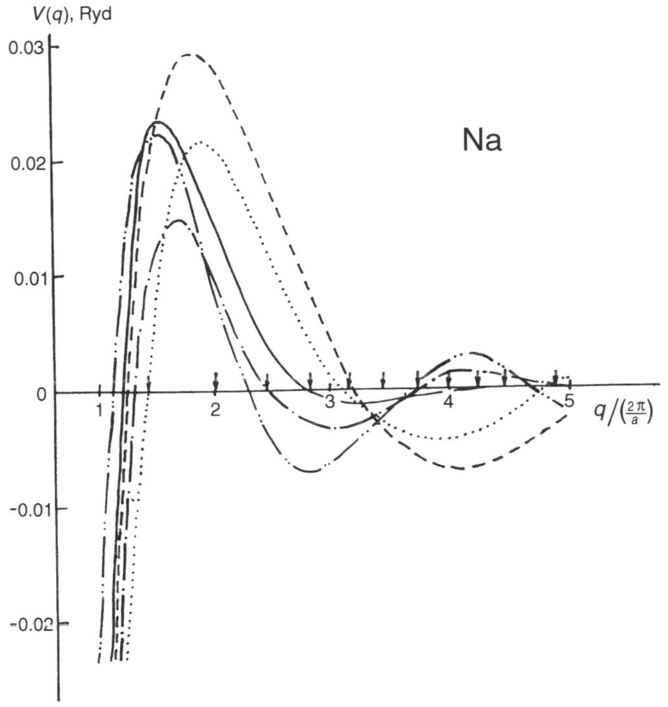

FIGURE 2.8 Form-factors of the local non-screened preudopotential for sodium for different models.

where γ is some arbitrary coefficient which, however, must be of the order one. We can prove this by comparison of (2.98) with the expression for the $X\alpha$-potential (2.94). Note that for fixed values of the electron spin, the $X\alpha$-potential has the form (2.94). Using the definition

$$\frac{4\pi}{3}(r_s)^3 = \frac{1}{n_0}, \tag{2.99}$$

it is possible to transform equation (2.94) to the form (2.98), in which

$$\gamma = \frac{3\alpha}{2\pi}\left(\frac{9\pi}{2}\right)^{1/3} \approx 1.154\alpha. \tag{2.100}$$

Taking into account that, in the framework of the $X\alpha$-method, the value of the parameter α varies from 2/3 to 1, we conclude that the region of variation of the parameter γ is 1.15–0.77. This transfer from v_{HF} to the local potential qualitatively agrees with the $X\alpha$-method. Note, however, that replacing v_{HF} with the local exchange potential v_{xc} makes its necessary to replace the Hartree–Fock expression for E_{xc}

$$E_{xc} = \frac{1}{2}\int \frac{|n^{(1)}(\mathbf{r}_1, \mathbf{r}_2)|^2}{|\mathbf{r}_1 - \mathbf{r}_2|} d^3r_1 d^3r_2 \tag{2.101}$$

by

$$E_{xc} = \frac{3}{4} \int n(\mathbf{r}) v_{xc}(\mathbf{r})\, d^3 r. \tag{2.102}$$

It is necessary because an application of the variation procedure to the total energy E must provide the respective one-particle equation of the $X\alpha$-method. It means that the energy E in the $X\alpha$-method, in contrast to the Hartree–Fock theory, does not correspond more to the mean value of the Hamiltonian H calculated using the wave function of the ground state Φ_0 (which is a determinant constructed from the one-particle wave functions φ_i). However, concrete calculations show that there is no large difference between the wave functions φ_i obtained in the HF theory and in the framework of the $X\alpha$-method.

As we already mentioned, expression (2.94) at constant α actually includes the exchange interaction between electrons in the framework of the HF method only. The many-electron effects (correlations) can be taken into account in the framework of the local field approximation [77], if we allow α to be dependent on r_s.

The values of α have been calculated for almost all elements of the Periodic Table [92]. In the theory of metals, it is useful to determine this parameter as a function of the mean electron density n_0 or the mean Fermi-momentum $k_F = (3\pi^2 n_0)^{1/3}$. To do this, let us use linear response theory and Fermi-liquid theory. Let us introduce some perturbation to the electron gas. It provides the variation of the electron density $\Delta n(\mathbf{r})$. Considering this perturbation to be small, we obtain the change of the exchange potential Δv_{ex} in the framework of the Slater approximation:

$$\Delta v_{ex} = v_{ex} - v_{ex}^0 \approx -\alpha \frac{3\pi}{2k_F^2} \Delta n(\mathbf{r}), \tag{2.103}$$

where v_{ex}^0 is determined by equation (2.94).

From the theory of Fermi-liquids [61], it is well-known that the Fourier-amplitude $\Delta v_{ex}(\mathbf{q})$ can be written as

$$\Delta v_{ex}(\mathbf{q}) = -\frac{4\pi}{q^2} G(\mathbf{q}) n(\mathbf{q}), \tag{2.104}$$

where $G(\mathbf{q})$ is the local field correction discussed in [95]. For the function $G(\mathbf{q})$ there now exist many approximative expressions which include the exchange and correlation contributions to screening the perturbation. The local field correction $G(\mathbf{q})$ determines the local screening $\tilde{\varepsilon}(\mathbf{q})$ (see section 1 and the text below).

2.3 Pseudopotentials Conserving the Norm

In [19], some non-singular pseudopotentials were introduced to provide calculations in the framework of the density functional formalism. These pseudopotentials have been obtained using results of many-electron atomic calculations and have the following properties:

1) They provide the same energy for the "model atomic configuration" under consideration for both the exact and the pseudopotential calculations;
2) The exact atomic and pseudopotential wave functions coincide outside the core radius r_c;
3) Integrals of the real and the pseudo-density over the interval $[0, r]$ have the same values at $r > r_c$ for all valent states (the norm conserving condition);
4) The logarithmic derivatives of the exact and the pseudo wave-functions, as well as their first derivatives with respect to the energy, must have the same values at $r > r_c$.

Further development of this method of construction of non-singular pseudopotentials was made in [7]. This work contains numerical values for pseudopotentials for a majority of the elements in the Periodic Table (from H to Pu). These pseudopotentials may be used in self-consistent calculations of the band structure. They also show a reliable reproduction of results of many-electron atomic calculations.

Below, we briefly discuss the method of constructing non-singular pseudopotentials supposed in [7].

This method of taking into account the effects of electron interaction in the framework of the pseudo-atom method is based on the local density functional theory (DFT) [24, 25] (see also [20]). The energy of the ground state of the system of interacting electrons in the external field (produced by atomic nucleus) is written as some functional of the electron density $n(\mathbf{r})$

$$E[n] = T[n] + E_{Coul}[n] + \int V_{ext}(\mathbf{r})\, n(\mathbf{r})\, d^3r + E_{xc}[n], \tag{2.105}$$

where $T[n]$ is the kinetic energy of non-interacting electrons, $V_{ext}(r) = -Z/r$ is the potential of the nucleus (core), $E_{xc}[n]$ is the exchange and correlation energy, $E_{\text{Coul}}[n]$ is the actual Coulomb energy of electrons

$$E_{Coul}[n] = \frac{1}{2}\iint \frac{n(\mathbf{r})n(\mathbf{r}')}{|\mathbf{r}-\mathbf{r}'|} d^3r\, d^3r'. \tag{2.106}$$

Following to Kohn and Sham [25], the variational solution of equation (2.105) can be obtained using the self-consistent solution of the next Schrödinger-type equation:

$$[T + V(\mathbf{r})]\psi_i(\mathbf{r}) = \varepsilon_i\psi_i(\mathbf{r}), \tag{2.107}$$

$$V(\mathbf{r}) = \int \frac{n(\mathbf{r}')}{|\mathbf{r}-\mathbf{r}'|} d^3r' + \frac{\delta E_{xc}[n]}{\delta n(\mathbf{r})} + V_{ext}(\mathbf{r}), \tag{2.108}$$

$$n(\mathbf{r}) = \sum_i |\psi_i(\mathbf{r})|^2. \tag{2.109}$$

Summation in the last equation is taken over all occupied states.

To describe the relativistic effects for all atoms in the Periodic Table, the Dirac expression for the kinetic energy [72] is used. The Schrödinger equation (2.107) is replaced with the system of two simultaneous equations for the large ("positive energy") $G_i(\mathbf{r})$ and small $F_i(\mathbf{r})$ ("negative" energy) components of the wave function. In spherical coordinates, these equations are (see [73])

$$\frac{dF_i(\mathbf{r})}{dr} - \frac{k}{r}F_i(\mathbf{r}) + \alpha[\varepsilon_i - V(\mathbf{r})]G_i(\mathbf{r}) = 0, \tag{2.110}$$

$$\frac{dG_i(\mathbf{r})}{dr} - \frac{k}{r}G_i(\mathbf{r}) + \alpha\left[\frac{2}{\alpha^2} + \varepsilon_i - V(\mathbf{r})\right]F_i(\mathbf{r}) = 0, \tag{2.111}$$

In (2.110) and (2.111), the atomic units were used: $\hbar = m = e = 1$, $c = \alpha^{-1} = 137.04$, and $k \neq 0$ is the quantum number (l is the orbital momentum)

$$k = \begin{cases} l, & \text{for } j = l - 1/2 \\ -(l+1), & \text{for } j = l + 1/2 \end{cases} \tag{2.112}$$

The charge density is a sum of both these components:

$$n(\mathbf{r}) = \sum_i \left[|G_i(\mathbf{r})|^2 + |F_i(\mathbf{r})|^2\right]. \tag{2.113}$$

For one-electron potentials, the Dirac equation includes all relativistic effects and provides the energy of the spin-orbit splitting. The most important problem for numerical calculations is the choice of reliable approximations for the exchange and correlation energy $E_{xc}[n]$ in (2.105) and (2.108).

In the framework of the local density approximation

$$E_{xc}[n] = \int n(\mathbf{r})\varepsilon_{xc}(n(\mathbf{r}))\, d^3r, \tag{2.114}$$

where $\varepsilon_{xc}(n)$ is the exchange and correlation energy (per one electron) for the electron system with the uniform density n. From many interpolation formulas for $\varepsilon_{xc}(n)$, in [7], the results of [74] were used, parametrized by Perdew and Zunger in [75].

Now let us construct the pseudopotentials. The construction of the norm-conserving potential includes the next five stages.

I. The Dirac equation (2.110, (2.111) is solved for some given electron configurations (marked by the index ν). As a result, we obtain the set of one-electron eigenvalues (for valent states), and the radial wave functions $F_k(\mathbf{r})$ and $G_k(\mathbf{r})$, as well as the self-consistent potential in the framework of the local density approximation. As has been shown in [73, 76], for valence electrons outside the core region, the Dirac equations can be formally replaced with the Schrödinger equation for the principal component of the wave function $G_k(\mathbf{r})$

$$\left[\frac{1}{2}\frac{d^2}{dr^2} - \frac{K(K+1)}{2r^2} + V^{\nu}(\mathbf{r})\right]G_k(\mathbf{r}) = \varepsilon\, G_k(\mathbf{r}). \tag{2.115}$$

This replacement is correct within the accuracy of the order of α^2. Any appreciable mixing of the second components $F_k(\mathbf{r})$ with $G_k(\mathbf{r})$ takes place only in the core region of heavy atoms. Hence, equation (2.115) may be considered as the first step in construction of the pseudopotential and the pseudo-wave functions.

II. In the next step, the pseudopotential $\hat{V}^{\nu}_{1j}$ is obtained from the screened core potential V^{ν} by cutting the singularity near to $r = 0$ (j denotes all electron quantum numbers)

$$\hat{V}^{\nu}_{1j}(\mathbf{r}) = V^{\nu}(\mathbf{r})[1 - f(r/r_{cj})] + c^{\nu}_{j} f(r/r_{cj}), \tag{2.116}$$

where $f(r/r_{cj})$ is some smooth cut-off function, while vanishes as $r \to \infty$. This function must tend to 1 as $r \to 0$ to cut the vicinity of $r = r_{cj}$. The constant c^{ν}_{j} was fitted using the condition that the solution of the radial Schrödinger equation $\{\omega^{\nu}_{1j}(\mathbf{r})$ (for the potential V^{ν}_{1j}) must provide the value of the energy equal to the true eigenvalue ε^{ν}_{j}. Outside the core r_{cj}, the normalized function $\omega^{\nu}_{1j}(\mathbf{r})$ must coincide with the valence wave function of core electrons times some constant factor γ^{ν}_{j}

$$\gamma^{0}_{j}\, \omega^{\nu}_{1j}(\mathbf{r}) \to G_{j}(\mathbf{r}), \quad r > r_{cj}.$$

As was shown in [7], the optimal choice of the cut-off function for many atoms provides

$$f(r/r_{cj}) = \exp[-(r/r_{cj})^{\lambda}] \tag{2.117}$$

with $\lambda = 3.5$. The cut-off radius r_{cj} determines the region in which the pseudo-wave and the wave functions are different. This algorithm cannot be considered as a fitting procedure, but, as was supposed by the authors of [7], it can be used in qualitative calculations of the pseudopotential. The optimal values of r_{cj} for many atoms are given in [7].

III. This stage includes the next modification of the intermediate pseudo-wave function $\omega^{\nu}_{1j}(\mathbf{r})$ in the region of small r

$$\omega^{\nu}_{2j}(\mathbf{r}) = \gamma^{\nu}_{j}[\omega^{\nu}_{1j}(\mathbf{r}) + \delta^{\nu}_{j} r^{l+1} f(r/r_{cj})].$$

The fitting of γ^{ν}_{j} uses the condition that the wave function ω^{ν}_{2j} must coincide with the valence function of the core at $r > r_{cj}$. This condition is satisfied if δ^{ν}_{j} is the lowest solution of the next quadratic equation

$$(\gamma^{\nu}_{j})^{2} \int [\omega^{\nu}_{1j}(\mathbf{r}) + \delta^{\nu}_{j} r^{l+1} f(r/r_{cj})]^{2} d^{3}r = 1.$$

IV. This stage yields the resulting screened pseudopotential $\hat{V}^{\nu}_{2j}(\mathbf{r})$ provided by the zero-less eigenfunctions $\omega^{\nu}_{2j}(\mathbf{r})$ corresponding to eigenvalues ε_{j}. Calculating this pseudopotential corresponds to inverting the radial Schrödinger equation. It can be written analytically for given $\hat{V}^{\nu}_{1j}(\mathbf{r})$ and $\omega^{\nu}_{2j}(\mathbf{r})$:

$$\hat{V}^{\nu}_{2j}(\mathbf{r}) = \hat{V}^{\nu}_{1j}(\mathbf{r}) + \frac{\delta^{\nu}_{j} r^{l+1} f}{2\omega^{\nu}_{2j}(\mathbf{r})} \left[\frac{\lambda^{2}(r/r_{cj})^{2\lambda} - [2\lambda l + \lambda(\lambda + 1)](r/r_{cj})^{2}}{r^{2}} + 2\varepsilon_{j} - 2\hat{V}^{\nu}_{1j}(\mathbf{r}) \right]. \tag{2.118}$$

The second term in (2.118) is a smooth cut-off correction factor proportional to $f(r/r_{cj})$ from (2.117).

V. The last stage is the calculation of the non-screened potential using the screened potential $\hat{V}_{2j}^{\nu}(\mathbf{r})$ and the zeroless pseudo-wave functions $\omega_{2j}^{\nu}(\mathbf{r})$

$$\hat{V}_{l,\pm 1/2}^{ion}(\mathbf{r}) = \hat{V}_{2j}^{\nu}(\mathbf{r}) - \int \frac{n^{\nu}(\mathbf{r}')}{|\mathbf{r}-\mathbf{r}'|}\, d^3r' - \frac{\delta E_{xc}[n^{\nu}]}{\delta n^{\nu}(\mathbf{r})}; \tag{2.119}$$

$$n^{\nu}(\mathbf{r}) = \sum_j \left| \frac{\omega_{2j}^{\nu}(\mathbf{r})}{r} \right|^2$$

The summation in the last equation is spread over all occupied valence states j.

The averaged pseudopotential, which takes into account the degeneration of states with $(l, \pm 1/2)$, can be determined as

$$\hat{V}_l^{ion}(\mathbf{r}) = \frac{1}{2l+1}\left[l\hat{V}_{l,-1/2}^{ion}(\mathbf{r}) + (l+1)\hat{V}_{l,+1/2}^{ion}(\mathbf{r}) \right]. \tag{2.120}$$

This averaged potential can be used in scalar-relativistic calculations. The potential describing large spin-orbital effects can be written as

$$\hat{V}_l^{so}(\mathbf{r}) = \frac{2}{2l+1}\left[\hat{V}_{l,+1/2}^{ion}(\mathbf{r}) - \hat{V}_{l,-1/2}^{ion}(\mathbf{r}) \right]. \tag{2.121}$$

Hence, the total pseudopotential, which can be used in relativistic calculations, is

$$\hat{V}_{ps}^{ion}(\mathbf{r}) = \sum_l |l> \left[\hat{V}_l^{ion}(\mathbf{r}) + \hat{V}_l^{so}(\mathbf{r})\mathbf{LS} \right] < l|. \tag{2.122}$$

In [7], methods of the calculation of relativistic pseudopotentials were considered in detail. This work also includes numerical tables of these pseudopotentials for elements in the Periodic Table: from H ($Z = 1$) to Pu ($Z = 94$).

2.4 Screening of the Potential in Metals

2.4.1 Linear screening

The screening properties of the electron system with electron–electron and electron–ion interactions can be described by the dielectric function $\varepsilon(\mathbf{q}, \omega)$ depending on the wave vector $\mathbf{q}$ and the frequency ω. Actually, in studying atomic properties of metals, we have dealt with the static dielectric function $\varepsilon(\mathbf{q}, \omega) = \varepsilon(\mathbf{q})$, because all characteristic frequencies determining the atomic properties are of the order of the Debye frequency ω_D, i.e., they are appreciably smaller than the characteristic electron frequencies. This statement is the applicability condition for the adiabatic approximation widely used in the theory of atomic properties of crystals (see Chapter 1).

The problem of calculation of $\varepsilon(\mathbf{q})$ is very simple in the case of a high-density plasma: $n_0 \gg 1, r_s \ll 1$. In this case, the dielectric function of a non-interacting electron gas can be used. Similary, it is relatively easy to calculate $\varepsilon(\mathbf{q})$ in the case of a small electron density: $n_0 \ll 1, r_s \gg 1$. But in real metals, there exists an intermediate case $1 < r_s < 6$ (see Tables 2.3 and 2.10) without any small parameter. During the last three decades, there have been many attempts to construct a consistent theory for the dielectric function (see [77]). However, even today there exists no exact solution to the problem. These difficulties are related to the calculation of the correlation energy. In this subsection, we will describe some of the most consistent and well-grounded, from the theoretical point of view, interpolation schemes for the calculation of $\varepsilon(\mathbf{q})$ for realistic values of the electron densities in metals. Some questions relating to the diagram technique and the definition of $\varepsilon(\mathbf{q})$ were considered in Chapter 1. More detailed reviews of the present state of the problem can be found in [77–79].

The exchange and correlation contribution to the adiabatic potential have already been discussed (Chapter 2). For this reason, we present here only some approximations for the polarization operator $\pi(\mathbf{q})$ and the local field correction $G(\mathbf{q})$, which we will use further in our calculations of concrete properties of simple metals. Application of many interpolation schemes for $\pi(\mathbf{q})$ is necessary to provide an optimal choice of the calculation scheme for exchange and correlation effects, which were the best from the point of view of the unified description of a maximally large number of metallic properties. The dielectric function $\varepsilon(\mathbf{q})$, describing the weakness of the external potential in metals, is related to the polarization operator $\pi(\mathbf{q})$ by

$$\varepsilon(\mathbf{q}) = 1 + v_c(\mathbf{q})\pi(\mathbf{q}), \tag{2.123}$$

where $v_c(\mathbf{q}) = 4\pi e^2/q^2$. If we suppose that the amplitude of the particle–hole interaction is a function of the momentum transfer $\mathbf{q}$ only, then

$$\pi(\mathbf{q}) = \pi_0(\mathbf{q})[1 - \tilde{\gamma}_4(\mathbf{q})\pi_0(\mathbf{q})]^{-1}. \tag{2.124}$$

In this equation, $\pi_0(\mathbf{q})$ is the polarization operator calculated neglecting the local field correction. It corresponds to the well-known Lindhard formula (see [5, 80])

$$\pi_0(\mathbf{q}) = \frac{3}{2}\frac{n_0}{E_{F0}}\left[\frac{1}{2} + \frac{1-x^2}{4x^2}\ln\left|\frac{1+x}{1-x}\right|\right], \tag{2.125}$$

where $x = 1/2k_F$. The irreducible vertex function $\tilde{\gamma}_4(\mathbf{q})$, in the framework of the momentum transfer approximation, is

$$\tilde{\gamma}_4(\mathbf{q}) = v_c(\mathbf{q})G(\mathbf{q}). \tag{2.126}$$

The only difference between all interpolations and approximations, used by describing the correlation effects in the electron gas, is the choice of this function $G(\mathbf{q})$. The form of this function depends on approximations made in the calculations (see [5, 77]).

2.4.2 The Correlation energy and the compressibility sum rule

Before discussing the approximations for the dielectric function (or for the polarization operator $\pi(\mathbf{q})$), let us consider briefly the problem of the calculation of the correlation energy E_c. In accordance with the compressibility sum rule [61]

$$B^{(0)} \equiv \Omega \frac{\partial^2 E_e^{(0)}}{\partial \Omega^2} = \frac{n_0^2}{\pi(0)}, \tag{2.127}$$

the correlation energy directly relates to the exchange and correlation corrections for the polarization operator. It is easy to show that, if $E_c \equiv 0$, then the sum rule (2.127) is satisfied for $\pi(0) = \pi_0(0) = 3n_0/2E_{F0}$. Hence, there exists a certain relation between the form for E_c in the expression for the energy of the uniform electron gas $E_e^{(0)}$ and the deviation of $\pi(0)$ from $\pi_0(0)$.

The simplest approximation for $\pi(\mathbf{q})$ can be obtained in the frame of the ring diagram approximation: the random phase approximation (RPA). For this approximation, the polarization operator $\pi_0(\mathbf{q})$ (2.125) corresponds to the simple loop. On the other hand, Gell-Mann and Brueckner have calculated the RPA correlation to the energy (see [61]) (here and below, all energies are given in Ry)

$$E_c^{RPA} = (0.622 \ln r_s - 0.142). \tag{2.128}$$

Note that for $\pi(\mathbf{q}) = \pi_0(\mathbf{q})$, any expression for E_c, except zero value for all densities, does not agree with the sum rule (2.127). For this reason, using the RPA approximations, we will assume $\pi(\mathbf{q}) = \pi_0(\mathbf{q})$ and $E_c = 0$.

As was already mentioned, calculation of E_c at typical metallic densities is a difficult problem. The most popular interpolation expression is the Nozieres–Pines formula [61] (in Ry per electron)

$$E_c^{NP} = 0.031 \ln r_s - 0.115. \tag{2.129}$$

Together with E_c^{NP}, we will also use the expression for E_c obtained by Vashishta and Singwi [44]

$$E_c^{VS} = -0.112 + 0.0335 r_s - \frac{0.02}{0.1 + r_s} \tag{2.130}$$

The Perdew–Zunger form [94] has already been discussed. Remember that, for the densities considered here ($r_s \geq 1$.)

$$E_c^{PZ} = -\frac{0.1471}{1 + 1.1581\sqrt{r_s} + 0.344 r_s}. \tag{2.131}$$

Remember also that, in the framework of the random phase approximation, $E_c^{RPA} = 0$.

2.4.3 Expressions for the polarization operator and the local field correction

The great majority of calculations for the polarization operator at typical metallic densities use the momentum transfer approximation for the amplitude of the particle–hole process $\tilde{\gamma}_4(\mathbf{q})$ (2.126). It immediately gives

$$\pi(\mathbf{q}) = \frac{\pi_0(\mathbf{q})}{1 - 4\pi e^2 (G(\mathbf{q})/q^2)\pi_0(\mathbf{q})}. \tag{2.132}$$

Hence, the problem relates to the calculation of $G(\mathbf{q})$ or $G(\mathbf{q})/q^2$.

The first attempt to escape the limits of the RPA approximation ($G(\mathbf{q}) = 0$) was performed by Hubbard [81]. He tried to calculate approximately the sum of the simplest class of exchange contributions to the polarization operator. He obtained an expression like (2.132) with

$$G_H(\mathbf{q}) = \frac{1}{2}\frac{q^2}{q^2 + k_{F0}^2} \tag{2.133}$$

However, the polarization operator (2.132) with $G_H(\mathbf{q})$ does not satisfy the identity (2.127) with the correlation energy E_c^{NP} for $E_e^{(0)}$. Later, Geldart and Vosko [42] obtained the expression

$$G_{GV}(\mathbf{q}) = \frac{1}{2}\frac{q^2}{q^2 + k_{F0}^2\,\xi}. \tag{2.134}$$

In this formula, the value of the parameter ξ has been obtained using the sum rule (2.127). Since the compressibility $\chi = B^{-1}$ can be calculated with rather good accuracy using the expression for the energy. The values of $\pi(0)$ and, consequently, $\varepsilon(\mathbf{q})$ at small $\mathbf{q}$, are also determined very accurately. Using E_c^{NP}(2.129) provides

$$\xi = \frac{2}{1 + (0.0155\pi/k_{F0})} \approx 2. \tag{2.135}$$

Apparently, Geldart and Vosko [42] were the first who paid special attention to satisfying the compressibility sum rule (2.127).

As was shown in many calculations, the most accurate approximation was found by Geldart and Taylor in [39, 40]. In [39], the attempt to select the most essential class of diagrams for the static polarization operator was undertaken. In these calculations, special attention was paid to the mutual cancellation of some classes of diagrams at each stage of the calculation. This cancellation is especially important to the Coulomb systems. Special attention was also paid to the identity (2.127). The expression for $\pi(\mathbf{q})$ depends on the electron density n_0

$$\pi(\mathbf{q}, n_0) = \pi_0(\mathbf{q})\left[1 + \frac{f(x, n_0)}{1 - c(n_0)\pi_0(\mathbf{q})}\right]; \quad x \leq 4,$$

$$\pi(\mathbf{q}) = \pi_0(\mathbf{q})\Big/\left(1 + \frac{\beta}{x^4}\right); \quad x \geq 4. \tag{2.136}$$

Here $x = q/k_{F0}$, and the function $f(x, n_0)$ was tabulated in [39]. These calculations were performed with using the second-order interpolation for given x and n_0. The parameter β provides a fitting of both equations (2.136) at $x = 4$; the function $c(n_0)$ describes the self-consistent contribution of diagrams of the order higher than one in $v_c(\mathbf{q})$.

In [40], Taylor used with great success (in our opinion) a very simple model for $G(\mathbf{q})/q^2$. It was mentioned that $G(\mathbf{q})/q^2$ has a nearly constant value at $q \leq 2k_{F0}$, and $\pi_0(\mathbf{q} \to 0) \sim q^{-2}$. Since $G(\mathbf{q})/q^2$ always participates as a combination $\pi_0(\mathbf{q})G(\mathbf{q})/q^2$ (see, for example, (2.132)), then it is convenient to approximate $G(\mathbf{q})/q^2$ with

$$\left(G(\mathbf{q})/q^2\right)_{q=0} = [1 - L]\frac{\pi}{4k_{F0}}. \tag{2.137}$$

From (2.127), it follows that $L = \pi_0(0)/\pi(0)$. Using the correlation energy in the Nozières–Pines form (2.129), it is easy to obtain

$$L^{NP} = 1 - \lambda - 0.1535\lambda^2, \tag{2.138}$$

where $\lambda = (\pi k_{F0})^{-1}$. Expression (2.137) can be used for all q, and results are very close to the results obtained by using (2.136) [40]. Further, we will use the same notation, GT, for both approximations (2.136) and (2.137).

Another approach used in [44] also provides similar results. The main idea of this approach is to take into account the short-range electron correlations by introducing the pair correlation functions. It allows us to provide self-consistent calculations of the local field correction (see also [20]). This approximation, known as the VS form, was obtained by using the correlation energy E_c^{VS}(2.130). Results of the calculations of $G(\mathbf{q})$ have been tabulated in [44].

The method based on the splitting of the Green's functions was put forward by Toigo and Woodruff [43] (the TW approximation). The results of the calculation have been presented in a tabulated form.

In the frame of the second-order model, there is no difference in using the dielectric function of the screening of the external test charge $\varepsilon(\mathbf{q})$ (2.123) or the dielectric function for the electron participating in the screening of $\tilde{\varepsilon}(\mathbf{q})$(2.141). It follows from the fact that $\pi(\mathbf{q})$ is given by equation (2.132) and

$$\pi(\mathbf{q})/\varepsilon(\mathbf{q}) = \pi_0(\mathbf{q})/\tilde{\varepsilon}(\mathbf{q}). \tag{2.139}$$

However, as we have already mentioned, in higher contributions to the energy, it is necessary to extract the value $\tilde{\varepsilon}(\mathbf{q})$, since the pseudopotential participates only as a combination

$$W(\mathbf{q}) = \frac{V_{\mathbf{q}}}{\tilde{\varepsilon}(\mathbf{q})} \tag{2.140}$$

$$\tilde{\varepsilon}(\mathbf{q}) = 1 + \left\{\frac{4\pi e^2}{q^2} - \left[\frac{1}{\pi_0(\mathbf{q})} - \frac{1}{\pi(\mathbf{q})}\right]\right\}\pi_0(\mathbf{q}). \tag{2.141}$$

The form-factor (2.140) satisfies the well-known condition [4]

$$W(0) = -\frac{2}{3}E_F. \tag{2.142}$$

However, representation (2.140) supposes that the potential acting on each quasi-particle is the mean local potential, i.e., formula (2.142) is a particular case of a more general result obtained using the Fermi-liquid formalism in [82]. This formalism provides for the Fourier component of the potential which corresponds to the scattering of a quasi-particle with changing the state from $|\,\mathbf{k}>$ to $|\,\mathbf{k}+\mathbf{q}>$:

$$W_{\mathbf{k},\mathbf{k}+\mathbf{q}}(E) = \frac{z(E)\hat{\Lambda}(\mathbf{k},\mathbf{q},\varepsilon)v_{\mathbf{k},\mathbf{k}+\mathbf{q}}}{\tilde{\varepsilon}(\mathbf{q})}. \tag{2.143}$$

Here, $\hat{\Lambda}(\mathbf{k},\mathbf{q},\varepsilon)$ is the irreducible vertex part, $z(E)$ is the quasi-particle renormalization constant (see [83]).

Taking into account that $E = \mu$ at the Fermi surface and

$$\lim_{\mathbf{q}\to 0}\hat{\Lambda}(\mathbf{k}_F,\mathbf{q},\mu) = z^{-1}(\mu)\frac{m}{m^*}\frac{B^{(0)}}{B} \tag{2.144}$$

(m and m^* are masses of free electrons and quasi-particles, respectively), as well as condition (2.127), we obtain a general result independent of the scattering mechanism

$$\lim_{\substack{\mathbf{q}\to 0\\ k=k_F}} W_{\mathbf{k},\mathbf{k}+\mathbf{q}}(\mu) = \frac{2}{3}\frac{m}{m^*}\mu. \tag{2.145}$$

For the scattering process, because the electron does not leave the Fermi surface, the next approximation was supposed in [80] for $\hat{\Lambda}$:

$$\hat{\Lambda}(\mathbf{q}) = \Lambda(k_F,\mathbf{q},\mu) \approx 1 - \frac{\pi(\mathbf{q})}{\pi_0(\mathbf{q})}\left\{1 - z^{-1}(\mu)\frac{m}{m^*}\frac{B^{(0)}}{B}\right\}. \tag{2.146}$$

In concrete calculations, as a rule, the approximation $m = m^*$ is used. The values of $z^{-1}(\mu)$ for some electron densities have been tabulated in [84].

Finally, let us note also the Gaspar–Kohn–Sham approximation (see [20]). It uses formula (2.132) for $\pi(\mathbf{q})$ and some constant values for $G(\mathbf{q})/q^2$ and $\tilde{\gamma}_6(\mathbf{q}_1,\mathbf{q}_2,\mathbf{q}_3)$:

$$G(\mathbf{q})/q^2 \approx \frac{r_s^2}{4(9\pi/4)^{2/3}}. \tag{2.147}$$

Using this formula and the relation between $[G(\mathbf{q})/q^2]_{\mathbf{q}\to 0}$ and $\tilde{\gamma}_6(0.0.0)$ (see [95]), we obtain

$$\tilde{\gamma}_6(0,0,0) = -\frac{4\pi r_s^5}{9(9\pi/4)^{2/3}}. \tag{2.148}$$

The correlation energy in the framework of this approximation is equal to zero.

2.5 Resumé

As was already emphasized in sections 1 and 2, the charge density distribution $n(\mathbf{r})$ in the coordinate space plays a fundamental role in the investigation of all properties of ordered, as well as disordered systems. On the other hand, since the pseudopotentials used here are determined by fitting to measured crystal properties, there exists a problem of how adequate the description using the scattering characteristics is. In this sense, *ab initio* pseudopotentials may serve as a test, for example, by comparing the phase shifts obtained in calculations (see the RT1 model). For this purpose, we have calculated the radial distribution of the electron density $n(\mathbf{r})$, induced by a singly ionized ion of alkali metal in the framework of the linear approximation

$$n(\mathbf{r}) = \frac{4\pi}{r}\int_0^\infty n(q)\sin(qr)q\,dq, \tag{2.149}$$

where

$$n(\mathbf{q}) = V(\mathbf{q})\left[1 - \frac{1}{\varepsilon(q)}\right] \Big/ \frac{4\pi e}{q^2}. \tag{2.150}$$

These calculations were made in the framework of the second-order model. A comparison (see Figure 2.9) of these results with results of similar calculations for Li, Na, and K made by Rasolt *et al.* in [13, 14, 85], and also with results of self-consistent calculations with nonlinear screening by Dagens [23], provide some conclusions.[4] Figure 2.9b shows that the local AHA pseudopotential, with fitting in accordance with (2.39) and with the GT approximation for $\pi(\mathbf{q})$ (see Tables 2.1, 2.2), provides the electron density distribution $n(\mathbf{r})$ around Na^+ and K^+ ions very close to the results of calculations by Rasolt, *et al.* [13, 14, 85] based on the non-local pseudopotential, with fitting to experimental data for phase shifts. Moreover, the curves obtained for the model considered here are very close to the self-consistent calculations by Dagens [23]. For Li (Figure. 2.9,a), the agreement is not so good.

The fact that values of $n(\mathbf{r})$ for Na and K obtained by us in the framework of the second-order (in the pseudopotential) model, i.e., within the linear screening approximation, agree well with the results obtained in [23] using the nonlinear screening, shows that, apparently, in these metals, high-order in the pseudopotential contributions are small (see also [87]). The disagreement of results for $n(\mathbf{r})$ in the case of Li, especially taking into account the fact that the Li^+ ion is a rather strong scatterer [17], definitely demonstrates the necessity of taking into account higher-order in $V(\mathbf{q})$ terms (although in [87], this conclusion was not confirmed), or to provide calculations in the frame of the DFT [20]. Therefore, the pseudopotential obtained in the frame of the second-order model is not good enough and the aim of

---

[4] Authors gratefully acknowledged Drs. R. Taylor and L. Dagens for detailed plots of $n(r)$ and results of calculations sent to us.

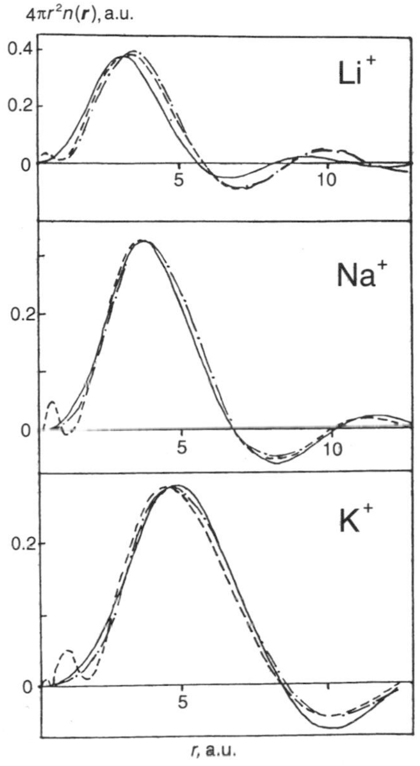

FIGURE 2.9 Calculated electron density, induced by an isolated ion of lithium, sodium, and potassium: — · — the phase shift method [80, 85]; – – – – self-consistent calculations [23]; ——— calculations using $V_{AHA}(\mathbf{q})$ (2.36), (2.39).

further calculations of properties of Li in the framework of the fourth-order model is to establish the qualitative role of higher approximations (see Chapter 3).

Hence, it would be expected that the model pseudopotential (2.36) obtained with the fitting of its two parameters to experimentally measured crystal properties, could also provide (at least in Na and K) an adequate description of the scattering by isolated atoms of alkali metal.

Now, we make some comments about approximation (2.73) for the multi-poles $\Lambda_0^{(3)}$ and $\Lambda_0^{(4)}$. Formula (2.76) for $\Lambda_0^{(3)}$ for different sets of $\boldsymbol{\tau}_1, \boldsymbol{\tau}_2$ was compared with results provided by the exact (in the framework of the ring diagram approximation) expression for $\Lambda_0^{(3)}$ (2.70) (Table 2.13). It turns out that, despite some disagreements, approximation (2.76) provides an adequate description of $\Lambda_0^{(3)}$ by the variation of its arguments. Starting from this fact, and taking into account that, in contrast with calculations of $E_e^{(3)}$, calculations of $E_e^{(4)}$ include an additional summation (this summation provides additional smoothing of the singularities), we can conclude that approximation (2.75)–(2.76) is good enough.

TABLE 2.13 Different approximations for $\Lambda^{(3)}(\boldsymbol{\tau}_1, \boldsymbol{\tau}_2, -\boldsymbol{\tau}_1 - \boldsymbol{\tau}_2)$ for aluminium

$\boldsymbol{\tau}_1$	$\boldsymbol{\tau}_2$ in units of $(2\pi/a)$	$\boldsymbol{\tau}_1 + \boldsymbol{\tau}_2$	$\Lambda_0^{(3)} \times 10^2$ eq. (2.70)	$\Lambda_0^{(3)} \times 10^2$, eq. (2.76) GT	RPA
	1 − 1 − 1	200	6.8	8.6	4.4
	11 − 1	220	2.6	6.6	2.7
	111	222	2.5	5.9	2.3
	200	311	3.1	3.5	2.1
111	−200	−111	8.2	8.5	4.4
	220	331	0.39	1.4	0.79
	−220	−131	1.2	1.7	1.0
	−2 − 20	−1 − 11	12.7	6.6	2.7
	31 − 1	420	1.2	1.1	0.59
	3 − 1 − 1	400	0.97	1.1	0.65
	−31 − 1	−220	1.4	1.7	1.0
	−3 − 1 − 1	−200	32.0	4.3	2.1
	020	220	2.3	3.5	2.1
	200	400	3.6	2.8	1.6
	220	420	3.6	0.68	0.68
	022	222	0.95	1.2	0.85
200	−220	020	7.3	3.5	2.1
	113	313	1.2	0.71	0.53
	−113	113	0.43	0.95	0.71
	−222	022	2.5	1.2	1.8
	−1 − 13	113	0.92	0.34	0.31
	−3 − 11	−111	−0.32	1.4	1.0
220	1 − 31	313	0.25	0.24	0.23
	−2 − 22	002	2.9	1.2	0.85
	2 − 22	402	0.22	0.22	0.21
	0 − 22	202	1.9	0.58	0.51

Note, additionally, one fact. As was shown in calculations, the AHA pseudopotential (with parameters obtained here, see Tables 2.1, 2.2, Figures 2.3, 2.5) is appreciably weaker than the *ab initio* pseudopotential obtained in the frame of the on-Fermi-sphere approximation. Especially large is this effect in light metals Li and Al. The analysis made in the present chapter shows that the localization of the model pseudopotentials using (2.32), i.e., the condition of a correct reproduction of the electron density, provides pseudopotential, of which the form-factors are very close together. The form-factors obtained using (2.32) are very close to the form-factor of the fitted potential $V_{pC}(\mathbf{q})$ (2.36) and are quite different from actually used potentials obtained in the frame of the on-Fermi-sphere approximation (2.30) for all metals considered (except Na). This difference is a consequence of using different fitting procedures for the V_{pC} potential and the localized model potentials RT and HA. These RT and HA potentials are calculated for an isolated ion, whereas the fitting parameters of V_{pC} (AHA) potential were determined for the ion in a crystal. Hence, this fitting also includes some information about the reconstruction of the wave function of core electrons under the influence of neighboring ions in crystal, as well as of valent electrons. Since the inclusion of all effects listed above to the model pseudopotentials needs relatively complicated and cumbersome calculations without simply controlled accuracy, then it seems to be reasonable to use, in quantitative calculations, the potentials with fitting parameters. In this case, many fine effects are automatically taken into account by the fitting procedure. Calculations of many properties of simple metals (see Chapter 3) confirm the conclusion that it is preferable to use the potentials with fitting parameters (although recently, calculations with the *ab initio* non-local pseudopotentials have been successfully performed, too; see [7–9] and references therein).

For any procedure determining the local pseudopotential, the transfer of this pseudopotential to another structure is a separate problem. It relates to the fact that the fitting of the pseudopotential parameters uses the cohesive energy (or its derivatives), i.e., the characteristics calculated with the averaging over the band spectrum. Additionally, it is necessary to prove the consistence of the pseudopotential obtained with the model used in calculating electronic properties. Nevertheless, this attempt was performed for alkali metals in [86]. Calculations of deformations of the Fermi-surface in the frame of the second-order model using the pseudopotential fitted using (2.39) were successful only for Na and K, whereas for Li and Cs, results of calculations demonstrate appreciable disagreement with experimental data. Some other properties (the temperature and volume dependencies of the electrical conductivity of solid and liquid phases, the thermoelectric potential, etc.) also have been successfully described for some metals. However, the accuracy of these calculations is much worse than the accuracy of calculations of atomic properties. Nevertheless, the comparison with results of other calculations (including calculations with *ab initio* potentials) [86] shows that the pseudopotential used provides the best description of electronic properties of alkali metals.

3 Atomic and Thermodynamic Properties of Simple Metals. Anharmonic Effects

Models with local pseudopotential described in section 2 of the previous chapter have been, with great success, used by many other authors to calculate the great number of atomic properties of the whole series of alkali metals (Li, Na, K, Rb, and Cs) and also alkaline-earth (Ca, Sr, Ba) and polyvalent (Al, Pb) metals. In agreement with experimental data, the following were calculated: the phonon spectra, specific heat [1–10], elastic constants of the second and third order [1, 11–15], mean quadratic displacements and the Debye–Waller factor [4, 6, 16], the coefficient of thermal expansion [6, 17], and temperature and volume dependencies of elastic moduli of the second [1, 9, 18–20] and third order [12, 13], equation of state at $T = 0$ and at finite temperatures [9, 20–23], the temperature shift of phonon frequencies and phonon attenuation [2], variation of the phonon spectrum under pressure [3, 24], and deviations from the Cauchy relations [25].[1]

In the present chapter, we will mainly deal with the anharmonic properties of alkali (and other) metals (predominantly, with those for which there are experimental data), using, if necessary, results concerning other properties calculated for models of the second, third and fourth order. Therefore, in the framework of a unified calculation method, the quantitative theory of anharmonic effects in alkali metals is presented. The agreement of theoretical calculations with experimental results, achieved by this procedure, demonstrates good reliability of the method. This agreement also demonstrates a high degree of predictability of the theory. However, we start this chapter with the comparison of energies for some structures. On the one hand, our aim is to demonstrate a high accuracy of the theory, which correctly reproduces the sequences of crystal phases of alkali metals at normal conditions

[1] For a review of advances and applications of the pseudopotential method see [26, 29].

observed experimentally.[2] On the other hand, we will show on this example the best approximations for the polarization operator $\pi(\mathbf{q})$ of those discussed in the last section of the previous chapter.

3.1 Cohesive Energy of Metals. Comparison of Energies of Different Crystal Structures

The cohesive energy of a metal in the framework of the second-order model was already discussed in Chapter 2. It has the form (see equation (2.46))

$$E = E_i + E_{sr} + E_e^{(0)} + E_e^{(1)} + E_e^{(2)}. \tag{3.1}$$

Table 3.1 shows the energy differences between the HCP (and FCC) phases of alkali metals and the BCC phase. These calculations were provided without taking into account the repulsive energy E_{sr} [1] (model A).

The comparison of the results for the energy of different structures shows that results of calculations are predominantly dependent on the approximation used for $\pi(\mathbf{q})$. Table 3.1 shows that only the TW and GT approximations can correctly reproduce the most energetically preferable structure in agreement with low-temperature experiments: the HCP structure for Li and Na and the BCC phase for K, Rb, Cs [27]. The most accurate (from the theoretical point of view) approximation GT predicts for K, Rb, Cs the FCC structure to be the next (in energy scale) after the BCC. This fact agrees also with experiment [27]. For Li and Na, also in accordance with experiments, the next in the energy scale structure is the HCP [27].[3] Ignoring the core energy E_{sr} at $p = 0$ provides no principle change in result, although this term provides contributions to the energy difference comparable to the difference in the energy between different structures ΔE [24] (see also [5, 29]).

Concrete calculations have shown that the value E_{sr} is very sensitive to the crystal structure and the pressure (as a consequence of a strong dependence of E_{sr} upon the interatomic distance). However, this dependence has an appreciable value only for metals with a large enough number of core electrons. For example, for Na and Li the effect of core overlapping remains small up to rather large compressions. On the other hand, for Cs, Rb, and K, the energy

$$E_{sr} = \frac{A}{2} \sum_{\mathbf{R}} \exp(-R/\rho) \tag{3.2}$$

tends (in absolute value) to $E_e^{(2)}$ with increasing pressure (see Table 3.2).

---

[2] The sequences of crystal phases in Li and Na under pressure will be discussed in Chapter 4.

[3] The effective pseudopotential and the structure-dependent part of the total energy E_{str} (3.3) was calculated in [28] using the non-local Animalu–Heine–Abarenkov model form factor (2.9) which includes the hybridization of s- and d-levels. The results of calculation of the total energy of alkali metals with the BCC and HCP structures predict an energetic preference for the BCC structure for all these metals. This fact contradicts experimental results. Hence, we show that the accuracy of present-day first-principle calculations is not good enough.

TABLE 3.1 Crystal structure parameters for alkali metals, calculated using the Animalu–Heine pseudopotential (2.36) and different approximations for the polarization operator: the difference in energy $\Delta E = E - E_{ECC}$ (Ry/atom), the difference in volume $\delta = 10^3(V/V_{BCC} - 1)$, and the ratio c/a

metal	structure	approx. for $\pi(q)$							experiment		
		RPA	GV	TW	VS	GT					
		$\Delta E = E - E_{BCC}$					δ	c/a	ΔE	δ	c/a
Li	FCC		−8.8	−8.8	−8.94	−9.3	2.0				
	HCP		−9.6	−10.0	−8.86	−10.8	2.7	1.629	−4.5	3.3 ± 4.1	1.637 ± 0.003
Na	FCC	−3.8	−3.3	−2.5	−2.5	−2.36	1.9				
	HCP	−3.6	−3.1	−2.7	−1.3	−2.48	2.4	1.633	−3.2	2.7 ± 1.6	1.634 ± 0.001
K	FCC	−1.1	−0.2	1.1	1.8	2.3	2.6				
	HCP	−0.8	0.1	0.8	3.0	2.5	3.0				
Rb	FCC	−0.3	0.7	2.0	2.8	3.3	2.9				
	HCP	−0.2	1.0	1.7	4.0	3.6	3.3				
Cs	FCC	1.0	1.9	3.1	4.0	4.6	3.6				
	HCP	0.7	2.1	2.9	5.2	4.8	4.3				

Comment: RPA, CT, TW, and GT approximations are described in section 2.4 and in references therein. Sources of experimental data see in [1]

The energy E_i can be calculated with arbitrary accuracy. Although this value is a characteristic of the structure only, it, nevertheless, depends weakly on the structure of the metal, and depends appreciably on the volume [29, 30]). The competition between the negative $E_i \sim \Omega^{-1/3}$ and positive $E_e^{(1)} \sim \Omega^{-1}$ (only at very high pressures p does the term $E_e^{(0)} \sim \Omega^{2/3}$ also become important) contributions to the energy predominantly determines the equilibrium density of the metal or the minimum of the cohesive energy $E(\Omega)$, where Ω corresponds to the unit cell volume of the stable phase. The realization of one or another structure depends not only on large volume-dependent terms, but also on small structure dependent terms, in particular, the electron energy $E_e^{(2)}$ and the core energy E_{sr}. Hence, the structural energy E_{str} of heavy alkali metals must also include the energy E_{sr} (Model M2B)

$$E_{str} = E_e^{(2)} + E_{sr}. \tag{3.3}$$

As will be shown in Chapter 4, the core polarization energy E_p must be taken into account to obtain a correct description of the phase (BCC–FCC) transition under pressure, experimentally observed in Cs at $p = 22\,\text{kbar}$ and $\Delta V/V_0 = 0.37$ ($T = 295\,\text{K}$).

Table 3.2 demonstrates all these characteristic features of the behavior of cohesive energy of alkali metals and different contributions to it as a function of the volume and the crystal structure. Note: small dependence of the energy $E_e^{(0)}$ upon Ω; the competition of energies $E_i(\Omega)$ and $E_e^{(1)}(\Omega)$; very small sensitivity of the energy E_i on the structure; and small value of E_{str} in comparison with E and E_i ($\sim 1\%$). The differences in the energies $E_e^{(0)}$, $E_e^{(1)}$, E_i between BCC and FCC phases relate to small differences between volumes of these phases, although the equations of state $p = p(\Omega, T)$ for both these phases are very close together (see also Chapter 4).

In Table 3.2 can be easily seen the role of the contribution E_{sr} (3.2) to the cohesive energy of heavy alkali metals can easily be seen.

1. For all compressions (and also at $p = 0$) the difference between energies of the BCC and FCC phases $|\Delta E_{sr}|$ is comparable with the difference $\Delta E_e^{(2)}$ and with the differences between values of the whole energy ΔE_{FB}.
2. A large dependence of the energy E_{sr} upon the volume approaches, at very high pressures, the value of E_{sr} to the absolute value of $E_e^{(2)}$. Hence, the role of E_{sr} increases with increasing p.
3. The contribution of the energy E_{sr} to the cohesive energy increases in the sequences from Li to Cs, whereas, for the electron contribution to energy $E_e^{(3)}$, the reverse dependence takes place.

All the facts listed above demonstrate clearly the necessity of expanding the model used in calculations by including the energy of the short-range repulsive core interaction. It is especially important in studying structural dependent properties of heavy alkali metals under pressure as well as all other properties under large compressions.

TABLE 3.2 The cohesive energy of heavy alkali metals and different contributions to it (in Ry/atom) for different structures (BCC – (B), FCC – (F)) and volumes

metal	phase	$\frac{\Delta V}{V_0}$	$-E$	$\Delta E_{FB} \cdot 10^5$	$-E_i$	$-E_e^{(0)}$	$E_e^{(1)}$	$-E_e^{(2)} \cdot 10^5$	$\Delta E_{FB}^{(2)} \cdot 10^5$	$E_{sr} \cdot 10^5$	$\Delta E_{sr}^{FB} \cdot 10^5$
K	B	0	0.389544	1.4	0.367588	0.160875	0.143988	510.6	2.2	3.7	0.7
	F	0	0.389530	1.4	0.367281	0.160849	0.143653	508.4	2.2	3.0	0.7
	B	0.3	0.379894	7.7	0.413905	0.163327	0.205562	855.1	13.5	33.7	4.6
	F	0.3	0.379817	7.7	0.413801	0.163326	0.205445	042.6	13.5	29.1	4.6
Rb	B	0	0.370164	0.6	0.344815	0.158413	0.136096	314.7	1.7	11.5	1.9
	F	0	0.370158	0.6	0.344556	0.158383	0.135814	313.0	1.7	9.6	1.9
	B	0.3	0.360193	9.5	0.389525	0.162459	0.196196	541.2	12.3	100.7	11.4
	F	0.3	0.360098	9.5	0.389636	0.162466	0.196401	528.9	12.3	89.3	11.4
Cs	B	0	0.346541	0.8	0.318576	0.154521	0.128475	206.6	2.1	14.9	2.5
	F	0	0.346533	0.8	0.318360	0.154488	0.128237	204.5	2.1	12.4	2.5
	B	0.3	0.337458	10.0	0.359050	0.160050	0.183925	352.4	10.7	124.1	13.5
	F	0.3	0.337358	10.0	0.359269	0.160075	0.184296	341.7	10.7	110.6	13.5

3.2 Lattice Dynamics of Alkali Metals

3.2.1 Basic equations

The phonon frequencies $\omega_{\lambda\mathbf{q}}$ and the microscopic Grüneisen parameters

$$\gamma_{\lambda\mathbf{q}} = -\partial \ln \omega_{\lambda\mathbf{q}} / \partial \ln \Omega, \tag{3.4}$$

at $p = 0$ calculated in the framework of the second order model for some different approximations for the exchange and correlation contributions to the polarization operator $\pi(\mathbf{q})$ were analyzed in [1] for alkali metals. The influence of the hydrostatic pressure upon the behavior of $\omega_{\lambda\mathbf{q}}$ and $\gamma_{\lambda\mathbf{q}}$ was also investigated in [24].

The dynamical matrix $D_{\alpha\beta}(\mathbf{q})$ in the case of mono-atomic crystals can be written as

$$D_{\alpha\beta}(\mathbf{q}) = \tilde{D}_{\alpha\beta}(\mathbf{q}) - D_{\alpha\beta}(0). \tag{3.5}$$

In the case of non-transition metals, in accordance with equation (3.1), the dynamical matrix $\tilde{D}_{\alpha\beta}(\mathbf{q})$ can be presented as the sum of three terms

$$\tilde{D}_{\alpha\beta}(\mathbf{q}) = \tilde{D}^{(i)}_{\alpha\beta}(\mathbf{q}) + \tilde{D}^{sr}_{\alpha\beta}(\mathbf{q}) + \tilde{D}^{(e)}_{\alpha\beta}(\mathbf{q}). \tag{3.6}$$

The terms in this equation have the following nature:

1. The contribution of direct Coulomb interaction between point-like cores (ions) is

$$\begin{aligned} \tilde{D}^{(i)}_{\alpha\beta}(\mathbf{q}) = \frac{4\pi(Ze)^2}{M\Omega} \Bigg\{ \sum_{\boldsymbol{\tau}} \frac{(\boldsymbol{\tau}+\mathbf{q})_{\alpha\beta}}{(\boldsymbol{\tau}+\mathbf{q})^2} \exp\left(-\frac{(\boldsymbol{\tau}+\mathbf{q})^2}{4\sigma^2} \right) \\ + \Omega \frac{2}{\pi^{3/2}} \sum_{\mathbf{R}\neq 0} \left[f_4(R) R_{\alpha\beta} - \frac{1}{2} f_2(R) \delta_{\alpha\beta} \right] \times \left(1 - \cos \widehat{\mathbf{qR}} \right) \Bigg\}, \end{aligned} \tag{3.7}$$

where

$$f_n(R) = \int_0^\infty \exp[-R^2 y^2] y^n \, dy.$$

The direct Coulomb interaction E_i between point-like ions Ze vanishes very slowly with the increase of the distance between ions. For this reason, calculation of the their contribution to the dynamical matrix $D^{(i)}_{\alpha\beta}(\mathbf{q})$ gives a very slowly converging sum over sites of the reciprocal lattice $\boldsymbol{\tau}$. To avoid these difficulties, Ewald (see for example [29, 32]) has suggested the procedure of transforming slowly converging sums over $\boldsymbol{\tau}$ to two rapidly converging sums over the sites of the reciprocal (the sum over $\boldsymbol{\tau}$ in (3.7)) and direct lattices (the sum over $\mathbf{R}$ in (3.7)). (see Appendix B.) From (3.7), we can see that both sums converge rapidly owing to the exponentially decreasing factors in each sum.

2. The repulsive core interaction E_{sr} (3.2) is

$$\tilde{D}^{sr}_{\alpha\beta}(\mathbf{q}) = \frac{A}{M}\sum_{\mathbf{R}\neq 0}\frac{R}{\rho}\exp(-R/\rho)\left[-\frac{\delta_{\alpha\beta}}{R^2}+\frac{R_\alpha R_\beta}{R^4}\left(1+\frac{R}{\rho}\right)\right]\times\left(1-\cos\widehat{\mathbf{qR}}\right), \tag{3.8}$$

3. The contribution of the electron subsystem [33] is

$$\tilde{D}^{(e)}_{\alpha\beta}(\mathbf{q}) = -\frac{\Omega}{M}\sum\frac{\pi(\boldsymbol{\tau}+\mathbf{q})}{\varepsilon(\boldsymbol{\tau}+\mathbf{q})}|V(\boldsymbol{\tau}+\mathbf{q})|^2(\boldsymbol{\tau}+\mathbf{q})_{\alpha\beta}. \tag{3.9}$$

In formulae (3.7)–(3.9) M is the ion mass, $V(\mathbf{q})$ is the local non-screened form factor (see Chapter 2). $\tau_{\alpha\beta\ldots} = \tau_\alpha\tau_\beta\ldots$, Z is the valence of ions, Ω is the volume per atom (in the case of mono-atom crystals with the BCC structure $\Omega = a^3/2$), $\mathbf{R}$ and $\boldsymbol{\tau}$ are vectors of the direct and reciprocal lattices. Other definitions are given in Chapter 2.

The frequency of the phonon with polarization λ and momentum $\mathbf{q}$ is determined by the equation of motion in the framework of the adiabatic approximations

$$\omega^2_{\lambda\mathbf{q}} = e^\alpha_{\lambda\mathbf{q}}D_{\alpha\beta}(\mathbf{q})\, e^\beta_{\lambda\mathbf{q}} = [D]_{\lambda\lambda}, \tag{3.10}$$

where $\mathbf{e}_{\lambda\mathbf{q}}$ is the polarization vector and the summation over repeating indices α and β from 1 to 3 is supposed. The microscopic Grüneisen parameters $\gamma_{\lambda\mathbf{q}}$ for the phonon $\lambda\mathbf{q}$, in accordance with definitions (3.4) and (3.10) (see also section 1, Chapter 1), arc

$$\gamma_{\lambda\mathbf{q}} = -\frac{\Omega}{2\omega^2_{\lambda\mathbf{q}}}\frac{\partial[D]_{\lambda\lambda}}{\partial\Omega}. \tag{3.11}$$

Let us consider concrete expressions for derivatives of the dynamical matrix $\tilde{D}_{\alpha\beta}(\mathbf{q})$ in (3.6). First, let us present $\tilde{D}_{\alpha\beta}(\mathbf{q})$ as

$$\tilde{D}_{\alpha\beta}(\mathbf{q}) = \tilde{T}_{\alpha\beta}(\mathbf{q}) + \tilde{R}_{\alpha\beta}(\mathbf{q}), \tag{3.12}$$

where $\tilde{T}_{\alpha\beta}$ ($\tilde{R}_{\alpha\beta}$) denotes all terms in equations (3.7)–(3.9), containing summations over the reciprocal (direct) lattice $\boldsymbol{\tau}(\mathbf{R})$. Now, it is necessary to recall that, in accordance with the results of Chapter 1, the dynamical matrix (in the same manner as the contributions to the energy of the metal) depends on the parameters of deformations u_{in} (in particular, on the volume) through the change of the volume of the unit cell Ω and also through the change of the vectors of the direct $\mathbf{R}$ and the reciprocal lattices $\boldsymbol{\tau}$, as well as the equilibrium electron density, i.e.,

$$\frac{\partial}{\partial u_{in}} = \frac{\partial\Omega}{\partial u_{in}}\left(\frac{\partial}{\partial\Omega}\right)_{n_0,\boldsymbol{\tau},\mathbf{R}} + \frac{\partial R_\alpha}{\partial u_{in}}\left(\frac{\partial}{\partial R_\alpha}\right)_{n_0,\Omega,\boldsymbol{\tau}}$$
$$+\frac{\partial\tau_\alpha}{\partial u_{in}}\left(\frac{\partial}{\partial\tau_\alpha}\right)_{n_0,\Omega,\mathbf{R}} + \frac{\partial n_0}{\partial u_{in}}\left(\frac{\partial}{\partial n_0}\right)_{\Omega,\mathbf{R},\boldsymbol{\tau}} \tag{3.13}$$

By applying these formula to the matrix $\tilde{D}_{\alpha\beta}(\mathbf{q})$, we obtain

$$\tilde{T}^{in}_{\alpha\beta}(\mathbf{q}) \equiv \partial \tilde{T}_{\alpha\beta}(\mathbf{q})/\partial u_{in} = -\sum_{\tau}\left(\frac{\partial}{\partial p_n} p_{\alpha\beta i} + \delta_{in}\, p_{\alpha\beta}\, n_0 \frac{\partial}{\partial n_0}\right)\psi(p) \qquad (3.14)$$

$$\tilde{R}^{in}_{\alpha\beta}(\mathbf{q}) \equiv \frac{\partial R_{\alpha\beta}(\mathbf{q})}{\partial u_{in}} = \frac{(Ze)^2}{2M}\sum_{\mathbf{R}}(1-\cos\widehat{\mathbf{qR}})$$
$$\times\left[\left(\delta_{\alpha n}R_{i\beta} + \delta_{in}R_{\alpha\beta} + \delta_{\alpha i}R_{n\beta}\right)F_1 + R_{\alpha\beta in}F_2\right], \qquad (3.15)$$

where $\mathbf{p} = \mathbf{q} + \boldsymbol{\tau}$; $p_{\alpha\beta} = p_\alpha p_\beta$; $R_{\alpha\beta} = R_\alpha R_\beta,$

$$\psi(p) = \frac{4\pi(Ze)^2 \exp(-p^2/4\sigma^2)}{M\Omega p^2} - \frac{\Omega\pi(\mathbf{p})}{\varepsilon(\mathbf{p})}|V(\mathbf{p})|^2,$$

$$F_m = \left(\frac{1}{R}\frac{d}{dR}\right)^{m+1}\left[\frac{2}{\sqrt{\pi}R}\int_{\sigma R}^{\infty}\exp(-x^2)\,dx + A\exp(-R/\rho)\right]. \qquad (3.16)$$

The derivatives participating in (3.11) can be calculated using the following equations

$$\frac{\partial[D]_{\lambda\lambda}}{\partial u_{in}} = e^{\alpha}_{\lambda\mathbf{q}}\frac{\partial D_{\alpha\beta}(\mathbf{q})}{\partial u_{in}}\, e^{\beta}_{\lambda\mathbf{q}} = [D^{in}(\mathbf{q})]_{\lambda\lambda,} \qquad (3.17)$$

as follows from (1.36)–(1.39). To obtain $\partial D_{\alpha\beta}(\mathbf{q})/\partial u_{in}$ equations (3.14)–(3.16) are used.

At small $\mathbf{q}$, the value of the small parameter for the expansion of the cohesive energy in a series of the perturbation theory is $\eta = V(\mathbf{q})/E_F \sim 1$. Therefore, as $\mathbf{q} \to 0$ some contributions to $D_{\alpha\beta}(\mathbf{q})$ that contain $V^3(\mathbf{q})$ and $V^4(\mathbf{q})$ are of the order η^2. However, concrete calculations demonstrate that this region of long-wave phonons is very small: $q \leq 10^2\pi/a$. Therefore, to compare the phonon spectrum $\omega_{\lambda\mathbf{q}}$ as well as its characteristics including integrals of $\omega_{\lambda\mathbf{q}}$ with experimental data, we can restrict our consideration of the matrix $D_{\alpha\beta}(\mathbf{q})$ to the terms of the second order in $V(\mathbf{q})$ only, in the same manner as in equation (3.16).

The main notations (needed to understand the text below) for the branches of the phonon spectrum for the BCC structures [35] are given in Table 3.3.

3.2.2 The phonon frequencies

The phonon frequency of all five alkali metals calculated using different approximations from $\pi(\mathbf{q})$, while neglecting E_{sr}, are shown in Table 3.4. This table shows that, after transfering to natural "ionic variables", i.e., after extracting the scaling factors, properties of all alkali matters, especially for K, Rb, Cs, become very close together (and monotonically change in the sequence from Li to Cs [1]).

The effect of the ion–core interaction upon the phonon frequencies $\omega_{\lambda\mathbf{q}}$ in Li is negligable to the pressure $p \sim 1$ Mbar. In the case of Na, this term affects some

TABLE 3.3 Notations for the phonon branches and the symmetric points in the Brillouin zone. Relations between the second-order Birch elastic moduli and the slope of the phonon dispersion curves as $\mathbf{q} \to 0$ [35]

direction	branch	polarization	point	$\mathbf{q}$ in units of $2\pi/a$	elastic moduli and their combinations
	L, Δ_1	(100)			B_{11}
[100]	T, Δ_5	(010)	H_{15}	(1; 0; 0)	B_{44}
	T, Δ_5	(001)			B_{44}
	L, Σ_1	(110)	N_1'		$(B_{11} + B_{12} + 2B_{44})/2$
[110]	T_1, Σ_4	$(1\bar{1}0)$	N_4'	(0.5; 0.5; 0)	B_{33}
	T_2, Σ_3	(001)	N_3'		B_{44}
	L, Λ_1	(111)			$(B_{11} + 2B_{12} + 4B_{44})/3$
[111]	T, Λ_3	$(1\bar{1}0)$	P_4	(0.5; 0.5; 0.5)	$(B_{11} - B_{12} + B_{44})/3$
	T, Λ_3	$(11\bar{2})$			$(B_{11} - B_{12} + B_{44})/3$

branches of the spectrum at very high p. However, in this subsection we consider, in general, moderate compressions up to about tenths of kilobars and can use the approximation of absolute hard cores.

In the case of the group of heavy alkali metals K, Rb, Cs, the similarities of many properties were observed experimentally and theoretically (see [1–18, 12, 13, 15–24]). Although taking into account the energy E_{sr} destroys the quantitative similarity of some characteristics, especially at large p, the qualitative similarity is not destroyed. It also relates to the phonon spectra $\omega_{\lambda\mathbf{q}}$. The phonon spectra $\omega_{\lambda\mathbf{q}}$ of Rb and Cs (in the units of the plasma frequency $\omega_{pl} = (4\pi e^2/M\Omega)^{1/2}$) calculated in [1] for the M2A model (without E_{sr}) at $p = 0$ are the same within the accuracy 1–2% (and also agree with experimental data [36–38]). Here, we present results of calculations of the phonon spectra of Rb and Cs in the framework of the M2B model (with E_{sr}). These spectra are also similar, but the difference between them is something larger and increases with increasing p.

Experimental results for the phonon spectra under external stresses have been obtained for K and Rb at small hydrostatic pressures only [36, 39, 40]. Concrete calculations show that the role of the short-range interaction at small p is small. Only analysis of all of the available experimental data for K and Rb allows us to draw general trends on the dependence of the short-range repulsive inter-core contribution to the phonon spectrum on p.

Tables 3.5 and 3.6 show that the model with the GT approximation for $\varepsilon(\mathbf{q})$ reproduces relatively well the spectrum $\omega_{\lambda\mathbf{q}}$ for K under pressure [39–40] and especially well for Rb [36]. The AHA pseudopotential V_{pC} provides the most accurate description of the low-frequency transversal modes $\omega_{\lambda\mathbf{q}}$, the slope of which as $\mathbf{q} \to 0$ relate to the elastic modulus B_{44}(branches Δ_5 and Σ_3). The AHA pseudopotentials

TABLE 3.4 Phonon frequencies for symmetric points of the Brillouin zone (in units of $\omega_{pl} = (4\pi e^2/M\Omega)^{1/2}$)

metal	approx. of $\pi(\mathbf{q})$	$\xi = (\frac{1}{2}, \frac{1}{2}, 0)$			$\xi = (\frac{1}{2}, \frac{1}{2}, \frac{1}{2})$	$\xi = (1, 0, 0)$
		N_1'	N_3'	N_4'	P_4	H_{15}
	GV	0.561	0.335	0.101	0.408	0.487
	TW	0.514	0.334	0.099	0.380	0.471
Li	VS	0.522	0.335	0.099	0.388	0.477
	GT	0.501	0.333	0.096	0.369	0.463
	exper.	0.514 ± 23	0.331 ± 12	0.104 ± 10		0.512 ± 23
	RPA	0.693	0.370	0.124	0.499	0.561
	GV	0.645	0.369	0.124	0.474	0.554
	TW	0.606	0.369	0.126	0.454	0.545
Na	VS	0.608	0.369	0.126	0.457	0.546
	GT	0.589	0.368	0.125	0.441	0.538
	TW, third order in V	0.602	0.371	0.123		0.548
	exper.	0.541 ± 10	0.364 ± 7	0.124 ± 3	0.406 ± 6	0.512 ± 6
	RPA	0.732	0.384	0.135	0.522	0.590
	GV	0.687	0.384	0.136	0.508	0.585
K	TW	0.655	0.384	0.139	0.493	0.582
	VS	0.654	0.385	0.141	0.495	0.582
	GT	0.633	0.384	0.141	0.479	0.572
	exper.	0.605 ± 10	0.378 ± 5	0.134 ± 5	0.449 ± 5	0.557 ± 5
	VS	0.565	0.385	0.144	0.498	0.584
Rb	GT,	0.635	0.384	0.144	0.480	0.572
	exper.	0.615 ± 8	0.394 ± 12	0.139 ± 8	0.464 ± 6	0.568 ± 6
	VS	0.655	0.384	0.147	0.498	0.582
Cs	GT	0.627	0.383	0.147	0.478	0.570
	exper.	0.620 ± 30	0.360 ± 20	0.130 ± 10	0.450 ± 20	0.550 ± 30
"gele" model		0.929	0.347	0.131	0.577	0.577

V_{ph}[1],[4] as was expected, provides a more accurate description of the middle- and high-frequency parts of the spectrum $\omega_{\lambda\mathbf{q}}$. The pseudopotential V_{pC} overestimate frequencies of longitudinal modes, that takes place in the M2A model [1]. For Rb,

[4] Two AHA pseudopotentials (2.36) V_{pC} and V_{ph} differ in the fitting conditions for parameters: (3.39) or (3.53).

there is practically no difference in values of $\omega_{\lambda\mathbf{q}}$ for both V_{pC} and V_{ph} pseudopotentials. Additionally, as we can see from Table 3.6, both these potentials provide results within the accuracy of experimental data [36]. Note also very precise results for the high-symmetry directions in the Brillouin zone (branches $\Delta_5, \Sigma_4, \Sigma_3, \Lambda_1$) in K (points 1, 4–7 in Table 3.5), as well as for low-symmetry directions ($F_1, G_{1,4}, D_4$ branches) (points 9–12). For branches Λ_3 and Σ_1, the accuracy increases with increasing pressure (points 8, 2).

Let us investigate the difference between calculating $\omega_{\lambda\mathbf{q}}$ for K (Table 3.5) (as well as for Rb (Table 3.6)) at $p = 0$ for two models: M2A and M2B.

The relative variation of $\omega_{\lambda\mathbf{q}}$ is smaller than 1%, but taking into account the core overlapping improves, as a rule, an agreement with experiment. For some points there was an exact agreement of the theory with experiment (within the limit of the experimental accuracy) and including core overlapping does not destroy this agreement. The same conclusion holds for the comparison of theoretical and experimental results for $\omega_{\lambda\mathbf{q}}$ in K and Rb at small p. Although only a few experimental measurements exists for the phonon spectra $\omega_{\lambda\mathbf{q}}$ in Rb for different λ and $\mathbf{q}$, these data allow us to confirm a conclusion about very good reproduction of the branches Δ_5, Σ_4 in calculations and provide additional information concerning the behavior of the branch Σ_1 at small p.

The transversal branch Σ_3 for Rb is not shown. For K, the variation of $\omega_{\mathbf{q}}(p)$ in experiment is much stronger than in theoretical calculations, although at the border

TABLE 3.5 Experimental ([39, 40] $T = 4.5$ K) and theoretical ($T = 0$) values of phonon frequencies $\omega_{\lambda\mathbf{q}}$ (in THz) for potassium at some pressures

point	wave vector	polar-ization	branch	$a = 0.5239$ nm; $p = 0$			
				ω_{exper}	ω_{theor}		
				± 0.01	V_{pC} [1]	V_{pC}	V_{ph}
1	(0; 0; 0.25)	(110)	T_1, Δ_5	0.838	0.838	0.839	0.831
2	(0.12; 0.12; 0)	(110)	L, Σ_1	0.831	0.916	0.914	0.892
3	(0.15; 0.15; 0)	(001)	T_2, Σ_3	1.046	1.137	1.134	1.105
4	(0.15; 0.15; 0)	(110)	T_1, Σ_4	0.706	0.692	0.694	0.688
5	(0.5; 0.5; 0)	(001)	T_2, N_3'	1.532	1.522	1.525	1.511
6	(0.5; 0.5; 0)	(110)	T_1, N_4'	0.550	0.559	0.550	0.534
7	(0.1; 0.1; 0.1)	(111)	L, Λ_1	0.942	0.969	0.968	0.946
8	(0.2; 0.2; 0.2)	(112)	T, Λ_3	0.765	0.923	0.819	0.802
9	(0.7; 0.7; 0.7)	(110)	L, F_1	1.022	1.019	1.019	1.006
10	(0.4; 0.4; 1)	(110)	G_1	0.815	0.837	0.830	0.807
11	(0.4; 0.4; 1)	(110)	G_4	2.380	2.486	2.473	2.395
12	(0.5; 0.5; 0.2)	(110)	D_4	0.965	1.019	1.008	0.976

(contd)

TABLE 3.5 (*contd*) The relation between the lattice constant a and the pressure p provided by the equation of state (see section 4)

point	wave vector	branch	$p = 2.05$ kbar				$p = 1.4$ kbar			
			ω_{exper} ± 0.01	ω_{theor}			ω_{exper} ± 0.01	ω_{theor}		
				V_{pC} [1]	V_{pC}	V_{ph}		V_{pC} [1]	V_{pC}	V_{ph}
1	(0; 0; 0.25)	$T_1,\ \Delta_5$	0.872	0.865	0.868	0.875	0.896	0.886	0.892	0.900
2	(0.12; 0.12; 0)	$L,\ \Sigma_1$	0.929	0.969	0.969	0.974	0.950	1.015	1.016	1.021
3	(0.15; 0.15; 0)	$T_2,\ \Sigma_3$	1.158	1.204	1.203	1.209	1.208	1.261	1.262	1.268
4	(0.15; 0.15; 0)	$T_1,\ \Sigma_4$	0.732	0.714	0.717	0.723	0.742	0.731	0.736	0.742
5	(0.5; 0.5; 0)	$T_2,\ N_3'$		1.569	1.576	1.590	1.607	1.618	1.618	1.633
6	(0.5; 0.5; 0)	$T_1,\ N_4'$	0.588	0.591	0.582	0.583	0.619	0.617	0.608	0.609
7	(0.1; 0.1; 0.1)	$L,\ \Lambda_1$	1.015	1.024	1.023	1.030	1.053	1.069	1.071	1.078
8	(0.2; 0.2; 0.2)	$T,\ \Lambda_3$	0.820	0.861	0.858	0.862	0.866	0.892	0.890	0.895
9	(0.7; 0.7; 0.7)	$L,\ F_1$	1.065	1.056	1.057	1.066	1.103	1.085	1.089	1.098
10	(0.4; 0.4; 1)	G_1	0.862	0.886	0.879	0.882	0.911	0.926	0.921	0.924
11	(0.4; 0.4; 1)	G_4		2.661	2.650	2.659		2.808	2.802	2.810
12	(0.5; 0.5; 0.2)	D_4	1.040	1.082	1.071	1.072	1.066	1.134	1.124	1.125
	a, nm			0.5152		0.511	0.506		0.508	0.504

Comments: V_{pC} [1] is the AHA form-factor (2.36) with parameters determined using (2.39) with $E_{sr} = 0$ (see [1]); V_{pC} and V_{ph} are the AHA form-factors fitted with taking into account E_{sr} with parameters fitted to (2.39) and (2.53) [21, 24] (see Table 2.1)

TABLE 3.6 The phonon frequencies (in THz) for Rb at small pressures. Experimental data are obtained by extrapolation of results of [36] to $T = 0$

point number	wave vector	polar-ization	branch	$a = 0.562$ nm; $p = -0.6$ kbar			$a = 0.5390$ nm; $p = 4.3$ kbar		
				ω_{exper} ± 0.02	ω_{theor}		ω_{exper} ± 0.02	ω_{theor}	
					V_{pC} [1]	V_{pC}		V_{pC} [1]	V_{pC}
1	(0.2; 0; 0)	(100)	L, Δ_1	0.50	0.53	0.51	0.60	0.62	0.61
2	(0.2; 0; 0)	(010)	T, Δ_5	0.42	0.41	0.41	0.45	0.44	0.45
3	(0.2; 0.2; 0)	(110)	L, Σ_1	0.85	0.91	0.88	1.02	1.03	1.03
4	(0.3; 0.3; 0)	(110)	T_1, Σ_4	0.28	0.28	0.26	0.32	0.32	0.30

* The value of the pressure $p = -0.06$ kbar has been calculated from the equation of state for Rb (see Section 4) because the value $a = 0.562$ nm [36] differs from the value used by the fitting of parameters of V_{pC} (see Table 2.1)

of the Brillouin zone (point N_3'), the theoretical value is higher (and inside the zone smaller) than the experimental value (it apparently, relates to a low accuracy of the experimental data). This fact confirms applicability of the M2B model used here. Figures 3.1 and 3.2 show that taking into account the short-range core interaction results in a more rapid increase of the frequency of the branch Σ_3 with increasing p (see also Figures 3.3–3.6). At $p = 0$, there is no difference (within the accuracy of the

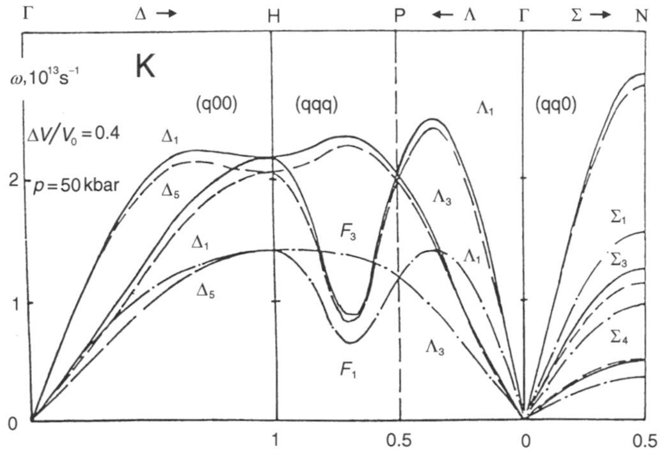

FIGURE 3.1 Phonon frequencies in potassium against the wave vector **q**. At the pressure $p = 50$ kbar (or the relative compression $\Delta V/V_0 = 0.4$): full lines, the M2B model (without taking into account the atomic polarization energy $E_p \approx E_{sr}$); dashed lines, the M2A model (without taking into account the polarization). Dashed–dotted lines correspond to $p = 0$ for both models.

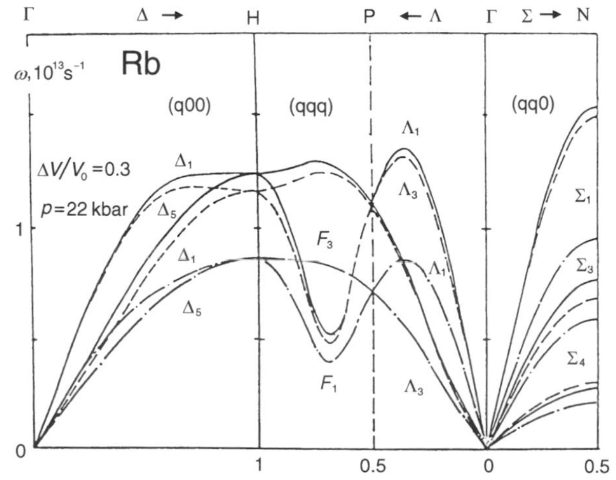

FIGURE 3.2 Phonon frequencies in rubidium against the wave vector **q** at $p = 22$ kbar $(\Delta V/V_0) \approx 0.3$. The same notations as in Figure 3.1.

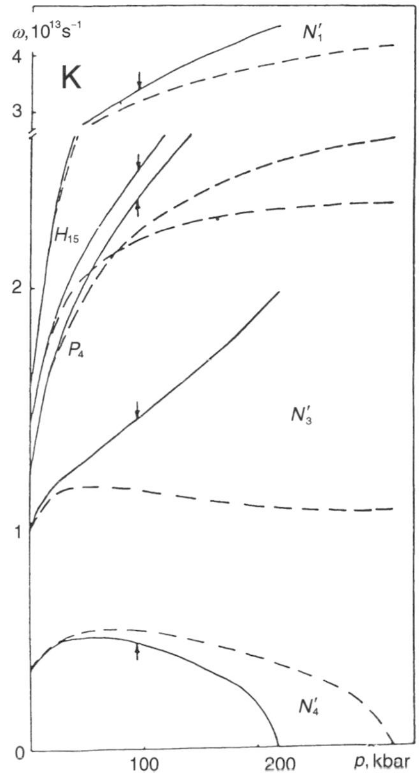

FIGURE 3.3 Phonon frequencies against pressure in the high-symmetry points of the Brillouin zone of potassium: solid line, the M2B model; dashed line, the M2A model. The arrows show the pressure of the BCC–FCC transition.

plot) between plots of $\omega_{\lambda\mathbf{q}}$ for Rb and K calculated in the framework of M2A and M2B models (see Figures 3.1–3.2). The short-range contribution E_{sr} to the increase of $\omega(N_3')$ is of the order of 30% for K at $\Delta V/V_0 = 0.4$ and $p = 50$ kbar. The same contribution for Rb at $\Delta V/V_0 = 0.3$ and $p = 22$ kbar is $\sim 50\%$.

Figures 3.1 and 3.2 show that this similarity of the phonon spectra $\omega_{\lambda\mathbf{q}}$ for K and Rb remains under pressure and taking into account the short-range core interaction does not destroy this similarity, too. The effect of p and the short-range term E_{sr} (which itself is increasing with increasing p) on $\omega_\lambda(\mathbf{q})$ has a different form for different λ and $\mathbf{q}$. A general feature of the model M2B is an increase of all phonon frequencies under pressure, except the branch Σ_4.

The dependence on p is maximal for longitudinal modes ω branches Σ_1, Λ_1 and Δ_1. The slope of the dispersion curves for these branches as $\mathbf{q} \to 0$ correspond, predominantly, to the modulus B (the slope of Δ_1 is determined exclusively by B, see Table 3.3). As can be seen from Figures 3.1 and 3.2, for K as well as for Rb at pressures considered here, values for the slopes of branches Σ_1, Λ_1 and Δ_1 as $\mathbf{q} \to 0$ are very close together for both models. This fact demonstrates once again a small

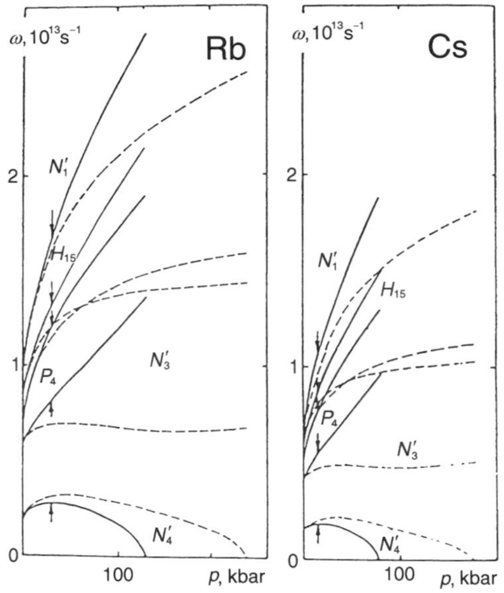

FIGURE 3.4 Phonon frequencies against pressure of the high-symmetry points of the Brillouin zone in rubidium and cesium. The same notations as in Figure 3.3.

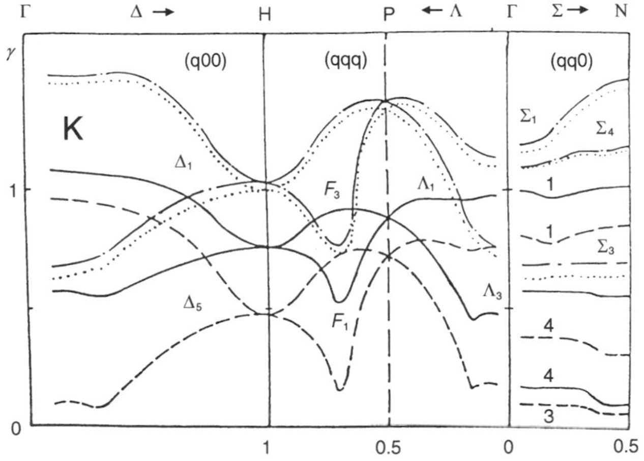

FIGURE 3.5 Microscopic Grüneisen parameter $\gamma_{\lambda\mathbf{q}}$ for potassium against the wave vector **q**. At $p = 50\,\text{kbar}$ ($\Delta V/V_0 \approx 0.4$): solid lines, the M2B model, dashed lines, the M2A model. At $p = 0$: dashed–dotted lines, M2B, dotted lines, M2A.

influence of E_{sr} on $B(p)$. However, for branches Σ_1 and Δ_1, this effect increases with increasing q. For K and Rb, the changes of ω at the border of the Brillouin zone are for $\omega(H_{15}) \sim 7\%$, for $\omega(N_1) \sim 3\%$. For the branch Λ_1, the change of ω is about 3%

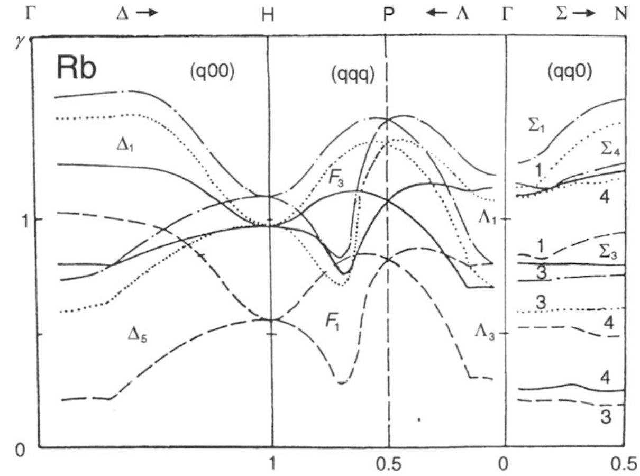

FIGURE 3.6 Microscopic Grüneisen parameter $\gamma_{\lambda\mathbf{q}}$ for rubidium against the wave vector $\mathbf{q}$. For $p = 22\,\text{kbar}$ ($\Delta V/V_0 \approx 0.3$) and $p = 0$. The same notations as in Figure 3.5.

near the border, but at the zone border in the point P_4 this change is smaller (see Figures 3.1, 3.2).

For a low-symmetry direction F_1, the frequency of the longitudinal mode is weakly dependent on p inside the zone. However, this dependence increases by approaching to the zone border. The value E_{sr} increases with increasing p for the longitudinal branch F_3, as well as for the transversal branch F_1.

Let us consider the influence of p and E_{sr} on the transversal branch $\omega_{\lambda\mathbf{q}}$ in K and Rb. The most remarkable feature is the increase of the frequency of the branches F_3 (inside the Brillouin zone) and Λ_3, Δ_5, Σ_3, Σ_4 (on the border) under pressure p. The long-wave behavior of both branches Δ_5 and Σ_3 determines the shear modulus $B_{44} = C_{44} - p$ (see Table 3.3). Figures 3.1 and 3.2 show that, in contrast to the case of the modulus B, the Born–Mayer interaction E_{sr} (3.2) has appreciable effects on the behavior of B_{44} under pressure. Taking into account this term provides a variation of the slope of branches Δ_5 and Σ_3 at $\mathbf{q} \to 0$ under action of p. Its influence is also important for the shear modulus $B_{33} = \frac{1}{2}(C_{11} - C_{12}) - p$, which determines the slope of the branch Σ_4 as $\mathbf{q} \to 0$. However, the effect of the short-range term E_{sr} on the shear moduli B_{44} and B_{33}, similar to the change of $\omega(\Sigma_3)$ and $\omega(\Sigma_4)$ near $q \to 0$ under pressure, has a different character for both models M2A and M2B. For B_{44}, it results in the increase of this modulus with increasing p, whereas in the case of B_{33}, it provides a more rapidly decreasing frequency. Here, it is necessary to consider the correlations in behavior of both branches at the border of the Brillouin zone and as $q \to 0$. Hence, some peculiarities in the pressure dependence of transversal branches (even at the border of the BZ) can affect volume dependences of shear elastic moduli.

Figures 3.3 and 3.4 show the pressure dependences of the phonon frequencies in the high-symmetry points of the Brillouin zone for K, Rb, and Cs up to the pressure, corresponding to the absolute instability of the BCC lattice for these metals, i.e., when the frequency of the branch Σ_4 is equal to zero at point N on the border of the

Brillouin zone. It is well known [41, 42] that the absolute instability of the phase means the necessity of a phase transition in the solid. Phase transitions under pressure are possible for all alkali metals [20], and the value of the short-range repulsion in heavy alkali metals effects only on the value of the critical compression, at which $\omega(\Sigma_4) = 0$. Phase transitions and absolute instabilities in heavy alkali metals will be considered in detail in Chapter 4. Here, we consider only the influence of high pressures and the short-range term E_{sr} on the phonon spectrum of K, Rb and Cs. However, to provide a complete and transparent discussion of this question from a physical point of view, and speaking in advance, we show values of the pressure p_c of the BCC–FCC phase transitions in Cs (experimentally observed), as well as K and Rb calculated in the frame of the M2B model at $T = 0$.

As follows from Figures 3.3 and 3.4, neither high pressures nor the Born–Mayer-repulsion affect the similarity of the phonon spectra of K, Rb and Cs. There is no reason to suppose that the structural phase transition under pressure in Cs can destroy this similarity. The contribution of the short-range interaction to the phonon frequencies for high-symmetry points in the BZ reaches its maximum for the value $\omega(N_3')$. In the frame of the M2B model, the value of $\omega(N_3')$ rapidly (almost linearly) increases with increasing p starting with small p. Other frequencies demonstrates similar behavior, except $\omega(N_4')$. For the M2A model, the value $\omega(N_3')$ changes little with increasing p from zero to the critical value.

Although a smooth decrease of $\omega(N_3')$ takes place at some region of p, nevertheless, the value of $\omega(N_3')$ is almost constant at high compressions (a rapid decrease of $\omega(N_3')$ under pressure discussed in [3, 20] is a consequence of using the dimensionless value ω/ω_p in these calculations [3]; however, the plasma frequency increases with increasing p as $\omega_p \sim \Omega^{-1/2}$). Hence, in the frame of the M2B model, all frequencies (excluding $\omega(\Sigma_4)$) increase with increasing p, in accordance with general physical considerations. In the frame of the M2A model, the frequency of the transversal branch Σ_3 in the [110] direction (nearest to the branch Σ_4) at large compressions neither increases (in contrast to all other branches) nor decreases (in contrast to $\omega(\Sigma_4)$) but remains almost constant. The difference between values of $\omega(N_3')$ calculated in the frame of the M2A and M2B models at $p \sim p_c$ is of the order 20–30% for K, Rb, and Cs. Consequently, experimental measurements of $\omega(\Sigma_3)$ at large p could provide information about the short-range repulsive interaction of cores in alkali metal and also about a degree of applicability of the approximation (3.2) for E_{sr}.

The behavior of $\omega(N_4')$ under pressure is qualitatively the same for both models M2A and M2B. However, the differences between results for K, Rb, and Cs at $p \sim p_c$ are about 10%. This difference in behavior of $\omega(p)$ for the transversal phonon branches Σ_3 and Σ_4 calculated in the framework of the M2B model must be important at large p for the shear elastic moduli B_{33} and B_{44}.

3.2.3 Microscopic Grüneisen parameter

A more important test of the model is a comparison with experimental data for the microscopic Grüneisen parameter $\gamma_{\lambda\mathbf{q}}$ for K and Rb [36, 39] (see Tables 3.7, 3.8 and

TABLE 3.7 Experimental ([39, 40] $T = 4.5\,K$) and theoretical values ($T = 0$) for the macroscopic Grüneisen parameters of potassium

point number	wave vector	polar-ization	branch	$p = 0$					$p = 4.1\,kbar$	
				γ_{exper}	γ_{theor}				γ_{theor}	
					V_{pC} [1]	V_{pC}	V_{ph}	[40]	V_{pC} [1]	V_{pC}
1	(0; 0; 0.23)	(110)	T, Δ_5	0.54 ± 0.15	0.65	0.69	0.73	0.79	0.62	0.64
2	(0.5; 0.5; 0)	(110)	T_1, N_4'	1.54 ± 0.13	1.16	1.18	1.27	1.30	0.99	1.01
3	(0.5; 0.5; 0)	(001)	T_2, N_3'	0.83 ± 0.16	0.64	0.69	0.72	0.81	0.61	0.63
4	(0.18; 0.18; 0)	(001)	T_2, Σ_3	0.49 ± 0.13	0.63	0.67	0.70	0.73	0.60	0.62
5	(0.13; 0.13; 0)	(110)	L, Σ_1	1.76 ± 0.10	1.15	1.19	1.24	1.35	1.10	1.12
6	(0.19; 0.19; 0)	(112)	T, Λ_3	1.23 ± 0.12	0.90	0.94	0.99	1.12	0.83	0.84
7	(0.1; 0.1; 0.1)	(111)	T, F_1	1.20 ± 0.10	1.10	1.14	1.18	1.24	1.06	1.10
8	(0.72; 0.72; 0.72)	(111)	L, Λ_1	0.95 ± 0.07	0.72	0.76	0.79	0.86	0.69	0.72
9	(0.4; 0.4; 1)	(110)	G_1	1.32 ± 0.12	1.39	1.42	1.48	1.32	1.28	1.30
10	(0.5; 0.5; 0.2)	(110)	D_4	1.09 ± 0.10	0.98	1.02	1.06	1.43	0.92	0.93

See comments to Table 3.5

TABLE 3.8 The microscopic Grüneisen parameters for rubidium. Experimental data after [36]

point number	wave vector	polar-ization	branch	$p = 0$ γ_{exper}	$p = -0.6$ kbar γ_{theor}			$p = 4.3$ kbar γ_{theor}	
					V_{pC} [1]	V_{pC}	V_{ph}	V_{pC} [1]	V_{ph}
1	(0.2; 0; 0)	(100)	L, Δ_1	2.15 ± 0.23	1.47	1.57	1.67	1.39	1.47
2	(0.2; 0; 0)	(010)	T, Δ_5	0.99 ± 0.13	0.63	0.76	0.81	0.71	0.74
3	(0.2; 0.2; 0)	(110)	L, Σ_1	1.65 ± 0.07	1.24	1.36	1.42	1.22	1.27
4	(0.3; 0.3; 0)	(110)	T_1, Σ_4	1.31 ± 0.25	1.19	1.24	1.14	0.93	0.91

Figure 3.7). Since $\gamma_{\lambda\mathbf{q}}$ is a derivative of the phonon frequency with respect to volume, then it is determined by the third coordinate derivatives of the inter-atomic potential and, hence, this value is more sensitive to its deviation from the true potential of the crystal lattice than the phonon frequency itself.

Experimental data for the microscopic Grüneisen parameters for K and Rb [36, 39] were obtained only at $p = 0$ by measuring the phonon frequencies at small p as was discussed above. Hence, comparison of experimental data with theoretical values for $\gamma_{\lambda\mathbf{q}}$ can provide conclusions about the influence of the short-range interaction E_{sr} on $\gamma_{\lambda\mathbf{q}}$ for K and Rb only without pressure.

Comparison of the calculated value of $\gamma_{\lambda\mathbf{q}}$ for the M2B model to experimental data, as well as with theoretical values of $\gamma_{\lambda\mathbf{q}}$ calculated in the framework of the M2A model [17], shows that, in contrast to $\omega_{\lambda\mathbf{q}}$, the contribution of the short-range core repulsion to $\gamma_{\lambda\mathbf{q}}$ is not more negligable even at $p = 0$. This term provides an

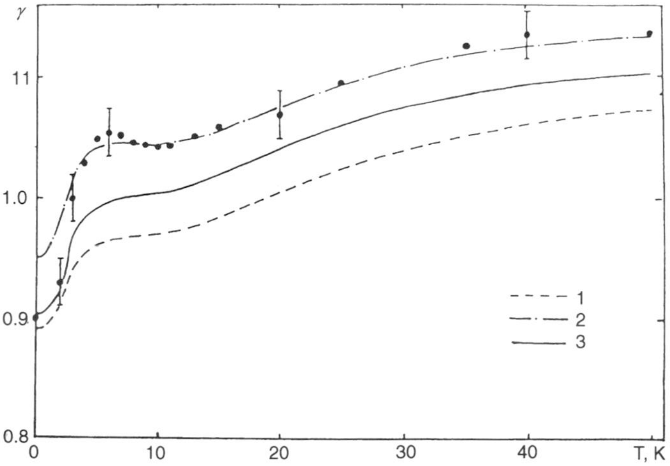

FIGURE 3.7 Macroscopic Grüneisen parameter $\gamma(T)$ for potassium. Experiment point from [35]. Theoretical calculation: 1 – V_{pC}, $E_p = 0$; 2 – V_{ph}, $E_p = 0$; 3 – V_{pC} with E_{sr} (see Table 2.1).

improvment of the agreement with experimental data for Rb and, generally speaking, for K to about 5–10% (see Tables 3.7 and 3.8). In the case of K, there exists some remarkable underestimation of experimental values for $\gamma(\Sigma_3)$ at small q (point 4 in Table 3.7). It relates to the uncertainty of experimental data for $\omega(\Sigma_3)$ inside the Brillouin zone mentioned above [39]. Additionally, taking into account the results of pseudopotential calculations by Taylor and Geldart [43] (see Table 3.7), as well as by other authors (see detailed analysis in [43]), and also by recalling the similarity of the phonon spectra of K and Rb, we can conclude that, at $q \leq 0.2$, the inequality $\gamma_{theor} < \gamma_{exper}$ takes place. For this reason, this disagreement between results of calculations and experimental data for the value γ for the branch Σ_3 as well as results of the calculation of the value $\gamma(\Delta_5)$ for K (point 1 Table 3.7) for transversal branches at small q (which, however, are within the limit of experimental errors) force us to conclude that these experimental data are doubtful. Hence, additional experimental tests are necessary to explain the question. Note that the relation $\gamma_{theor} < \gamma_{exper}$ as $\mathbf{q} \to 0$ for longitudinal phonon branches in K and Rb follows from peculiarities of the pseudopotential calculation scheme for long-wavelength phonons.

Table 3.9 shows results of calculation and experimental data for $\gamma_{\lambda\mathbf{q}}$ as $\mathbf{q} \to 0$. The limiting values $\gamma_{\lambda\mathbf{q}\to 0}$ are determined to be

$$\gamma_{\lambda\mathbf{q}\to 0} = -\frac{1}{2}\frac{d\ln C_\lambda(\mathbf{n})}{d\ln\Omega} - \frac{1}{6}, \tag{3.18}$$

where $C_\lambda(\mathbf{n})$ are some combinations of elastic constants for the branch λ and the direction $\mathbf{n}$. The elastic constants C_{ik} and their derivatives with respect to the volume have been calculated at $T = 0$ in the framework of the M2A model using the V_{pC} pseudopotential (see Table 1 in Ref. [1]). Experimental values C_{ik} were obtained at $T = 0$, their derivatives with respect to the volume change were measured at room temperature (see [1]). The experimental data for γ, shown in Table 3.9, were obtained

TABLE 3.9 Theoretical and experimental values (see text) for the microscopic Grüneisen parameters of potassium and rubidium as $\mathbf{q} \to 0$

direction	branch	polarization	potassium			rubidium		
			γ_{theor}		γ_{exper}	γ_{theor}		γ_{exper}
			$\gamma(st)$	$\gamma(dyn)$		$\gamma(st)$	$\gamma(dyn)$	
[001]	L, Δ_1	(001)	1.63	1.31	1.65 ± 0.14	1.66	1.32	1.74 ± 0.12
	T, Δ_5	(110)	0.63		1.13 ± 0.32	0.73		1.01 ± 0.22
	L, Σ_1	(110)	1.25	1.06	1.46 ± 0.23	1.30	1.05	1.47 ± 0.22
[110]	T_1, Σ_4	(110)	1.09		1.19 ± 0.20	1.10		1.17 ± 0.20
	T_2, Σ_3	(001)	0.67		1.13 ± 0.32	0.73		1.01 ± 0.31
[111]	L, Λ_1	(111)	1.18	1.01	1.43 ± 0.24	1.23	1.00	1.42 ± 0.23
	T, Λ_3	(112)	0.76		1.14 ± 0.26	0.80		1.04 ± 0.25

from experimental results using these "mixed-determined" values of C_{ik} and $d\ln B_{ik}/d\ln\Omega$. However, in [13], it was theoretically shown that the derivatives $dB_{ik}/d\Omega$ decrease with decreasing T from room temperature to $T = 0$ at 8 (7) and 26 (24) per cent for elastic constants B_{11} and B_{44} of K (Rb), respectively. The derivative $dB_{33}/d\Omega$ remains practically constant by the variation of T. Using these data for $d\ln B_{ik}/d\ln\Omega$, errors in experimental data for $\gamma(\mathbf{q}\to 0)$ in Table 3.9 were evaluated.

Table 3.9 shows that, for the transversal branches Σ_3 and Δ_5, in contrast to Table 3.8, γ_{theor} is smaller than γ_{exper}. This relation takes place for other pseudopotential models, too (see [43]). Hence, the experimental results for γ for the transversal branches Σ_3 and Δ_5 at small $\mathbf{q}$ [39] contradict experimental data for C_{ik} and their derivatives with respect to the volume. Hence, further improvement of these measurements would be necessary (note a rather good description of the low-lying branch Σ_4, which determines the thermodynamic properties of alkali metals at low T [17], as well as of the shear module C' under pressure).

The relation $\gamma_{theor} < \gamma_{exper}$ obtained (but not explained) in [43] for small q in the case of longitudinal branches (actually seen as the main difficulty in the pseudopotential description of phonon spectra [43]) is a consequence of the not fully correct (also in our model) use of the dynamical matrix $D_{\alpha\beta}(\mathbf{q})$ (calculated only to the second order in the pseudopotential $V(\mathbf{q})$) in describing the longitudinal branches of the phonon spectrum of alkali metals at $\mathbf{q} = 0$. The values $\gamma(dyn)$ obtained in the frame of this approach as $\mathbf{q}\to 0$ are smaller than experimental data (Table 3.9). As was shown in [33, 34], as $\mathbf{q}\to 0$ it is necessary to keep some terms of the order of $V^3(\mathbf{q})$ and $V^4(\mathbf{q})$ in the expression for $D_{\alpha\beta}(\mathbf{q})$ (3.15), to consider correctly the terms of the order $(V(\mathbf{g})/E_F)^2$. These terms do not affect the spectra of the transversal branches $\omega_{\lambda\mathbf{q}}$ and $\gamma_{\lambda\mathbf{q}}$. The values $\gamma(st)$ calculated with this improvement (see Table 3.9) are in good agreement with experimental data for both K and Rb.

The dependencies $\gamma(\mathbf{q})$ for $p = 0$ and $p \neq 0$, calculated in the framework of the M2A and M2B models for K and Rb, are shown in Figures 3.5 and 3.6. In contrast to the spectra $\omega_{\lambda\mathbf{q}}$, calculated spectra $\gamma_{\lambda\mathbf{q}}$ at $p = 0$ are different for both models. This difference depends on λ and $\mathbf{q}$ and increases in the sequences from K to Cs. For example, for the branches Δ_5 and Σ_3, this differences between values of γ obtained in the frame of the M2A and M2B model is about 20%. Note also that taking into account the short-range core repulsion result in the increase of calculated values for all shown in Figure 3.5 and 3.6 branches for K and Rb. This correction improves the agreement with experimental data at $p = 0$ [36, 39] and, probably, allows more accurate description of characteristic features of the thermal expansion for heavy alkali metals.

This inclusion of the short-range core repulsion results in less rapidly decreasing $\gamma_{\lambda\mathbf{q}}$ with increasing p for all branches (except Σ_4, for which values of γ appreciably decrease)[5] (see Figures 3.5 and 3.6) and at large p, these values become negative. This behavior of $\gamma(\Sigma_4)$ relates to a more drastic decrease of the phonon frequency of the branch Σ_4 at large p as a consequence of large short-range core repulsion mentioned

[5] Decreasing $\gamma_{\lambda\mathbf{q}}$ with increasing pressure for the branch $\Sigma_4 S$ was obtained also in [3], where the core polarization was not taken into account.

above. This behavior can be considered to be a consequence of the "softening" of this phonon mode. As a consequence of a weak pressure dependence of $\gamma_{\lambda\mathbf{q}}$ for the majority of branches of the phonon spectrum, we expect, for calculations in the frame of M2B model, some weak variation of the macroscopic parameter Grüneisen under pressure.

The effect of the short-range term E_{sr} on the dependence $\omega(p)$ for the branch Σ_3 in heavy alkali metals can more clearly be observed in the behavior of the value $\gamma(p)$ for this branch. As we can see from Figures 3.5 and 3.6, the Born–Mayer contribution to the pressure dependence of $\gamma(\mathbf{q})$ for the branch Σ_3 has the opposite sign to the value for the branch Σ_4. For example, for Rb the value $\gamma(\Sigma_3)$, in contrast to all other branches of the spectrum $\gamma_{\lambda\mathbf{q}}$ shown in Figure 3.6, increases with increasing p starting from $\Delta V/V_0 = 0.3$. For K, it takes place starting from $\Delta V/V_0 = 0.4$. Although at these compressions the values $\gamma(\Sigma_3)$ are smaller than at $p = 0$, they are large enough to provide increasing Σ_3 almost linearly in p (Figures 3.3 and 3.4). Some differences between spectra of $\gamma_{\lambda\mathbf{q}}$ in K and Rb (see Figures 3.5 and 3.6) relate to different deviation in the critical compression or in the phase transition point. Nevertheless, even for K (whose ions have the smallest short-range repulsive interaction provided by the overlapping electron shells in comparison with other heavy alkali metals in a non-stressed state) far from eventual phase transition, the values $\gamma(\Sigma_3)$ calculated using the M2A and M2B models differ by a factor of 10 (see Figure 3.5). The short-range term E_{sr} also affects the value $\gamma(p)$ for another phonon branches, especially on transversal phonons, changing their values many times as large.

For compressions larger than those shown in Figures 3.5 and 3.6, the M2B model provides negative values of γ for the branch Σ_4, whereas γ increases for all branches $\Delta_{1,5}$, Σ_1, and $\Lambda_{1,3}$ in the same manner as the value $\gamma(\Sigma_3)$. However, for the M2A model, values of γ for the transversal branches Δ_5, F_1, and Λ_3 decrease and become negative with increasing p (see Figures 3.5 and 3.6). This difference in behavior of $\gamma_{\lambda\mathbf{q}}$ under pressure is rather well observable for K at relatively small compressions ($p < p_c$). While neglecting E_{sr} for K, the value $\gamma(\Sigma_3)$ is approximately zero at $p = 50$ kbar, at $p = 60$ kbar $\gamma(\Sigma_3) < 0$, and at $p = 80$ kbar γ becomes negative for all transversal branches inside the Brillouin zone. Hence, the short-range repulsion of cores at large enough p affect not only absolute value of $\gamma_{\lambda\mathbf{q}}$, but also its sign, which could undoubtedly be observed in experiment.

Note also, that the curves $\gamma(\mathbf{q})$ shows no appreciable dependence on small variations of the pressure as well as on the short-range term E_{sr}. For this reason, the similarity of the phonon spectra of K, Rb, and Cs remains under the action of external stresses on the metal.

3.2.4 Frequency shift of phonons and damping

Further development of the theory of anharmonic effects is one of the most important problems in solid state physics. This problem was considered in many publications. However, until recently, the possibility of constructing any quantitative theory of these effects was thought to be a problem of the distant future. Since experimental study of anharmonic effects is also very difficult, there exists only a small amount of dubious experimental data relating to these effects and their temperature behavior. In experiments where these effect were observed, there was often no clear enough interpretation of their physical nature. For example, in measurements of the specific heat and the thermal expansion at high temperature T, actually it is very difficult to separate contributions provided by genuinely anharmonic effects from contributions provided by the vacancy creation [4, 5, 16, 19, 44]. By studying the temperature shift of the phonon frequency $\Delta\omega_{\lambda\mathbf{q}}(T)$ and the damping $\Gamma_{\lambda\mathbf{q}}(T)$, it is not clear wheter these effects can be described in the framework of actual lowest order in anharmonicity approximation [45] or whether this analysis must also take into account the highest order terms. In particular, this problem is especially important if T is of the order of the melting temperature T_m. Discussing these questions in the present subsection we will follow works [2, 3].

Alkali metals in many respect are simple metals: they have the lowest values of $T_m \sim 300-400\,\mathrm{K}$, and, hence, they can be considered as one of the most suitable objects for studying anharmonic effects. The anharmonic shifts $\Delta\omega_{\lambda\mathbf{q}}(T)$ were experimentally studied in Na [46, 47], K [48], Rb [49], Li [50]; the damping $\Gamma_{\lambda\mathbf{q}}(T)$ was measured in K [48] and Li [50]. However, experimental errors and spreading of experimental data actually are rather appreciable, especially for measurements of $\Gamma_{\lambda\mathbf{q}}$. Only recently were good enough quantitative data for $\Delta\omega_{\lambda\mathbf{q}}$, and $\Gamma_{\lambda\mathbf{q}}$ for Li obtained (see [50] and text below).

Many pseudopotential calculations of values $\Delta\omega_{\lambda\mathbf{q}}(T)$ and $\Gamma_{\lambda\mathbf{q}}(T)$ for alkali metals have been done (see Refs 2, 3, 5–8 in [2]). However, these calculations did not result in more clarity in the questions considered. For example, although, for the majority of the calculations the values of $\Delta\omega_{\lambda\mathbf{q}}(T)$ agree well with experimental data (within the limits of experimental errors, which, however, are rather large), calculated values of $\Gamma_{\lambda\mathbf{q}}$, as a rule, appreciably disagree with experimental data (are 2–5 times smaller than measured values). As a consequence of these difficulties, the conclusion about insufficiency of actually used lowest order anharmonic approximation for the damping at $T \sim T_m$ and about the necessity to use high-order approximations or methods of computer simulation was made in [51, 52]. Moreover, temperature dependency of $\Gamma_{\lambda\mathbf{q}}(T)$ (as well as, in many cases, $\Delta\omega_{\lambda\mathbf{q}}(T)$) calculated in [49, 51–54], in many cases appreciably differ from temperature dependency calculated using the first-order anharmonic perturbation theory [45]. Finally, there exists a big spread of results in the calculation of $\Delta\omega_{\lambda\mathbf{q}}$ and $\Gamma_{\lambda\mathbf{q}}$ for different alkali metals in contrast to a well-known, both theoretical and experimental (see [1–8, 12, 13, 15–24]) similarity of dynamic and static properties of these metals. As will be shown below, there is no physical reason for a conclusion about the inapplicability of actual anharmonic perturbation theory for these metals, all disagreements, mentioned above, are, probably, a consequence of a inadequate application or small accuracy of models used and (or) methods of calculation.

The anharmonic shift $\Delta\omega_{\lambda\mathbf{q}}$ and the damping $\Gamma_{\lambda\mathbf{q}}$ are actually calculated [2, 3] for all alkali metals from Li to Cs using the second-order pseudopotential model, described in Chapter 2. There are reasons to suppose that, for the values $\Delta\omega_{\lambda\mathbf{q}}(T)$ and $\Gamma_{\lambda\mathbf{q}}(T)$, the accuracy of this model is large enough and the results of calculation will provide reliable information about values and the physical nature of anharmonic effects. Simultaneous analysis, in the frame of a unified scheme of a series of metals, simplifies the interpretation of results, and, in particular, makes the comparison with experiment more transparent and reliable.

General formulae needed for the calculation of the shift of the phonon frequency $\Delta\omega_{\lambda\mathbf{q}}$ and the damping $\Gamma_{\lambda\mathbf{q}}$, were discussed in Chapter 1. Now we rewrite these formulae especially for metals considered here.

Let us introduce the dimensionless values $\tilde{\mathbf{q}}$ and $\tilde{\omega}_{\nu\mathbf{k}}$, for $\mathbf{q}$ and $\omega_{\nu\mathbf{k}}$ expressing $\mathbf{q}$ through the parameter a and $\omega_{\nu\mathbf{k}}$ in units of the ion plasma frequency $\omega_p^2(\Omega_0) = 4\pi(Ze)^2/M\Omega_0$, i.e., $\tilde{\mathbf{q}} = \mathbf{q}a/2\pi$; $\tilde{\omega} = \omega/\omega_p$. Values of a and $\nu_p = \omega_p(\Omega_0)/2\pi$ used in calculations are listed in Table 3.3. Similary, let us introduce dimensionless values

$$\tilde{V}_{\nu\lambda\mu,} = V_{\nu\lambda\mu}\frac{a}{2\pi}\frac{\Omega}{(Ze)^2}, \qquad \tilde{V}_{\nu\nu\lambda\lambda} = V_{\nu\nu\lambda\lambda}\frac{a^2}{4\pi^2}\frac{\Omega}{4\pi(Ze)^2} \tag{3.19}$$

instead of $V_{\nu\lambda\mu}, V_{\nu\nu\lambda\lambda}$ in (1.52), (1.53). Now, formulae (1.48), (1.49) for $\Delta^{(3)}_{\nu\mathbf{k}}$ and $\Delta^{(4)}_{\nu\mathbf{k}}$ have a final form used in calculations

$$\tilde{\Delta}^{(3)}_{\nu\mathbf{k}} = \frac{\Delta^{(3)}_{\nu\mathbf{k}}}{\omega_p} = -\frac{1}{\tilde{\omega}_{\nu\mathbf{k}}}\sqrt{\frac{\hbar^2}{Ma(Ze)^2}}\left(\frac{\pi}{8}\right)^{3/2}\times\sum_{\lambda,\mu}\int d^3\tilde{q}\,\frac{\tilde{L}^{\mathbf{k},\mathbf{q},\mathbf{k}+\mathbf{q}}_{\nu\lambda\mu}}{\tilde{\omega}_{\lambda\mathbf{q}}\tilde{\omega}_{\mu,\mathbf{k}+\mathbf{q}}}\left(\tilde{V}^{\mathbf{k},\mathbf{q},\mathbf{k}+\mathbf{q}}_{\nu\lambda\mu}\right)^2; \tag{3.20}$$

$$\tilde{\Delta}^{(4)}_{\nu\mathbf{k}} = \frac{\Delta^{(4)}_{\nu\mathbf{k}}}{\omega_p} = \frac{4}{\tilde{\omega}_{\nu\mathbf{k}}}\sqrt{\frac{\hbar^2}{Ma(Ze)^2}}\left(\frac{\pi}{8}\right)^{3/2}\times\sum_{\lambda}\int\frac{d^3\tilde{q}}{\tilde{\omega}_p}\left(n_{\lambda\mathbf{q}}+\frac{1}{2}\right)\tilde{V}^{\mathbf{kkqq}}_{\nu\nu\lambda\lambda}; \tag{3.21}$$

where all integrals over $\tilde{\mathbf{q}}$ are taken over the Brillouin zone (BZ) and the value $\tilde{L}_{\nu\lambda\mu}$ is obtained from (1.50) by the substitution of all $\omega_{\lambda\mathbf{q}}$ with the values $\tilde{\omega}_{\lambda\mathbf{q}}$.

The damping or the total width of the phonon peak $\Gamma_{\nu\mathbf{k}}$, in accordance with [45, 48, 49], is taken as double[6] expressions like (1.48). Only in formula (1.50), which determines $L_{\nu\lambda\mu}$, all principal values $\hat{\mathcal{P}}(1/x)$ must be substituted with the factor $-\pi\delta(x)$ (1.51), and the first (non-singular) term must be omitted. As a result, we obtain instead of (3.20), the dimensionless damping $\tilde{\Gamma}_{\nu\mathbf{k}} - \Gamma_{\nu\mathbf{k}}/\omega_p$

$$\tilde{\Gamma}_{\nu\mathbf{k}} = \frac{2\pi}{\tilde{\omega}_{\nu\mathbf{k}}}\sqrt{\frac{\hbar^2}{Ma(Ze)^2}}\left(\frac{\pi}{8}\right)^{3/2}\times\sum_{\lambda,\mu}\int d^3\tilde{q}\,\frac{S^{\mathbf{k},\mathbf{q},\mathbf{k}+\mathbf{q}}_{\nu\lambda\mu}}{\tilde{\omega}_{\lambda\mathbf{q}}\,\tilde{\omega}_{\mu,\mathbf{k}+\mathbf{q}}}[\tilde{V}^{\mathbf{k},\mathbf{q},\mathbf{k}+\mathbf{q}}_{\nu\lambda\mu}]^2; \tag{3.22}$$

$$\begin{aligned} S^{\mathbf{k},\mathbf{q},\mathbf{k}+\mathbf{q}}_{\nu\lambda\mu} &= \left(1+N_{\lambda\mathbf{q}}+N_{\mu,\mathbf{k}+\mathbf{q}}\right)\delta\left(\tilde{\omega}_{\nu\mathbf{k}}-\tilde{\omega}_{\lambda\mathbf{q}}-\tilde{\omega}_{\mu,\mathbf{k}+\mathbf{q}}\right) \\ &+ \left(N_{\lambda\mathbf{q}}-N_{\mu,\mathbf{k}+\mathbf{q}}\right)\left[\delta\left(\tilde{\omega}_{\nu\mathbf{k}}+\tilde{\omega}_{\lambda\mathbf{q}}-\tilde{\omega}_{\mu,\mathbf{k}+\mathbf{q}}\right)-\delta\left(\tilde{\omega}_{\nu\mathbf{k}}-\tilde{\omega}_{\lambda\mathbf{q}}+\tilde{\omega}_{\mu,\mathbf{k}+\mathbf{q}}\right)\right] \end{aligned} \tag{3.23}$$

Note, that the formula for $\tilde{\Gamma}_{\nu\mathbf{k}}$ (3.22) describes the Landau damping of phonons [55]. In this case, the condition $\omega_{\nu\mathbf{k}}\Gamma^{-1}_{\nu\mathbf{k}} \gg 1$ is satisfied. The Akhiezer mechanism [56]

[6] Note that our $\Gamma_{\nu\mathbf{k}}$ denotes the total width but not the half-width Γ_{half} as in [45.48.51–54], i.e., our $\Gamma_{\nu\mathbf{k}} = 2\Gamma_{half}$.

(internal friction), for which it is possible to consider phonons to be in a uniform deformation field, is not considered here.

The first term in (3.23) determines, as is well known, the "decay" damping of phonons, which exists even at $T=0$. The second term relates to the scattering by other phonons; this term vanishes at $T=0$.

The expression for the high-temperature damping $\tilde{\Gamma}_{\nu\mathbf{k}} = \tilde{\Gamma}^{\infty}_{\nu\mathbf{k}}(T)$, which, as will be show below, can actually be used for all T, is

$$\tilde{\Gamma}^{\infty}_{\nu\mathbf{k}}(T) = k_B T \frac{\pi^2}{32}\frac{a}{(Ze)^2}\sum_{\lambda,\mu}\frac{d^3\tilde{q}}{\tilde{\omega}_{\lambda\mathbf{q}}\,\tilde{\omega}_{\mu,\mathbf{k}+\mathbf{q}}}\left(\tilde{V}^{\mathbf{k},\mathbf{q},\mathbf{k}+\mathbf{q}}_{\nu\lambda\mu}\right)^2$$
$$\times\left[\delta\left(\tilde{\omega}_{\nu\mathbf{k}} - \tilde{\omega}_{\lambda\mathbf{q}} - \tilde{\omega}_{\mu,\mathbf{k}+\mathbf{q}}\right) + \delta\left(\tilde{\omega}_{\nu\mathbf{k}} + \tilde{\omega}_{\lambda\mathbf{q}} - \tilde{\omega}_{\mu,\mathbf{k}+\mathbf{q}}\right)\right.$$
$$\left. + \delta\left(\tilde{\omega}_{\nu\mathbf{k}} - \tilde{\omega}_{\lambda\mathbf{q}} + \tilde{\omega}_{\mu,\mathbf{k}+\mathbf{q}}\right)\right]. \tag{3.24}$$

Since the dimensionless frequencies $\tilde{\omega}_{\lambda\mathbf{q}}$ have close values for different alkali metals (see Table 3.10), using similarity of expressions for the vertices $\tilde{V}_{\nu\lambda\mu}$ and the dynamical matrix $D_{\alpha\beta}/\omega_p^2$, which determines these $\tilde{\omega}_{\lambda\mathbf{q}}$, we can expect from (3.24) that values of $\tilde{\Gamma}^{\infty}_{\nu\mathbf{k}}$ for different metals with close values of a, i.e. for K, Rb, and Cs, would also be close together. As will be shown below, this is also actually true for Na and Li, for which decrease of a is compensated with some decrease of the frequency $\tilde{\omega}_{\lambda\mathbf{q}}$.

These formulae correspond to the lowest order in anharmonicity. They are valid when the relative shift and the relative damping

$$\delta_{\nu\mathbf{k}} = \frac{\Delta_{\nu\mathbf{k}}}{\omega_{\nu\mathbf{k}}}; \quad \eta_{\nu\mathbf{k}} = \frac{\Gamma_{\nu\mathbf{k}}}{\omega_{\nu\mathbf{k}}} \tag{3.25}$$

are small:

$$\delta_{\nu\mathbf{k}} << 1; \quad \eta_{\nu\mathbf{k}} << 1. \tag{3.26}$$

For metals considered here, these conditions are fulfilled to $T \approx T_m$ for all phonon branches, except the critical, anomaly "soft" branch Σ_4, for which $\delta \sim 0.2$–0.3 and $\eta \sim 0.3$––0.4 at $T \sim T_m$. For non-small δ and η, in describing the dynamic properties instead of simple formulae (3.20), (3.21), we must, generally speaking, use a more general formalism, calculating the phonon spectrum as poles of the phonon Green's function $K(\mathbf{k},\omega)$ (see, e.g., [45, 42]). However, in our case, large values of the ratios $\delta = \Delta/\omega$ and $\eta = \Gamma/\omega$ are not a consequence of large values of the anharmonic contributions $\Delta_{\nu\mathbf{k}}$ and $\Gamma_{\nu\mathbf{k}}$. These large values relate to anomalically small values of the harmonic frequencies themselves: $\tilde{\Delta}_{\nu\mathbf{k}}, \tilde{\Gamma}_{\nu\mathbf{k}} \sim 0.05$ and $\tilde{\omega}(\Sigma_4) \sim 0.1$, instead of actual values $\tilde{\omega}_{\nu\mathbf{k}} \sim 0.5$ (see Table 3.10). For this reason, similar to the formalism used for the "soft modes" in the theory of displacement-type phase transitions [42], expressions (3.20)–(3.23) for $\Delta_{\nu\mathbf{k}}$ and $\Gamma_{\nu\mathbf{k}}$ are actually valid for the critical branch Σ_4 even at $T \approx T_m$. Only in calculations of observable values, e.g., the cross-section of inelastic neutron scattering, these $\Delta_{\nu\mathbf{k}}$ and $\Gamma_{\nu\mathbf{k}}$ must be used not in simple formulae of the lowest anharmonic approximation (like (11) and (12) from [48]), but in more

TABLE 3.10 Phonon frequencies, anharmonic shift and damping at $T = 0$ for symmetric points of the Brillouin zone

$\tilde{\mathbf{k}}$	branch	metal	$\tilde{\omega}_{\lambda\mathbf{q}}(0)$ theory	$\tilde{\omega}_{\lambda\mathbf{q}}(0)$ experiment	$\Delta\tilde{\omega}^{zp}_{\lambda\mathbf{q}} \cdot 10^2$	$\Delta\tilde{\omega}^{zp}_{\lambda\mathbf{q}}/\tilde{\omega}_{\lambda\mathbf{q}}(0) \cdot 10^2$	$\tilde{\Gamma}_{\lambda\mathbf{q}}(0) \cdot 10^2$	$\eta_{\lambda\mathbf{q}}(0) \cdot 10^2$
		Li	0.487	0.512 ± 32	2.6	5.4	1.04	2.14
		Na	0.538	0.512 ± 6	1.5	2.8	0.40	0.74
(1,0,0)	H_{15}	K	0.573	0.557 ± 5	1.0	1.7	0.25	0.43
		Rb	0.574	0.568 ± 6	0.6	1.1	0.15	0.26
		Cs	0.571	0.550 ± 30				
		Li	0.416		1.6	4.1	0.56	1.34
(1/2,		Na	0.448	0.406 ± 6	1.0	2.1	0.23	0.50
1/2,	P_4	K	0.483	0.449 ± 5	0.6	1.3	0.16	0.34
1/2)		Rb	0.484	0.464 ± 6	0.4	0.8	0.11	0.22
		Cs	0.479	0.450 ± 20	0.3	0.6	0.08	0.17
		Li	0.572	0.514 ± 23	2.1	3.7	1.19	2.07
		Na	0.597	0.541 ± 10	1.2	2.0	0.45	0.75
	N_1'	K	0.639	0.605 ± 10	0.8	1.2	0.26	0.40
		Rb	0.639	0.615 ± 8	0.5	0.8	0.16	0.26
		Cs	0.629	0.620 ± 30	0.3	0.5	0.12	0.19
		Li	0.337	0.331 ± 12	1.4	4.0	0.15	0.45
(1/2,		Na	0.369	0.364 ± 7	0.7	1.9	0.07	0.20
1/2,	N_3'	K	0.384	0.378 ± 5	0.4	1.2	0.05	0.14
0)		Rb	0.385	0.394 ± 12	0.3	0.8	0.03	0.09
		Cs	0.383	0.360 ± 20	0.2	0.5	0.03	0.07
		Li	0.112	0.104 ± 10	−0.12	−1.1	0.008	0.07
		Na	0.130	0.124 ± 3	0.00	0.0	0.002	0.02
	N_4'	K	0.143	0.134 ± 5	0.02	0.1	0.002	0.01
		Rb	0.145	0.139 ± 8	0.02	0.1	0.001	0.01
		Cs	0.148	0.130 ± 10	0.02	0.1	0.001	0.01

Comments: calculated values after [2], experimental data for Cs after [37]. Experimental errors in units of 10^{-3}.

general relations like formula (10) from [48]. This formalism needs solutions of equations like

$$2\omega_{\lambda\mathbf{k}}\Sigma_{\lambda\lambda}(\mathbf{k},\omega) = \omega[2\Delta_{\lambda\mathbf{k}}(\omega) - i\Gamma_{\lambda\mathbf{k}}(\omega)],$$

where $\Sigma_{\lambda\lambda}(\mathbf{k},\omega)$ is the mass-operator for the phonon Green's function $K(\mathbf{k},\omega)$.

Calculations of $\Delta\omega_{\nu\mathbf{k}}$ and $\Gamma_{\nu\mathbf{k}}$ were provided in [2, 3] using the second- order model with the AHA pseudopotential (see Chapter 2) with $E^{zp} \neq 0$, $E_p = 0$. Since in [2, 3] the short-range core repulsion E_{sr} was not taken into account, we can conclude that, although the results for $p = 0$ [2] are adequate, the accuracy of the description of the effect of the pressure on $\Delta\omega_{\nu\mathbf{k}}$ and $\Delta_{\nu\mathbf{k}}$ [3] is unknown, we cannot provide any conclusion about the quality of these calculations because experimental data are absent. For this reason, we will consider [3] only briefly. For the polarization operator $\pi(\mathbf{q})$ for all metal (except Li) the Geldart–Taylor (GT) (for Li Geldart–Vosko (GV)) approximation was used in [2]. As we have already emphasized, these approximations for $\pi(\mathbf{q})$ provide the most accurate description of the properties of these metals. We will not consider here any details of calculation and will refer the reader to work [2].

Let us make some comments relating to the experimental data and their errors, shown in Figures 3.8, 3.9, 3.12, 3.13. For Na the data for $\Delta\omega(T)$ and estimations of errors of those are taken from [53] (where a number of unpublished results were quoted). For other metals, we use the data from [48–50]. For the symmetry points H_{15} and P_4 in Figures 3.8 and 3.12, in many cases we have plotted many data, obtained from experiment by approaching these points from different directions in the BZ. Since, factually, the spectrum is triply degenerate in these points, the spread of results characterize the value of experimental errors.

Values $\Delta\omega(T) = \omega(T) - \omega(0)$ in Figures 3.8–3.10 were obtained from experimental data for the frequency shift $\Delta\omega_{obs}(T, T_1) = \omega(T) - \omega(T_1)$ using the following procedure. If the lowest temperature in the experiment T_1 was small, as in [48] for K ($T_1 = 9\,\mathrm{K}$) and in [49] for Rb ($T_1 = 12\,\mathrm{K}$), then we have the assumptions $\omega(T_2) = \omega(0)$ and $\Delta\omega_{obs}(T, T_1) = \Delta\omega(T)$. But if T_1 is not small, as for Na ($T_1 = 90\,\mathrm{K}$) and for Li ($T_1 = 100\,\mathrm{K}$), then the values of the shift $\Delta\omega(T_1)$ were supposed to be equal to their theoretical value $\Delta\omega_{calc}(T_1)$ from [2]. It gives

$$\Delta\omega(T) = \Delta\omega_{obs}(T, T_1) + \Delta\omega_{calc}(T_1).$$

For Li, the phonon frequencies for the branches Σ_4 and Δ_5 were measured in [50] at $T = 295\,\mathrm{K}$ only, whereas data for $\omega_{\nu\mathbf{k}}$ at $T = 98\,\mathrm{K}$ for the same corresponds to slightly different values of the wave vector $\mathbf{k}$ (see [2, 50]). To obtain from these data $\Delta\omega_{\nu\mathbf{k}}(T)$, an extrapolation procedure was used in [2]. For each value of $\mathbf{k}$, this procedure uses the experimental data [50] for $\omega_{\nu\mathbf{k}}$ (98 K) for the points most closely located to $\mathbf{k}$ ($\Delta\tilde{k} \leq 0.02$). Then these values were extrapolated to $\mathbf{k}$ using theoretically calculated values of $\partial\omega^0_{\nu\mathbf{k}}/\partial\mathbf{k}$. Resulting values of $\Delta\omega_{\nu\mathbf{k}}(T)$ are shown in Figures 3.8, 3.9, the experimental error was evaluated as a sum of errors provided by both the measurements and interpolations.

The measurements of the damping $\Gamma(T)$ in Li in [50] were perfomed using the assumption that the observed width W_{obs} relates to the intrinsic width Γ and the resolution W_{res} by the relation $B_{obs} = (W^2_{res} + \Gamma^2)^{1/2}$. It was also supposed, that at the lowest temperature achieved in the experiment, $T = 100\,\mathrm{K}$, $\Gamma(T)$ has negligible value, i.e. $W_{res} = W_{obs}(100\,\mathrm{K})$. But Figure 3.11 shows that at $T = 100\,\mathrm{K}$, the damping Γ in Li has, generally speaking, a remarkable value. For this reason, in

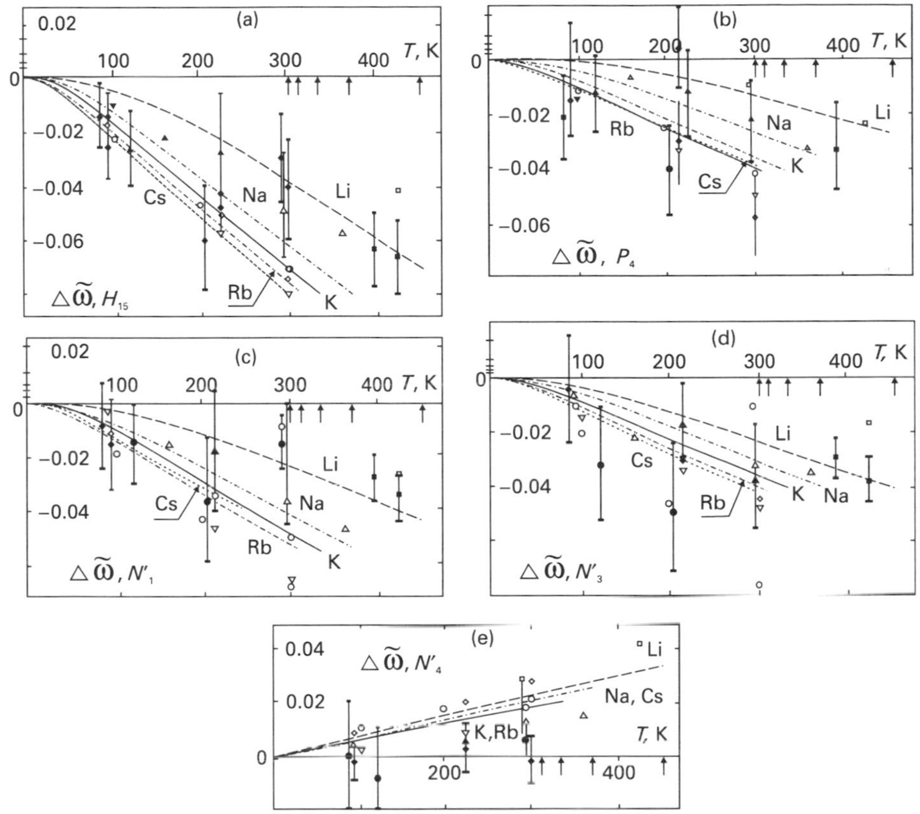

FIGURE 3.8 Temperature dependencies of the frequency shift $\Delta\tilde{\omega}(T) = \Delta\omega(T)/\omega_p$ for the high-symmetry points of the Brillouin zone (BZ): H_{15}, P_4, N_1', N_3', N_4'. The curves are results from [2] for: – – – Li; –·–·– Na; —— K; —··— Rb; ······ Cs. For each metal the curve was drown to the melting point T_m marked by an arrow in the T axes. Heavy symbols denote experiment data. Light symbols are results of calculations of other authors. Horizontal lines on the ordinate show values of $\Delta\tilde{\omega}^{zp}$ (the anharmonic shift at $T = 0$). The order of those correspondes to increasing of $|\Delta\tilde{\omega}^{zp}|$ from Cs to Li (Figures 3.8–3.13 after [2, 3]).

proceeding with the experimental data [50] for $W_{obs}(T)$, we have used in evaluating W_{res} theoretically calculated values for $\Gamma_{calc}(100\,\mathrm{K})$ [2], i.e., the experimental values $\Gamma_{\exp}(T)$ for Li, shown in Figures 3.11 and 3.12, were obtained from the data [50] as

$$\Gamma_{\exp}(T) = \left[\, W^2_{obs}(T) - W^2_{obs}(100\,\mathrm{K}) + {}^{2}_{\mathrm{calc}}(100\,\mathrm{K}) \right]^{1/2}.$$

Figures 3.9, 3.12 show both dependences $\Delta\omega_{\nu\mathbf{k}}$ and $\Gamma_{\nu\mathbf{k}}$ on ν and $\mathbf{k}$ inside the BZ. Experimental data are shown only for Li (after [50]). Experimental errors in measurements [50] are appreciably smaller than in measurements [48] for K and [49] for Rb.

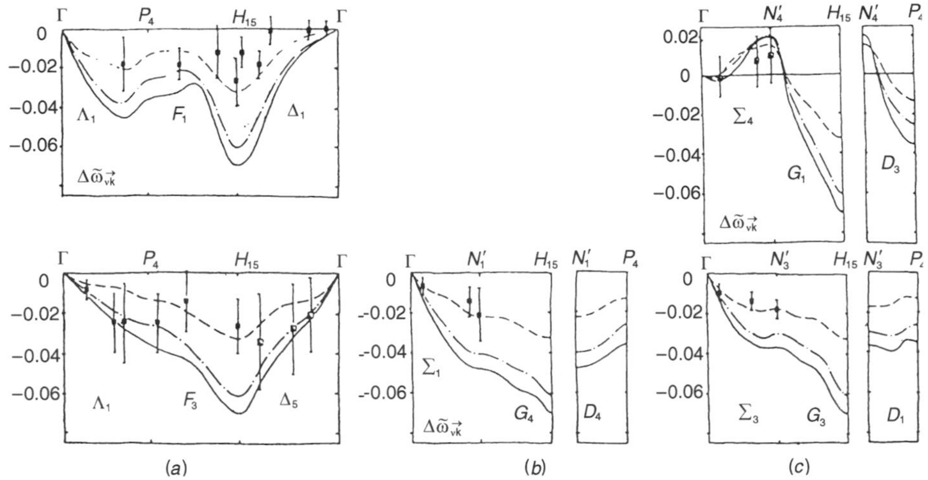

FIGURE 3.9 Variation of $\Delta\tilde{\omega}_{\nu\mathbf{q}}(T)$ in the BZ as a function of ν and $\mathbf{q}$. The curves are results of calculations in [2]: the difference $\Delta\tilde{\omega}_{\nu\mathbf{q}}(T, T_1) = \Delta\tilde{\omega}_{\nu\mathbf{q}}(T) - \Delta\tilde{\omega}_{\nu\mathbf{q}}(T_1)$ for $T = 295$ K, $T_1 = 100$ K (dashed line) for Li; $\Delta\tilde{\omega}_{\nu\mathbf{q}}(T)$, $T = 295$ K (dotted line) for Na; $\Delta\tilde{\omega}_{\nu\mathbf{q}}(T)$, $T = 299$ K (full line) for K. Experimental results are shown only for Li (after [2]).

Table 3.10 shows low-temperature values for the shift $\Delta\omega^{zp}_{\nu\mathbf{k}}$ and the frequency $\omega_{\nu\mathbf{k}}(0) = \omega^0_{\nu\mathbf{k}} + \Delta\omega^{zp}_{\nu\mathbf{k}}$. We can see that, although the relative shift $\Delta\omega^{zp}_{\nu\mathbf{k}}/\omega(0)$ is small, nevertheless it reaches the rather remarkable values 4–5% for Li, 2–3% for Na and 1–2% for K. These values can be considered to be a characteristic of the contribution of zero oscillations to all thermodynamic quantities.

As was mentioned in [1, 4, 17] and is demonstrated in Table 3.10, this model provides the most accurate description of the lowest and middle branches of the phonon spectrum. But, these low-lying branches actually provide the main contribution to anharmonic effects considered here. It follows from the fact that the denominator of expressions (3.20)–(3.23) includes the powers of the frequency $\omega_{\lambda\mathbf{q}}$ (for $\Delta^{(3)}_{\lambda\mathbf{q}}$ and $\Gamma_{\lambda\mathbf{q}}$ these powers are rather high) and at large $T \geq \theta_D$ an additional power of $\omega^{-1}_{\lambda\mathbf{q}}$ arises from the power series expansion of $N_{\lambda\mathbf{q}} + 1/2 \approx k_B T/\hbar\omega_{\lambda\mathbf{q}}$. Hence, we conclude that, for the properties considered here, the accuracy of the model is good enough.

The results of calculation of the temperature behavior of $\Delta\omega_{\lambda\mathbf{q}}(T)$ for high-symmetry points in the BZ are shown in Figure 3.8 for all five metals together with experimental data and results of other calculations. Figure 3.9 shows results of calculations of $\Delta\omega_{\lambda\mathbf{q}}(T)$ as a function of λ and $\mathbf{q}$ at constant T. We see results of [2] to be in good agreement with results of the most precise measurements [50] for Li for all points and direction in the BZ. For Na, K, and Rb, these results also agree well with the theory within the limits of experimental error (except, maybe, the N_4' point). However, large values of experimental errors make it impossible to provide any quantitative comparison.

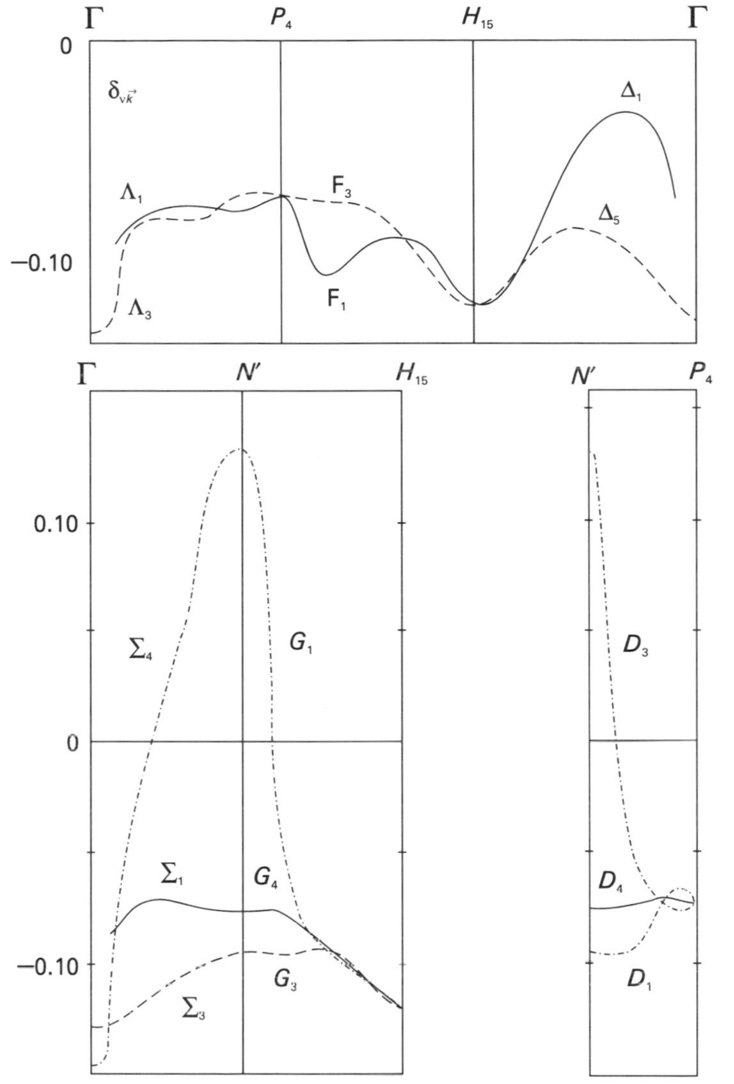

FIGURE 3.10 Variation of the relative frequency shift $\delta_{\nu\mathbf{q}}(T) = \Delta\,\omega_{\nu\mathbf{q}}(T)/\omega_{\nu\mathbf{q}}(0)$ in the BZ for potassium at $T = 299\,\mathrm{K}$ (after [2]).

Especially interesting seem to be the results for the point N in the Brillouin zone (Figures 3.8, 3.9) corresponding to the critical branch Σ_4. This value, $\Delta\,\omega_{\lambda\mathbf{q}}$, demonstrates in all theoretical calculations an anomalous behavior: at non-small $T \sim \theta_D$, this frequency increases with increasing T instead of the actual observed decrease. We see that for Li, this effect is confirmed by the experimental data [54].

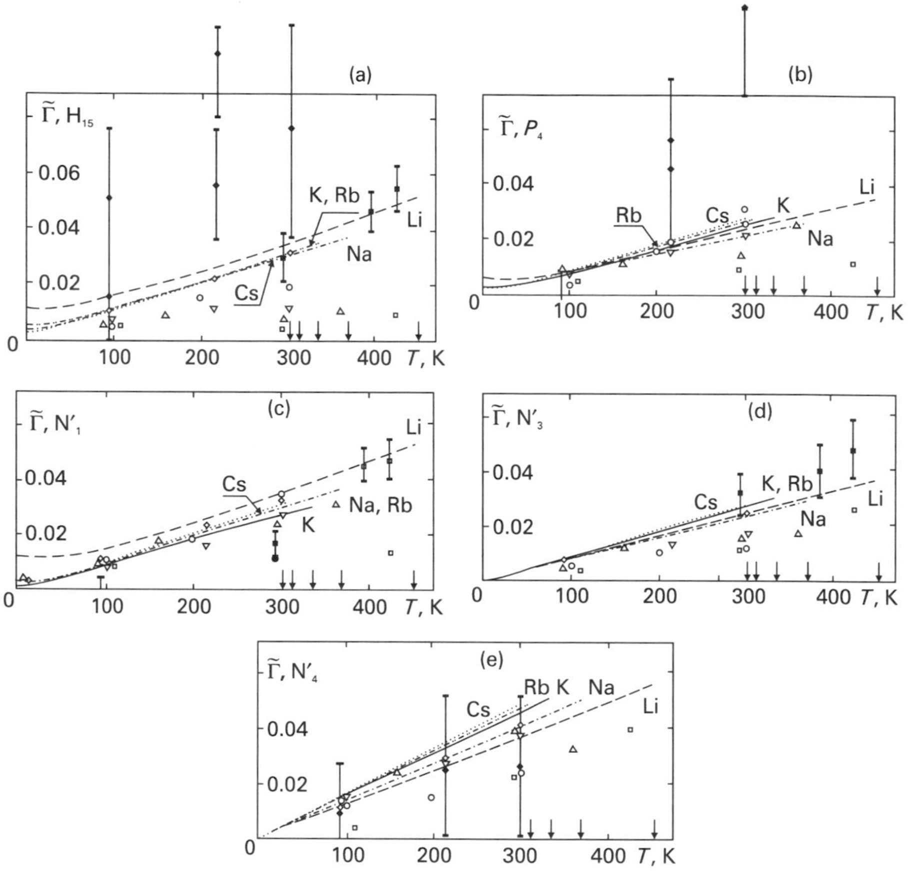

FIGURE 3.11 Temperature dependence of the damping $\tilde{\Gamma}(T) = \Gamma(T)/\omega_p$ in the high-symmetry points of the BZ. The same notations as in Figure 3.8 (after [2]).

Above we have seen that the accuracy of the model used in calculations of $\Delta\,\omega_{\lambda\mathbf{q}}$ in Li is large enough for all points inside the BZ and there exists some similarity of dynamic properties of all alkali metals (see, e.g., tables and figures of the present subsection). Hence, we can expect that, in other alkali metals, $\omega(N_4')$ also increases with T and this effect can be observed in more precise experiments.

The temperature behavior of $\Delta\,\omega_{\lambda\mathbf{q}}(T)$ (Figure 3.8), calculated in the framework of the lowest-order anharmonic approximation, demonstrates a simple monotonic behavior, which is determined by the factors $2N_{\lambda\mathbf{q}} + 1 = \coth(\hbar\omega_{\lambda\mathbf{q}}/2k_BT)$. The high-temperature linear dependence of $\Delta\,\omega_{\lambda\mathbf{q}}$ on T for all metals, except Li, takes place starting from rather low values of T. For the quasi-harmonic contribution $\Delta_{\lambda\mathbf{q}}^{(qh)}$ in (1.56) as well as for the specific heat, it takes place at $T \geq \theta_D/2$ [4, 6], whereas for $\Delta_{\lambda\mathbf{q}}^{(3)}$ and $\Delta_{\lambda\mathbf{q}}^{(4)}$ at even much lower T. This fact is a consequence of large contributions

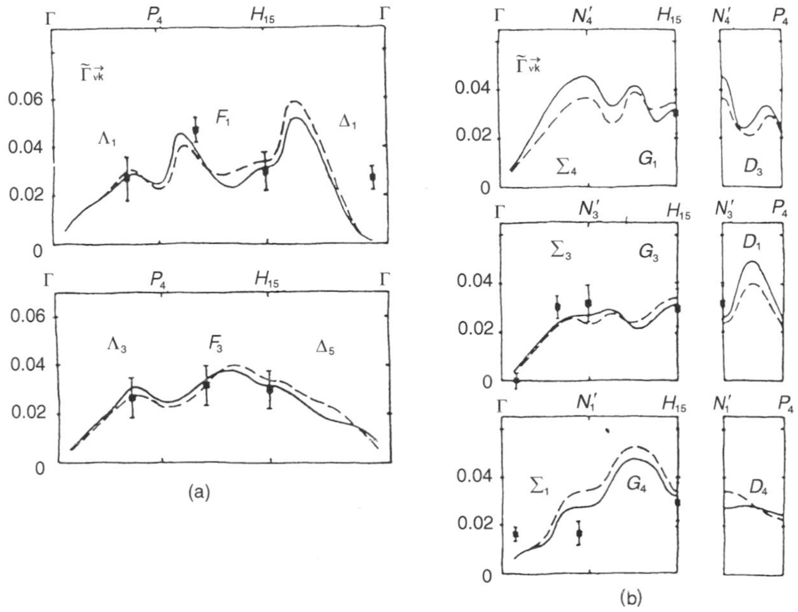

FIGURE 3.12 Variation of $\tilde{\Gamma}_{\nu q}(T)$ in the BZ of lithium at $T = 295\,\mathrm{K}$ and potassium at $T = 299\,\mathrm{K}$. The same notations as in Figure 3.8 (after [2]).

provided by the region of small $\omega_{\lambda\mathbf{q}}$ to the integrands in (3.20)–(3.21), mentioned above. Hence, relative shifts of the frequency $\delta_{\lambda\mathbf{q}}(T)$ at large T can be written as

$$\delta_{\lambda\mathbf{q}}(T) = \frac{\Delta\,\omega_{\lambda\mathbf{q}}(T)}{\omega_{\lambda\mathbf{q}}(0)} = \frac{k_B(T^{\omega}_{\lambda\mathbf{q}} - T)}{E^{\omega}_{\lambda\mathbf{q}}}, \tag{3.27}$$

where $T^{\omega}_{\lambda\mathbf{q}}$ and $E^{\omega}_{\lambda\mathbf{q}}$ are some characteristics of the metal. The energy $E^{\omega}_{\lambda\mathbf{q}}$ is a characteristic of the cohesion in the metal, which lowers the anharmonic shifts, and the values $T^{\omega}_{\lambda\mathbf{q}} = E^{\omega}_{\lambda\mathbf{q}}\Delta\,\omega^{zp}_{\lambda\mathbf{q}}/\omega_{\lambda\mathbf{q}}(0)$ characterize the temperature region, where the simple relation (3.27) is valid. Table 3.11 shows that the values $T^{\omega}_{\lambda\mathbf{q}}$ for metals considered are really small: $T^{\omega}_{\lambda\mathbf{q}} \sim (0,3\text{--}-0.5)\theta_D$ (values of θ_D are given in Table 3.3). Hence, relations (3.25) can be used in a wide interval of T. The energy $E^{\omega}_{\lambda\mathbf{q}}$ (see Table 3.11) actually have close values for different metals (as well as the characteristic values E_δ of amplitudes of thermal oscillations [4]) and are about 0.2–$-0.1\,\mathrm{eV}$, i.e., these values are of the same order of magnitudes as the energy of vacancy creation for these metals (see [7]).

Variations of the relative shifts $\delta_{\lambda\mathbf{q}}$ by changing λ and $\mathbf{q}$ inside the BZ are shown in Figure 3.10 for K at $T = 299\,\mathrm{K}$. Similar results were obtained for other metals and temperatures.

The plots for the longitudinal acoustic branches Λ_1, Δ_1, Σ_1 in Figure 3.10 are drawn only for $\tilde{q} \geq 0.05$. It is a consequence of the fact that at smaller q the frequencies $\omega_{\lambda\mathbf{q}}$ correspond to the so-called "hydrodynamic" regime, where these

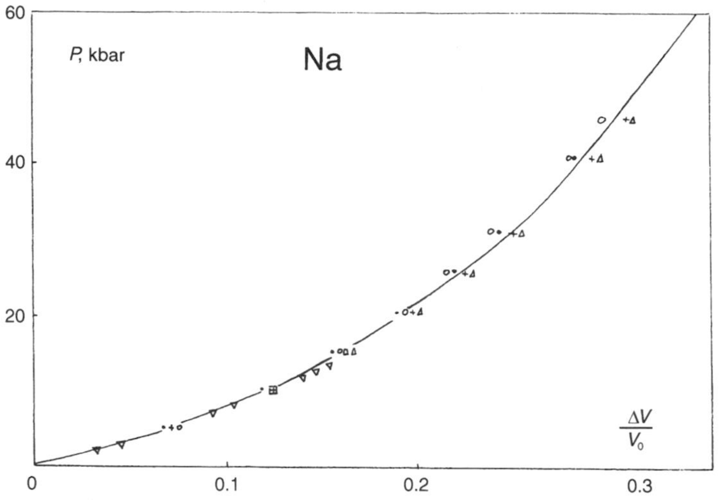

FIGURE 3.13 The $T = 295\,\mathrm{K}$ isotherm $p(\Delta V/V_0)$ of sodium (after [20]).

frequencies become of the order of the mean phonon damping: $\omega \leq \langle\Gamma\rangle$. Further decreasing q and ω results in a well-known hydrodynamic dispersion of the sound velocity (and the damping, see below). It transforms the isothermic oscillations to adiabatic ones. To describe this region, the kinetic description should be used [57]. Since this hydrodynamic dispersion does not take place for the transversal branches (see [57]), then for the branches $\Lambda_3, \Delta_5, \Sigma_3$ and Σ_4 values $\delta_{\lambda\mathbf{q}}$ as $q \to 0$ (Figure 3.10) were obtained using the isothermic elastic constants $C_{ik}(T)$ calculated in [18] (see section 4 of this chapter).

The values $\delta_{\lambda\mathbf{q}}$ rapidly and non-trivially change within the BZ, especially in the vicinity of the critical branch Σ_4. This variation of $\delta_{\lambda\mathbf{q}}$ is more rapid than the variation of the Grüneisen parameters $\gamma_{\lambda\mathbf{q}}$ [5, 17], which determine, in accordance with (1.57), the "quasi-harmonic" shift $\Delta_{\lambda\mathbf{q}}^{(qh)}$ (see details in [2, 3]).

Table 3.12 shows the role of different contributions to the total value $\delta_{\lambda\mathbf{q}}$ (1.56) for high-symmetry points in the BZ and $T = T_m$. Contributions $\Delta_{\lambda\mathbf{q}}^{(3*)}$ and $\Delta_{\lambda\mathbf{q}}^{(4*)}$ are frequenly mutually compensated, e.g., in the points P_4 and N_1', but sometimes these contributions have the same sign, as, e.g., in H_{15} and N_3'. The quasi-harmonic contribution $\Delta^{(qh*)}$ provides a major contribution (80–90%) to the total value $\Delta\,\omega_{\lambda\mathbf{q}}$ only in the case of the compensation mentioned. In other cases, it provides about one third of the total effect.

Table 3.12 shows also relative and absolute values of frequency shifts at $T = T_m$, i.e., their maximal values for metals considered. The relative shifts $\delta_{\lambda\mathbf{q}}(T)$ remain small even at $T = T_m$: $|\delta_{\lambda\mathbf{q}}| \leq 0.1{-}{-}0.15$ for all phonon (except the critical modes Σ_4 in Na and Li, where $\delta_{\lambda\mathbf{q}}(T_m) \sim 0.2{-}{-}0.3$). Hence, approaching the melting point does not give any appreciable change of phonon frequencies.

TABLE 3.11 Parameters $E_{\lambda\mathbf{q}}^{\omega,\Gamma}$ and $T_{\lambda\mathbf{q}}^{\omega,\Gamma}$ for symmetric points in the Brillouin zone

branch	metal	$E^{\omega}, 10^3$K	T^{ω}, K	$E^{\Gamma}, 10^3$K	T^{Γ}, K
	Li	2.27	122	4.20	90
	Na	2.09	59	5.30	39
H_{15}	K	2.16	36	5.39	23
	Rb	2.11	22	5.47	15
	Cs	2.05	15	5.53	10
	Li	4.31	178	5.39	72
	Na	3.71	79	6.44	33
P_4	K	3.38	46	5.82	20
	Rb	3.35	28	5.55	12
	Cs	3.53	20	5.25	9
	Li	3.94	147	4.91	102
	Na	3.38	66	6.01	45
N_1'	K	3.43	41	6.95	28
	Rb	3.34	25	6.46	17
	Cs	3.72	17	6.09	12
	Li	2.82	114	4.25	19
	Na	2.93	56	4.82	10
N_3'	K	2.82	33	4.37	6
	Rb	2.71	20	4.32	4
	Cs	2.61	14	4.15	3
	Li	−1.43	15	0.91	0.64
	Na	−1.87	0.3	0.97	0.14
N_4'	K	−2.26	3	0.94	0.13
	Rb	−2.35	4	0.92	0.06
	Cs	−2.17	4	0.91	0.05

Table 3.12 shows that for each mode λ, $\mathbf{k}\ \delta_{\lambda\mathbf{k}}$ at $T = T_m$ has similar values for all metals (except the critical branch Σ_4). This proximity of anharmonic properties at $T = T_m$ is quite similar to "the melting criteria" by Lindeman [32] and Varshni [58] discussed above. This fact illustrates once again the similarity of properties of all alkali metals in solid, as well as in liquid, phases.

Table 3.10 shows values of both the absolute and relative damping, $\Gamma_{\lambda\mathbf{q}}$ and $\eta_{\lambda\mathbf{q}}$ (3.25), at $T = 0$. As was already mentioned, this damping is provided by the decay from the one-phonon to the two-phonon state with smaller energies. Hence, this damping has significant value only for upper branches of the spectrum, in particular,

TABLE 3.12 Frequency shifts Δ and the damping Γ at the melting temperature T_m

branch	metal	$10^2 \cdot \tilde{\Delta}^{(gh*)}$	$10^2 \cdot \tilde{\Delta}^{(3*)}$	$10^2 \cdot \tilde{\Delta}^{(4*)}$	$10^2 \cdot \Delta\tilde{\omega}(T_m)$	$\delta(T_m)$	$10^2 \cdot \tilde{\Gamma}(T_m)$	$\eta(T_m)$	η^m
	Li	−2.69	−4.15	−0.26	−7.10	−0.146	5.26	0.108	0.125
	Na	−3.06	−4.35	−0.60	−8.01	−0.149	3.76	0.070	0.082
H_{15}	K	−2.84	−4.10	−1.06	−8.00	−0.139	3.58	0.062	0.073
	Rb	−2.76	−4.04	−1.10	−7.90	−0.138	3.28	0.057	0.066
	Cs	−2.55	−4.26	−1.18	−7.99	−0.140	3.04	0.055	0.063
	Li	−2.68	−1.80	1.82	−2.66	−0.064	3.49	0.084	0.090
	Na	−3.25	−1.86	1.58	−3.53	−0.079	2.58	0.058	0.063
P_4	K	−3.18	−2.05	1.19	−4.04	−0.084	2.79	0.058	0.063
	Rb	−3.14	−2.08	1.11	−4.11	−0.085	2.73	0.056	0.062
	Cs	−2.80	−2.17	1.14	−3.83	−0.080	2.69	0.057	0.062
	Li	−3.80	−2.24	1.59	−4.45	−0.078	5.28	0.092	0.100
	Na	−4.56	−2.38	1.55	−5.39	−0.090	3.69	0.062	0.068
N_1'	K	−4.49	−2.16	1.15	−5.50	−0.086	3.09	0.048	0.053
	Rb	−4.43	−2.16	1.09	−5.50	−0.086	3.09	0.048	0.053
	Cs	−3.63	−2.35	1.17	−4.81	−0.076	3.05	0.050	0.054
	Li	−1.50	−1.41	−1.15	−4.06	−0.121	3.59	0.107	0.121
	Na	−1.46	−1.18	−1.33	−3.97	−0.107	2.84	0.077	0.086
N_3'	K	−1.21	−1.28	−1.64	−4.13	−0.108	2.96	0.077	0.087
	Rb	−1.15	−1.32	−1.69	−4.16	−0.108	2.78	0.072	0.081
	Cs	−1.00	−1.39	−1.83	−4.22	−0.110	2.72	0.073	0.082
	Li	−0.80	−4.56	8.80	3.44	0.306	5.60	0.499	0.382
	Na	−0.80	−3.43	6.79	2.56	0.198	4.58	0.384	0.321
N_4'	K	−0.79	−3.48	6.42	2.15	0.151	5.14	0.360	0.313
	Rb	−0.82	−3.28	6.04	1.94	0.134	4.93	0.340	0.299
	Cs	−0.86	−2.95	5.89	2.08	0.141	4.80	0.332	0.291

in the points H_{15}, N_1', and P_4. However, even for Li values of $\eta_{\lambda\mathbf{q}}(0)$ are small and do not exceed 1–2%.

The results of calculations of the temperature dependences of $\Gamma_{\lambda\mathbf{q}}(T)$ are given in Figure 3.11. Variations in $\Gamma_{\lambda\mathbf{q}}$ with λ and $\mathbf{q}$ within BZ are shown in Figure 3.12. Experimental data for $\Gamma_{\lambda\mathbf{q}}$ are available only for K [48] and Li [50]. Similar to the case of $\Delta\omega_{\lambda\mathbf{q}}(T)$, the theory agrees well with the majority of existing experimental data [50] for Li. Significant disagreement takes place for branches Δ_1 and Σ_1 at small q only (see Figure 3.12). For this region, measurements [50] also show rather large values of Γ instead of theoretically predicted values $\Gamma \approx 0$. However, from general physical reasons, it follows that, at small q the values $\Gamma_{\lambda\mathbf{q}}$, must tend to zero and these disagreements are, probably, the consequences of experimental errors only. For K, our values of $\Gamma_{\lambda\mathbf{q}}$ are close to theoretical values of $\Gamma_{\lambda\mathbf{q}}$ from [48] and, in the same manner as these last, are remarkably smaller than the experimentally measured values of $\Gamma_{\lambda\mathbf{q}}$ in [48]. However, since experimental errors for the damping in [48] are very large, then these disagreements probably do not exist at all. At the same time, the similarity of properties of alkali metals makes any appreciable difference between values of $\tilde{\Gamma}_{\lambda\mathbf{q}}(T)$ for Li and K less probable (see Figure 3.11).

The temperature dependences $\Gamma_{\lambda\mathbf{q}}(T)$ (see Figure 3.11) are more smooth than $\Delta\omega_{\lambda\mathbf{q}}(T)$, and the interval of linear dependence on T starts earlier. This fact relates to a large contribution to (3.22) provided by the region of small $\omega_{\lambda\mathbf{q}}$, as was already mentioned. For this reason, in describing the temperature behavior of the damping, it is more useful to use the high-temperature expansion (3.24), which (together with (3.25)) gives

$$\eta_{\lambda\mathbf{q}}(T) = \frac{k_B T}{E^{\Gamma}_{\lambda\mathbf{q}}}, \tag{3.28}$$

This relation is valid at $T \ggg T^{\Gamma}_{\lambda\mathbf{q}} = E^{\Gamma}_{\lambda\mathbf{q}}\eta_{\lambda\mathbf{q}}(0)$. The values of the parameters $T^{\Gamma}_{\lambda\mathbf{q}}$ and $E^{\Gamma}_{\lambda\mathbf{q}}$ for the high-symmetry points in the BZ are given in Table 3.11. We can see, that values of $T^{\Gamma}_{\lambda\mathbf{q}}$ for the upper branch of the spectrum are smaller than $T^{\omega}_{\lambda\mathbf{q}}$ (3.27) by a factor of 1.5–2, and by a factor of 5–10 for the middle and lower branches. Hence, for the middle and lower branches formula (3.28) is valid practically for all T (see Figure 3.11). We see also, that the energies $E^{\Gamma}_{\lambda\mathbf{q}}$ are actually 1.5–2 times higher than $E^{\omega}_{\lambda\mathbf{q}}$, at about 0.4–0.6 eV. These energies also have close values for different metals. The critical branch Σ_4, as usual, demonstrates a very anomalous behavior, which results in extremely small $E^{\Gamma}_{\lambda\mathbf{q}} \sim 0.1$ eV, i.e., in very rapidly increasing damping with T.

Both the dampings $\tilde{\Gamma}_{\lambda\mathbf{q}}(T)$ and $\eta_{\lambda\mathbf{q}}(T)$ have similar values for different alkali metals and for all T (Figures 3.11 and 3.12). Even for Li, whose properties mostly differ from other alkali metals, values of $\tilde{\Gamma}_{\lambda\mathbf{q}}$ at room temperature are close to $\tilde{\Gamma}_{\lambda\mathbf{q}}$ for K. For other metals, the calculated values $\tilde{\Gamma}_{\lambda\mathbf{q}}(T)$ are even closer to the results obtained for K. The proximity of values $\tilde{\Gamma}_{\lambda\mathbf{q}}(T)$ for different alkali metals for all non-small temperatures is a very important prediction made by the theory.

The three last columns in Table 3.12 show characteristics of the damping at $T = T_m$. Together with $\eta(T_m) = \Gamma(T_m)/\omega(0)$, we also give values of $\eta^m = \Gamma(T_m)/\omega(T_m)$, where $\omega(T_m) = \omega(0) + \Delta\omega(T_m)$. From the formal point of view, this correction to the frequency ω by calculating the relative damping is rather an overestimation of the accuracy limit of the lowest-order perturbation theory (PT) used. Table 3.12 shows that the difference between η^m and $\eta(T_m)$ usually does not exceed 10–15%, i.e. has, generally speaking, the same order of magnitude as the highest order terms of the PT (omitted here). But, for the critical branch Σ_4 (for Li and Na), the difference between η^m and $\eta(T_m)$ can be of about 30% and, as was mentioned in the begining of the present subsection, the usage of the PT for calculating this quantity can be aproved despite not very small $\delta_{\lambda\mathbf{q}}$ and $\eta_{\lambda\mathbf{q}}$. In this case,

η^m are more precise characteristics of the damping and the width of the phonon peaks than $\eta(T_m)$.

Table 3.12 shows that, for all phonons, except for the critical branch Σ_4, the relative damping remains small, $\eta \leq 0.05$–0.1 even at T_m. This very important result means that, at temperatures up to the melting point, the conception of phonons has a full physical sense from both the thermodynamical, and the kinetic points of view. Hence, this concept can be used for the description of properties of metals at all $T \leq T_m$.

The similarity in behavior of the damping at $T = T_m$, i.e. the proximity of the values $\eta_{\lambda\mathbf{k}}(T_m)$ or $\eta_{\lambda\mathbf{k}}^m$ for different metals, takes place for K, Rb, and Cs only. For Li, the values η^m are 1.5–2 times smaller than for other metals. Hence, calculations in [2] do not confirm the conclusion made in [51] that the anharmonicity in Li is smaller than in Na and K. Figures 3.11 and 3.12 show that the damping in Li is actually not smaller than in other alkali metals, even at room temperature (and, naturally, at $T \sim T_m = 453\,\mathrm{K}$).

Let us discuss the results of previous calculations of $\Delta\omega$ and Γ in alkali metals [48, 49, 51–54], shown in Figures 3.8 and 3.11 together with the results of [2]. The closest to [2] are the results obtained in [48] for K. The values $\Gamma_{\lambda\mathbf{k}}(T)$ are, practically, the same for all λ, $\mathbf{k}$ and T. For $\Delta\omega_{\lambda\mathbf{k}}(T)$, a significant difference 30% exists only for the critical branch N_4'. This calculation seems to be the most precise from [48, 49, 51–54] from the point of view of the model, as well as the methods of calculation.

The results of calculations in [49] for Rb disagree with results of [2] more significantly. This disagreement is especially large for the value $\Delta\omega(T)$ for the branch N_3', as well as for the damping $\Gamma_{\lambda\mathbf{q}}(T)$ of all branches (except N_1'). For this last quantity, rather strange temperature dependences were obtained in [49]: they do not tend at large T to the correct asymptotic form $\Gamma_{\lambda\mathbf{q}}(T) = \mathrm{const}\,T$ which follows from the general structure of the perturbation theory used in [49]. In contrast to [48] and [2], the calculations [49] were performed not in the momentum space, but in the coordinate space, with truncation of the summation over $\mathbf{R}$ on the second coordination sphere. This procedure can provide very large errors since the inter-ionic potential slowly decreases with $\mathbf{R}$. Estimations made in [49] show that these errors could reach, for some branches, 20–30%.

Especially large is the disagreement between results of calculations by Taylor *et al.* [51, 53, 54] for Na, K, and Li and results of Vaks *et al.* [2]. In particular, it relates to the damping, which (see Figure 3.11) is actually many times smaller than the results of [2] and even demonstrates a non-monotonic temperature behavior, in contrast to the temperature dependence obtained in the framework of the PT: $\Gamma_{\lambda\mathbf{q}} = \mathrm{const}\,T$. In many cases, results of [2] appreciably differ from results of [51, 53, 54] for $\Delta\omega_{\lambda\mathbf{q}}(T)$.

Pseudopotentials, used in [51, 53, 54], describe the spectra with approximately the same accuracy, as in [2, 3]. Hence, this disagreement, especially for the temperature behavior, relates either to the accuracy of calculations or to usage of the "self-consistent phonons with cubic anharmonicity" (SCP + C) approximation by Taylor *et al.* (instead the actual perturbation theory in the anharmonicity parameter [2, 3]). Formally speaking, the (SCP + C) approximation is not any power series expansion in the anharmonicity parameter ε, because, besides the terms first-order in ε, it also includes a sum of some arbitrary chosen sequences of highest order terms, containing only an even number of anharmonic vertices. These terms correspond to four-, six-, eight-, etc. phonon contributions. If these contributions are small, then this "over-estimation of the accuracy" is not essential and provides an error $\sim \varepsilon^2$. But, if the results become remarkably different from the results of the perturbation theory (this is the unique possibility to explain this large disagreement for the temperature behavior), then the physical sense and the accuracy of these approximations become absolutely indefinite.

In reality, all existing results of theoretical and experimental studies show relatively small values of anharmonic effects for metals considered, as for the mean value of the shift $\langle \delta_{\lambda \mathbf{q}} \rangle$, as for the damping $\langle \eta_{\lambda \mathbf{q}} \rangle$. Thus, there are many reasons to suppose that the factual anharmonicity parameter ϵ is small. At the same time, Table 3.12 and Figure 3.12 show that the characteristic feature of the theory is an appreciable compensation of contributions of even and odd anharmonic vertices $V^{\mathbf{k},\mathbf{q},\mathbf{k}+\mathbf{q}}_{\nu\lambda\mu}$ and $V^{\mathbf{kkqq}}_{\nu\nu\lambda\lambda}$, i.e., contributions $\Delta^{(3)}_{\lambda\mathbf{q}}$ and $\Delta^{(4)}_{\lambda\mathbf{q}}$ for the majority of phonons in the BZ (see respective columns for the modes P_4 and N_1'). This compensation probably takes place for highest order terms of the PT, and, maybe, the compensation degree even increases. In this event, approximations like (SSP + C), which take into account only even-order contributions (corresponding to even numbers of vertices in high-order diagrams), would totally destroy the results. This pecularity of the compensation would explain, why these high-order terms can appreciably change the result obtained in the framework of the first-order PT (despite small values of these terms, estimated above). If the compensation degree were increasing with increasing the order of the PT series expension in ε, then the usage of this approximation could appreciably overestimate the role (which, factually, is small) of high-order contributions.

Hence, if these disagreements are really caused by the usage of the (SSP + C) approximation in [55–57] instead of actual PT, then the comparative study of results shown in Figures 3.8 and 3.11 (including a comparison with experimental data for Li) can be considered as a witness of a general inapplicability of approximations like the "self-consistent-phonons" at least for metals considered.

To conclude this subsection, let us briefly consider the results of the work [3]. Here, the results of [2] were generalized to allow the consideration of the case of non-zero pressures $p \neq 0$ and the volume behavior of the anharmonic characteristics $\Delta\,\omega_{\lambda\mathbf{q}}$ and $\Gamma_{\lambda\mathbf{q}}$ were considered. This discussion is interesting from many points of view. First, from the general physical point of view, it is interesting to establish either an increase or a decrease in these anharmonic effects with increasing the compression and which parameter is the measure of this variation. We have neither experimental nor theoretical information about this problem for any solid. Hence, calculations for alkali metals can provide information about the value of the dependence of $\Delta\,\omega_{\lambda\mathbf{q}}$ and $\Gamma_{\lambda\mathbf{q}}$ on the pressure. Note, that calculations of the volume dependence of $\Delta\,\omega_{\lambda\mathbf{q}}$ and $\Gamma_{\lambda\mathbf{q}}$ in [3] were performed (as in [2]) while neglecting the contributions E_{sr} (3.2). However, from the first subsection of the present section, it follows that the role of E_{sr} increases with increasing compression, and, without taking this term into account, it is impossible to obtain any accurate quantitative description. For this reason, conclusions made in [3] seem to be very interesting, but only as the qualitatively correct description of some trends in behavior of $\Delta\omega_{\lambda\mathbf{q}}$ and $\Gamma_{\lambda\mathbf{q}}$ under pressure.

In [3], the change of the anharmonic frequency shift $\Delta\omega_{\lambda\mathbf{q}}(T)$ and the phonon damping $\Gamma_{\lambda\mathbf{q}}(T)$ for the interval of compression $\Delta\Omega/\Omega_0 = 0 \div 0.6$ were considered. The method of calculation and the pseudopotential model were the same as used in [2] in calculating $\Gamma_{\lambda\mathbf{q}}$ and $\Delta\omega_{\lambda\mathbf{q}}$ at $p = 0$.

It was found, that the relative shifts $\delta_{\lambda\mathbf{q}}$ (3.25) rapidly decrease at relatively small compressions. The relative damping $\eta_{\lambda\mathbf{q}}$ (3.25) change weakly, but also significantly.

From general considerations, it would be natural to expect that $\eta_{\lambda\mathbf{q}}$ and $\delta_{\lambda\mathbf{q}}$ will decrease with increasing p as a consequence of an increase of the hardness of the lattice. However, in [3] was shown that this conclusion is valid only for the relative frequency shifts $\delta_{\lambda\mathbf{q}}$, whereas the relative damping $\eta_{\lambda\mathbf{q}}$ increases for some branch with increasing p.

The role of anharmonic temperature effects for the microscopic Grüneisen parameters $\gamma_{\lambda\mathbf{q}}(T)$ was also investigated. It was obtained that the anharmonic effects result in remarkable temperature dependences of $\gamma_{\lambda\mathbf{q}}(T)$, especially in the case of $\gamma_{\lambda\mathbf{q}}(T)$ for the critical branch Σ_4, where $\gamma_{\lambda\mathbf{q}}(T)$ decreases many times with increasing temperature on the interval $T = 0 \div 300\,\mathrm{K}$. However, for all other branches, $\gamma_{\lambda\mathbf{q}}$ increases with increasing T.

Details of calculations and results can be found in [3].

Finally, let us make just one more comment. It is well known (see [59]) that increasing temperature is accomplished by increasing the number of point defects in the crystal $n_{imp} \sim \exp(-U/k_B T)$, where U is the formation energy for defects. These defects are, in accordance with Frenkel [59], either interstitial atoms or vacancies. Naturally, these point defects can coagulate in dislocation loops and lines and, consequently, affect elastic properties as well as the phonon damping. Any consistent theory of these phenomena needs microscopic calculations of the fields induced by dislocations and point defects. As we know, at present, this theory does not exist in any consistent form.

3.2.5 Main results

Let us emphasize the most important results obtained in this section and discuss those experimental studies, which, from the point of view of the problems considered above, seem to be of great importance to further development of the theory.

First, the possibility of neglecting the core polarization energy E_p by calculating the phonon frequencies in non-stressed alkali metals was clearly demonstrated. Hence, we can expect that many properties of alkali metals, depending on the phonon frequencies at zero pressure only, are, practically, independent of both M2A and M2B models which were used at $p = 0$. These properties are: the specific heat at constant volume, the free and internal energies, the entropy, mean atomic displacements, etc. [4–7, 16]. However, the rather large sensitivity of volume derivatives of the phonon frequencies to the accuracy of the model is important for non-stressed heavy alkali metals K, Rb, and Cs, too. For this reason, the effect of the Born–Mayer interaction E_{sr} on the microscopic Grüneisen parameters has a remarkable value even at $p = 0$. This is natural, because anharmonic properties of the lattice in contrast to the value

$\omega_{\lambda\mathbf{q}}$, are determined by the third coordinate derivatives of the inter-atomic potential and, hence, are more sensitive to the approximations used.

Complete analysis of experimental data and their comparison to theoretical values of $\omega_{\lambda\mathbf{q}}(\Omega, T)$ and $\gamma_{\lambda\mathbf{q}}(\Omega)$ at $p = 0$, obtained in the present subsection, as well as with calculations performed by other authors, show that our model provides the most accurate description of the microscopic Grüneisen parameters of Rb and K. Since an inclusion of the short-range term E_{sr} results in increasing $\gamma_{\lambda\mathbf{q}}$ for non-stressed metals, then we can expect that taking this term into account would improve the agreement of the theory with experiment for temperature behavior (at $p = 0$) those anharmonic characteristics of K, Rb, and Cs, which are, predominantly, determined by $\gamma_{\lambda\mathbf{q}}$. Especially important is this fact for the macroscopic Grüneisen parameters, the thermal expansion, and the thermal pressure. We will consider these properties below.

At large p, the core polarization in K, Rb, and Cs is not more negligible and provides more intensive changes to the spectrum $\omega_{\lambda\mathbf{q}}$ under pressure. For some phonon branches, especially for Σ_3 (Figures 3.3–3.6), this increase of the contribution of core interaction to $\omega_{\lambda\mathbf{q}}$ increases when p reaches the value of about one half of the total increase of $\omega_{\lambda\mathbf{q}}$ under pressure. But, actually, the relative changes of the phonon frequencies due to the ion interaction in K, Rb, and Cs remain for different phonon branches at moderate compression in the limits of 2–20%. Hence, the anharmonic changes of phonon frequencies by lattice deformation can reach rather large values because these changes are not restricted to any small parameter.

A more complete picture of the influence of the core interaction (polarization) on the phonon spectrum provides the dependence $\gamma_{\lambda\mathbf{q}}(p)$. For example, the behavior of $\gamma_{\lambda\mathbf{q}}$ of the branch Σ_3 under pressure clearly demonstrates a qualitative change in the volume dependence of $\omega_{\mathbf{q}}(\Sigma_3)$ at large p caused by the Born–Mayer interaction. We see (Figures 3.5–3.6) that there exists not only an appreciable difference between absolute values of $\gamma_{\mathbf{q}}(\Sigma_3)$ calculated for the M2A and M2B models, but also for the signs. For this reason, experimental investigations of the volume dependences of $\gamma_{\lambda\mathbf{q}}$ for the transversal branch Σ_3 and Σ_4 (and their further verification at $p = 0$) can provide important information about the role of the short-range core repulsion in these metals. Such experiments would be very important for understanding the nature of interatomic interaction in crystals and the microscopic structure of ion cores. They would allow us to provide a test for the applicability of one or another model and would accomplish a further development of the theory of metals. The most convenient object for these studies is K. For this metal the short-range core repulsion is much lower then for Cs and Rb. The pressure of the phase transitions in K is a few times larger, than in Cs and Rb (see Chapter 4). For this reason, the influence of the short-range interaction on the phonon spectrum can be studied on a wide region of p. Results of such measurements under pressure would provide very important information about the dependence of the energy E_p on the inter-atomic distance in the crystal. From the theoretical point of view, in the case of K, neglecting the influence of the empty d-zone located near the Fermi level as well as the "hard core" approximation (valid for K up to $p \sim 200$ kbar) seem to be mostly well-founded.

Our analysis of the behavior of phonon spectra under pressure and the effect of the core polarization provides a supposition that these peculiarities and differences in behavior of $\omega_{\lambda\mathbf{q}}(p)$ for transversal branches of the phonon spectra of heavy alkali metals can also affect the pressure dependence of the second order shear elastic moduli. This question will be considered below in the present chapter.

The results obtained for the anharmonic shift $\Delta\omega_{\lambda\mathbf{q}}$ and the damping $\Gamma_{\lambda\mathbf{q}}$ in the frame of the M2A model have a numerical accuracy of about a few per cent. On the other hand, the accuracy of calculations of the shift and damping in alkali metals in the framework of this model is about 10–20%. This value is much higher than the accuracy of the majority of existing experimental data. The theory agrees well with the most precise experimental data [50] for Li. For other metals, the accuracy of experimental data is rather bad, but, as a rule, they also agree well with the theory within the limit of experimental errors.

Since existing experimental data for $\Delta\omega_{\lambda\mathbf{q}}$ and $\Gamma_{\lambda\mathbf{q}}$ are rare and less-accurate, the theory considered can, apparently, be used to obtain some qualitative and quantitative information about anharmonic effects in real metals. It appears that the relative frequency shift $\delta_{\lambda\mathbf{q}}$ and the relative damping $\eta_{\lambda\mathbf{q}}$ (3.25) remain small up to $T = T_m$: $|\delta_{\lambda\mathbf{q}}(T_m)| \leq 0.1$–$0.15$; $\eta_{\lambda\mathbf{q}}(T_m) \leq 0.05$–$0.1$, for all phonons, except for some in the vicinity of the critical branch Σ_4, where $\delta(T_m) \sim 0.15$–0.3 and $\eta(T_m) \sim 0.3$–0.5. Thus, anharmonic effects remain small and actual anharmonic perturbation theory can be used to calculate these values. Hence, the phonon concept can, with confidence, be used in describing dynamic and thermodynamic properties of these metals for all $T \leq T_m$.

The values $\delta_{\lambda\mathbf{q}}(T)$ and $\eta_{\lambda\mathbf{q}}(T)$ demonstrate at non-small $T \sim T_m$ simple linear dependences on T. In particular, this fact must result in close values for $\eta_{\lambda\mathbf{q}}(T)$ at given T for the whole series of alkali metals. The values $\delta_{\lambda\mathbf{q}}(T)$ and $\eta_{\lambda\mathbf{q}}(T)$ appreciably change within the BZ. In particular, they increase rapidly in the vicinity of the critical branch Σ_4. The shift of the phonon frequencies for this branch at non-small $\mathbf{q}$ is positive, i.e. the frequencies are increasing, but not decreasing, with increasing T. It would be very interesting to test these theoretical predictions experimentally.

Finally note that, although we have considered here alkali metals only, it seems reasonable to suppose that all the main conclusions about the value and temperature behavior of anharmonic effects, made here, can, in general, be valid for other metals, too. For example, the "melting criteria" by Lindeman and Varshni, mentioned above, provide close values for the anharmonicity parameter at $T = T_m$ not only for alkali, but for other metals [16, 18], too. If we suppose that other (relative) anharmonic characteristics, like $\delta_{\lambda\mathbf{q}}(T)$ and $\eta_{\lambda\mathbf{q}}(T)$, have also close values for different metals at $T \sim T_m$, then the results obtained can provide some estimations of the characteristic scale of the values $\delta_{\lambda\mathbf{q}}(T)$ and $\eta_{\lambda\mathbf{q}}(T)$ for non-alkali metals, too.

3.3 Anomalies in the Temperature Behavior of the Macroscopic Grüneisen Parameter

Actually, it is assumed that anharmonic effects are of the most important at large temperatures ($T > \theta_D$, θ_D is the Debye temperature), because the majority of anharmonic contributions increase proportional to some powers of the temperature.

Here, and in section 4 of this chapter we will show that for the thermal expansion and elastic properties of alkali metals anharmonic effects are also important at low temperatures. In the calculation of the macroscopic Grüneisen parameter $\gamma(T)$ its "step-like" variation at low temperatures (see Figure 3.7) was established. This behavior is a consequence of the anisotropy of the phonon frequencies $\omega_{\lambda\mathbf{q}}$ themselves as well as their derivatives with respect to the volume $\partial\omega_{\lambda\mathbf{q}}/\partial\Omega$ [17, 19]. This anisotropy becomes apparent in the temperature dependences of elastic properties [19]. We emphasize that the possibility of discussing properties of metals at $T \leq \theta_D/10$ becomes possible only due to the high-precession method of integration over the Brillouin zone.

The values of the thermal expansion coefficient $\alpha(p, T)$, the specific heat at constant volume $C_V(p, t)$ and at constant pressure $C_p(T)$, calculated in the framework of the M2A and M2B models described above [5, 6], the macroscopic Grüneisen parameter $\gamma(T)$ [32], the elastic constants $C_{ik}(T)$ [18, 19] and their derivatives with respect to T and the volume Ω (or the pressure p), considered here, determine all thermodynamic properties of anharmonic crystals. In particular, these characteristics determine the relations between derivatives of adiabatic and isothermic elastic constants with respect to external stresses for different combinations of external conditions [60, 62]. These relations provide information about the general equation of state. They are also used by proceeding experimental data for the penetration of shock waves in crystals.

The results of calculations of $\gamma(T)$ are shown here for K only (results for other metals see in [17]). The choice of this metal to illustrate the results obtained is a consequence of the fact that (in contrast to Li and Na) this metal has no martensitic transformations at low temperatures. For this metal, there exists a complete set of experimental data, in particular, results of very precise measurements of the Grüneisen parameter [63]. These calculations were performed in the framework of the second-order in the pseudopotential and the first-order anharmonicity approximation. The effect of the highest orders in anharmonicity was estimated using the quasiharmonic approximation (i.e. taking into account the temperature dependence of the cell volume). For the polarization operator $\pi(\mathbf{q})$, the best approximation (GT) [64] was used.

Anharmonic characteristics of the crystal, like the macroscopic Grüneisen parameter $\gamma(T)$, the volume expansion coefficient $\alpha(T)$, the thermal pressure $p^*(T)$, as was already mentioned in Chapter 1, are determined by the phonon frequencies $\omega_{\lambda\mathbf{q}}$ and the microscopic (one-phonon) Grüneisen parameters[7]

$$\gamma_{\lambda\mathbf{q}}^{ij} = -\frac{\partial \ln \omega_{\lambda\mathbf{q}}}{\partial u_{ij}} = -\frac{D_{\lambda\lambda}^{(ij)}}{2\omega_{\lambda\mathbf{q}}^2}, \tag{3.29}$$

[7] The contribution of the energy of thermal electron excitations is subtracted from all experimental data (see discussion in [18] and in section 4 of the present chapter) and, for this reason, this value is not considered here.

where u_{ij} are the components of the distortion tensor, $u_{ij} = \partial u_i / \partial x_j$ and

$$D^{(ij)}_{\lambda\lambda'} = e^{\alpha}_{\lambda\mathbf{q}} \frac{\partial D_{\alpha\beta}(\mathbf{q})}{\partial u_{ij}} e^{\beta}_{\lambda'\mathbf{q}}, \tag{3.30}$$

$D_{\alpha\beta}(\mathbf{q})$ is the dynamical matrix (3.5), $\mathbf{e}_{\lambda\mathbf{q}}$, $\omega^2_{\lambda\mathbf{q}}$ are its eigenvectors and eigen-values, respectively. (We assume overall summation over pairs of identical Greek indexes, except indices λ, λ', which always will denote the index of the phonon branch).

The thermal expansion characteristics $\alpha(p, T)$ and $\Delta V/V_0$ were considered in the framework of the quasi-harmonic approximation in the first order anharmonicity in [17], and in the framework of the whole quasi-harmonic approximation in [6]. A very interesting result obtained is an unusual low-temperature behavior of the macroscopic Grüneisen parameter (was calculated in [17] neglecting the volume change with temperature: $\Omega = \Omega_0$). In the present section we will consider the low-temperature behavior of $\gamma(T)$, and the derivatives of $\gamma(T)$ with respect to Ω and T [19].

Figure 3.7 shows the results of calculation of $\gamma(T)$ in the framework of the quasi-harmonic approximation, which agree well with modern high-precession experiments [63]. These calculations were performed using two approximation for the AHA pseudopotential $V(\mathbf{q})$ (2.36): for parameters fitted with and without taking into account $E_p \approx E_{sr}$ (3.2). The best results for all static and thermodynamic properties were obtained using the second order model and with the fitting of the parameters of V_{pC} using the condition $p = 0$, $C_{44}(T = 0) = C^{exp}_{44}(T = 0)$ in the framework of the Geldart–Taylor approximation (GT) [64] for the polarization operator. But in some cases, more accurate results were obtained using the fitting of the parameters to the phonon spectrum (we denote this pseudopotential as V_{ph} (2.53)).

Despite that the pseudopotential V_{ph} seems to be more accurate in describing the temperature behavior of the Grüneisen parameter in potassium (Figure 3.7), it is more preferable to use the results obtained using the conditions $p = 0$, $C_{44} = C^{exper}_{44}$, but taking into account the repulsive term E_{sr} (see Table 2.1), too. As was already mentioned, the choice of V_{ph} does not satisfy the condition $p = 0$. The difference between the results obtained for the pseudopotential V_{pC} with and without taking into account the E_{sr} term is a consequence of the difference between values of the microscopic parameters $\gamma_{\lambda\mathbf{q}}$. The mean value of this difference is about 5%.

Note that the difference $\Delta\gamma$ between the quasi-harmonic value of $\gamma(T)$ and the value calculated for $\Omega = \Omega_0$, does not exceed 1% for the temperature interval shown in Figure 3.7. This value increases with increasing T and near the melting temperature T_m $\Delta\gamma \sim$ 6–7%. This behavior agrees rather well with the results of calculations $\Delta V(T)/V_0$ (remember, that $V = N\Omega$) for K [6]. Hence, taking into account the temperature variation of $\Omega(T)$ provides a substitution of $\gamma_{\lambda\mathbf{q}}$ for the value $\gamma_{\lambda\mathbf{q}} + (\partial\gamma_{\lambda\mathbf{q}}/\partial \ln \Omega)\Delta V/V_0$. Since $\partial\gamma/\partial \ln \Omega \sim 1$ then at $T \sim T_m$ $\Delta\gamma \sim \Delta V/V_0$ $\sim 6\%$ [6].

The most important characteristic feature of the value $\gamma(T)$ for potassium (as well as for other alkali metals) is a step-like variation of $\gamma(T)$ at low temperatures $T \sim 10\,\mathrm{K}$. At these T, the temperature behavior of $\partial\gamma/\partial \ln \Omega$ also has anomalies relating to anomalies of $\gamma(T)$ and $\partial \ln \gamma/\partial T$. Anomalies of $\gamma(T)$ and its derivatives with respect to the distortion tensor also become apparent in the low-temperature region for $C_{ik}(T)$ (see below).

The "step-like" increase of $\gamma(T)$ in the region 5–10 K in alkali metals is a consequence of the existence of the low-lying transversal branch Σ_4 of the phonon spectrum, which provides a dominant contribution at low T (since $\omega_{max}(\Sigma_4) = \omega(N_1) \sim 20\,\mathrm{K}$) to all integrals (1.59)–(1.69) which contain the functions $N_{\lambda\mathbf{q}}$, in particular, to $\gamma(T)$. For this branch, $\gamma_{\lambda\mathbf{q}}$ increases with increasing q, whereas for all other branches, this value remains almost constant within the so-called "working" phase volume (where $\omega_{\lambda\mathbf{q}} \sim T$) [17]. Hence, increasing $\gamma_{\lambda\mathbf{q}}(\Sigma_4)$ with increasing T provides a rapid increase of the macroscopic Grüneisen parameter. With further increasing T (with no further increase of the contribution from the branch Σ_4) the contribution of branches $G_1, D_{3,4}$ near the point N on the border of the Brillouin zone, where $d\gamma_{\lambda\mathbf{q}}/dq < 0$, become significant. This results in a small "depth" in the dependence $\gamma(T)$, which corresponds to a negative value of the derivative $\partial \ln \gamma / \partial T$. The further smooth increase of $\gamma(T)$ with increasing T is a consequence of a positive value of the derivative $d\gamma_{\lambda\mathbf{q}}/dq$ for the majority of branches.

This "step-like" behavior of $\gamma(T)$ is a characteristic feature of many cubic metals, e.g., copper, silver, gold [65], as well as taking place also in alkali-halide crystals: KBr, KCl, KI, NaCl, NaI [66]. In the case of alkali-halide crystals, this "step" in behavior of $\gamma(T)$ is well distinguished in the region 100–200 K. Note that this be havior is also determined by $\gamma_{\lambda\mathbf{q}}$ for the low-lying transversal [111] acoustic branch. If in the phonon spectrum there exist branches with large values of $d\gamma_{\lambda\mathbf{q}}/dq$ and $\omega_{\lambda\mathbf{q}}$, then, generally speaking, the "step-like" variation of $\gamma(T)$ is possible to observe at $T \sim \theta_D$ also.

Hence, our consideration of $\gamma(T)$ gives reasons to suppose that the step-like variation of $\gamma(T)$ can take place in other crystals also, and that it is not an experimental error, as was declared in [63]. For this reason, systematic experimental investigations of the dependence $\gamma(T)$ for all crystals, in particular, for alkali metals, would be very useful in testing and improving the models used in calculations of lattice dynamics.

3.4 Equation of State at Small Compressions

Preparing the detailed consideration of martensitic transitions in Li and Na and polymorphic BCC–FCC transitions in K, Rb, Cs (see Chapter 4), we outline the equations of state for alkali metals. Our main aim here is to explain some general features in the whole series of these metals and to demonstrate the accuracy of calculations, which can be achieved in the framework of the second-order model.[8]

[8] In 1986 the work [67] was published, in which the cohesive energy, the equation of state, and the pressure dependence of elastic constants were obtained in the frame the second-order model with $E_{sr} = 0$, the same as used here, and the GT approximation for the polarization operator and the HA pseudopotential. Naturally, these results reproduce with marvellous accuracy the results of calculations, discussed here, which were performed in 1978–79.

Since the dependence $p(\Omega)$ is weakly dependent on details of the model, we neglect here the repulsion energy E_{sr} [20]. As we will see in Chapter 4, this term only weakly affects the isotherms of polymorphic phase transitions. However, our analysis will show the necessity of taking into account the energy E_{sr} by describing the peculiarities of the BCC–FCC transitions.

The equations of state $p(\Omega, T)$ of alkali metals at moderate pressures $p \leq 50\,\text{kbar}$ were investigated in a lot of publications [68–71]. Some interesting phenomena discovered were, in particular, the similarity of the "zero-temperature" isotherms $p(\Omega, 0)$ of different metals and rather weak volume dependences of the thermal pressure $p^* = p(\Omega, T) - p(\Omega, 0)$ [69]. However, the number of theoretical works relating to these questions is rather small. In [30] the dependences $p(\Omega, 0)$ were calculated for Na and K and some reasons for the similarity mentioned were discussed. In [73], the isotherms $p(\Omega, T)$ at $T = 100\,\text{K}$ and 300 K for Na and K were calculated. The results of these calculations agree well with experimental data (in spite of rather rough approximations for the pseudopotential used, in particular, for electron correlations). However, these calculations do not provide good accuracy in describing other properties [1, 4, 5, 17].

In this section, we restrict our consideration to the BCC-phase. We will not consider here any phase transitions under compression. As we will see below, apparently, the accuracy of calculations of $p(\Omega, T)$ for this theory is good enough to provide a successful usage of these results for improving experimental data at moderate compression as well as for some qualitative estimations at large p.

The pressure p will be calculated using

$$p(\Omega, T) = p_{st}(\Omega) + p^{zp}(\Omega) + p_e^*(\Omega, T). \tag{3.31}$$

Here, Ω is the volume of the unit cell; $p_{st} = -\partial E/\partial\Omega$ is the "static" pressure at $T = 0$ calculated using E from (3.1), i.e. without taking into account the zero-oscillation term; p^{zp} and the thermal (i.e. vanishing at $T = 0$) phonon pressure p^* are calculated here in the framework of the harmonic approximation; p_e^* is the thermal pressure of electron excitations. Expressions used for p^{zp}, p^* and p_e^* were given in Chapter 1. We will neglect anharmonic contributions to p and, having the same order of magnitude, deviations of p_e^* from the low-temperature behavior $\lambda_e T^2/3\Omega$ (see (1.95)).

In the same manner as above, we will consider the ion pseudopotential $V(\mathbf{q})$ as being "hard", i.e. volume-independent, which corresponds to the neglect of the core deformations (i.e. the repulsive term E_{sr}). At large compressions $u = \Delta V/V_0 \geq 0.5$, the validity of this supposition is not clear. However, the role of the core deformations, as well as the Born–Mayer repulsion, need additional consideration (see Chapter 4). A comparison with experimental isotherm shows that, for the accuracy achievable in present-day measurements of $p(\Omega)$, these effects are not completely negligible.

The sensitivity of results in the form of the polarization operator $\pi(\mathbf{q})$ was already discussed in Chapter 2. It was found that, for the calculating scheme used, the best results for Na, K, Rb, and Cs were obtained using the Geldart–Taylor (GT) approximation for $\pi(\mathbf{q})$ whereas, for Li, the Geldart–Vosko (GV) approximation is better. The dependence $p(\Omega, T)$ considered here, is weakly sensitive to the form of $\pi(\mathbf{q})$, and we will discuss the results for "the best" approximations for $\pi(\mathbf{q})$: GV for Li and GT for other metals.

The values of parameters of the AHA pseudopotential (2.36) were taken from Table 2.1 for Na, K, Rb, and Cs (GT) and from Table 3.2 for Li (GV) for $E_p \equiv 0$, $E^{zp} \neq 0$.

The methods of calculation $p_{st}(\Omega)$ and $p^*(\Omega, T)$ differ significantly. The static contribution to the pressure $p_{st}(\Omega)$ was calculated with direct application of the relation $p_{st} = -\partial E/\partial\Omega$ to the next expression

$$E(\Omega) = E_e^{(0)} + E_e^{(1)} + E_e^{(2)} + E_i, \tag{3.32}$$

where (in Ry/atom)

$$E_e^{(0)} = Z\left[\frac{2.21}{r_s^2} - \frac{0.916}{r_s} - 0.115 + 0.031 \ln r_s\right], \tag{3.33}$$

$$E_e^{(1)} = \frac{Zb}{\Omega}, \quad b = \frac{4\pi}{3} Z r_0 (3 + 2u), \tag{3.34}$$

$$E_e^{(2)} = -\Omega \sum_{\boldsymbol{\tau}} \frac{\pi(\boldsymbol{\tau})}{\varepsilon(\boldsymbol{\tau})} |V(\boldsymbol{\tau})|^2; \quad \varepsilon(\mathbf{q}) = 1 + \frac{4\pi}{q^2}\pi(\mathbf{q}), \tag{3.35}$$

$$(r_s a_B)^3 = \frac{3}{4\pi n_0}, \quad n_0 = \frac{Z}{\Omega}, \tag{3.36}$$

$$E_i = Z\left\{\sum_{\boldsymbol{\tau}} \frac{4\pi}{\Omega\tau^2} \exp\left(-\frac{\tau^2}{4\sigma}\right) - 2\sqrt{\frac{\sigma}{\pi}} + \sum_{\mathbf{R}} \frac{1}{NR} G\left(\sqrt{\sigma}R\right) - \frac{\pi}{\sigma\Omega}\right\} \tag{3.37},$$

$$G(X) = \frac{2}{\sqrt{\pi}} \int_X^\infty \exp(-t^2)\, dt.$$

In these formulae, all energies are given per unit cell (atom), Z is the valency (for alkali metals $Z = 1$), Ω is the cell volume (for alkali metals $\Omega = a^3/4$, where a is the lattice constant), u and r_0 are the parameters of the Animalu–Heine–Abarenkov pseudopotential (2.36) determined using (2.39) and listed in Table 2.1 for the case V_{pC}, $E_{sr} = 0$, $E^{zp} \neq 0$. The Ewald parameter σ in (3.47) was taken to be π/a [2].

The differentiation of E in (3.32) with respect to the volume is performed using equation (3.13), where we must suppose $u_{in} = \Omega$. The energy depends on Ω directly, as well as through n_0 (see (3.36)), $k_F = (3\pi^2 n_0)^{1/3}$, r_s (see (3.36)), and vectors of the reciprocal $\boldsymbol{\tau} = \mathbf{n}2\pi/a$ and direct $\mathbf{R} = a\,\mathbf{m}$ lattices. The parameters σ in the expression for E_i and the cut-off parameter ξ for $V(\mathbf{q})$ are not differentiated. The respective formulae for $p_{st}(\Omega)$ will not be shown here because they are very cumbersome.

Note that, at first sight, it would be possible to calculate the equation of state $p = p(\Omega)$ using the series expansion of the free energy in powers of the change in volume $\Delta V/V_0$. However, in fact, this procedure can describe (using known values of elastic constants at $p = 0$) only a small ($\Delta V/V_0 \leq 0.01$) part of the isotherm, where $p \approx \chi\Delta V/V_0$ (χ is the compressibility). At large $\Delta V/V_0$, the dependence $p = p(\Omega)$ becomes appreciably non-linear and its description needs calculations of the whole power series in $\Delta V/V_0$ that is equivalent to formulae (3.31)–(3.38).

The isotherms $p(\Omega, T)$ were calculated as functions of $\Delta V/V_0 = 1 - V/V_0(T)$, where $V_0(T)$ is the equilibrium crystal volume at $p = 0$. The energy of zero oscillations also provides some contribution $p^{zp}(\Omega)$ to the equation of state (3.31). This contribution was obtained by differentiation of the energy of zero oscillation E^{zp}, which includes the thermal pressure of phonons $p^*(\Omega)$, i.e., the pressure of harmonic phonons is

$$p_h = p^{zp}(\Omega) + p^*(\Omega, T) = -\frac{\partial}{\partial\Omega}[E^{zp}(\Omega) + F^*(\Omega, T)] \tag{3.38}$$

(see formula (1.255)). The calculation of the phonon frequencies was considered in the previous section.

The zero-temperature isotherms $p(\Omega, 0) = p_{st}(\Omega)$ for small compressions $p \leq 20\,\text{kbar}$ ($\Delta V/V_0 = 0.35$) were calculated without taking into account zero oscillations [1]. It turns out that taking into account the term p^{zp} almost does not change the isotherms $p(\Omega, T)$ for Na, K, Rb, and Cs. Only for Li is a new plot of $p(\Omega, 0)$ 1.5–2% below the results calculated without p^{zp}.

Comparison of calculated zero isotherms with experiment is difficult, since there were only a few low-temperature measurements of $p(\Omega)$ [69]. The differences between calculated and measured isotherms do not exceed 5–10%. The spread of results obtained by different authors is of the same order or smaller than the disagreements between theory and experiment mentioned above. For this reason, the true value of this disagreement seems to be rather unknown. More definite conclusions relating to the accuracy of the theory can provide a comparison with experimental isotherms at room temperature.

This comparison is made here for Na (see Figure 3.13) to demonstrate the accuracy of the theory only. The equations of state for other alkali metals were discussed in detail in [20].

Figure 3.13 shows that for Na the agreement of the theory with result [71] is much better, than the comparison of the results for zero-temperature isotherms with the data from [69]. In particular, there exists no over-estimation of p_{exper} in comparison with p_{theor} for Na, which could be considered a consequence of neglecting the Born–Mayer repulsion. This agreement with experimental data demonstrates good accuracy of calculations of $p(\Omega)$ and the volume elastic moduli $B(\Omega) = -\partial p/\partial \ln \Omega$ for all alkali metals (even for Li, for which the accuracy of calculation of the volume dependence of shear moduli and the phonon frequencies is not good enough). For Rb, some small over-estimation of p_{exper} in comparison with p_{theor} at large compression would show, for example, the value of the Born–Mayer repulsion discussed above. The influence of this term on the equation of state of K, Rb, and Cs will be discussed in Chapter 4.

Although here we do not consider the phase transitions under pressure, in fact, this transition takes place (see Chapter 4). For example, for Cs the structural phase transition from the BCC to FCC phase takes place at $\Delta V/V_0 \approx 0.37$. Further compression (to $\Delta V/V_0 \approx 0.45$) induced the electronic $s-d$ transition [70]. It seems to be natural to expect the existence of such transitions in other alkali metals, too, in the first place in Rb and K, although even today there is no unambiguous confirmation of the existence of these transitions [27].

3.5 Temperature Dependences of the Second-Order Elastic Constants of Alkali Metals

3.5.1 Present state of the problem

Calculation of the temperature dependence of the elastic constants $C_{ik}(T)$ is a classic problem in the theory of anharmonic effects in crystals. From the point of view of a quantitative theory, this problem seems to be more difficult than the calculation of the thermal expansion $\Delta\Omega(T) = \Omega(T) - \Omega(0)$, since $\Delta\Omega(T)$ is determined only by the first derivatives of phonon frequencies $\omega_{\lambda\mathbf{q}}$ with respect to volume, whereas $C_{ik}(T)$ depend on the second derivatives of $\omega_{\lambda\mathbf{q}}$ and not with respect to the volume only, but with respect to other deformations, too. For this reason, the comparison of the calculated temperature dependences with experiment is a sensitivity test of the accuracy of the model. Any success in the description of temperature dependences of elastic properties can be considered to be a definite confirmation of maturity of the theory.

At first, qualitative evaluations of the dependences $C_{ik}(T)$ for alkali metals were discussed in [74], where the quasi-harmonic change in $C_{ik}(\Omega)$, relating to the thermal expansion $\Omega(T)$, were calculated for Na and K.[9] Taking this value from experiment, the author of [74] has found that, in the case of the elastic modulus B, the quasi-harmonic formulae describe the experimental data for $B(T)$ well whereas the experimental values of the shear moduli C_{44} and $C' = (1/2)(C_{11} - C_{12})$ decrease with increasing T more rapidly than the prediction of the quasi-harmonic approximation, i.e., apparently, for these values, intrinsic anharmonic contributions are also important. The authors of [53, 54] have calculated a total set of $C_{ik}(T)$ for Na and K in the frame of the pseudopotential model [76], using the "self-consistent phonon with cubic anharmonic" (SCP + K) approximation discussed in section 3 of this chapter. Elastic constants were calculated using formulae for the long-wavelength limit of the dynamical matrix, so that the values of the elastic moduli B_{dyn} obtained differ from the static B_{st} values obtained by differentiation of the free energy F. Moreover, in the case of Na, the authors of [53] have expressed an uncertainty in the accuracy of

[9] Note that the values of $C_{ik}(T)$ at $p = 0$ were calculated in [75] using methods of molecular dynamics in the framework of the second-order model in the pseudopotential.

results because of poor convergence of these numerical calculations. The authors of [73] discussed the results of calculations of $C_{ik}(T)$ for Na and K using the Walles model [35] but by replacing the phonon spectrum with the spectrum for the Einstein model with a single "average" frequency [32]. In [77, 78], $C_{ik}(T)$ were also calculated for Na and K using the pseudopotentials from [79], but, instead of regular calculations, the Monte Carlo method was used for the description of the ion movement in the lattice.

Hence, neither of the works listed here contains consistent calculations of $C_{ik}(T)$ in the framework of the actual perturbation theory in the anharmonicty and the pseudopotential. Moreover, the fact that calculated values of $C_{ik}(T)$ in these works appreciably disagree with experimental data is very important especially for the shear modulus C' (see below). Hence, not one of these models can be considered as a suitable model for the description of the temperature dependencies of elastic moduli discussed.

In this section following [18], we investigate in detail the temperature and volume dependence of isothermal elastic constants (EC) of alkali metals in the framework of the second-order model M2A (see section 2 Chapter 2) in the first order in the anharmonicity parameter $\varepsilon = \langle x^2 \rangle / A^2$ [45] without taking into account the polarization energy. We will show that, for the first time, good agreement with experiment has been reached. Additionally, we will investigate here the role of each anharmonic contribution to $C_{ik}(T)$ and the effect of the phonon spectrum $\omega_{\lambda \mathbf{q}}$ and $\gamma_{\lambda \mathbf{q}}$ on the low-temperature dependence of $C_{ik}(T)$ [19].

Before we begin calculation of EC, let us remember that the name "isothermal elastic constants" does not mean that these values are temperature-independent, but rather relates to methods of measurements of these values. These measurements are actually performed at constant temperature and, for example, under static compression. Then the elastic constants are determined as derivatives of the free energy F. The next measurement is provided at some other temperature, etc. This procedure yields the dependence $C_{ik}(T)$. The adiabatic elastic constant (the derivatives of the internal energy U) are measured in experiments, where the entropy remains constant during the measurement (e.g., investigations of a sound penetration through the media). These elastic constants are also temperature-dependent.

If we want to calculate $C_{\alpha\beta\gamma\delta}(T)$ in the first order in the anharmonicity parameter ε then, for alkali metals, for which $\varepsilon \leq 10^{-2}$, it seems to be reasonable to calculate the temperature-dependent part of the free energy using the harmonic approximation, i.e., using $F^*(\Omega, T)$ from (1.14) [80]. In this section, we will mainly consider the temperature dependence $C_{\alpha\beta\gamma\delta}(T)$ with no pressure $p = 0$, when the unit cell volume $\Omega = \Omega(T)$ changes only due to the thermal expansion. In this case, we denote the elastic constants simply as $C_{\alpha\beta\gamma\delta}(T)$ and $\Omega(T = 0)$ as Ω_0. In the first order in the anharmonicty, we obtain

$$C_{\alpha\beta\gamma\delta}(T) = C^0_{\alpha\beta\gamma\delta}(\Omega) + \Omega \frac{dC^0_{\alpha\beta\gamma\delta}(T)}{d\Omega} \frac{p^*(\Omega, T)}{B^{(0)}} + C^{zp}_{\alpha\beta\gamma\delta}(\Omega) + C^*_{\alpha\beta\gamma\delta}(\Omega, T). \tag{3.39}$$

Here

$$p^* = -\left[\frac{\partial F^*}{\partial \Omega}\right]_T \tag{3.40}$$

is the thermal pressure (1.62),

$$B^{(0)} = \Omega_0 \left[\frac{d^2 E}{d\Omega_0^2}\right] \tag{3.41}$$

is the compressibility modulus at $T = 0$ if E is determined by (3.1);

$$\frac{p^*}{B^{(0)}} = \frac{\Omega(T)}{\Omega_0} - 1 \tag{3.42}$$

is the thermal expansion (1.64) in the lowest order in anharmonicity.

The second term in (3.39), which it is possible to write as $-p^* dC^0_{\alpha\beta\gamma\delta}/dp$, we will denote as the quasi-harmonic contribution $\Delta C^{qh}_{\alpha\beta\gamma\delta}$. The last term $C^*_{\alpha\beta\gamma\delta}$, we denote as the intrinsic anharmonic contribution. For a cubic crystal with no stress, we have

$$\frac{1}{\Omega}\frac{\partial F^*}{\partial \bar{u}_{\alpha\beta}}\bigg|_{\bar{u}=0} = \delta_{\alpha\beta}\left[\frac{\partial F^*}{\partial \Omega}\right]_T = -\delta_{\alpha\beta} p^*(\Omega, T), \tag{3.43}$$

where $\bar{u}_{\alpha\beta}$ is the deformation tensor.

Let us introduce the Voigt notations[10] C_{ik} for the elastic constants $C_{\alpha\beta\gamma\delta}$. Instead of C_{11}, C_{12}, C_{44}, we use the Fuchs moduli B_{11}, B_{33}, B_{44} (1.12). These last have a transparent physical meaning as the compressibility B_{11} and shear moduli B_{33}, B_{44} under pressure p. They simply relate to C_{ik} and the Birch moduli $\mathcal{B}_{ik}$ (see section 3 Chapter 1).

For all T, the electronic contribution F_e^* (1.3) to the elastic moduli is relatively small in comparison with the phonon contribution F^* (1.14) Only this value becomes important for the derivatives $dC_{\alpha\beta\gamma\delta}/dT$ at low T only. For all other T, this contribution has the same order of magnitude as the highest order contributions in the anharmonicity parameter which we neglect here. For this reason, we will use for F_e^* the low-temperature formula $\lambda_e T^2/2$, where $\lambda_e T = C_e$ is the electronic specific heat measured experimentally. To obtain the contribution of F_e^* to the values A_{ij}^* (1.83), we must know the behavior of λ_e under deformations, which is a rather complicated problem. Let us suppose, for simplicity, this dependence to be the same form as in the case of the free electron gas: $\lambda_e = \lambda_e(\Omega) \sim \Omega^{2/3}$. It means that we suppose the electron Grüneisen parameter γ_e to be equal to 2/3 (the free-electron-gas value). This supposition agrees with results obtained in [63] for K. However, generally speaking,

[10] The second-order elastic constants have four Cartesian indices $C_{\alpha\beta\gamma\delta}$, needed for theoretical investigations. Experimentalists use the Voigt indices C_{ik}, of which the correspondence to the Cartesian ones is well known [35]: $11 \to 1$, $22 \to 2$, $33 \to 3$, $23 \to 4$, $13 \to 5$, $12 \to 6$.

this supposition can provide only a rough estimate of the order of the contribution F_e^* to elastic moduli. For this reason, we will discuss here the lattice contributions B_{ik}^{lat} only. For the experimental values $B_{ik,exp}^{lat}$, we will take the differences between observed values B_{ik}^{obs} and the contributions from F_e^* obtained as described above:

$$B_{11,exper}^{lat} = B^{obs} + p_e^* \left[\frac{dB^{(0)}}{dp} - \frac{1}{3} \right]; \quad B_{33,\exp}^{lat} = B_{33}^{obs} + p_e^* \frac{dB_{33}^{(0)}}{dp},$$

$$B_{44,\exp}^{lat} = B_{44}^{obs} + p_e^* \frac{dB_{44}^{(0)}}{dp}; \quad p_e^* = \lambda_e T^2 / 3\Omega. \tag{3.44}$$

Note once again that "electronic" corrections in (3.44) are small and have an approachable value only for dB_{ik}/dT at low T. For example, at the melting point $T = T_m$, the contribution of p_e^* in (3.44) is less than 0.3% for Li and decreases in a sequence from Li to Cs. For brevity, we will drop the index "lat" in B_{ik}^{lat}, i.e., we will below consider the values $B_{ik}^{lat} = B_{ik}$ only.

In accordance with the definitions (1.74) and (1.83), we obtain

$$A_{ij}^* = A_{ij}^*(3) + A_{ij}^*(4) = \frac{1}{\Omega} \left[\frac{\partial^2 F^*}{\partial u_i \partial u_j} \right], \tag{3.45}$$

where u_i are the components of the distortion tensor: $u_1 = u_{11},\ u_2 = u_{22},\ u_3 = u_{33}$. The values A_{ij}^* are:

$$A_{ij}^*(3) = -\frac{\hbar^2}{N\Omega k_B T} \sum_{\lambda \mathbf{q}} N_{\lambda \mathbf{q}} (N_{\lambda \mathbf{q}} + 1) \omega_{\lambda \mathbf{q}}^2 \gamma_{\lambda \mathbf{q}}^i \gamma_{\lambda \mathbf{q}}^j - R_{ij}, \tag{3.46}$$

$$A_{ij}^*(4) = \frac{\hbar}{N\Omega} \sum_{\lambda \mathbf{q}} N_{\lambda \mathbf{q}} \, \omega_{\lambda \mathbf{q}} \left[\gamma_{\lambda \mathbf{q}}^i \gamma_{\lambda \mathbf{q}}^j - \frac{\partial \gamma_{\lambda \mathbf{q}}^i}{\partial u_j} \right] + R_{ij}, \tag{3.47}$$

$$R_{ij} = \hbar \sum_{\lambda \mathbf{q}} \frac{N_{\lambda \mathbf{q}}}{\omega_{\lambda \mathbf{q}}} \left[D_{\alpha\beta} - \omega_{\lambda \mathbf{q}}^2 \delta_{\alpha\beta} \right] e_{\lambda \mathbf{q}}^{\alpha,i} e_{\lambda \mathbf{q},}^{\beta,j} \quad e_{\lambda \mathbf{q}}^{\alpha,i} = \frac{\partial e_{\lambda \mathbf{q}}^{\alpha}}{\partial u_i} \tag{3.48}$$

To obtain the contribution of zero oscillations to A_{ij}^{zp}, we must keep in $A_{ij}^*(3)$ R_{ij} only and substitute in R_{ij} and $A_{ij}^*(4)$ the Planck function $N_{\lambda \mathbf{q}}$ by 1/2. The formulae for A_{ij} have been written in a form that clearly shows the dependence of B_{ik}^* on the microscopic Grüneisen parameters $\gamma_{\lambda \mathbf{q}}$, as well as on their derivatives $\gamma_{\lambda \mathbf{q}}^i = \partial \gamma_{\lambda \mathbf{q}} / \partial u_i$. Note that the dependences of $A_{ij}^*(3)$ and $A_{ij}^*(4)$ on T are determined not by the factor $N_{\lambda \mathbf{q}}$ only, but also by the derivatives of eigen-vectors of the dynamical matrix $\mathbf{e}_{\lambda \mathbf{q}}^i$ with respect to deformations, although in C_{ik}^* the term R_{ij} has been cancelled.

The temperature dependences of the Fuchs elastic moduli $B_{11}(\Omega, T)$, $B_{33}(\Omega, T)$ and $B_{44}(\Omega, T)$ were calculated as

$$B_{11}(\Omega, T) = B(\Omega, T) = B^{(0)}(\Omega) + B^{zp}(\Omega) + B^{*}(\Omega, T) - p^{*}(\Omega)\frac{dB^{(0)}(\Omega)}{dp}, \tag{3.49}$$

$$B_{33}(\Omega, T) = C'(\Omega, T) = B_{33}^{(0)}(\Omega) + B_{33}^{zp}(\Omega) + B_{33}^{*}(\Omega, T) - p^{*}(\Omega)\frac{dB_{33}^{(0)}(\Omega)}{dp}, \tag{3.50}$$

$$B_{44}(\Omega, T) = C_{44}(T) = B_{44}^{(0)}(\Omega) + B_{44}^{zp}(\Omega) + B_{44}^{*}(\Omega, T) - p^{*}(\Omega)\frac{dB_{44}^{(0)}(\Omega)}{dp}. \tag{3.51}$$

Here, $B^{*} = \frac{1}{3}\left[A_{11}^{*} + 2A_{12}^{*} + 2p^{*}\right]$;

$$\begin{aligned} B_{33}^{*} &= \frac{1}{2}\left[A_{11}^{*} - A_{12}^{*} - p^{*}\right] = C'^{*} - p^{*}; \\ B_{44}^{*} &= A_{44}^{*} = C_{44}^{*} + p^{*}. \end{aligned} \tag{3.52}$$

The functions A_{ij}^{*} have been determined above.

Formulae (3.39), (3.45)–(3.52) are exact in the first order in anharmonicity. The effect of higher order terms is possible to estimate using the quasi-harmonic approximation. It corresponds to using in all the terms in equations (3.49)–(3.51) the value $\Omega(T)$ (the volume calculated taking into account a thermal expansion [6, 17] in the framework of the same model) instead of Ω, but the phonon contributions B_{ij}^{zp}, B_{ij}^{*} will be calculated as before using equations (3.45)–(3.52). Usage of this substitution, which we denote as the quasi-harmonic perturbation theory (QHPT), will be specially marked.

The values E, B_{ik}^{0}, dB_{ik}^{0}/dp and the dynamical matrix $D_{\alpha\beta}(\mathbf{q})$ were calculated using actual formulae of the second-order model (see section 3 of the present chapter). The values of the parameters u and r_0 of the pseudopotential were fitted using the conditions $p(\Omega_0) = 0$, $C_{44}(\Omega_0) = C_{44,exp}$ ($E_{sr} \equiv 0, E^{zp} \neq 0$). They are listed in Tables 2.1–2.2. Parameters of the potential were considered as invariant with respect to variations of the volume and deformations ("hard core"). The following approximations were used for the polarization operator $\pi(\mathbf{q})$: RPA [1]; Geldart–Vosko [81] GV; Toigo–Woodruff [82, 83] TW; Vashishta–Singwi [84] VS, and Geldart–Taylor [64] GT. As we have already mentioned many times, the last approximation, GT, seems to be the most accurate, at least in describing the properties of Na, K, Rb, and Cs. The results for other approximations of $\pi(\mathbf{q})$ will be shown here only for Na as a methodological illustration. The GV approximation provides a better description of properties of Li in the framework of the scheme with local pseudopotential used. Probably this approximation corresponds to some imitation of non-local effects and highest order terms in $V(\mathbf{q})$. Here, we will show results for Li for $\pi(\mathbf{q})$ calculated in the framework of the GV approximation only.

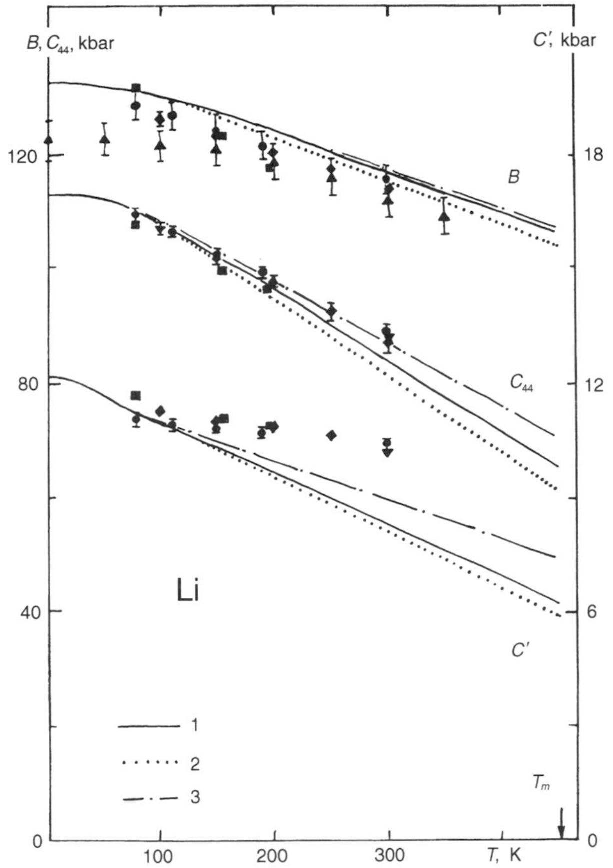

FIGURE 3.14 Temperature dependence of the elastic constants $C_{ik}(T)$ at $p = 0$ for lithium. Notations: 1 – the first-order perturbation theory (3.39); 2 – the first-order quasi-harmonic perturbation theory C_{ik}^{qh} (see text); 3 – calculation using (3.39) and experimental data [87] for dB_{ik}^{0}/dT. Experimental points: ● – [88]; ▲ – [89]; ■ – [90]; ▼ – [87]; ◆ – [91].

3.5.2 Temperature dependences of B, C_{44}, and C'

Figures 3.14–3.18 show results of the calculations for $C_{ik}(T)$ (3.49)–(3.51) at $p = 0$ ($\Omega = \Omega_0$). Here, the experimental values $C_{ik} = C_{ik}^{lat}$ were obtained using formulae (3.44) from experimental data for C_{ik}^{obs} from publications cited, the values λ_e were taken from [85] and the experimental data for dB_{ik}/dp. Because no experimental data exist for dB_{ik}^{0}/dp in Cs, we use here the theoretical values (at $T = 0$). If the experiment provides the adiabatic elastic modulus B_s then the value of the isothermal modulus $B_T = B$ can be calculated from B_s using a well-known thermodynamic relation [86]:

$$B - B_s = \left[\partial \ln \Omega / \partial T\right]_p^2 \cdot \frac{B_s \Omega}{C_p^2}, \tag{3.53}$$

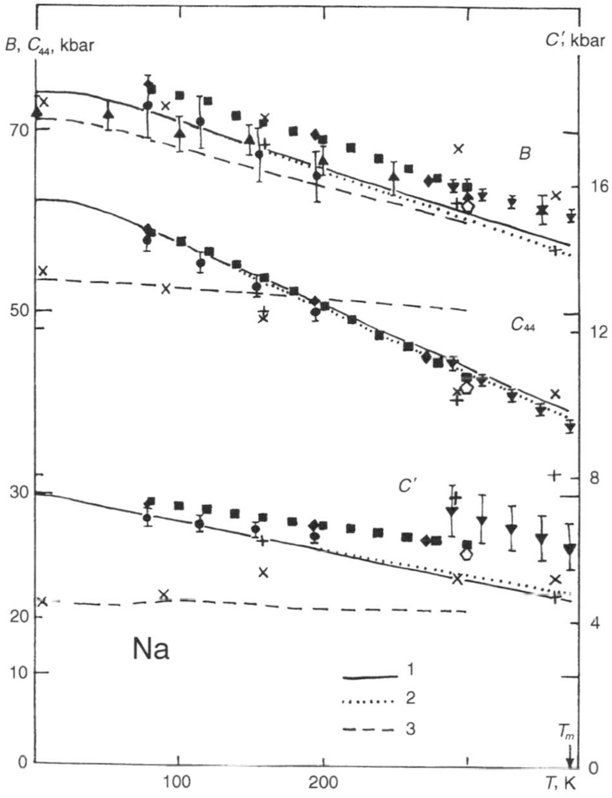

FIGURE 3.15 $C_{ik}(T)$ for sodium. Curves 1 and 2 in Figures 3.15–3.18 show the same values as in Figure 3.14. Curve 3 – calculations [73]. Heavy symbols are experimental data: • — [92]; ▲ – [93]; ■ – [94]; ▼ – [95]; ◆ – [96], ⌂ – [97]. Crosses are results of calculations of other authors: × – [53]; + – [78].

where the experimental values from Table 2 from [4] were used on the right-hand side.

Figures 3.14–3.18 show that the theory provides a good description of $B(T)$ and $C_{44}(T)$ for all metals through the whole temperature region from zero to the melting temperature T_m. Even for Li, the disagreement with the experimental data for $B(T)$ and $C_{44}(T)$ does not exceed the limits of the spread and errors of the experimental data.

The results for the shear modulus C' are less accurate: the calculated C' decreases with T more rapidly than in experiment. On the other hand, for all metals, except Li, this disagreement for C' does not exceed the limits of experimental error and the differences between values obtained in different publications. However, we can see (Figures 3.14–3.18 and Table 3.13) that, for the value dC'/dT, this disagreement takes place not only for Li, but also for Na and K (and, maybe, for other metals). Note also, that the accuracy of the measurements of the modulus C' in Rb in [98] was, probably, overestimated by the authors and, factually, the experimental values

TABLE 3.13 Temperature derivatives of elastic constants at $p = 0$ (in 0.01 kbar/K)

1	metal	Li			Na		
2	temperature, K	295			195		
3	value	$-\frac{dB}{dT}$	$-\frac{dC_{44}}{dT}$	$-\frac{dC'}{dT}$	$-\frac{dB}{dT}$	$-\frac{dC_{44}}{dT}$	$-\frac{dC'}{dT}$
4	PT	7.0	12.5	1.4	5.0	6.7	0.79
5	QHPT	7.5	13	1.45	5.9	7.05	0.70
6	experiment	7.0	9.3	0.30 [88]	5.0	7.5	0.54 [94]
		6.0	9.8	0.42 [91]	6.0	7.1	0.47 [92]

1	K			Rb			Cs		
2	195			195			78		
3	$-\frac{dB}{dT}$	$-\frac{dC_{44}}{dT}$	$-\frac{dC'}{dT}$	$-\frac{dB}{dT}$	$-\frac{dC_{44}}{dT}$	$-\frac{dC'}{dT}$	$-\frac{dB}{dT}$	$-\frac{dC_{44}}{dT}$	$-\frac{dC'}{dT}$
4	2.6	3.1	0.45	2.1	2.5	0.39	1.4	1.9	0.33
5	2.8	3.3	0.41	2.3	2.6	0.35	1.4	1.6	0.29
6	3.0	3.5	0.27 [97]	3.3	2.9	0.28 [98]	1.8	1.8	0.20 [103]
	2.8		[99]				1.0		[69]

Comments: PT – perturbation theory (3.49)–(3.51); QHPT – quasi-harmonic perturbation theory.

of C' at low T would be much closer to the theoretical curve (Figure 3.17). This conclusion follows from an excellent agreement with theory [4] for the results obtained in [85] for the low-temperature specific heat and the Debye temperature θ_D. These results were 6 times larger than the experimental error estimated in [85].

Table 3.14 shows relative values of both the quasi-harmonic and intrinsic harmonic contributions to ΔB_{ik}^{qh} and B_{ik}^{*} (the contribution $C_{ik}(0)$ at $T = 0$ is subtracted), as well as the total relative change in $B_{ik}(T)$ through the interval from $T = 0$ to T_m. As we can see, for the elastic moduli B, the thermal contribution ΔB is small as has been already mentioned in many experimental studies [89, 69, 104]. But, at the same time, the intrinsic anharmonic contributions play the main role for $\Delta B_{33}(T)$ and $\Delta B_{44}(T)$. Hence, the quasi-harmonic approximation cannot be used in calculations of the temperature dependence of shear modulus [74]. Taking into account Figures 3.14–3.18 and formulae (3.45)–(3.48), we can conclude that the values B_{44}^{*} (derivatives of the frequency $\omega_{\lambda\mathbf{q}}$ with respect to $u_4 = u_{32}$) have been calculated with good accuracy, but the calculations of B_{33}^{*} were less accurate. The values of the difference $A_{11}^{*} - A_{12}^{*}$, participating in the expression (3.52) for B_{33}^{*}, actually are of an order of magnitude smaller than p^{*}. Since the thermal pressure p^{*} and the value $dB_{33}^{(0)}/dp$ for Na and K have been calculated here with rather good accuracy, then the

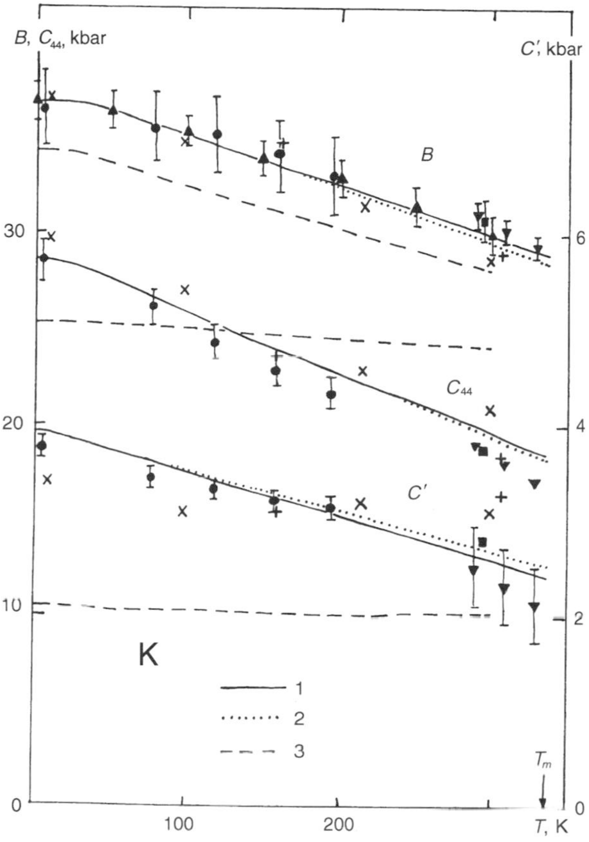

FIGURE 3.16 Modules $C_{ik}(T)$ for potassium. Curve 3 – results of calculations [73]. Experimental data: • – [63]; ▲ – [99]; ■ – [100]; ▼ – [101]; crosses are calculation of other authors: × [54]; + – [77].

TABLE 3.14 Calculated values of the ratio of the quasi-harmonic contribution to the module $\Delta B_{ik}^{qh} = B_{ik}^{qh}(T) - B_{ik}(0)$ to the total temperature contribution $\Delta B_{ik}(T) = \Delta B_{ik}^{qh} + B_{ik}^{*}$ at the melting temperature T_m, and the total temperature contribution $\Delta B_{ik}(T_m)/B_{ik}(0)$

metal	Li	Na	K	Rb	Cs
approx. of $\pi(q)$	GV	GT	GT	GT	GT
$\Delta B_{11}^{qh}/\Delta B$	0.96	0.91	0.89	0.88	0.93
$\Delta B_{44}^{qh}/\Delta C_{44}$	0.26	0.26	0.23	0.22	0.20
$\Delta B_{33}^{qh}/\Delta C'$	0.29	0.35	0.32	0.31	0.29
$\Delta B_{11}(T_m)/B(0)$	0.196	0.220	0.214	0.212	0.181
$\Delta B_{44}(T_m)/C_{44}(0)$	0.425	0.365	0.344	0.338	0.341
$\Delta B_{33}(T_m)/C'(0)$	0.485	0.380	0.377	0.385	0.408

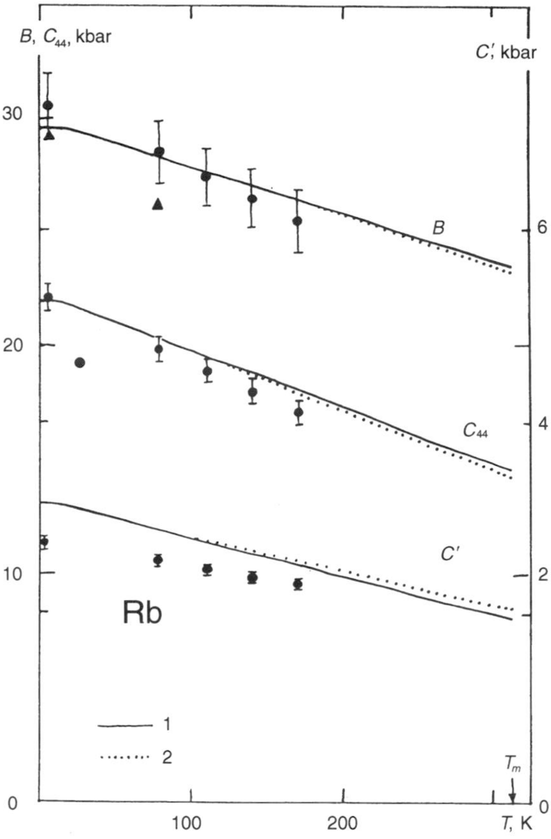

FIGURE 3.17 Moduli $C_{ik}(T)$ for rubidium. Experimental data: • – [98]; ▲ – [102].

disagreement with experimental data for dC'/dT would demonstrate rather bad accuracy in calculations of $A^*_{11} - A^*_{12}$, i.e., low accuracy in calculations of the derivatives of phonon frequencies.

But, probably, this disagreement is, at least partially, a consequence of contributions of next-order terms in the anharmonicity. Typical values of C' are an order of magnitude smaller than B and C_{44}. Hence, highest order contributions to this value would be much more important. To illustrate this conclusion, let us consider the results obtained in the frame of the quasi-harmonic perturbation theory (see Table 3.13 and Figures 3.15–3.17). We can see that the results obtained taking into account this correction (i.e., some part of higher-order terms) show relativly larger variations of C' and dC'/dT in comparison with B and C_{44}. For all metals except Li (Figure 3.14), the results obtained are rather close to experimental data. Thus, if the relation between quasi-harmonic and intrinsic anharmonic contribution to C' for higher-order terms were approximately the same as for the first-order terms considered here, then the disagreement with experiment for $C'(T)$ would decrease

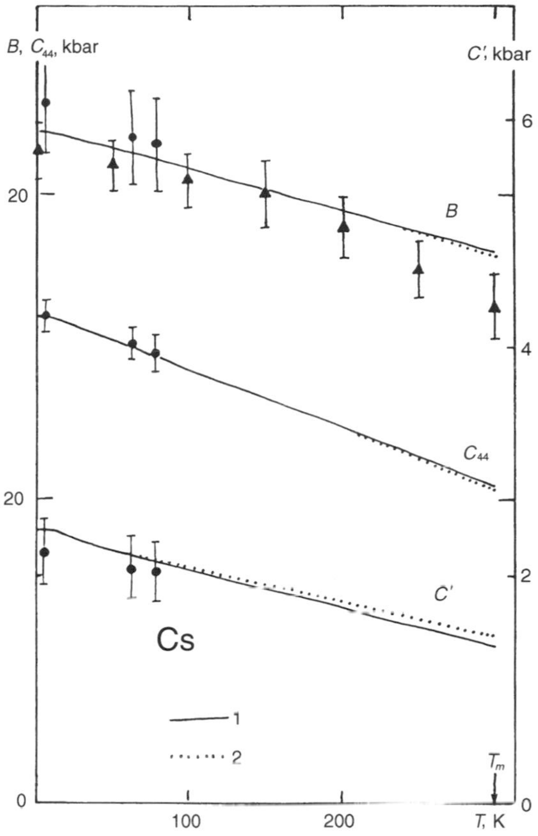

FIGURE 3.18 Modules $C_{ik}(T)$ for cesium. Experimental data: • – [103]; ▲ – [69].

remarkably. Note also that the importance of inclusion of anharmonic high-order contributions to calculations of the temperature dependences of the phonon frequencies for the branch Σ_4 (the long-wavelength limit that is determined by the modulus B_{33} or C' at $p = 0$) was discussed in [2] and earlier in [53] for Na and in [49] for Rb and K.

For Li, the taking into account of the quasi-harmonic high-order contributions (Figure 3.14 and Table 3.13) makes the description worse. This demonstrates once again the non-applicability of the model in the case of Li. It is possible to ask the question if this disagreement is a consequence of a large disagreement between theoretical results obtained in the framework of the second order model theory and experimental data for the static characteristics $dB_{33}^{(0)}/dp$ and $dB_{44}^{(0)}/dp$

To investigate this question, we have also provided for Li the calculations using the experimentally measured values dB_{ik}/dp from [87] (which are the same as the results of calculations for the fourth-order model (see [149, 95]) in formulae (3.50)–(3.51) instead of results for the second-order model dB_{ik}/dp (Table 3.15). In the frame

TABLE 3.15 Temperature dependence of the derivatives of the elastic moduli with respect to pressure, dB_{ik}/dp

	dB_{11}/dp			dB_{44}/dp			dB_{33}/dp		
	theory		experiment	theory		experiment	theory		experiment
T, K	0	295	295	0	295	295	0	295	295
Li	3.69	3.79	3.6 ± 0.25	1.74	1.90	1.03 ± 0.03	0.255	0.264	0.081 ± 0.002
			3.56 ± 0.1			1.08 ± 0.05			0.08
Na	3.85	4.07	3.63 ± 0.07	1.53	1.82	1.99 ± 0.02	0.254	0.262	0.261 ± 0.004
			3.84 ± 0.2			1.74 ± 0.08			0.258 ± 0.015
K	3.84	4.10	4.00	1.23	1.51	1.62	0.261	0.242	0.251
			3.85 ± 0.2						
T, K	0	195	195	0	195	195	0	195	195
Rb	3.78	3.99	3.96	1.15	1.30	1.48	0.264	0.242	0.229
Cs	3.59	3.77	3.54	1.02	1.20		0.275	0.260	

Comment: sources of experiment data see in [18]

TABLE 3.16 Contributions of zero oscillations to the static characteristics of metals (in %, except E^{zp} and p^{zp})

metal	E^{zp}[K/atom]	p^{zp}, kbar	$B^{zp}/B^{(0)}$	$B_{44}^{zp}/C_{44}^{(0)}$	$B_{33}^{zp}/C'^{(0)}$
Li	436	3.26	3.65	−4.25	3.23
Na	195	0.86	1.30	−2.16	1.10
K	116	0.25	0.59	−1.48	0.36
Rb	71.3	0.13	0.33	−0.98	0.12
Cs	50.5	0.06	0.32	−0.76	−0.03

metal	$\left(\frac{dB^{zp}}{dp}\right)/\left(\frac{dB^{(0)}}{dp}\right)$	$\left(\frac{dB_{44}^{zp}}{dp}\right)/\left(\frac{dB_{44}^{(0)}}{dp}\right)$	$\left(\frac{dB_{33}^{zp}}{dp}\right)/\left(\frac{dB_{33}^{(0)}}{dp}\right)$
Li	1.65	−0.29	1.78
Na	0.42	−1.49	−0.29
K	0.07	−0.105	−0.88
Rb	0.02	−0.62	−0.87
Cs	0.01	−0.45	$-3\cdot10^{-5}$

of this procedure, the disagreement with experimental data for $C'(T)$ remains small (Figure 3.14). It meanes that for Li, the derivatives of $\omega_{\lambda\mathbf{q}}$ with respect to deformations cannot be satisfactorily described in the framework of this model at $q \neq 0$. Perhaps, the fourth-order model should be used in calculations of $C_{ik}(T)$ for Li.

The results for the elastic moduli $B(T)$ and $C_{44}(T)$ (Figures 3.14–3.19) have a good enough accuracy up to the melting point T_m. For this reason, the calculated values $\Delta B_{44}(T_m)/C_{44}(0)$ (Table 3.14) can be used, in particular, in discussing the supposition by Varshni [58] that the melting of metals with the same structure and bonding type corresponds to close values of $f_m = C_{44}(T_m)/C_{44}(0)$ for all these metals. For the FCC metals, Pb, Al, Cu, Ag and Au, he had found $f_m \approx 0.55$. This supposition is very close to the Lindeman criterium [32], which is rather good for alkali metals [21, 4]. Hence, it is possible to examine for these metals this hypothesis, too. From Figure 3.14, we can see that the Lindeman criterium [16] is almost satisfied for K, Rb, and Cs: $f_m = 0.695 \pm 0.03$. This value is 4% smaller than for Na and 13% smaller than for Li. This fact reflects once again the similarity of the metallic properties considered (in solid, as well as in liquid, phases). The values $\Delta B_{33}(T_m)/C'(0)$ (Table 3.14) are also close, but the accuracy of those calculations is less known.

Table 3.16 shows contributions to the elastic moduli provided by zero oscillations. Even in Li, these contribution do not exceed 3–4%, and, hence, the anharmonic effects cannot explain the disagreement with experiment for Li mentioned above.

The results for Na demonstrate (Figures 3.19–3.20) the sensitivity of $C_{ik}(T)$ to the form of $\pi(\mathbf{q})$ and $V(\mathbf{q})$ used in calculations. In the same manner as in describing other properties, the GT approximation for $C_{ik}(T)$ is more accurate than other

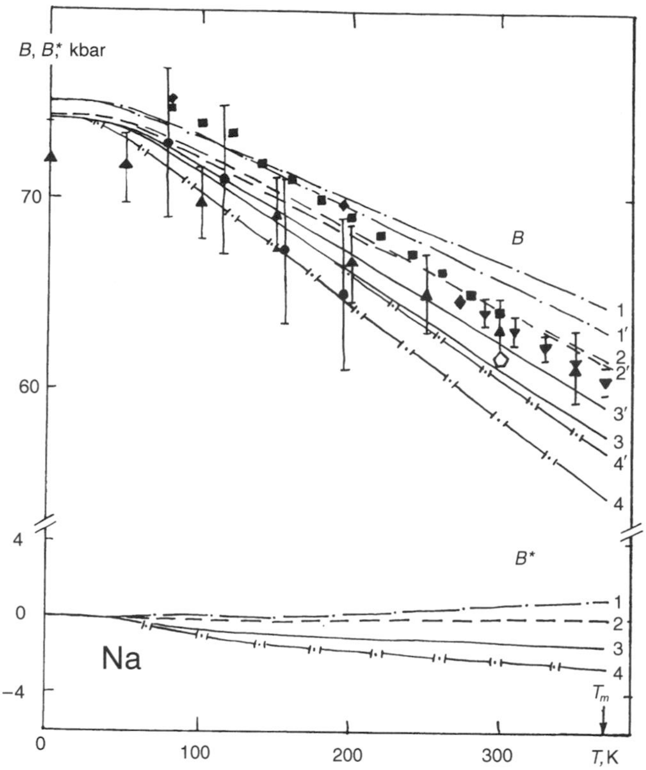

FIGURE 3.19 Sensitivity of results of calculations of $B(T)$ and $B^*(T)$ for sodium to the approximation used for $\pi(\mathbf{q})$ and for the pseudopotential $V(\mathbf{q})$: 1 – RPA; 2 – GV; 3 – GT; 4 – GT, V_{ph}. The curves for the TW and VS approximation are almost the same. Both those are not plotted for clarity of the figure, For both these approximations values of all $C_{ik}(T)$ would be located between curves 2 and 3 (GV and GT). For each approximation primed numbers show the quasi-harmonic result.

forms of $\pi(\mathbf{q})$. The fitting of the pseudopotential parameters using the phonon spectrum (V_{ph}, see equation (2.42)) results in worse accuracy for these elastic moduli. The quasi-harmonic contribution ΔB_{ik}^{qh} plays a dominant role for $B(T)$ only. Hence, since the thermal expansion $\Omega(T)$ is very sensitive to the form of $\pi(\mathbf{q})$ and $V(\mathbf{q})$, then the total variation of $B(T)$, provided by changing the form of $\pi(\mathbf{q})$ and $V(\mathbf{q})$, is rather remarkable. The value $B^*(T)$ is also very sensitive to the form of $\pi(\mathbf{q})$ and $V(\mathbf{q})$ and, depending on the approximation used, can increase or decrease with T or even be non-monotonic (Figure 3.19). But, as a consequence of a small value of $B^*(T)$, this value affects only a little the total modulus $B(T)$. For the shear moduli, the role of quasi-harmonic terms is small, and the sensitivity to the form of $\pi(\mathbf{q})$ and $V(\mathbf{q})$ is smaller. Note also, that the zero-oscillation contributions B_{ik}^{zp} depend on the form of $\pi(\mathbf{q})$ and $V(\mathbf{q})$ more drastically than the total value B_{ik}. For example, for the TW approximation the values $B_{33}^{zp}/C'(0)$ for K, Rb, and Cs are equal to 0.46, −0.21, and 0.05%, respectively, instead of results shown in Table 3.16 for the GT approximation.

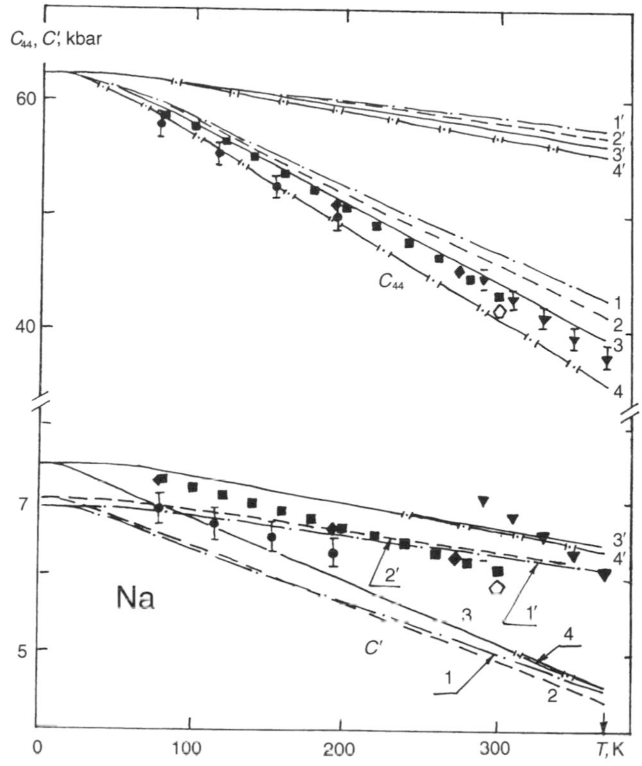

FIGURE 3.20 The same plots as in Figure 3.19 for the moduli $C_{44}(T)$ and $C'(T)$.

Now let us discuss the volume dependences of the moduli $B_{ik}(\Omega, T)$. The dashed line in Figure 3.21 shows $B_{ik}(\Omega, 0)$ for Na. Full lines correspond to $T = 273.2\,\mathrm{K}$. The temperature does not affect the character of the volume dependence of $B_{ik}(\Omega, 0)$ [20]. Remember that these results are valid only within the region of existence of the BCC phase, i.e. at pressures smaller than the pressure of phase transition to any other phase (see Chapter 4). These results were obtained neglecting the term E_{sr} in the adiabatic potential. This last fact gives a large over-estimation of the compression at which the value B_{33} vanishes.

Finally, let us briefly discuss the results of other calculations of the temperature dependences $C_{ik}(T)$, shown in Figures 3.15–3.16. The modulus $B(T)$ is described well, for $C_{44}(T)$, the results are less satisfactory, but the behavior of $C'(T)$ totally disagrees with experimental data. It is difficult to state any conclusions about reasons for these disagreements, since, besides the differences in the pseudopotential and in the methods of calculation used, some other approximations with indefinite accuracy were used in this work. Nevertheless, we can conclude that the results for $B(T)$ are rather good even for rough models, especially if the fitting of constants uses the condition $p = 0$, as was made in [78, 73, 77]. This circumstance relates to the

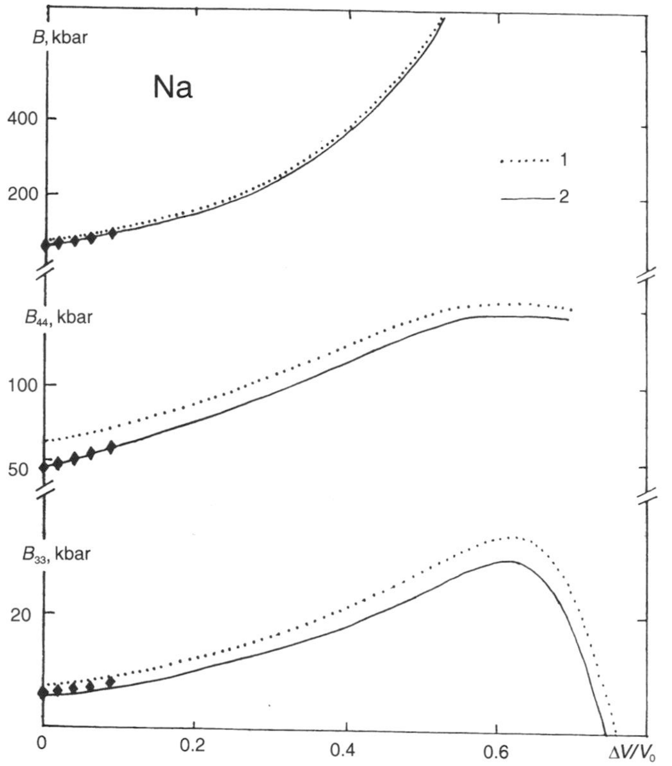

FIGURE 3.21 Volume-dependence of the elastic moduli $B_{ik}(T)$ of sodium: 1 – $T = 0$; 2 – $T = 273.2\,K$; ◆, experiment data from [96]

quasi-harmonic character of this dependence mentioned above and to the usage of the experimental values for $\Omega(T)$. But the problem of the description of $C'(T)$ is a more complicated problem. Some success reached in the work [18], which is the basis of our analysis, shows that this description in the framework of the perturbation theory with a simple pseudopotential is possible. It appears from our analysis of the problem that the application of rather artificial methods like (SCP + C) [53, 54] or the Monte Carlo method [105] seem to be well-founded. First, they provide additional difficulties in calculations and secondly, the analysis of results and the comparison of different pseudopotential models become more difficult. Beyond considering temperature dependences of the frequency shift and the phonon damping (see section 2), we have already demonstrated the inapplicability of these methods of investigation of anharmonic affects.

3.5.3 Anharmonic behavior of $C_{ik}(T)$ at low temperatures

In the begining of section 3, we have mentioned that some low-temperature anomalies of elastic moduli exists. The nature of these anomalies is the same as in the case of the low-temperature "step" in the temperature dependence of the macroscopic Grüneisen parameter $\gamma(T)$ and relates to the increase of $\gamma_{\lambda\mathbf{q}}$ for the branch Σ_4 with q. Formulae (3.45)–(3.48) show, that similar anomalies must take place in the temperature dependences of the intrinsic anharmonic contributions $B^*_{ik}(T)$ to the elastic constants.

Figures 3.22 and 3.23 show the results of calculations of isothermal values $C_{ik}(T)$, $C^*_{ik}(T)$, $B^*_{ik}(T)$, $A^*_{ij}(3)$, $A^*_{ij}(4)$ (for potassium), as well as the difference between the adiabatic and isothermal modules.

$$\Delta(T) = C^S_{12} - C^T_{12} = T\gamma^2(T)C_V(T) \tag{3.54}$$

First, let us pay attention to the behavior of $\Delta(T)$ on the region near the low-temperature step-like variation of $\gamma(T)$ (section 3, Figure 3.7). Here, the curve $\Delta(T)$ at low T is much lower than the asymptotic behavior $\Delta(T \to \infty)$. If there were $\gamma(T) = \text{const}$, then the asymptotic value would be lower than the curve $\Delta(T)$ (3.54). A similar, but larger, effect was observed in NaI theoretically [106] and experimentally [107], but its nature was not explained in these works.

Near $\theta_D/4$, a "hump" in the plot of $C_{12}(T)$ exists (Figure 3.22a) and a "depth" in the plot of $C'(T)$. Since these peculiarities are consequences of the low-temperature

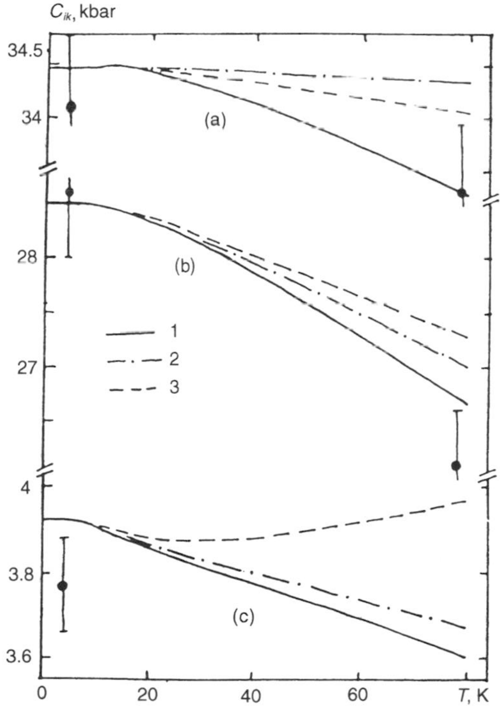

FIGURE 3.22 Temperature dependence of the elastic constants of potassium: (a) – C_{12}, (b) – C_{44}, (c) – C' (3.52). 1 – $C_{ik}(T)$ (3.49)–(3.51), 2 – $B^*_{ik}(\Omega_0, T)$ (3.52), 3 – $C^*_{ik}(T)$. Experiment data from [97].

behavior of the intrinsic anharmonic contributions C^*_{ik} (or B^*_{ik}) (Figure 3.23b,c), then the quasi-harmonic contribution $C_{ik}(T) - C^{(0)}_{ik} - C^{zp}_{ik} - C^*_{ik}$ depends on the temperature through the pressure $p^*(T)$ only, which is a monotonic function of T.

Figures 3.23 b, c, d illustrate the formation of $C^*_{ik}(T)$ from $A^*(3)$ and $A^*(4)$. Usually, opposite sign values $A^*(3)$ and $A^*(4)$ have to be supposed. However, this is true for C'^* (and B^*) only. For these moduli, a large mutual compensation of the contributions $A^*(3)$ and $A^*(4)$ exists. As a result, the "depth" in the plot of C'^* takes place. Moreover, $A^*_{12}(3)$ changes the sign that results in the "hump" in the plot of C_{12} (Figure 3.22). This peculiarity of the low-temperature behavior of $C_{ik}(T)$ is possible to explain only when taking into account the peculiarities of the spectra of $\omega_{\lambda\mathbf{q}}$, $\gamma^i_{\lambda\mathbf{q}}$ and $\partial\gamma^i_{\lambda\mathbf{q}}/\partial u_j$ discussed. Indeed, if the values $\gamma^i_{\lambda\mathbf{q}}$ and $\partial\gamma^i_{\lambda\mathbf{q}}/\partial u_j$ were constant, then the values A^*_{ij} (3.42) would be proportional to the specific heat and the phonon energy, and, hence, they would not show any anomalous behavior.

As we have already mentioned, the temperature dependences $B(T)$ and $C_{44}(T)$ agree well with experiment, whereas the accuracy in calculations of the shear modulus $C'(T)$ is worse. We have already mentioned that this disagreement probably relates

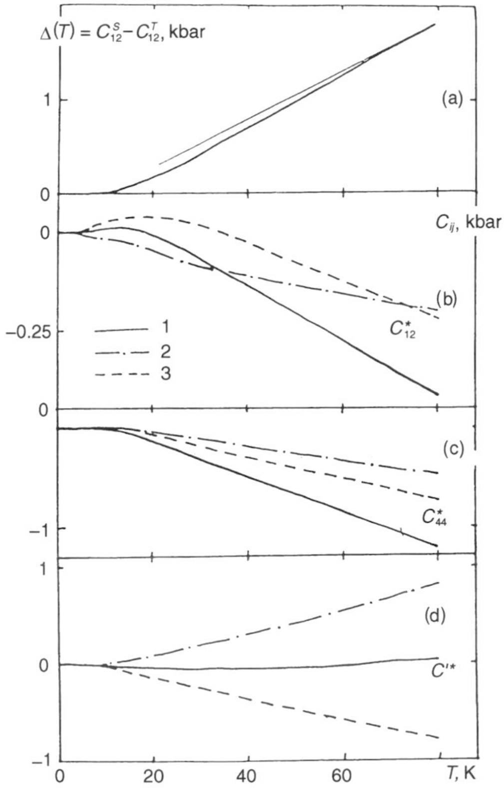

FIGURE 3.23 (a) – the difference $\Delta(T) = C^S_{12} - C^T_{12}$; (b), (c), (d) show different contributions to $C^*_{ik}(T)$ for potassium: 1 – $C^*_{ik}(T)$; 2 – $A^*_{ik}(4)$; 3 – $A^*_{ik}(3)$.

to the influence of high-order anharmonic terms. The results of calculation $A^*(3)$ and $A^*(4)$ confirm this supposition. For $C'^*(T)$, almost total mutual compensation of contributions of the two first anharmonic corrections $A^*(3)$ and $A^*(4)$ takes place (Figure 3.23), whereas in the case of C^*_{12} and C^*_{44}, the contributions $A^*(3)$ and $A^*(4)$ participate with the same signs.

Let us discuss the possibility to observe in experiment the non-smoothe behavior of elastic moduli C_{12} and C' at low T discussed above. First note that in alkali metals (and, apparently, in other crystals, too (see [107])), these anomalies become apparent in a temperature region which is very inconvenient for experimental studies: between the helium (4.2 K) and the nitrogen (78 K) temperatures. There exist only a few measurements at these temperatures. The first measurements in this interval have shown some unusual behavior by $\Delta(T)$ in NaI [107]. In the case of C_{12} and C', the situation is more complicated, because the anomalous behavior of these moduli is camouflaged by the smooth decrease in the quasi-harmonic contribution to the value $C_{ik}(T)$ with increasing T. For this reason, it would be very useful to perform experiments at pressure p equal to the thermal pressure p^*. In this case, the moduli $B_{ik}(p^*, T)$, measured in ultra-sound experiments, would include the contributions $C^{(0)}_{ik}$, C^*_{ik}, and p^* only. Hence, we can simply obtain the value $C^*_{ik}(T)$ using the fact that p^* is known with very good accuracy experimentally and theoretically (see [6]). The anomalies in dependences $C^*_{12}(T)$ and $C'^*(T)$ are of the order 0.2–0.3%, whereas the accuracy of modern experiments can be not worse than 0.1% [108].

3.6 Temperature Dependence of the Third-Order Elastic Moduli of Alkali Metals

In the previous section, we have constructed, in good agreement with experiment for the first time, the quantitative theory of the temperature dependence of the second-order elastic constants B, C', and C_{44} in alkali metals. However, although decreasing the elastic constants B, C', and C_{44} with increasing temperature follows from rather general physical reasons, even qualitative prediction of the behavior of the moduli of the third (and higher) order C_{ikl} with increasing T is very difficult and even today this problem was never considered theoretically. However, since the accuracy of modern measurements of C_{ikl} is good enough, this problem seems to be opportune.

However, one comment is needed. It is well known (see [33, 41, 72]) that the supposition about the adiabatic movement of electrons following the ion movement is valid up to the fourth-order terms in the anharmonicity parameter ε (see section 3 Chapter 1). Hence, we can, strictly speaking, consider the elastic moduli up to the fourth order at $T = 0$, as well as the temperature dependences of the first and second order elastic moduli in the lowest non-zero order in anharmonicity. Elimination of these restrictions needs to escape the adiabatic approximation and (as we know), until today, there were no attempts to do it.[11] However, as immediately follows from

[11] Recently in the work [109] the effect of the non-adiabaticity on the phonon spectrum of normal metals was investigated.

the result of [33] (see section 3 Chapter 1), taking into account the electron transitions would result in some additional terms in the expression for the energy E determined above. The role of these terms requires additional investigation. Hence, the calculation of the temperature dependence of highest, for example third, order elastic moduli in the frame of the models with the adiabatic potential E is a separate problem whose solution would be very useful in considering other non-adiabatically effects. Moreover, the analysis of systematical differences between the calculated "adiabatic" values $C_{ikl}(T)$ and experimental data would provide some conclusions about values of the non-adiabatic contributions to the crystal energy.

In calculations of the temperature dependence of elastic constants, the second-order model for non-transition metal was used. This model, as was shown above, provides a good description of all main lattice properties of alkali metal, including anharmonic ones. For the dielectric (static) susceptibility $\varepsilon(q)$, we use the GT approximation [64] for all alkali metal, except Li, for which the CV approximation [81] was used.

In the present section, the temperature dependences of the isothermal Fuchs modules B_{111}, B_{144}, B_{133}, as well as the derivatives $dB_{ik}/dp = (dB_{ik}/dp)_{p=0,T}$ (which are close relating to these modules (see equations (1.12)–(1.17)) are considered. By definition, the n-th order isothermal elastic module is the n-th derivative of the free energy F with respect to the deformation parameters, normalized to the initial volume of the system [86]. Taking the components of the Lagrange tensor to be the deformation parameters, we obtain the Bragger modules $C_{\alpha\beta\gamma\ldots}$ actually used (section 2 Chapter 1). However, for cubic crystals, it is more convenient to use the parameters γ_i which have more transparent physical sense: they correspond either to the volume change (γ_1) or the pure shear deformations $(\gamma_2, \ldots, \gamma_6)$ (see section 2 Chapter 1).

3.6.1 Calculation of $B_{ikl}(T)$ and $dB_{ik}(T)/dp$

Calculations of $dB_{ik}(T)/dp$ have been performed using a numerical differentiation of the second order isothermal elastic modules $B_{ik}(T)$ calculated analytically (see (3.49)–(3.51)). These derivatives can be written as

$$\frac{dB_{ik}(T)}{dp} = \left[\frac{dB_{ik}}{dp}\right]^{qh} + \left[\frac{dB_{ik}}{dp}\right]^{*} + \left[\frac{dB_{ik}}{dp}\right]^{zp} \tag{3.55}$$

For $(dB_{ik}(T)/dp)^{qh}$, theoretical values for $\Delta\Omega(T)/\Omega$ were used, calculated in the frame of the same model [6, 17]. The results for all alkali metals are shown in Table 3.15. The GT approximation was used for $\pi(\mathbf{q})$ for all metals, except Li, for which the GV approximation was used.

Before starting a discussion of the results obtained, note that experimental data for $dB_{ik}(T)/dp$ for many temperatures are available only for Li and Na, whereas for K and Rb, the pressure derivatives $dB_{ik}(T)/dp$ have been measured only at 295 K and 198 K, respectively. For Cs, there exist experimental results for $dB_{11}(T)/dp$ only. A deficit of available experimental information and large (in comparison to the

TABLE 3.17 Relative change of some values with temperature (for each parameter X, the value $\delta X = X(T_m)/X(0) - 1$ is shown)

metal	δB_{11}	δB_{44}	δB_{33}	$\delta\left(\frac{dB_{11}}{dp}\right)$	$\delta\left(\frac{dB_{44}}{dp}\right)$	$\delta\left(\frac{dB_{33}}{dp}\right)$	δB_{111}	δB_{144}	δB_{133}
Li	−20	−42	−42	6.0	16	1	−16	1	1
Na	−22	−36	−38	8.0	22	2	−17	3	−8
K	−21	−34	−38	8	26	−6	−15	9	−17
Rb	−21	−34	−38	7	24	−9	−17	8	−23
Cs	−18	−34	−41	7	32	−6	−13	21	−11

results for the temperature variation of $dB_{ik}(T)/dp$) experimental errors (6–10%) do not allow us to make any unambiguous conclusion about the possibility of calculating theoretically the variation of $dB_{ik}(T)/dp$ with temperature. Comparing experimental data, we subtract the contribution of electron excitations from experimental values of $dB_{11}(T)/dp$. For all T, this contribution is small in comparison with the lattice one ($\sim 0.1\%$). Note also that in actual experiments, the pressure derivative of the adiabatic modulus was measured. To obtain the values $(dB_{ik}(T)/dp)_T$ used here, formulae and estimations from [61, 62] were used.

Tables 3.15 and 3.17 show that the pressure derivatives change with temperature appreciably less than the respective moduli $B_{ik}(T)$. Small variation in dB_{ik}/dp gave the a conclusion [69, 110] that the pressure derivative of the isothermal compressibility is temperature independent. Calculations [12, 13] show a weak ($\sim 7\%$) in comparison with 20% variation of the modulus B_{11} itself) increase of this derivative with increasing T up to the melting temperature T_m. This fact agrees with experimental data for Li [91], for which this change is 8%.

Calculated values of dB_{ik}/dp are in good agreement with experiment for all alkali metals, except Li and Cs.

For Li, as was already mentioned in discussing other properties, the overestimation of theoretical values of dB_{44}/dp and dB_{33}/dp, is, apparently, a consequence of the usage of the local pseudopotential and the neglect of terms of order $V^3(\mathbf{q})$, etc. Non-applicability of the second-order model with local pseudopotential was strictly demonstrated for results of the calculation of high-order elastic moduli of Li, whereas the phonon frequencies, the specific heat, $B_{ik}(T)$ for Li, generally speaking, can be well described in the framework of this model. This fact confirms an extreme sensitivity of the values of B_{ikl} to details of the model used and the necessity of further experimental studies of the problem.

For Cs, any analysis of the situation is more difficult because experimental data are absent. The only available experimental values of $dB_{11}/dp = 2.84 \pm 0.05$ [69] for the region $0 < T < T_m$ are appreciably smaller than the theoretical results: $dB_{11}(0)/dp = 3.59$ and $dB_{11}(T_m)/dp = 3.86$. This fact makes questionable the results of [69], because values obtained in this work qualitatively disagree with the results of measurements of $B_{11}(T)$ (at $p = 0$, $B_{11} = B$) for Cs. The experimental values of $|dB/dT|$ near T_m from [103] and [69] are equal to 2.4 and 2.6, respectively (in units of

TABLE 3.18 The ratio $\Delta X = \frac{X^{qh}(T_m)-X(0)}{X(T_m)-X(0)}$; X^{qh} is the quasi-harmonic contribution to X at the melting temperature T_m

metal	Li	Na	K	Rb	Cs
ΔB_{11}	0.96	0.91	0.89	0.88	0.93
ΔB_{44}	0.26	0.26	0.23	0.22	0.20
ΔB_{33}	0.29	0.35	0.32	0.31	0.29
$\Delta(dB_{11}/dp)$	0.62	0.72	0.81	0.79	1.04
$\Delta(dB_{44}/dp)$	0.84	0.83	0.79	0.75	0.75
$\Delta(dB_{33}/dp)$	4.76	6.90	−3.15	−1.85	−3.43

10^7kbar/(cm^{2}K)), i.e. remarkably larger than our theoretical value 1.4 (see Table 3.13). At the same time, the derivative

$$\frac{dB}{dT} = \frac{dB_{11}}{dT} = -\left[\frac{dp^*}{dT}\right]_\Omega \frac{dB_{11}}{dp} - p^* \left[\frac{d}{dT}\left(\frac{dB_{11}}{dp}\right)_\Omega\right] + \frac{dB^*}{dT}$$
$$\approx -\left[\frac{dp^*}{dT}\right]_\Omega \frac{dB_{11}}{dp} \tag{3.56}$$

is predominantly determined by the first term. In accordance with experimental [69] and theoretical estimations, the second term is close to zero, as well as the third term. Calculated values for the thermal expansion $(dp^*/dT)_\Omega$ Cs are in good agreement with experiment [17]. Hence, a smaller value of $(dB_{11}/dT)_{exper}$ must correspond to a smaller (in comparison with theoretical values) value of $(dB_{11}/dT)_{exper}$. But the experimental results from [69] demonstrate the opposite behavior. Hence, more precise investigations of the elastic moduli of Cs and their pressure derivatives would be very useful.

The pressure derivative of the shear modulus B_{33} depends on temperature even less than dB_{11}/dp (see Figure 3.24 and Table 3.17) whereas the variation of the value B_{33} itself on the interval 0–T_m is about 35–40%.

As an example, Figure 3.24 and Table 3.18 show contributions to $dB_{ik}(T)/dp$ for Na. About 90% of the value $d[dB_{11}(T)/dp]/dT$ is provided by the quasi-harmonic term which itself weakly depends on T. A weak temperature dependence of dB_{33}/dp is determined by the competition of $(dB_{33}/dp)^{qh}$ and $(dB_{33}/dp)^*$, both of which are not small, but intensively cancel each other (Table 3.18, Figure 3.24). This fact results in a good agreement of the calculated values $dB_{33}(T)/dp$ with precise experimental data for Na [94]. For Li, the experimental points [54] lie appreciably lower than the theoretical curve, although calculated values of the derivatives $d[dB_{11}(T)/dp]/dT$ [54] agree well with experiment. For other metals (K, Rb, Cs), the intrinsic anharmonic contribution "over-compensates" the contribution of thermal expansion and provides $d[dB_{11}(T)/dp]/dT$ to be negative (Tables 3.17, 3.18). All calculated values of the derivatives of elastic moduli for K and Rb agree well with

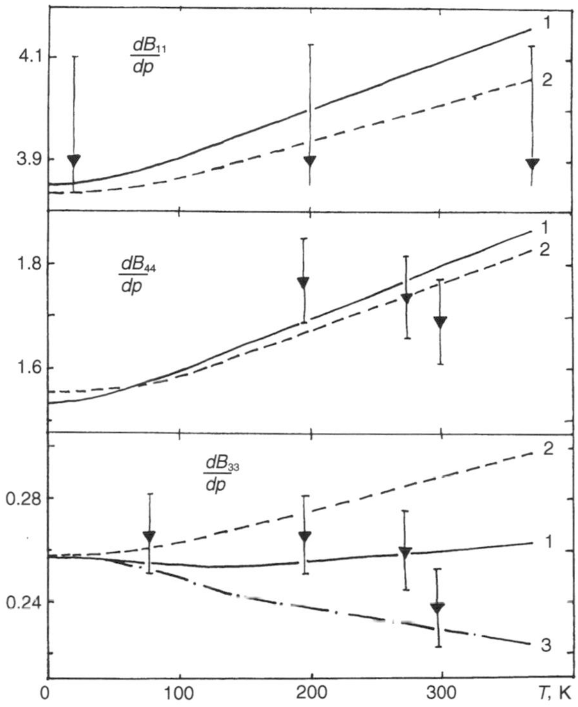

FIGURE 3.24 Different contributions to the temperature dependence of $dB_{ik}(T)/dp$ for sodium. 1 – the first-order perturbation theory, 2 – the quasi-harmonic approximation, 3 – intrinsic anharmonic contributions. The different between curves 1 and 2 at $T = 0$ relates to the contribution of zero oscillations. Experimental data from [94, 100].

experiment. The larger role of the intrinsic anharmonic contributions to $dB_{33}(T)/dp$ would be an additional argument for a conclusion about high sensitivity of the modulus B_{33} to high-order anharmonic terms.

Taking into account the temperature dependence of dB_{44}/dp completely eliminates the disagreement between experimental and theoretical values of dB_{44}/dp observed in earlier publications [1], but for K only (and maybe for Rb, for which experimental errors are unknown). For Na, calculated values of $(dB_{44}/dp)_{theor}$ agree with experimental data only in the region of 200–300 K and do not reproduce the rapid increase near 100 K [94] (Figure 3.24). As we can see from Table 3.17, the most intensive change with temperature demonstrates the value dB_{44}/dp. Its variation, as well as the variation in dB_{11}/dp, is predominately determined by the quasi-harmonic term, but the changes in dB_{11}/dp are small and have almost the same values for all metals, whereas the change in dB_{44}/dp increases in a sequences from Li to Cs (Table 3.17) and reaches for the latter the value 32%. Variations of the modulus B_{44} demonstrate similar behavior, although these variations have opposite signs.

Zero oscillations practically does not affect the value dB_{ik}/dp for Na, K, Rb, and Cs ($\sim$ 1–1.5%). For Li, contributions of zero oscillations to dB_{ik}/dp reach 10%.

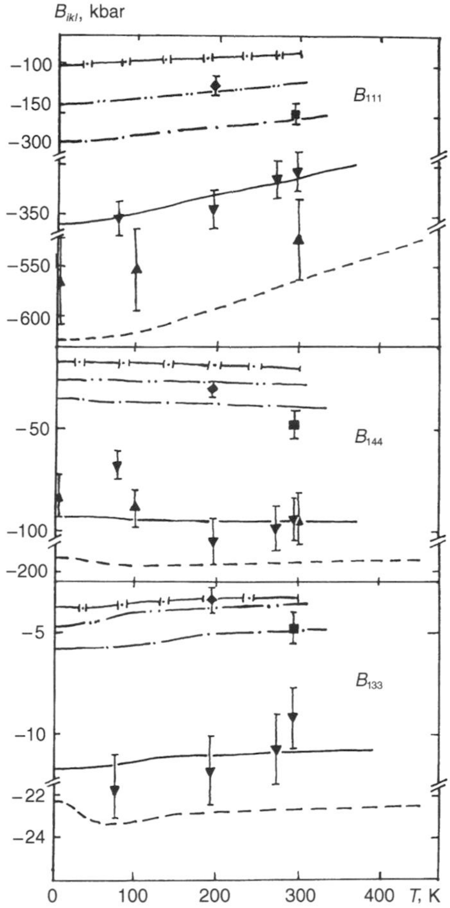

FIGURE 3.25 Temperature dependence of B_{111}, B_{144}, B_{133} for alkali metals: – – – – Li; —— Na; – · – K; – ·· – Rb; ⊢—⊣ Cs. Experimental data see in [13].

Despite the fact that the fitting of parameters with taking into account zero oscillations does not eliminate disagreement between theory and experiment for Li, it is clear that any quantitative theory of properties of this metal must include effects of zero oscillation.

Here, we will not provide any detailed discussion of results of calculations of B_{ikl} at $T = 0$, an influence of different approximations for the dielectric susceptibility, and the comparison with results obtained by other authors. These questions have been considered in detail in [12]. Now we consider the temperature dependences of B_{ikl}.

Figure 3.25 and Table 3.17 show the results of calculations of the temperature dependence of the isothermal third-order moduli B_{111}, B_{144}, B_{133} for alkali metals obtained from the dependences $B_{ik}(T)$ calculated in the framework of the same model as well as from theoretical values of $dB_{ik}(T)/dp$. Even today, there exist no experimental data for the moduli B_{ikl} for alkali metals. For this reason, the points plotted in Figure 3.25 were obtained by processing the data for $(dB_{ik}/dp)_S$ and B_{ik}^S.

The third-order elastic moduli for alkali metals are negative [12]. B_{111} and B_{133} increase and B_{144} decreases with increasing temperature (Table 3.17). The relative variations of B_{111} for different metals have almost the same value of about 15%. The changes of the moduli B_{144} and B_{133} increase from 0.6% for Li to 21–23% for Cs and Rb. Note, that the change in B_{144} (and B_{133}) has remarkable values up to $T \approx T_m/3$, while for higher temperatures this variation is much weaker. This behavior of $B_{ikl}(T)$ appreciably differs from the temperature variation of $B_{ik}(T)$ (Figures 3.14–3.18). Generally speaking, variations of the third-order elastic moduli with temperature are appreciably weaker than variations of the second-order elastic moduli.

The third-order moduli are very sensitive at $T = 0$ to the details of the model (the choice of the pseudopotential and the approximation for the polarization operator $\pi(\mathbf{q})$) [12]. Naturally, this fact is a consequence of the definition of these values as the third derivatives of the energy with respect to the deformations u_i. Temperature dependences of these values, as was already mentioned, are determined by higher derivatives, up to the fifth order. For this reason, to provide a conclusion about applicability of one or another model, accurate experimental data for C_{ikl} are needed. Such data can be obtained in present-day experiments.

3.6.2 *The Cauchy relations for anharmonic crystals*

The suppositions that the atoms (ions) in the lattice interact through pair central forces and each atom is a center of symmetry results in an additional symmetry of the second-order elastic constants $C_{\alpha\beta\gamma\delta}$, which provides the so-called Cauchy relations (CR) [41]:

$$\delta = C_{\alpha\beta\gamma\delta} - C_{\alpha\gamma\beta\delta} = 0. \tag{3.57}$$

Note, that since here we use the Brugger elastic moduli (section 2 Chapter 1), then expression (3.57) is valid (by suppositions listed above) for an arbitrary, even stressed state (see [86]).

However, it is well-known from experiment that the CR (3.57) are not satisfied for all crystals experimentally investigated. Moreover, large values of $\delta \sim (0.1 \div 1)C_{12}$ were observed for metals (see [111], Table 15) and for A_3B_5 semiconductors [112], as well as for some binary compounds with the NaCl structure (e.g., AgCl, MgO, GeTe [113]).

Actually, the main reason for the breakdown of the CR is considered to be non-pair interactions in crystals. Many attempts have been made to reach an agreement with experiment for theoretical values δ calculated in the frame of phenomenological, as well as microscopic description of the non-pair forces (see publications cited in [114]). These models relate to ionic or van der Waals crystals at $T = 0$. For metals besides the many-ion interaction, described by terms like $E_e^{(3)}$, etc., two reasons for the breakdown of the CR at $T = 0$ [33] (even in the case of the pair approximation for E (3.1)) have been established. The first reason is that the equilibrium state of the lattice in metals depends on the forces provided by the electron liquid. The second reason is a non-pair character of the indirect inter-ion interaction described by the term $E_e^{(2)}$ in (3.1). Generally speaking, this term is also provided by the electron liquid, i.e., by the dependence of the pair potential $E_e^{(2)}$ on the electron density n_0.

At finite temperatures, there exists one additional reason giving the breakdown of the CR: the anharmonicty of the inter-atomic potential (see [111]). In fact, the anharmonicity affects the CR like the many-particle nature of the inter-ion interaction. Even in the case of the crystal with a pair interaction only, the contributions of each interacting pair of atoms to the oscillation part of the free energy F^* (1.14) are not additive because they participate in the argument of the logarithm. Strictly speaking, even at absolute zero, the CR would be destroyed as a consequence of anharmonic zero oscillations, but actually these effects are small [25, 12].

In the present subsection we consider in detail the breakdown of the CR provided by the anharmonicity of the inter-atomic potential. To do this, we use a rather general model of the crystal with the energy, containing not only the pair potential, but also depending on the cell volume as parameter (see section 1 Chapter 1). This can be used as for dielectrics for which, possibly, the dependence of the energy on the cell volume Ω is negligeable, as for metals, where the crystal energy directly depends on the equilibrium electron density $n_0 = z/\Omega$.

We will show general relations between the deviations of $\delta(T)$ from the CR at finite temperatures and the derivatives of the inter-atomic potential. For non-transition cubic metals, the temperature dependence of δ calculated in the framework of the second-order pseudopotential model (see Chapter 2) will be compared with experiment. The effects of the electron liquid on $\delta(T)$ will be discused.

Let us obtain the formulae describing the deviations from the Cauchy relations. For cubic crystals, it is convenient to write

$$\delta = C_{12} - C_{44} = \frac{1}{6}\left[C_{\alpha\alpha\beta\beta} - C_{\alpha\beta\alpha\beta}\right] \tag{3.58}$$

(Here and below, we suppose a summation over repeated Greek indices, except λ, λ' which numerate the phonon branches). On the other hand, taking into account the formulae for the contributions to $C_{\alpha\beta\gamma\delta}(T)$ (3.39), it is possible to write

$$\delta(\Omega, T) = \delta^0(\Omega) + \delta^{qh}(\Omega, T) + \delta^{zp}(\Omega, T) + \delta^*(\Omega, T). \tag{3.59}$$

The first term in this formula determines the deviation from the CR at $T = 0$, the second term is a contribution of the thermal expansion $\Omega(T)$, the third and the fourth terms directly relate to the anharmonicty and have been obtained from $C^{zp}_{\alpha\beta\gamma\delta}$ and $C^*_{\alpha\beta\gamma\delta}$.

Taking into account (1.12), (1.4), we obtain that

$$\delta(\Omega) = \left(2\frac{\partial}{\partial\Omega} + \Omega\frac{\partial^2}{\partial\Omega^2}\right)\left[w(\Omega) + \frac{1}{\Omega}\sum_{\boldsymbol{\tau}}\chi(\boldsymbol{\tau}, \Omega)\right] - \frac{2}{3\Omega}\sum_{\boldsymbol{\tau}}\frac{\partial^2\chi}{\partial\Omega\partial\tau_\alpha}\tau_\alpha. \tag{3.60}$$

This formula describes both $\Delta(\Omega(T))$ and $\delta(\Omega(p))$. In accordance with our consideration, the value $\delta(\Omega)=0$, if the cohesive energy does not depend on the volume, i.e., if

$$w(\Omega)=0 \quad \text{and} \quad \chi=\chi(\boldsymbol{\tau}). \tag{3.61}$$

Formula (3.60) can be used in different concrete models of the cohesive energy. In particular, this equation is valid at $T=0$. For example, for a non-transition metal with the electron density $n_0=z/\Omega$, we have to substitute in (3.60)

$$\begin{aligned} w(\Omega) &= E_e^{(0)}+E_e^{(1)}, \\ \chi(\mathbf{q},\Omega)\equiv\chi(\mathbf{q},n_0) &= \frac{1}{2}\left[\frac{4\pi(Ze)^2}{q^2}-\frac{\pi(\mathbf{q})(\Omega V_{\mathbf{q}})^2}{\varepsilon(\mathbf{q})}\right]. \end{aligned} \tag{3.62}$$

The energy of the free electron gas $E_e^{(0)}$ (3.33) depends on n_0 only. The energy of the non-point-like structure of ions $E_e^{(1)}\sim 1/\Omega$ (3.34) and the Fourier components of the pair potential $\chi(\mathbf{q},\Omega)$ intrinsically depend on Ω through the density n_0, participating in the polarization operator $\pi(\mathbf{q})$, and the dielectric susceptibility $\varepsilon(\mathbf{q})$ (see (3.35), (3.37)). On the other hand, the product of the form factor and the volume, $V_{\mathbf{q}}\Omega$, does not explicitly depend on the volume.

It is easy to see that, if the free energy contains terms proportional to $1/\Omega$ only, then these terms would not destroy the CR for the elastic moduli of any order. In the case of metals, the energy of non-point-like structure of ions $E_e^{(1)}$ demonstrates this peculiarly. Hence

$$\delta(\Omega)=-2p^{(0)}+B^{(0)}+\frac{1}{\Omega^2}\sum_{\tau}\left[2n_0\frac{\partial}{\partial n_0}+n_0^2\frac{\partial^2}{\partial n_0^2}+\frac{2}{3}n_0\tau\frac{\partial^2}{\partial n_0\partial\tau}\right]\chi(\boldsymbol{\tau},n_0), \tag{3.63}$$

where $p^{(0)}$ and $B^{(0)}$ are the pressure and the compressibility modulus obtained from the energy $E_e^{(0)}$. Formula (3.63) has been for the first time obtained by Brovman and Kagan [33], who considered it to be valid at $p=T=0$ only. However, this formula can be used at arbitrary $\Omega(p,T)$ [25].

Strictly speaking, the quasiharmonic correction to δ in (3.59) is

$$\delta^{qh}(\Omega,T)=-p^*(\Omega)\,\frac{d\delta(\Omega)}{dp}. \tag{3.64}$$

Using (1.116) and [12, 13], we obtain

$$\frac{d\delta(\Omega)}{dp}=-\frac{1}{3B}[\delta(\Omega)+2(K_1(\Omega)+K_2(\Omega))], \tag{3.65}$$

where

$$K_1=C_{112}-C_{166};\quad K_2=C_{123}-C_{144}. \tag{3.66}$$

are deviations from the CR for the third-order elastic constants C_{ikl}.

Let us now consider the intrinsic anharmonic contribution to δ: $\delta^{ph} = \delta^{zp} + \delta^*$. The expressions for $C^{ph}_{\alpha\beta\gamma\delta}$ were discussed above. Note only that the term δ^{ph} includes all terms except contributions of the thermal expansion. Hence,

$$\delta^{ph}(\Omega, T) = \frac{\hbar}{16\pi^3}\sum_{\lambda}\int d^3q\left\{\frac{N_{\lambda\mathbf{q}}+0.5}{\omega_{\lambda\mathbf{q}}}J^{(4)}_{\lambda\mathbf{q}} + \sum_{\lambda'}(\omega^2_{\lambda'\mathbf{q}} - \omega^2_{\lambda\mathbf{q}})^{-1}\right.$$

$$\left.\times\left[\frac{N_{\lambda\mathbf{q}}+0.5}{\omega_{\lambda'\mathbf{q}}} - \frac{N_{\lambda\mathbf{q}}+0.5}{\omega_{\lambda\mathbf{q}}}\right]J^{(3)}_{\lambda\lambda'\mathbf{q}}\right\}. \tag{3.67}$$

The first term in the braces is provided by the four-anharmonic contributions (in the first order) and the second term relates to the triple-anharmonic contributions (in the second order). Additionally, $J^{(4)}_{\lambda\mathbf{q}}$ includes the thermal pressure. Hence, using (3.58), we obtain

$$J^{(4)}_{\lambda\mathbf{q}} = \frac{1}{6}\left\{\left[D^{\alpha\alpha,\beta\beta}\right]_{\lambda\lambda} - \left[D^{\alpha\beta,\alpha\beta}\right]_{\lambda\lambda} + 2\left[D^{\alpha\alpha}\right]_{\lambda\lambda}\right\}$$

$$J^{(3)}_{\lambda\lambda'\mathbf{q}} = \frac{1}{6}\left\{\left[D^{\alpha\alpha}\right]_{\lambda\lambda'}\left[D^{\beta\beta}\right]_{\lambda'\lambda} - \left[D^{\alpha\beta}\right]_{\lambda\lambda'}\left[D^{\alpha\beta}\right]_{\lambda'\lambda}\right\}, \tag{3.68}$$

where (see (1.37))

$$\left[D^{\alpha\beta}(\mathbf{k})\right]_{\lambda\lambda'} = e^{\gamma}_{\lambda\mathbf{k}}\left(\partial D_{\gamma\delta}(\mathbf{k})/\partial\bar{u}_{\alpha\beta}\right)e^{\delta}_{\lambda'\mathbf{k}}.$$

Using general properties of the dynamical matrix, we obtain (see [25])

$$J^{(4)}_{\lambda\mathbf{q}} = -2X^{(1,1)}_{xx\lambda\lambda} + X^{(2,0)}_{\lambda\lambda} - \frac{4}{3}X^{(1,0)} + \left(\omega^2_{\lambda\mathbf{q}} - \frac{1}{3}\sum_{\lambda'}\omega^2_{\lambda'\mathbf{q}}\right)$$

$$J^{(3)}_{\lambda\lambda'\mathbf{q}} = \left(X^{(0,1)}_{xx,\lambda\lambda} + X^{(1,0)}_{\lambda\lambda'}\right)^2 - \frac{1}{3}X^{(0,1)}_{\lambda\lambda\lambda'\lambda'}\left(\omega^2_{\lambda\mathbf{q}} + \omega^2_{\lambda'\mathbf{q}}\right)$$

$$+\frac{1}{3}\left[2\left(6X^{(0,1)}_{xx\lambda\lambda} - 5X^{(1,0)}_{\lambda\lambda}\right) + 8\omega^2_{\lambda\mathbf{q}}\right]\omega^2_{\lambda\mathbf{q}}\delta_{\lambda\lambda'} - \frac{1}{6}\left(\omega^4_{\lambda\mathbf{q}} + \omega^4_{\lambda'\mathbf{q}}\right). \tag{3.69}$$

Here, we denote

$$X^{(r,s)}_{\alpha\ldots} \equiv X^{(r,s)}_{\alpha\ldots}(\mathbf{q}) = \tilde{X}^{(r,s)}_{\alpha\ldots}(\mathbf{q}) - \tilde{X}^{(r,s)}_{\alpha\ldots}(0),$$

$$\tilde{X}^{(r,s)}_{\alpha_1\ldots\alpha_m}(\mathbf{q}) = \frac{1}{M}\sum_{\tau}\left\{\Omega^{r-1}\frac{\partial^r}{\partial\Omega^r}\left[\chi^{(s)}(\mathbf{p},\Omega)p_{\alpha_1}\ldots p_{\alpha_m}\right]_{\mathbf{p}}\right\} \tag{3.70}$$

$$\chi^{(s)}(\mathbf{p},\Omega) = \left[\left(\frac{1}{p}\frac{\partial}{\partial p}\right)^s\chi(\mathbf{p},\Omega)\right]_{\Omega}; \quad X_{\lambda\lambda'\ldots} = e^{\alpha}_{\lambda\mathbf{q}}X_{\alpha\beta\ldots}e^{\beta}_{\lambda'\mathbf{q}},$$

$$\mathbf{p} = \boldsymbol{\tau} + \mathbf{q}; \quad m = 2(s+1).$$

Note, that

$$\tilde{X}^{(0.0)}_{\alpha\beta}(\mathbf{q}) = \tilde{D}_{\alpha\beta}(\mathbf{q})$$

The formulae (3.69) contain rather interesting results. For the system with pair interaction only, the deviation $\delta^{ph}(T)$ from the CR is independent of the four-phonon processes and is determined by the three-phonon vertex (proportional to $X^{(0.2)}$), the phonon frequencies $\omega_{\lambda\mathbf{q}}$ and their partial derivatives with respect to volume only (see [25]).

All these conclusions about the CR are valid for every model of the cohesion (without explicit contribution of many-particle interaction), i.e., for crystals with different types of chemical bonding.

For ion and van der Waals crystals, the deviations from the CR, provided by three-particle forces, were considered in [114]. These terms must be added to the terms considered above in the present review. Additionally, we must take into account that for such crystals the cohesive energy does not explicitly depend on the volume (see (3.61)). From (3.69)–(3.70) we can see that, if the cohesive energy (exactly speaking $\chi(\mathbf{q})$) does not explicitly depend on Ω, then $X^{(1.1)}$, $X^{(2.0)}$, and $X^{(1.0)}$ must be identically equal to zero and expressions for $J^{(4)}_{\lambda\mathbf{q}}$ and $J^{(3)}_{\lambda\lambda'\mathbf{q}}$ are simplified.

In the case of simple metals, the cohesive energy depends explicitly on Ω only through n_0. Hence, we can use in (3.70) a simple substitution $(\partial/\partial\Omega)_{\mathbf{p}} \rightarrow -n_0\Omega^{-1}(\partial/\partial n_0)_{\mathbf{p}}$.

Finally, let us mention the Cauchy relations for the third-order elastic constants. If each atom is a center of symmetry and the interatomic interaction can be described by central forces, then the next relations must be satisfied at all pressures:

$$C_{\alpha\beta\gamma\delta\mu\nu} = C_{\alpha\gamma\beta\delta\mu\nu} = C_{\alpha\beta\gamma\mu\delta\nu}. \tag{3.71}$$

These relations provide three conditions for C_{ikl}:

$$C_{123} = C_{144} = C_{456}; \quad C_{112} = C_{166}. \tag{3.72}$$

There are no principal difficulties in providing the analysis of the breakdown of the Cauchy relations for C_{ikl}, similar to that made above for the value δ, but the respective formulae are very cumbersome. We restrict our consideration to the case of a simple metal at zero temperature and, as an example, show the expression for $C_{144} - C_{456}$ only

$$K_3 = C_{144} - C_{456} = -\delta + n_0 \frac{d}{dn_0}(C_{44} - p). \tag{3.73}$$

This simple formula demonstrates clearly the fact that the breakdown of the CR for C_{ikl} in metals has the same nature as was established for the breakdown of the CR for $C_{\alpha\beta}$: both of these are provided by an explicit dependence on the density n of the electron liquid, as well as by the non-pair character of the inter-ion interaction.

The formulae for K_1 and K_2 in (3.66) are qualitatively similar to the expression for K_3, but they are more cumbersome because of high-order derivatives with respect to n_0.

Figure 3.26 and Table 3.19 show the results of respective calculations for alkali metals. The second-order (see Chapter 2) pseudopotential model was used.

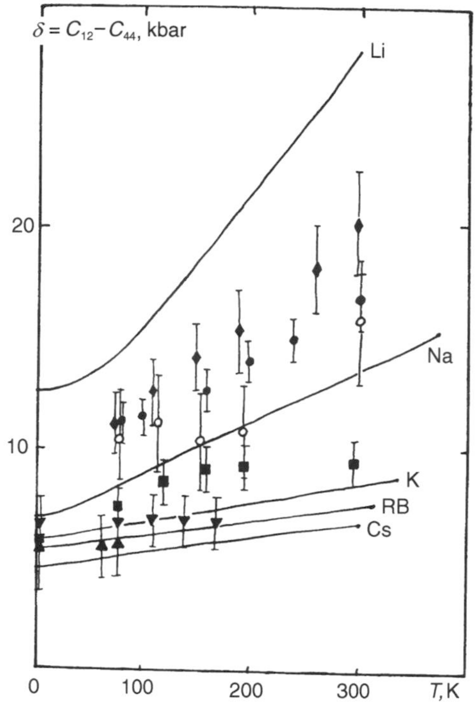

FIGURE 3.26 Deviation from the Cauchy relations for alkali metals. Experimental data: ◆ – Li [88]; ● – Na [94]; ○ – Na [92]; ■ – K [97]; ▼ – Rb [98]; ▲ – Cs [103].

Let us consider the results of calculations of $\delta(T)$. First note that, for alkali metals, in agreement with experiment, $\delta(0)/C_{12}(0)$ (at $T = 0$) is about $0.1 \div 0.3$.

From Figure 3.26 and Table 3.19 two general conclusions emerge. First, in light metals (Li, Na), the values $\delta^*(T)$ are comparable with $\delta(0)$. Second, the derivative $\partial \ln \delta / \partial T$, which characterizes the slope of the curve $\delta(T)$, decreases with increasing the atomic number (to be precise the atomic mass). This fact can by simply explained if we take into account that $\delta^* \sim 1/M$.

The comparison of calculated temperature dependence $\delta(T)$ with experiment is less informative because of the small accuracy of experimental data. Nevertheless, the theory correctly reproduces trends in experimental data for all alkali metals, except Li. The behavior of $\delta(T)$ Li calculated by us appreciably disagrees with experimental data. In particular, the theoretical value of $\partial \ln \delta / \partial T$ is almost two times larger than the experimental one. As we have already mentioned in discussions of many other properties, this disagreement relates to the non-applicability of the second-order model with the local pseudopotential in the case of Li. For Na, the calculated curve $\delta(T)$ does not coincide with the results of very accurate measurements [94], although the values of the derivative $\partial \ln \delta / \partial T$ are very close to experimental data.

TABLE 3.19 Deviations from the Cauchy relations for the second-order elasticity moduli of alkali metals

metal		Li	Na	K	Rb	Cs
$\delta(0)$, kbar	a	11.7	7.0	6.0	5.5	4.6
	b	10 ± 5	10.6 ± 5.7	5.9 ± 1.8	6.7 ± 3.0	5.6 ± 1.7
$\delta(0)/C_{12}(0)$	a	0.094	0.101	0.173	0.200	0.223
	b	0.082	0.146	0.171	0.223	0.259
$\frac{\delta(T_m) - \delta(0)}{\delta(0)}$	a	2.21	1.19	0.47	0.34	0.46
T, K		295	195	195	195	78
$\frac{\partial \ln \delta}{\partial T}$; 10^{-3}/K	a	2.3	2.0	1.1	1.0	1.4
	b	1.2	2.1	0.7	−0.3	0.3
exper.		[88]	[94]	[97]	[98]	[103]

Comment: a – theoretical, b – experimental results.
Experimental values for $\delta(0) = \delta(T = 0)$ and $C_{12}(0)$ are obtained by extrapolation to $T = 0$.
Approximations used for $\pi(q)$: Li – GV [81]; Na, K, Rb, Cs – GT [64].

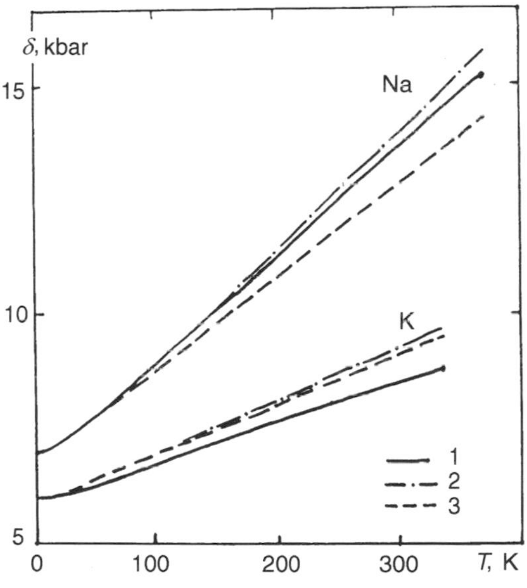

FIGURE 3.27 Contributions to $\delta(T)$ (see text) for sodium and potassium: 1 – the total $\delta(T)$, calculated in the first order in the anharmonicity; 2 – the intrinsic anharmonic contribution $\delta^*(T)$; 3 – $\delta_{qh}(T)$ calculated at constant electron density.

Figure 3.27 shows the role of different contributions to $\delta(T)$ (3.59) and the role of the dependence of χ in (3.62) on n_0. The difference between curves 1 and 2 is the quasi-harmonic contribution δ^{qh} to $\delta(T)$ calculated using (3.64). Using the results of

TABLE 3.20 Deviations from the Cauchy relations for the third-order elastic moduli of alkali metals

metal	approx. for $\pi(q)$	$\dfrac{C_{112}-C_{166}}{C_{112}}$	$\dfrac{C_{123}-C_{144}}{C_{123}}$	$\dfrac{C_{144}-C_{456}}{C_{144}}$
Li	GV [81]	−0.089	−0.109	−0.041
Na	GT [64]	0.084	−0.012	−0.083
K	GT	0.236	0.048	0.118
Rb	GT	0.254	0.067	0.110
Cs	GT	0.271	0.027	0.124

[33, 12], it is possible to establish that $\delta + 2(K_1 + K_2)$ and, consequently, the value $\delta^{qh}(T)$ are positive for Li and negative for Na, K, Rb, Cs.

The dominant role in $\delta(T)$ plays the intrinsic anharmonic term $\delta^*(T)$ (Figure 3.27). The role of the derivatives with respect to the density n_0 is relatively small, of the same order or somewhat larger than $\delta^{qh}(T)$. It is very interesting that the sign of contributions of these derivatives to $\delta^*(T)$ changes irregularly from metal to metal.

Table 3.20 shows deviations from the CR for C_{ikl} for alkali metals. The values $K_{1.2.3}$ are determined by the terms $E_e^{(0)}$ and $E_e^{(2)}$. Contributions of both these terms have the same order of magnitude. For this reason, the deviations from the CR for C_{ikl} and C_{ik} are of the same order of magnitude.

Deviation from the CR depends remarkably on the approximation used for $\pi(\mathbf{q})$. Generally speaking, these deviations, even for alkali metals, can be described by a theory much worse than for the values $C_{ik}(T)$ themselves. There is no difficulty in understanding that, if we take into account the fact that, in contrast to the majority of properties for which the role of the pseudopotential (or $E_e^{(2)}$) is small in comparison with the contribution E_i, these deviations from the CR are directly determined by the term $E_e^{(2)}$.

3.7 The Core Polarizability

In the pseudopotential approach, which is intensively used in the microscopic theory of atomic and thermodynamic properties of metals, there are two main directions: *ab initio* calculations and calculations using model potentials. The advantage of the first direction (see, for example, [26, 152–157]) is that is does not use any experimental information relating to metallic properties. But, as a consequence of extremely sophisticated calculations, it is impossible to reach a high enough accuracy. As a consequence, these calculations are restricted to a relatively small number of properties in a few metals. The second direction provides adequate description of a large number of properties in many metals. However, until now, for a majority of polyvalent metals, results obtained by this approach cannot be considered as successful (see the present review and [170]). This is a consequence of using simple perturbation theory, neglecting the non-locality effects, the supposition of the "rigidity" of ion cores, and so on. For alkali metals, the accuracy of describing a very large number of metallic properties in the framework of the rather simple pseudopotential scheme can reach the level of accuracy of modern experiments (see preceding sections of the present chapter). However, some characteristics, which are very sensitive to the form of the inter-ion interaction, like the Grüneisen parameter (and consequently the thermal expansion), could not be described well for K, Rb, and Cs (see [158]). To provide more accurate calculations, we have to adopt the model for the terms providing maximum contributions to the volume-dependent part of the total energy of the metal, namely, the energies of the van der Waals (E_W) and Born–Mayer (E_{sr}) interactions.

The quantitative investigation of the van der Waals interaction (V_W) in metals was initiated by the work [158], where, for the first time, the expression for the pair potential of the dispersion forces, provided by the polarization of ions in the electron gas, was obtained

$$V_W(R) = -\int_0^\infty \frac{d\omega}{2\pi}\alpha^a(i\omega)\alpha^b(i\omega)\left[\left(\frac{\partial^2 w_c(R, i\omega)}{\partial R^2}\right)^2 + \frac{2}{R^2}\left(\frac{\partial w_c(R, i\omega)}{\partial R}\right)^2\right]. \tag{3.74}$$

Here, $\alpha^a(i\omega)$ is the ion polarizability for an imaginary frequency, a denotes the ion sort, R is the distance between ions a and b,

$$w_c(R, i\omega) = \int \frac{d^3q}{(2\pi)^3} w_c(q, i\omega)\exp(i\mathbf{qR}), \quad w_c(\mathbf{q}) = \frac{v_c(\mathbf{q})}{\varepsilon(\mathbf{q}, i\omega)} = \frac{4\pi e^2}{q^2\varepsilon(\mathbf{q}, i\omega)} \tag{3.75}$$

is the screened Coulomb potential, $\varepsilon(\mathbf{q}, i\omega)$ is the dielectric function of free electrons. The contribution of (3.74) to the energy of K, Rb, Cs was investigated in [159]. It was found to be appreciable. Studies of expression (3.74) made in [160] have shown that, taking into account the spatial dispersion in $\varepsilon(\mathbf{q}, i\omega)$ provides some additional terms which are small in comparison with actual $\sim 1/R^6$ contributions. One of them is proportional to $\exp(-R/R_d)/R^{5/2}$ ((R_d is some value of order of the screening radius), and the second is of the order $\cos^2(2k_F R)/R^7$ ($\hbar k_F$ is the Fermi momentum). However, the frequency dependence of $\varepsilon(0, i\omega) = 1 + \omega_{pl}^2/\omega^2$ (ω_{pl} is the electron plasma frequency) is important for noble metals. Similar conclusions were also made in [159], where V_W was calculated numerically. The dependence of the ion polarizability on $\mathbf{q}$ was considered in [151], but not exactly. In [162, 163], the conclusion was made that V_W is important not only for noble metals but also for heavy alkali metals (K, Rb, Cs). Using an erroneous expression for the energy, a wrong conclusion about an appreciable change in the equation of state for K at $T = 0$ due to the core polarizability was made in [164].

Hence, we can see that in calculations of V_W mentioned above, only a few terms in the expansion of the total energy (in powers of $\kappa = 4\pi\alpha(0)/\Omega_0$ and $\eta = \omega_{pl}/\omega_0$, where $\alpha(0)$, Ω_0, and ω_0 are the static polarizability of the ion, the volume per atom, and the characteristic energy of core excitations, respectively) were taken into account. This inconsistency has provided serious errors. For example, the importance of the screening of the Madelung interaction between polarized ions was manifested in [164]. However, as will be shown below (see [151]), this effect does not exist at all.

The consistent procedure of calculation of the power series expansion of the total energy in parameters κ and η was obtained in works [150, 151], which we will follow in this section.

The case $\eta \ll 1$ will be considered in subsections 1 and 2, and the case $\eta \geq 1$ will be briefly outlined in subsection 3.

3.7.1 The total energy of metals calculated taking into account the "non-point-like" structure of ions

Let us consider the expression for the total energy of the metal in the case of small ion radius and large energy of excitation, i.e., we will consider both the ratio of the core radius to the inter-ion distance and the ratio of the characteristic energy of conducting electrons to the energy of excitation of core electrons as small parameters:

$$\kappa = 4\pi\alpha(0)/\Omega \sim k_F r_0 \ll 1, \tag{3.76}$$

$$\eta = \omega_{pl}/\omega_0 \sim E/\hbar\omega_0 \ll 1, \tag{3.77}$$

where r_0 is of the order of the ion radius. Let us write the total Hamiltonian of an electron–ion system as

$$\hat{H} = \hat{H}_0 + \hat{H}', \quad \hat{H}_0 = \hat{H}_{0e} + \sum_{\mathbf{R}} \hat{h}_{\mathbf{R}}, \quad \hat{H}' = \hat{H}_{ee} + \hat{H}_{ii} + \hat{H}_{ei}. \tag{3.78}$$

Here, $\hat{h}_{\mathbf{R}}$ is the Hamiltonian of the $\mathbf{R}$-th ion, $\hat{H}_{0e}$ is the free electron Hamiltonian

$$\hat{H}_{ee} = \frac{1}{2}\sum_{\mathbf{q}} v_c(\mathbf{q})\rho_e(-\mathbf{q})\rho_e(\mathbf{q}),$$

$$\hat{H}_{ii} = \frac{1}{2}\sum_{\mathbf{q},\mathbf{R}} v_c(\mathbf{q})\rho_i^{\mathbf{R}}(-\mathbf{q})\rho_i^{\mathbf{R}}(\mathbf{q}),$$

$$\hat{H}_{ei} = \frac{1}{2}\sum_{\mathbf{q},\mathbf{R}} v_c(\mathbf{q})\rho_i^{\mathbf{R}}(-\mathbf{q})\rho_e(\mathbf{q})$$

are Hamiltonians of the electron–electron, ion–ion and electron–ion interactions, respectively. The values $\rho_i^{\mathbf{R}}(-\mathbf{q})$ and $\rho_e(\mathbf{q})$ denote the Fourier components of the charge density of $\mathbf{R}$-th ion and conducting electrons, respectively. As a consequence of a small value of the ion radius, the value $\rho_i^{\mathbf{R}}$ can be written as

$$\rho_i^{\mathbf{R}}(\mathbf{q}) = \exp(i\mathbf{q}\mathbf{R})(Z + i\mathbf{q}\hat{\mathbf{d}}_{\mathbf{R}}) + \cdots) \tag{3.79}$$

where Z is the ion charge and $\hat{\mathbf{d}}_{\mathbf{R}}$ is the dipole moment operator. We will consider $\hat{H}_0$ in (3.78) as the zero-order Hamiltonian, and $\hat{H}'$ as a perturbation. The electron Green's function will be plotted as a solid line and $v_c(\mathbf{q})$ as a dashed line. If the source of the Coulomb field is an ion, the dashed line has to start from a point. An additional element introduced in [159] is the dipole vertex $i\mathbf{q}\mathbf{d}_{\mathbf{R}}$. We will denote it as a white circle. Giving the averaging of diagrams with these circles over the ground state of $\hat{H}_0$, we must take into account that $<\mathbf{d_R}> = 0$, and the averaging of two operators $\mathbf{d_R}$ at the same site given their advanced Green's function [159,165]

$$\alpha(t) = -\langle 0|\hat{T}\,\mathbf{d_R}(t)\,\mathbf{d_R}(0)|0\rangle. \tag{3.80}$$

Thus, the rectangle with two dashed lines

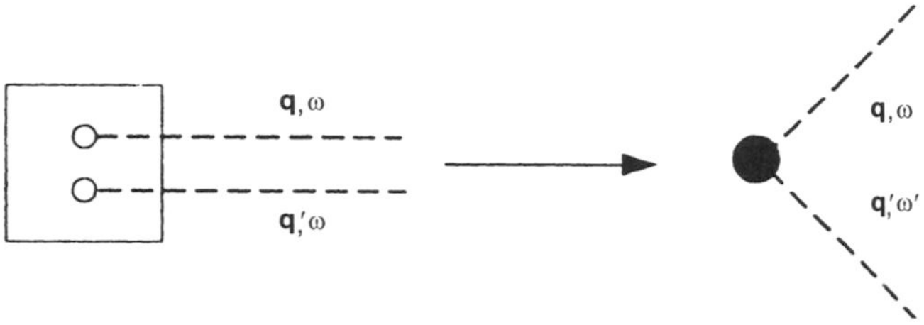

(3.81)

denotes the averaging of the next expression (over the ground state of $\hat{h}_\mathbf{R}$)

$$(-i\mathbf{q}')(i\mathbf{q})\frac{\alpha(\omega)}{\Omega_0}\exp[i(\mathbf{q}-\mathbf{q}')\mathbf{R}]\delta(\omega-\omega'), \tag{3.82}$$

where $\alpha(\omega)$ is the Fourier transform of the function (3.80). After summation over $\mathbf{R}$, the value $\exp(i(\mathbf{q}-\mathbf{q}')\mathbf{R})$ in (3.82) is replaced by the structure factor. Further, we will denote this expression as a black circle (like the left side of (3.81)). Let us consider the zero-, first-, and second-order in κ contributions to the total energy.

The zero-order approximation corresponds to diagram "a" in (3.83)

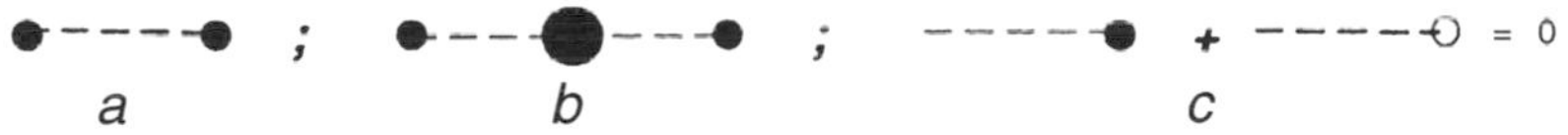

(3.83)

Its analytical representation is

$$E_M = \frac{1}{2}\sum_{\mathbf{g}\neq 0} Z^2 v_c(\mathbf{g}) = -\alpha_M/\Omega^{1/3}, \tag{3.84}$$

where α_M is the Madelung constant, $\mathbf{g}$ is the vector of the reciprocal lattice (for simplicity, we allow the lattice to be cubic). The contribution of the first order in κ corresponds to diagram "b" in (3.83). It is

$$\Delta E_M = \frac{Z^2}{2}\sum_{\mathbf{g}\neq 0}\sum_{\mathbf{g}'\neq 0}\mathbf{g}\mathbf{g}' v_c(\mathbf{g}) v_c(\mathbf{g}) - \int_{-\infty}^{\infty}\frac{d\omega}{2\pi}\alpha(i\omega) = 0. \tag{3.85}$$

The contribution of $\mathbf{g} = 0$ vanishes as a consequence of the charge neutrality condition (see diagram "c"). Thus, for a system with a center of symmetry, the screening of the direct inter-ion interaction (E_M) by the polarized ion does not take place. Similar contributions, provided by the quadrupole polarizability, do not vanish for low-symmetry lattices. However, in this case, it is also possible to neglect this contribution as a consequence of a small value of the structure-dependent part of E_M [1, 149]. It is not difficult to show that all higher-order in κ contributions to E_M are also equal to zero.

Taking into account the ion polarizability in calculating the energy of the electron gas $E_e^{(0)}$ provides diagrams with black circles inserted in the dashed lines. For example, diagram "a"

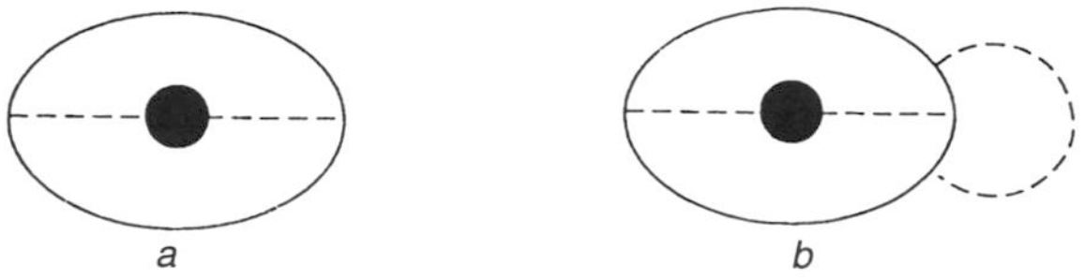

(3.86)

corresponds to the first-order in κ correction to the Fock energy. A summation of all these inclusions to the dashed lines (provided as electrons and as ions) replaces $v_c(\mathbf{q})$ by $v_c(\mathbf{q})/\varepsilon_{tot}(\mathbf{q}, i\omega)$, where

$$\varepsilon_{tot}(\mathbf{q}, i\omega) = \varepsilon(\mathbf{q}, i\omega) + \frac{4\pi}{\Omega_0}\tilde{\alpha}(i\omega) \tag{3.87}$$

is the total dielectric function and $\tilde{\alpha}(i\omega)$ is the ion polarizability calculated taking into account the local field correction. For cubic lattices

$$\tilde{\alpha}(i\omega) = \alpha(i\omega)\left[1 - \frac{4\pi}{3\Omega_0}\alpha(i\omega)\right]^{-1}. \tag{3.88}$$

We suppose κ and η to be small and restrict our consideration to the diagrams with one black circle only. As a consequence of $\eta \ll 1$, it is possible to neglect in the upper line of diagram "b" in (3.86) the energy ε in comparison with $\omega \sim \omega_0$. We see also that the main contribution to the integral over $\mathbf{q}$ provides the vicinity of region $q \sim \sqrt{\omega_0}$. Hence, the expression, corresponding to diagram "a" in (3.86), taking into account the screening $v_c(\mathbf{q})$, has the form

$$\begin{aligned}
E_a &= \frac{1}{2}\sum_{\sigma}\int\frac{d^3k}{(2\pi)^3}\int_{-\infty}^{\infty}\frac{d\varepsilon}{2\pi i}e^{i\varepsilon 0}G_0(i\varepsilon, \mathbf{k}) \\
&\quad\times\int_{-\infty}^{\infty}\frac{d\omega}{2\pi}\int_{-\infty}^{\infty}\frac{d^3q}{(2\pi)^3}\left(\frac{v_c(\mathbf{q})}{4\varepsilon(0, i\omega)}\right)^2\frac{q^2\alpha(i\omega)}{i\omega + q^2/2} \\
&= 2\delta b\int\frac{d^3k}{(2\pi)^3}\int_{-\infty}^{\infty}\frac{d\varepsilon}{2\pi}e^{i\varepsilon 0}G_0(i\varepsilon, \mathbf{k})
\end{aligned} \tag{3.89}$$

Here, $G_0(i\varepsilon, \mathbf{k})$ is the non-perturbed Green's function and

$$\begin{aligned}
\delta b &= \frac{1}{2}\int_{-\infty}^{\infty}\frac{d\omega}{2\pi}\frac{\alpha(i\omega)}{\varepsilon^2(0, i\omega)}\int_{-\infty}^{\infty}\frac{d^3q}{(2\pi)^2}\left(\frac{4\pi}{q^2}\right)^2\frac{1}{i\omega + q/2} \\
&= -2\int_0^{\infty}\frac{d\omega\,\alpha\,(i\omega)}{\sqrt{\omega}}\left(\frac{\omega^2}{\omega^2 + \omega_{pl}^2}\right)^2.
\end{aligned} \tag{3.90}$$

In equations (3.89) and (3.90), we keep the dependence of $\varepsilon(\mathbf{q}, i\omega)$ on ω because it provides contributions of the order $\sim \eta^{1/2}$, whereas taking into account the dependence on $\mathbf{q}$ and keeping arguments $(\varepsilon, 0)$ (which are small in comparison with (ω, q)) both provide contributions $\sim \eta$. Every insert into diagram "a" in (3.86) (see, for example, diagram "b") results in small factors $\sim \eta$, but we will neglect such diagrams. As a result, the lowest orders in κ and η contribution of the ion polarizability to the structure-independent part of the energy is

$$E_e^{(1)} = \delta b Z/\Omega_0. \tag{3.91}$$

We start the calculations of the energy of interaction of a pair of non-point-like ions with diagram "a" in the next expression

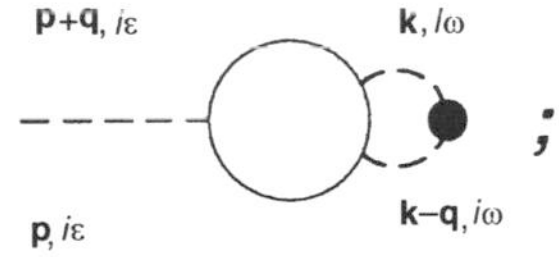

(3.92)

a

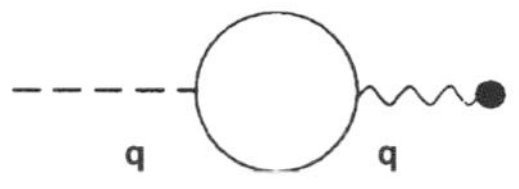

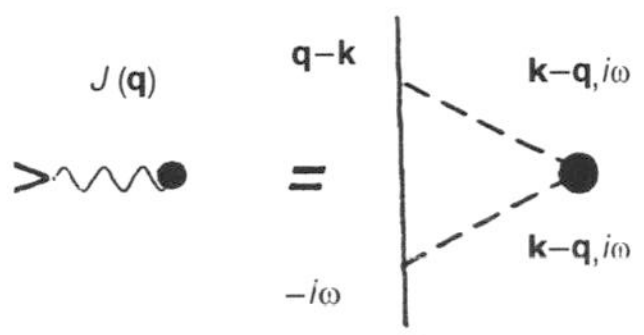

b c

In the lowest order in η, the function $G_0(i\varepsilon - i\omega, \mathbf{p} + \mathbf{q} - \mathbf{k})$ can be replaced by $G_0(-i\omega, \mathbf{q} - \mathbf{k})$ (since $p \ll q$ [54]) and the diagram is transformed into diagram "b". In this last case, the wave corresponds to the block $J(\mathbf{q})$ (diagram "c")

$$J(\mathbf{q}) = -\frac{1}{2} \int_{-\infty}^{\infty} \frac{d\omega}{2\pi} \alpha(i\omega) \int \frac{d^3k}{(2\pi)^3} \frac{v_c(\mathbf{k})}{\varepsilon(0, i\omega)} \frac{v_c(\mathbf{k} - \mathbf{q})}{\varepsilon(0, i\omega)} \frac{\mathbf{k}(\mathbf{k} - \mathbf{q})}{-i\omega + 1/2(\mathbf{q} - \mathbf{k})^2}, \tag{3.93}$$

and the respective correction to the energy is

$$\Delta E = -\frac{Z}{\Omega_0} \sum_{\mathbf{g} \neq 0} w_c(\mathbf{g}) J(\mathbf{g}) \pi_0(\mathbf{g}, 0), \tag{3.94}$$

where $\pi_0(\mathbf{g}, 0)$ is the Lindhard polarization operator at zero frequency. Here and below, the bold dashed line denotes the screened Coulomb potential $v_c(\mathbf{q})/\varepsilon_e(\mathbf{q}, i\omega)$. Since all complications of the internal electron line of block "c" provide additional factors $\sim 1/\omega$, then using the bold lines for the electron–electron interaction in diagram "a" provides, in the lowest order in η, the replacement of $\pi_0(\mathbf{g}, 0)$ in (3.94) by the exact operator $\pi(\mathbf{g}, 0)$.

Let us now calculate the second-order in κ contributions to the pair inter-ion interaction. The simplest diagram "a" is

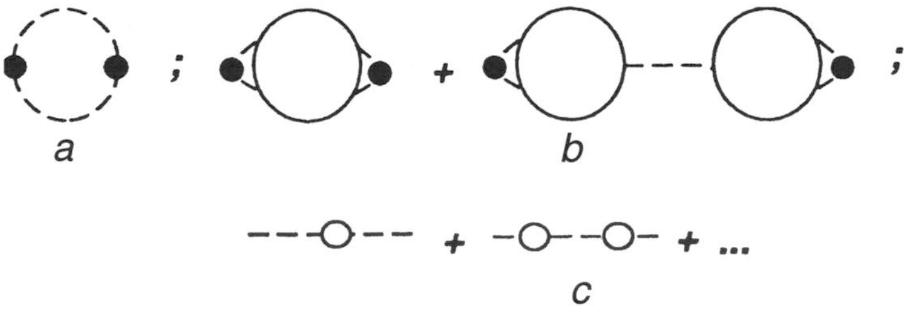

(3.95)

This diagram corresponds to actual van der Waals energy [159, 165]

$$E_W = -\frac{1}{2\Omega_0}\sum_{\mathbf{g}}\int\frac{d^3q}{(2\pi)^3}\int_{-\infty}^{\infty}\frac{d\omega}{2\pi}\alpha^2(i\omega)[\mathbf{q}\,(\mathbf{q}+\mathbf{g})]^2\frac{v_c(\mathbf{q})}{\varepsilon(\mathbf{q},i\omega)}\frac{v_c(\mathbf{q}-\mathbf{g})}{\varepsilon(-\mathbf{q},i\omega)}, \tag{3.96}$$

Eliminating the self-interaction, we obtain

$$E_W = \frac{1}{2}\sum_{\mathbf{R}\neq 0} V_W(\mathbf{R}) \tag{3.97}$$

where $V_W(\mathbf{R})$ is determined by (3.74). Taking into account, as we have manifestly done at in the beginning of this section, the dependence of $\varepsilon(\mathbf{q}, i\omega)$ on ω only, we obtain

$$V_W(\mathbf{R}) = -c/R^6, \quad c = \frac{3}{\pi}\int_0^{\infty} d\omega\,\alpha^2(i\omega)\left(\frac{\omega^2}{\omega^2+\omega_{pl}^2}\right)^2. \tag{3.98}$$

To calculate the total energy of the pair interaction, we must calculate the contributions of diagrams "b" and "c" in (3.95). The sum of these contributions, together with contributions of the third order, is equal to the actual second-order in the pseudopotential contribution in [148, 149, 33]

$$E^{(2)} = -\frac{\Omega_0}{2}\sum_{\mathbf{g}\neq\mathbf{0}}\frac{\pi(\mathbf{g})}{\varepsilon(\mathbf{g})}|V(\mathbf{g})|^2, \tag{3.99}$$

if the form factor is taken to be

$$V(\mathbf{g}) = \frac{1}{\Omega_0}[Zv_c(\mathbf{g}) + J(\mathbf{g})] \tag{3.100}$$

Hence, the pseudopotential concept is a natural consequence of a consistent procedure based on the expansion of the energy in powers of κ and η.

If we eliminate the dipole approximation (3.79) and use the perturbation theory in the Hamiltonian $\hat{H}_{ei}$, then the expression $\mathbf{q}\mathbf{q}'\alpha(\omega)$ in (3.81) must be replaced with

$\chi(\mathbf{q}, \mathbf{q}'; \omega)$ — the Fourier transform of the advanced Green's function corresponding to the total ion density ($\hat{\rho}_i$)

$$\chi(\mathbf{q}, \mathbf{q}'; \omega) = -i \sum_{\mathbf{R}} \int_{-\infty}^{\infty} dt e^{i\omega t} \langle 0 | \hat{T} \delta\rho_i^{\mathbf{R}}(\mathbf{q}, t) \delta\rho_i^{\mathbf{R}}(-\mathbf{q}, t) | 0 \rangle, \tag{3.101}$$

$$\delta\rho_i^{\mathbf{R}}(\mathbf{q}, t) = \rho_i^{\mathbf{R}}(\mathbf{q}, t) - F^{\mathbf{R}}(\mathbf{q}), \quad F^{\mathbf{R}}(\mathbf{q}) = \langle 0 | \rho_i^{\mathbf{R}}(\mathbf{q}, t) | 0 \rangle$$

and Z with $Z - F^{\mathbf{R}}(\mathbf{q})$ ($\hat{F}$ is the form factor for the R-th ion). It provides some additional contributions to E_M as a consequence of the non-point-like structure of ions. These contributions to the energy of inter-ion interaction are [166] of the order $\sim \exp(-\gamma R)$ (for spherically symmetrical ions). The corrections, provided by the exchange interaction of electrons located at different lattice sites, are of the same order of magnitude. These corrections can be calculated using, for example, the Heitler–London method [166]. It provides an additional term in (3.84)

$$E_{sr} = \frac{1}{2} \sum_{\mathbf{R}}{}' v_{sr}(\mathbf{R}). \tag{3.102}$$

Hence, the total energy can be expressed as

$$E = E_{\Omega} + E_M + E_e^{(2)} + E_W + E_{sr}, \quad E_{\Omega} = E_e^{(0)} + E_e^{(1)} \tag{3.103}$$

Here, $E_e^{(1)}$ corresponds to expression (3.91) where δb is determined by an expression similar to (3.90) (replacing $\alpha(i\omega) \to \chi(\mathbf{q}, \mathbf{q}'; i\omega)/q^2$) and there exists an additional term

$$\delta b_{\mathbf{R}} = \lim_{\mathbf{q} \to 0} v_c(\mathbf{q})[Z - F_{\mathbf{R}}(\mathbf{q})] = \frac{2\pi e^2}{3} \langle 0 | r^2 | 0 \rangle = 2 \int_0^{\infty} d\omega \alpha(i\omega). \tag{3.104}$$

The characteristic parameter of the vanishing of $\chi(\mathbf{q}, \mathbf{q}'; i\omega)$ as a function of $\mathbf{q}$ is the ion radius $\mathbf{R}_i$. If the parameter

$$\lambda = \frac{R_i}{R_0}$$

(R_0 is the inter-ion distance) is not small and $\eta \ll 1$ then in diagram "a" in (3.86) it is possible to neglect the value of ε as small in comparison with ω. However, it is impossible to neglect $\mathbf{k}$ in comparison with $\mathbf{q}$. Now, instead of (3.91), we obtain

$$E_e^{(1)} = \Omega_0 \int \frac{d^3k}{(2\pi)^3} n_{\mathbf{k}} V_{\mathbf{k},\mathbf{k}'}, \tag{3.105}$$

$$V_{\mathbf{k},\mathbf{k}'} = \frac{1}{2\Omega_0} \int_{-\infty}^{\infty} \frac{d\omega}{2\pi} \int \frac{d^3q}{(2\pi)^3} \chi(\mathbf{q}, \mathbf{q}; i\omega) \left[\frac{v_c(\mathbf{q})}{\varepsilon(\mathbf{q}, i\omega)} \right]^2 G_0(\mathbf{k} - \mathbf{q}, -i\omega) + \frac{\delta b_R}{\Omega_0}. \tag{3.106}$$

Equation (3.105) corresponds to the general expression for $E_e^{(1)}$ in the case of the non-local pseudopotential [33]. Formula (3.106) provides microscopic expressions for the diagonal matrix elements of this pseudopotential. Thus, we see that the local character of the pseudopotential is a consequence of small values of the parameters λ, η. Hence, equation (3.90) can be applied if $\eta^{1/2} \ll \lambda$.

As was mentioned in [150, 151], taking higher-order terms in the power series expansion in η (as $\mathbf{q} \to 0$) into account in equation (3.106) naturally introduces the "orthogonalization hole" [149], i.e., replacing $Z \to Z(1+\delta)$, where δ relates to the energy dependence of the potential. To demonstrate this, let us consider the following terms in the series expansion in ε/ω, corresponding to diagram "a" in (3.92). Now, there is an additional term in the block "a" corresponding to the block "c" (in diagram (3.107) this term is denoted as a cross, remember that $\mathbf{p} \ll \mathbf{q}$). We denote this new block as $W(\mathbf{q})$; its diagrammatic representation is

$w(\mathbf{q}) = i\varepsilon$ 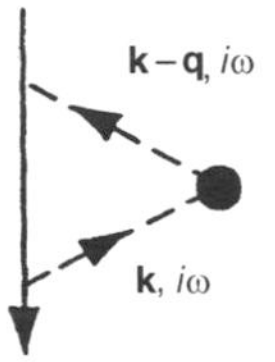

(3.107)

The analytical expression for this correction to diagram "a" in (3.92), multiplied by $i\varepsilon$, is

$$\frac{Z}{2}\frac{v_c(\mathbf{q})}{\varepsilon(\mathbf{q},0)} W(\mathbf{q}) \int_{-\infty}^{\infty} \frac{d\varepsilon}{2\pi} \int \frac{d^3p}{(2\pi)^3} G_0(\mathbf{p}+\mathbf{q}, i\varepsilon) G_0(\mathbf{p}, i\varepsilon) i\varepsilon \approx \frac{Z^2}{2\Omega_0} \frac{v_c(\mathbf{q})}{\varepsilon(\mathbf{q},0)}\bigg|_{\mathbf{q}\to 0} W(0),$$

where the last equality is valid as $q \to 0$. Here the electroneutrality condition has been used. Similar corrections to diagram "b" in (3.95), together with diagram "a" in (3.83), provide the substitution $Z \to Z(1 + W(0)/2)$ in the Coulomb potential of ion–ion interaction at small $\mathbf{q}$ (or at large R). The value $\delta = W(0)/2$ corresponds to the "orthogonalization hole" [149] *and* relates to the derivative of the pseudopotential (strictly speaking, the self energy part $M_{\mathbf{kk}'}$) with respect to the energy.

Note that an introduction of $V_{\mathbf{k},\mathbf{k}'}$ at $\eta \geq 1$ is, generally speaking, impossible.

The case of ions with low energy of electron excitations ($\eta \gg 1$) will be considered in subsection 3.

3.7.2 Van der Waals and Born–Mayer interactions of ions (the "hard core" case)

Our analysis in the previous section shows, in particular, that for $\lambda, \eta \ll 1$ all effects, provided by the non-point-like structure of the ions, can be considered in the framework of the general local pseudopotential approach with two additional terms: E_{sr} (3.102) and E_W (3.97). To provide a qualitative estimation of the role of these interactions, calculations of atomic properties of K, Rb, and Cs were provided in [150, 151]. These calculations used the well-known interpolation formula for E_{sr} (see [141, 166]). The values of the parameters A and ρ for Na, K, and Rb are taken from [167]. The van der Waals constant is determined by the general expression (3.98), in which it is necessary to use a concrete expression for the dependence $\alpha(i\omega)$. This dependence was calculated in [159, 161, 163] using the dispersion relations, which allows us to express $\alpha(i\omega)$ through the experimentally measured value $-\text{Im}\,\varepsilon(\omega)$. In this review, values of C from [163] are used. The possibility of using some simple approximations for $\alpha(i\omega)$ in calculating C was discussed in [184].

As above, these calculations have been performed using the model pseudopotential of the Animalu–Heine–Abarenkov type

$$V(\mathbf{q}) = -\frac{4\pi}{q^2}\left[Z\cos(qr_0) + V_0\,r_0\left(\frac{\sin(r_0)}{qr_0} - \cos(qr_0)\right)\right]\exp\left[-\xi(q/2k_{F0})^4\right], \quad (3.108)$$

where k_{F0} is the Fermi momentum at zero temperature and pressure, $\xi = 0.03$, and the parameters r_0, V_0 have been fitted to conditions $P = 0$, $C_{44} = (C_{44})_{exper}$ in the same manner, as in Chapter 2 (see Table 3.21), with the GT type of screening [80]. It provides the best description of properties of alkali metals. In contrast to Chapter 2 and sections 1–6 of the present chapter, where, predominantly, the Nozieres–Pines expression (2.129) for the correlation energy E_c was used, in the present subsection, we use the more accurate expression by Perdew and Zunger (2.131).

The values of the parameters of E_{sr} were calculated in [168] (for Cs they were obtained in [163]) using an extrapolation of the data for Na, K, Rb from [168]. These values are slightly different from the values given in Table 2.3. Note that it is convenient to rewrite the slowly converged sum of $V_W(R)$ as two rapidly convergent sums (see Appendix B).

Table 3.22 shows (see [150]) that the elastic moduli B_{ik} and the Debye temperature (which may be expressed through these models) are practically insensitive to aiding E_{sr} and E_W. As it was be expected, these contributions actually have opposite signs and, hence, are partially mutually cancelled. The difference between the calculated and experimental values of the sublimation energy E_{subl} is remarkable (10 mRy). This difference does not relate to the inclusion of E_{sr} and E_W, whose contributions to E are small. It is a consequence of the difference between E_c^{PZ} and E_c^{NP}. If we determine E_{subl} as the difference between the calculated energy and the experimental value of the ionization potential [171], then the agreement between the theory and experiment becomes much worse in comparison to [169]. However, the use of the ionization potential calculated in [172] remarkably improves this agreement. The authors of [150, 151] have supposed that the second method of calculation of E_{subl} provides a compensation of errors of calculations of the total energies of both the metal and the

TABLE 3.21 Parameters used in calculations (from [150], in a.u.)

metal	A	ρ	U	r_0	C
K	1510	0.4365	0.2013	3.125	26.86
Rb	1918	0.4876	0.2057	3.571	42.47
Cs	2157	0.5102	0.1923	4.051	76.40

TABLE 3.22 Atomic and thermodynamic properties of K, Rb, Cs (after [150]) (the PZ formula for correlation energy [147] was used): 1 – calculation without E_w; 2 and 3 – calculations [150] with and without E_w, respectively; 4 – experimental data.

		E_{subl}	mRy	B_{11}, kbar	B_{33}, kbar	B_{44}, kbar	$\frac{\partial B_{11}}{\partial p}$	$\frac{\partial \ln B_{33}}{\partial \ln \Omega}$	$\frac{\partial \ln B_{44}}{\partial \ln \Omega}$	Θ, K	γ_∞
K	1	70.1	79.2	37.2	3.82	28.6	3.79	2.48	1.57	91.0	1.074
	2	60.1	69.2	37.0	3.79	28.6	3.91	2.43	1.64	90.8	1.109
	3	60.6	69.7	36.9	3.71	28.6	3.84	2.48	1.65	90.4	1.128
	4	69 ± 0.2		37 ± 1	3.77 ± 0.07	28.6 ± 0.7	3.97	2.7	2.6	$90.6 \pm ^{1.4}_{0.3}$	1.15 ± 0.02
Rb	1	61.9	73.7	29.7	3.09	22.1	3.76	2.49	1.51	56.5	1.062
	2	52.5	64.3	29.5	2.89	22.1	3.98	2.41	1.73	55.8	1.183
	3	53.3	65.1	29.4	2.80	22.1	3.89	2.47	1.74	55.3	1.218
	4	64 ± 1.6		30.6 ± 2	2.74 ± 0.08	22.1 ± 0.5	3.96	2.67	2.35	56.5 ± 0.2	1.13 ± 0.1
Cs	1	59.1	73.5	22.3	2.350	16.0	3.71	2.46	1.44	40.6	1.048
	2	49.3	63.6	22.1	2.30	16.0	3.94	2.35	1.59	40.5	1.117
	3	50.3 ± 0.7	64.7	22.0	2.19	16.0	3.81	2.44	1.62	40.0	1.163
	4	61.0 ± 0.9		23.1 ± 0.9	2.2 ± 0.3	16 ± 0.6	3.5	–	–	40.5 ± 0.3	1.04 ± 0.10
				21.5							

ion. The derivatives of the elastic moduli with respect to volume and the Grüneisen parameters, which relate to these derivatives, appreciably changed by including E_{sr} and E_W (see Table 3.22). For K and Rb, these terms improve agreement of the theory with experiment for dB_{ik}/dp and γ_∞. For Cs, taking into account E_{sr} and E_W aggravates the agreement of the theory with experiment for the majority of properties, in particular, for γ_∞.

The phonon frequency $\omega_{\lambda\mathbf{q}}$ and the microscopic Grüneisen parameter $\gamma_{\lambda\mathbf{q}}$ for the symmetry points of the Brillouin zone are presented in [150, 151].

The effect of E_{sr} and E_W on $\omega_{\lambda\mathbf{q}}$ is small: about $1 \div 2\%$ for all these points excluding the point N_4' on the critical branch Σ_4, where this change is about 3%. For $\gamma_{\lambda\mathbf{q}}$, these effects are much higher: 5% for K, 25% for Rb, and 16% for Cs. Note, that taking into account E_{sr} and E_W provides the increase of $\gamma_{\lambda\mathbf{q}}$. The values of $\Delta\Omega(T)/\Omega_0$, calculated taking into account E_{sr} and E_W, are appreciably closer to the experimental data for all heavy alkali metals at high temperatures than the results of [83]. This fact also shows an improvement of calculations of $\gamma_{\lambda\mathbf{q}}$.

The results of calculations depend relatively weakly (taking into account the refitting of the pseudopotential parameters) on the constant C from (3.98). However, they are appreciably dependent on variations of the parameter ρ in the expression for E_{sr} (3.2). The aggravation of the description of properties of Cs relates, as was supposed by the authors of [150, 151], to the fact that for this metal there was no macroscopic calculation of v_{sr}, and the values A and ρ in [163] have been obtained using a not sufficiently clear extrapolation procedure. Probably, calculations for Cs should be provided with the values of A and ρ from Chapters 2–4 of this book (see Table 2.3), as well as from [173, 174].

Thus, simultaneous inclusion of both terms E_{sr} and E_W to the calculations at $p = 0$ provides a rather small variation of calculated properties of alkali metals, except the Grüneisen parameters, which are especially sensitive to the large volume dependence of these energies. This dependence is also very important for phase transitions and for the equation of state at large pressures.

3.7.3 The "soft-core" case

Now we, following [151, 174–188], briefly consider the interaction of conducting electrons with excitations of the ion core for the "soft-core" case, i.e., when the inequality opposite to (3.77) is valid

$$\eta \geq 1 \tag{3.109}$$

The diagram describing the contribution to the self-energy part of electrons, similar to "c" in (3.92), is

$$\Sigma(\mathbf{k}, \varepsilon) = \frac{1}{\Omega_0} \int \frac{d^3p}{(2\pi)^3} \int_{-\infty}^{\infty} \frac{d\omega}{2\pi i} \left(\frac{4\pi e^2}{\mathbf{p}^2 + \lambda^2} \right)^2 \frac{\chi(\mathbf{p}, \mathbf{p}; \omega)}{\varepsilon - \omega - \xi_{\mathbf{k}-\mathbf{p}} + i\delta \operatorname{sgn} \xi_{\mathbf{k}-\mathbf{p}}} \tag{3.110}$$

where $\varepsilon = \varepsilon(\mathbf{p}, \omega) \approx 1 + \lambda_0^2/\mathbf{p}^2$ because only small ω are important here. Let us suppose, for simplicity, that the condition (3.76) is satisfied. Then the function $\chi(\mathbf{p}, \mathbf{p}; \omega)$ can be approximated by its value at $k_F r_0 \ll 1$ [8,19]

$$\chi(\mathbf{p}, \mathbf{p}; \omega) \approx \mathbf{p}^2 \alpha(\omega) = \mathbf{p}^2 \frac{A\omega_0}{\omega_0^2 - \omega^2 - i\delta} \tag{3.111}$$

where $\alpha(\omega)$ is the dipole polarizability and the contribution of transitions with small ω_0 are taken. The constant A is proportional to the respective oscillatory force. Suppose, for clarity, that $\varepsilon \geq 0$. The integral over ω in (3.110) is

$$\begin{aligned} I(\varepsilon, \xi) &= \int_{-\infty}^{\infty} \frac{d\omega}{2\pi} \cdot \frac{1}{\varepsilon - \omega - \xi + i\delta \operatorname{sgn} \xi} \cdot \frac{1}{\omega_0^2 - \omega^2 - i\delta} \\ &= \frac{1}{2\pi} \frac{1}{\varepsilon - \xi - \omega_0 \operatorname{sgn} \xi}. \end{aligned} \tag{3.112}$$

Substituting this value in (3.110), we obtain at $\beta = me^2/\pi k_F \leq 1$

$$\Sigma(\mathbf{k}, \varepsilon) \approx \frac{me^2}{k} \frac{A}{\Omega_0} \ln \left[\frac{(k + k_F)^2 + \lambda_0^2}{(k - k_F)^2 + \lambda_0^2} \right] \ln \left| \frac{\varepsilon - \omega_0}{\varepsilon + \omega_0} \right|. \tag{3.113}$$

In the lowest order of the perturbation theory, the value of Σ diverges at $\varepsilon = \pm\omega_0$. Moreover, the gap Z_0 of the distribution function at the Fermi level [179, 192] is

$$Z_0 = \left[1 - \frac{\partial \Sigma(k_F, \varepsilon)}{\partial \varepsilon} \right]_{\varepsilon=0}^{-1} \approx \left[1 + \frac{2me^4}{k_F \omega_0} \frac{A}{\Omega_0} \ln \left(1 + \frac{1}{\beta} \right) \right]^{-1}. \tag{3.114}$$

As a consequence, the effective mass of conducting electrons $m^*/m \sim Z_0^{-1}$ contains some corrections $\sim \omega_0^{-1}$, which could be large at small ω_0.

The respective contributions to the total energy (strictly speaking, to the potential Ω) are

$$\delta\Omega = \frac{m^2 e^2}{\pi^2} A \ln \left(1 + \frac{1}{\beta} \right) \omega_0 \ln \left| \frac{E_F}{\omega_0} \right|. \tag{3.115}$$

Hence, the total energy is a non-analytical function of ω_0. The renormalization of the effective electron mass ($m^*/m \sim \omega_0^{-1}$) in the case of a small splitting of the energy levels of the ion in the crystal field was considered in detail in [183]. This effect is very important, for example, for Pr [183].

3.8 Resumé

In this chapter, the second-order model, described in Chapter 2, has been used for studying numerous dynamic, static and anharmonic properties of alkali metals. In particular, the temperature dependence of the second- and third-order elastic constants has been investigated. This problem is very non-trivial since, to describe the temperature dependences of C_{ik} and C_{ikl}, the model must provide a correct description of high-order derivatives of the inter-atomic potential: up to the fifth.

The successful description of $C_{ik}(T)$ for alkali metals, as well as an adequate description of practically all other atomic (elastic, thermodynamic) properties of whole series of the metals, clearly demonstrates that the calculation scheme used is well founded and has a good predictability. The pseudopotential approach is especially fruitful in the microscopic theory of non-transition metals.

However, it is very important that, although this calculation scheme was (partially) used earlier, the successive usage of this scheme becomes possible only after the intricate optimization of all elements of this method. This includes the choice of the pseudopotential parameters, the approximation of the screening function, the choice of the form of the pseudopotential.

From the point of view of description of all properties of the crystal lattice of alkali metals, we can conclude that the best form of the potential is the Animalu–Heine–Abarenkov one, the best form of the polarization operator gives the Geldart–Taylor (GT) approximation which is also the best from the theoretical point of view. However, we must fully understand, that this successful description of properties of an alkali metals used the fact that, for all properties of an alkali metal, the Coulomb contribution plays the dominant role which can be calculated exactly. This fact results in the similarity of properties of alkali metals mentioned here as well as by other authors. But by studying the properties determined by higher derivatives of the inter-atomic potential (e.g., C_{ikl}), the band-structure contribution $E_e^{(2)}$ increases and this similarity disappears. Unfortunately, experimentally these properties have been studied insufficiently. This deficit of information does not allow us to make a definite conclusion about the applicability of the theory to this problem.

Numerous investigations of anharmonic properties have shown some interesting effects, the main one being the step-like variation of the macroscopic Grüneisen parameter $\gamma(T)$ at low temperatures. We emphasize that the discovery of these effects was impossible in the framework of the phenomenological theory. The "step" in $\gamma(T)$ was experimentally observed in potassium [63]. and this effect is not an experimental error, as was supposed by authors of this work. Moreover, this anomalous (from the traditional point of view) behavior is, apparently, a rather general feature of crystal properties. This effect is a consequence of the anomalies in the dependence of the microscopic Grüneisen parameters $\gamma_{\lambda\mathbf{q}}$ on the wave vector $\mathbf{q}$. These anomalies in $\gamma(T)$ yield anomalies in the temperature dependences $C_{ik}(T)$ and $C_{ikl}(T)$, which could be observed experimentally.

The temperature dependence of the third-order elastic constants $C_{ikl}(T)$, for the first time analyzed by us, is appreciably different from the behavior of $C_{ik}(T)$. Experimental measurements qualitatively support results of our calculations. Unfortunately, low accuracy of measurements does not allow us to make a definite conclusion about the role of the non-adiabatic effects.

Note also that deviations from the Cauchy relations are very sensitive to the details of the model. The accuracy of present-day experiments is large enough to provide a comparative study of the deviations from the Cauchy relations. These measurements would be a good test for theory and could help to improve that. Even in its present-day version, the microscopic theory of alkali metals correctly describes the main feature of metal properties. This fact can be considered to a demonstration of some degree of ripeness of the theory.

Let us mention some modern trends is the theory of alkali metals. First of all, this is the usage of non-local pseudopotentials [28, 142, 143]. Moreover, in [28, 142] an attempt to take into account the effect of the proximity of *d*-zones to the Fermi level in heavy alkali metals was performed. In these works, the energies of different structures (BCC, FCC, and HCP) were calculated and the preference of the BCC structure for K, Rb, and Cs [28] as well as the HCP structure in [143] have been demonstrated. These results should be considered to be a demonstration of some definite progress in the theory with non-local *ab initio* potential.[12]

Using the density functional formalism in the theory of lattice dynamics [144] demonstrates the equivalence of the dynamic and static elastic constants, although the calculation scheme [144] itself is totally equivalent to that considered here.

Formula (3.82) expresses the exchange and correlation contributions to the energy $E_e^{(3)}$ through the amplitude of three-particle processes $\tilde{\gamma}_6$. This expression is exact, but the concrete form for $\tilde{\gamma}_6 = \tilde{\gamma}_6^0$ (3.76), (3.77) is an approximation. The necessity of such approximations follows from the fact that for real metals the value r_s changes from 1.8 to 6. For these values of r_s, the kinetic and potential energies of electrons have comparable values and any exact consideration is impossible. For this reason, the theory of $\tilde{\gamma}_6$ (Chapter 2) submitted here should be considered, using the Nozieres–Pines classification [127], as the second-type theory for which the errors provided by the approximations cannot be estimated. This comment relates to all elements of the pseudopotential approach in the theory of metals.

The usage of *ab initio* (model) potentials gives the conclusion that the taking into account of the exchange and correlation effects in high-order contributions to the adiabatic potential is necessary for developing a quantitative theory. Application of the local form of the model potential does not provide the pseudopotential to be independent of the model used. Each form factor is a function of the electron density n_0 and the number of terms in the power series expansion of E_e, although, in contrast to the case of fitted potentials, this value is independent of the structure of the crystal. This is a preference of the model potential approach in the consideration of theoretical problems, but, at the same time, this feature is a weakness of that from the point of view of the quantitative description of crystal properties. The breakdown of the equilibrium condition $p = O$ demonstrates this conclusion. On the other hand, the proximity of the localized model form factors to the fitted ones (Chapter 2), firstly, justifies the usage of the model potential approach and, secondly, shows the importance of taking into account the change of the form factor by the insertion of the ion into the crystal provided by the change of the wave functions of the core electrons.

Owing to this fact, it should be noted that the "hard core" approximation (see section 7), i.e., the usage of constant values for the pseudopotential parameters,

---

[12] Recently, a series of works (see [29]) were performed in which the effects of *d*-zone in heavy alkali and alkaline-earth metals were considered in detail (see also Chapter 4).

actually postulated for all the model form factors, is unsatisfactory, especially for calculations of high-order derivatives of the adiabatic potential and the properties of the metal under pressure. This conclusion mostly relates to polyvalent metals with large atomic number. However, the use of the form factor for Al [123, 124] which is linearly dependent on the pressure (the lattice constant) gave no clear success.

This great sensitivity of the calculated properties (for example, in the cases of Al or Pb) to the details of the model (see [170]) forces us to ask whether is possible, at the present time, to reach good agreement between theory and experiment using the *ab initio* pseudopotentials, obtained from the consideration of an isolated ion. Maybe some progress in this direction could be reached using the multiparameter form factors with adjustable parameters which are more flexible than the forms by Animalu–Heine–Abarenkov, Harrison, etc. However, we think this approach is very cumbersome and non-constructive. Possibly, a successful description of properties of polyvalent metals would be obtained using the theory of the non-uniform electron gas [126] (the density functional theory), but this question needs separate investigations.

Finally, let us consider the situation taking place in the theory of the static properties of lead and lithium. As was shown in [170], power series expansions of the perturbation theory in the pseudopotential diverge in calculations of p, B_{ik}, C_{ikl} for these metals. Maybe, for Li, a more accurate consideration of the non-locality of the pseudopotential is needed. In the case of Pb, the covalent effects become important. There are two possible ways to improve these calculations. It is necessary either to modify the summation of the power series in the pseudopotential [146], or to introduce the covalent effects in Pb phenomenologically, for example, as a charge on the bond.

There is no doubt that any quantitative microscopic theory of polyvalent metals (in particular, Al and Pb) must take into account the covalent interaction (the terms $E_e^{(3)}$, $E_e^{(4)}$, etc.). However, as reviews of current publications show, such calculations (except [5]) provide no improvement to the accuracy in comparison with calculations neglecting this many-particle interaction. This result is, predominantly, a consequence of the fact that, for these last calculations, there exist simple opportunities to optimize the model (the choice of the pseudopotential, a special choice of the fitting conditions using the sum rule for the compressibility, a possibility of controlling the accuracy of different approximations). We have tried to use these possibilities for the fourth-order model. However, this model, as was mentioned above, cannot provide any quantitative accuracy, and our aim was only to demonstrate the difficulties in using the perturbation theory for polyvalent metals.

The main defect of the second-order model is a bad description of the volume elasticity modulus B [11]. Our investigations show that this difficulty relates to low accuracy of the Nozieres–Pines formula for the electron correlation energy E_c in the region of electron densities $r_s \sim 1$ (as for Al and Pb). This fact was discussed earlier. The Vashishta–Singwi formula for E_c appreciably improves the situation for Al [170].

Hence, there are many possible directions for the future development of the theory. Possibly, the more accurate formula for the correlation energy E_c^{zp} (3.78), and for the values relating to it $\tilde{\gamma}_6(0.0.0)$ and $\tilde{\gamma}_8(0.0, 0.0)$, would be the first step in the right direction.

The unsatisfactory results obtained for the volume elasticity moduli B and C_{111} in the framework of the second order model [11], the relatively large core radius r_0 in Al ($r_0/r_a \approx 0.5$, where r_a is the radius of the atomic sphere), and large spatial spreading of core functions force us to include the energy of the core polarization in E. Such attempts have been performed for Al in [36], where the Born–Mayer repulsion term $\alpha_B \exp(-r_B/R)$ was included in the energy E. The fitting of the pseudopotential parameters and α_B to the phonon spectrum had shown that in the framework of this model $\alpha_B = 0$ and the Animalu–Heine–Abarenkov pseudopotential corresponds to the "empty core" potential [145], i.e., the case $u = 0$ in our notations. We have used the same expression (4.2) for the energy of the repulsive interaction E_{sr} with parameters from Table 3.1. The result was that, for Al, contributions E_{sr} at $p = 0$ are not important. But for Pb, the terns E_{sr} must be taken into account because its contribution is comparable with the contribution provided by $E_e^{(2)}$.

The question about the necessity of using non-local pseudopotentials only (this statement was made in some publications, see, for example, [115]) cannot be definitely solved in the framework of the present-day theory of polyvalent metals and this problem needs additional investigation. At least for Al and Pb, the usage of such pseudopotentials does not, even today, provide any success [120]. The existing calculations for Li and Al do not definitely demonstrate the necessity of using nonlocality pseudopotentials in calculations of static properties. Possibly, the other situation takes place for Pb for which the fourth-order model describes static properties worse than the second-order model [11].

The main technical result of section 7 is a consistent proof of the expression for the total energy of the metal for the case of a small radius of ions and large excitation energies ($\eta \ll 1$). It turns out that the result obtained can be presented as a power series of the perturbation theory in the pseudopotential together with the actual contributions provided by the van der Waals E_W and the Born–Mayer E_{sr} interactions. The results of calculation of some atomic and thermodynamic properties of K, Rb, Cs have shown that it is necessary to calculate both the contributions E_W and E_{sr} because, sometimes, contributions of these two terms are mutually cancelled. The most sensitive characteristics depending on these interactions are the Grüneisen parameters.

The results of calculation of the ion polarizability for the first two groups of the Periodic Table [184] show that, for K^+, Ca^{++} ions as well as for more heavy alkali and alkaline-earth metals, there exists an appreciable contribution provided by $np \to n'd$ transitions (where n is the shell number of the last closed shell). For these metals, the value $\omega_0 = \omega_{npn'd}$ satisfies the condition (3.77), $\omega_{pl}/\omega_0 \ll 1$ (for example, for K $\omega_0 = 0.753$ a.u., $\omega_{pl} = 0.162$ a.u.). This fact approves the usage of the pseudopotential model for these elements. On the other hand, for noble metals $\omega_0 \leq \omega_{pl}$ (for example, for Cu^+ the energy of $3d \to 4p$ providing the main contribution to the ion polarizability is 0.279a.u. and $\omega_{pl} = 0.399$a.u.). Therefore, for copper, the pseudopotential can only be introduced if we exclude the $3d$ shell from the core.

We have shown that, for the description of interaction of inner and outer electron shells, the most important role plays the parameter $\eta = \omega_{pl}/\omega_0$. If $\eta \geq 1$ (the "soft-core" case), then a large renormalization of the electron spectrum and non-analytical (in η) contributions to the thermodynamic potential takes place. For $\eta \ll 1$ (the "hard-core" case), some non-local (but energy-independent) pseudopotential can be used.

In conclusion, let us note that the consistent method of calculation of the energies of the short-range repulsion E_{sr} and the van der Waals attraction for crystals of noble gases was developed for the first time in [92–94]. Although atoms of these elements can be considered as cores for the metals investigated here, there is no screening by nearly free electrons and these concentrations were provided using the tight-binding approximation. The other difference is the fact that in metals E_{sr} and E_W are corrections to the main interaction $E_e^{(2)}$, whereas for crystals of noble gases, these terms are the main contributions to the adiabatic potential.

4 Phase Transitions in Simple Metals

The progress in the quantitative theory of crystal structures and phase transitions is one of the most important problems of the microscopic theory of the solid state, in particular, of the theory of metals. Progress in this problem is very important for the development of metals and alloys with controlled properties as well as for general studies in the physics of metals. Until today, problems like this have actually been solved empirically, although now it is clear that theoretical studies can effectively accomplish solutions of these problems. But to realize this possibility, the accuracy of the theory has to be good enough to provide qualitative calculations and to predict new effects.

Naturally, the progress in the theory of structures and phase transitions was always considered as one of the most important aims of the pseudopotential theory. The interest in these questions demonstrates two peakes in past decades. The first peak relates to the 1970s (see, for example, [1–15, 34–36]). However, in these years, there was no great success, and results obtained were, at best only qualitative. For example, the phase diagram of strontium calculated in [6] demonstrates qualitative agreement with experiment, whereas the temperature and the pressure of the phase transitions differ from experimental values by a factor of 5–7. Even for alkali metals, for which the pseudopotential theory demonstrates the highest accuracy, results of calculations in the energy differences between different structures ΔE and the temperature of phase transitions T_c show appreciable disagreement between results obtained by different authors as well as between theory and experiment (see [7, 8] and the present chapter below).

The reason for these difficulties is the fact that the difference between energies of different phases ΔE, which determines T_c, is very small, of the order of 10^{-5} Ry/atom for alkali metals, i.e., about 10^{-4} of the total cohesive energy E. For this reason, calculations of ΔE and T_c, especially for low-temperature phase transitions with small ΔE, require great accuracy of the model. Small variations in the form $V(\mathbf{q})$ and

$\pi(\mathbf{q})$ can drastically affect ΔE. Note that, in many publications, in particular in [2–4], the possibility of providing any quantitative calculations of T_c for alkali metals was considered as rather questionable, and results for T_c, which were a few times greater than experimental ones, were considered successful.

The next peak in the interest in the studying of the nature and the properties of concrete phase transitions was in the 1980s. In theoretical studies, the density functional method [9], as well as the *a priori* Schluter pseudopotential [10], which is based on this method (see section 3 Chapter 2), became very popular in recent times. The calculations of structures and structural transitions [11, 12] performed without any adjustable parameters reproduce correctly the sequences of phase transitions in simple metals. The universal phase diagram of these metals has been constructed in [11].

In the present chapter, we will consider phase transitions in simple metals. Lithium and sodium have simple electron structures and they are very useful objects for studies of characteristic features and the nature of martensitic phase transitions in metals. Decreasing the temperature T in these metals gives the phase transition from the BCC to other close-packed phases like HCP or to HCP-like structures (4H, 6H, 9R) [25]. These phase transitions discovered by Barret in 1956 [17] have been studied since the 1980s in many experimental and theoretical investigations, which are still going on today [16–30].

As we have shown in previous chapters, the accuracy of the model used in the description of atomic properties is, as a rule, very high. This takes place especially for the static properties, like the energy, the pressure, the elastic constants as well as the low-temperature thermodynamic properties determined by these values (see Chapter 3). But the description of the thermodynamics of phase transitions needs higher accuracy (see below) than can be provided by the second-order model used above.

In section 1, we will discuss the electronic theory of phase transitions in lithium and sodium using pseudopotential as well as *ab initio* calculations. We will show, in particular, that the difference between the phase diagrams of lithium and sodium relates to a qualitatively different variation in the band structure of both these metals under pressure. Actually, adequate calculations of the energy differences between different structures of metals require (as was already mentioned) very accurate methods. For intermetallic compounds and alloys, these calculations are especially complicated and there were only a few calculations performed for these objects. At the same time, the density $\nu(E)$ and the number of states $N(E)$ [31] actually could be calculated with good enough accuracy using relatively simple methods, for example, the LMTO–ASA method [2]. The same accuracy would be also expected in calculations of the shear moduli $C_{ss'}$. Hence, the application of these methods to the evaluation of the instability region ($C_{ss'} < 0$) or the region of rapid softening of $C_{ss'}$, would allow in many cases estimates of the region of structural stability more simply and reliably in contrast to direct calculations of differences in thermodynamic potentials. The effect of the peculiarities of the band structure on shear moduli will be considered in Subsection 1.3 taking as an example the behavior of $C_{ss'}(p)$ in alkali metals. In all these metals (except sodium), increasing p results in the phase transitions from the BCC to other close-packed structures, which were studied by many authors.

In the present chapter, we will consider phenomena relating to external stresses: first-order structural phase transitions in alkali metals (martensitic transformations in Li and Na [13, 14], polymorphic BCC–FCC transitions in K, Rb and Cs [15, 31–32]) and the absolute instability of the BCC crystal lattice in alkali metals by decreasing the volume, discovered in numerical calculations.

In the second section, polymorphic transitions in heavy alkali metals K, Rb, and Cs will be considered using the pseudopotential approach for the calculation of the energy of different structures. As we have already mentioned (this will be demonstrated once again in section 1), the pseudopotential approach (in particular, the second-order model) does not always result in a successful description of phase transitions. This fact should be taken into account when reading section 2. Nevertheless, this model can be considered as a starting point for the theory and we will use the second-order model adopted (in contrast to section 1, where it was successfully applied to Na [13, 14]) by the core interaction. In considering the role of the core polarization, we will use the Born–Mayer form of the core interaction term E_{sr} [15, 38, 39]. In the case of small pressures considered in [13, 14], neglecting E_{sr} provides apparently no effect on the final results. However, as will be shown in section 2, investigation of phase transitions under pressure without taking into account the term E_{sr} yields wrong results.

4.1 Martensitic Phase Transitions in Lithium and Sodium

4.1.1 Experimental results concerning the martensitic transitions in lithium and sodium

Before starting our consideration of results of calculations of martensitic transitions, let us recall the main experimental facts about the phase transitions in Li and Na. It should be done since actual theories ignore many of these facts and, as a consequence, the comparison of results of calculations with experiment could often be either inaccurate or incorrect.

The characteristic feature of the martensitic phase transition, in contrast to other structural transitions of the first order, is the fact that the martensitic transition appreciably changes the crystal structure. In Li and Na, a transformation from the BCC to FCC takes place. This transition has a diffusionless nature in contrast, for example, to the transitions in self-ordering alloys. In other words, the growth of the region of the new phase within the old one has the velocity of the order of the velocity of sound. As a result, the growth of regions of the new phase takes place predominantly along so-called "martensitic" crystal planes. This growth provides additional stresses on the old lattice. As a rule, these stresses overdraw the plasticity threshold. These stresses stop the growth of the new phase despite its thermodynamic preference, so that martensitic transformations are almost never complete. For example, estimations made in [17, 42] show that the contest of the HCP phase x_0 as $T \to 0$ is about 0.8–0.9 in Li, and $x_0 \sim 0.5$ in Na.

Hence, the hysteresis and the irreducible nature play more important roles for the martensitic transformations in contrast, for example, to ferroelectrics [113]. A typical

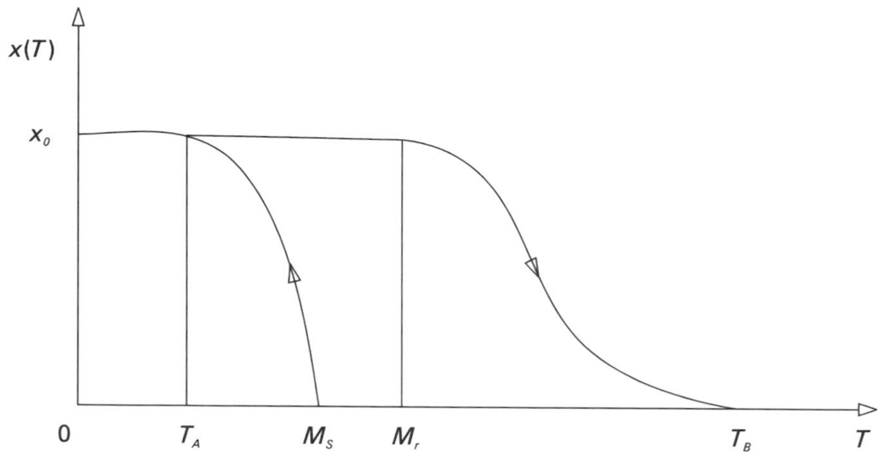

FIGURE 4.1 Temperature-dependence of the fraction of the low-temperature phase $x(T)$ for the martensitic transformation. M_s and M_r are the temperatures of the beginning of the phase transition by cooling and heating, respectively; T_A is the temperature below which there is no increasing of x, T_B is the disappearing temperature of the low-temperature phase.

curve for the temperature dependence of the fraction of the new phase $x(T)$ is shown in Figure 4.1. Below the point T_A, decreasing T does not provide any increase in $x(T)$ since the difference in the thermodynamic potentials $\Delta\Phi(T)$ tends to some constant value and cannot compensate more for the increase of the elastic energy provided by the expansion of the region of the new phase within the old one. The dependence $x(T)$ appreciably depends on the mechanical and thermal history of the sample, the size of grains, residual starless etc. [17, 40, 42]. Hence, neither the point M_s nor M_r are not equal, generally speaking, the temperature of the thermodynamic equilibrium between two phases, which is determined by the condition

$$\Delta\Phi_{ij}(T_c) \equiv \Phi_i(T_c) - \Phi_j(T_c) = 0 \tag{4.1}$$

It is clear that at least $M_s < T_c < T_B$. However, since the cold work, i.e. large mechanical deformations of the sample, reduces the residual stresses, then it has to give some increase of x within the region of thermodynamic preference of the low-

TABLE 4.1 Experimental data for characteristic temperatures (in degrees Kelvin) of the martensitic phase transitions in Li and Na (see Figure 4.1)

metal	T_A	M_s	M_r	M_d	T_B	sources
Li		70–80	90	105	165	[17, 18, 40]
	12	32	43		63–70	[41]
Na		35	45	51	65–70	[17, 40]
		35–50				[18]

temperature phase. For this reason, one should propose in experimental investigations (see [40, 41, 44]) to take as T_c the value of M_d, above which cold work provides no increase in the fraction of the low-temperature phase $x(T)$.

Experimental values of the characteristic temperatures for Li and Na are shown in Table 4.1. The value M_d, as we have seen above, is one of the most important characteristics of the transition. However, in experiment this value has been measured with minimal accuracy. For example, for M_d in Na only the interval $51\,\mathrm{K} \leq M_d < 62\,\mathrm{K}$ has been determined in [17] and in [40] M_d is taken to be the lower border of this interval.

Now it is clear that experimental values of T_c are actually not well-determined. In any case, this value is not M_s but rather M_d. This fact disagrees with results of works [3, 5].

The martensitic transition in Li has an additional important feature. A free crystal of Li, as well of Na, obeys by cooling the transformation from the BCC to the HCP phase. However, cold work of Li at all temperatures $T < M_d$ to very low temperatures, transforms it to the FCC phase, the growth of which is achieved by decreasing the part of the BCC, as well as the HCP phases [17]. This fact demonstrates extreme proximity of the thermodynamic potentials of the FCC and HCP phases. While taking into account the influence of cold work discussed above, it becomes not completely clear which phase must be considered in experiment as the most preferable. We see, that even small stresses or the history of the sample can change the sign of the difference $\Delta\Phi$ (4.1). This feature of the phase transition in Li was often totally ignored in early theoretical publications [2, 5].

In Na, the FCC phase does not appear by cold working (although, similar to Li, the number of stacking faults, demonstrating the proximity of energies of the FCC and the HCP phases, is rather large [77]).

Let us discuss the latent heat of the transition

$$q = T\Delta S_{ij} = T(S_i - S_j), \tag{4.2}$$

where S_i is the entropy (per atom) of phase i. The theoretical value of the latent heat $q = q_c$ is determined as the value of $T\Delta S$ at $T = T_c$. However, the experimentally [40] measured latent heat Q has been determined as the integral of the "excess" specific heat (the difference between values of the specific heat by heating and cooling) over the interval from $T \approx M_r$ to T_B (see Figure 4.1)

$$Q = \int_{M_r}^{T_B} (C_{heating} - C_{cooling})\,dT \equiv \int_{M_r}^{T_B} \Delta C_{ij}\,dT. \tag{4.3}$$

If we denote the entropy and the specific heat of pure HCP and BCC phases as S_H, C_H and S_B, C_B, respectively, and take into account that $C_{cooling} = C_B(T)$ then ΔC_{ij} in (4.3) becomes

$$\begin{aligned}\Delta C_{BH} &= T\frac{d}{dT}[(1-x)S_B + xS_H] - C_B(T)\\ &= T(S_B - S_H)\left|\frac{dx}{dT}\right| - x(C_B - C_H).\end{aligned} \tag{4.4}$$

Here we have taken into account that $dx/dT < 0$, i.e.

$$\frac{dx}{dT} = -\left|\frac{dx}{dT}\right|$$

If the spread of the transition region were small, then it would be possible to neglect the second term in (4.4) within the region of the phase transition $T|dx/dT| \geq x$, and to substitute the coefficient at $|dx/dT|$ by the value q_c. In this case, the integration of the experimental value $\Delta C_{BH}(T)$ from M_r to T would provide the dependence $x(T)$ on the interval from M_r to T_B. This procedure has been used in [40]. Using the results of (not very accurate) estimations of $x(T_A) = x_0$, the values $q_{exper} = Q/x_0$ for Li and Na, were also determined in [40] in the framework of the same approximation (see Table 4.2). But from the plots of $\Delta C_{BH}(T)$ in [19] and from values M_r and T_B in Table 4.1 we can see that the curves $x(T)$, obtained using this procedure, are appreciably spread out, especially for Li, for which the temperature spread of $\Delta C_{BH}(T)$ is almost 50 K. Hence, the main suppositions made in [40] about the sharpness of the transition and, consequently, about the small value of the second term in (4.4) and the constant value of ΔS_{BH}, are not fulfilled very well. For this reason, the values q_{exper} in Table 4.1 cannot be very accurate, especially for Li.

Even if the transition interval were narrow, this value could nevertheless not be identified with the difference in the energy of these structures $\Delta E(0)$ at $T = 0$, as it has been made in [2, 3]. The condition

$$\Delta\Phi_{ij}(T_c) = \Delta F_{ij}(T_c) = 0 \tag{4.5}$$

provides $q_c = \Delta E_{ij}(T_c)$ but not $q_c = \Delta E_{ij}(0)$. As a consequence, this identification would be possible only if the temperature dependence of $\Delta E_{ij}(T)$ were neglected.

TABLE 4.2 Values of the latent heat $q(T) = T\Delta S_{ij}(T)$ (in 10^{-5} Ry/atom) and the differences in the specific heat $\Delta C_{ij}(T)$ (in 10^{-2} cal/Kmol)

Li					Na				
T, K	BCC–HCP		BCC–FCC		T, K	BCC–HCP		BCC–FCC	
	q	ΔC_{BH}	q	ΔC_{FB}		q	ΔC_{BH}	q	ΔC_{FB}
70	4.09	5.10	3.80	5.01	30	1.43	4.98	1.30	4.83
80	4.99	3.97	4.66	3.93	40	2.19	2.80	2.01	2.79
90	5.87	3.05	5.49	3.03	50	2.91	1.55	2.68	1.57
100	6.73	2.33	6.31	2.32	51	3.00			
110	7.58	1.79	7.11	1.79	60	3.59	0.89	3.33	0.91
120	8.41	1.37	7.90	1.38	70	4.27	0.53	3.96	0.55
150	10.90	0.65	10.27	0.66					
experim. [40, 42]	4.5					3.2			

However, the difference $\Delta E_{ij}(T) - \Delta E_{ij}(0)$ has the same order of magnitude as $T\Delta S_{ij}$. As will be shown below, actually in Li and Na

$$\Delta E_{ij}(T_c) = q_c \approx 2\,\Delta E_{ij}(0). \tag{4.6}$$

4.1.2 The Pseudopotential theory. Results of calculations for the second-order model

In the same manner as in Chapters 1 and 3 (see, for example, formulae (3.32)–(3.37)), in calculating the thermodynamic potential $\Phi(p, T)$, we restrict our consideration to the harmonic approximation for ionic oscillations:

$$\begin{aligned} \Phi(p, T) &= \Phi_{st} + F_{ph} + F_e^*; \\ \Phi_{st} &= E_{st}(\Omega) + p\Omega, \quad E_{st}(\Omega) = E_i + \sum_{m=0}^{2} E_m(V^m); \\ F_{ph} &= E^{zp}(\Omega) + F^*. \end{aligned} \tag{4.7}$$

In the lowest-order approximation in the pseudopotential, the contribution of electron excitations F_e^* depends only on the volume Ω. Its change by the phase transition is $\Delta F_e^* = p_e^* \Delta\Omega_{ij}$. Estimations which use the data about the electronic specific heat (see Chapter 3) show that this value is three orders of magnitude smaller than other contributions to $\Delta\Phi$, and can be neglected.

Since the contributions of zero oscillations have been included in equations (4.7), the parameters of the pseudopotential $V(\mathbf{q})$ will be taken from Tables. 2.1 and 2.2. For $\pi(q)$, we will use, as usual, the CV approximation for Li, as well as the GT approximation for other metals (Na and K).

Values of the volume $\Omega(p, T)$ at $p \neq 0$ have been obtained, in the same manner as for the thermal expansion (see Chapter 3) using the perturbation theory in the ratio of F_{ph} to E_{st}. First, the values of the volume $\Omega_0(p)$ neglecting the phonon contribution were calculated for given p from the condition $(\partial\Phi_{st}/\partial\Omega) = 0$. The next step was the calculation of the correction, $\Omega_{ph}(p, T)$, using formulae (1.64) and (1.66). For the HCP phase, the value $c/a = c/a\,(p, T)$ has also been calculated in [13, 14].

Table 4.3 shows the results of calculation of the cell parameters for HCP and FCC phases from [13, 14]. For comparison, the results of calculations by other authors [2, 3] are also shown. By comparison of these results with experimental data for Li, we must take into account that, as a consequence of non-perfect structure of samples, the value of the lattice constant a_0 for the BCC phase was something larger than the value for pure samples [17]. At $T = 78$ K, the value $a_0(T)$ in [17] is equal to 3.491 Å. Taking into account the data for the thermal expansion from [45], we obtain $a(0) = a_0 = 3.490$ Å instead of for example, the value $a_0 = 3.478$ Å used in calculations [13, 14].

This general feature of the thermal expansion of non-ideal samples probably remains in HCP and FCC phases, too. For this reason, the comparison of the results of calculations with the experimental data [18], the cell parameters a, c, Ω for Li in

TABLE 4.3 Lattice parameters of the HCP and FCC phases

metal	T, K	source	HCP					FCC
			a	c	c/a	Ω	$10^3 \frac{\Omega-\Omega_0}{\Omega_0}$	$10^3 \frac{\Omega-\Omega_0}{\Omega_0}$
		[3]	3.784	6.171	1.631	38.255	14.5	
Na	0	theory	3.7666	6.15252	1.6334	37.796	2.30	1.81
	5	experim.	3.767 ± 1	6.154 ± 1	1.6337 ± 7	37.814 ± 26	2.77 ± 70	
		[2]	0.8931	1.4565	1.631	1.0062	6.2	
Li	0	theory	0.8919	1.4556	1.6321	1.0028	2.75	1.48
	78	theory	0.8923	1.4563	1.6325	1.0041	4.12	4.14
		experim.	0.8915 ± 3	1.4595 ± 26	1.6371 ± 34	1.0045 ± 34	4.54 ± 2.44	

Comments: a_0 is the lattice parameter for the BCC phase at $T = p = 0$ (see Table 2.3), $\Omega_0 = a_0^3/2$ is the volume of the elementary cell of this lattice. Experimental errors for all values (except $\Delta\Omega/\Omega_0$ for Li) are given in units for the last decimal digit of the mean value shown. The values a and c are given in units of 10^{-10} m, Ω_0 in 10^{-30} m^3.

Table 4.3 are given as the ratios a/a_0, c/a_0, Ω/Ω_0 to exclude the common "non-ideality factor" mentioned above. For Na, the value $a_0 = 4.225$ Å from [17] was used in calculations [13, 14] and the re-calculation mentioned is not necessary.

These calculated values of the lattice parameter of HCP phase agree well with the data from [17] (see Table 4.3), especially for Na for which these data themselves were the more accurate. Note, in particular, a very good agreement with experimental data for the value $\Delta\Omega = \Omega - \Omega_0$. This high accuracy is very important for the correct description of the phase diagrams (p, T). The thermal expansion appreciably affects the value $\Delta\Omega$ for Li in the vicinity of the phase transition. Taking into account this expansion, the results of the calculations of $\Delta\Omega$ for Li [2], as well as for Na [3], would remarkably disagree with experiment.

Table 4.4 shows the results of calculations of the differences between energies of different structures ΔE_{ij} as well as the temperature and the pressure of the phase equilibrium, T_c and p_c, respectively. To illustrate the zero-point contribution ΔE^{zp} to ΔE_{ij} as well as the redefinition of the parameters of $V_{pC}(\mathbf{q})$ provided by $E^{zp} \neq 0$, the third line in Table 4.4 shows the values of $\Delta E_{0ij}^{(0)}$ calculated without taking into account the contributions ΔE_{ij}^{zp} and with the non-redefined $V(\mathbf{q})$ (see Tables. 3.1, 3.2, $E_{sr} = 0$). The terms ΔE_{ij}^{zp} have approachable values for both Li and Na. These contributions remarkably affect ΔE_{ij} and, consequently, T_c. The results for K show that the role of ΔE_{ij}^{zp} decreases for heavier metals. Table 4.5 demonstrates the spread of the result obtained by different authors for the values ΔE_{ij} and T_c, mentioned above. Note, in particular, an incorrect result [4] for the structure of K. In this work, instead of the BCC phase observed experimentally at low T, calculations have yielded the HCP phase.

Figures. 4.2 and 4.3 show the phase diagrams of Li and Na calculated in [13, 14]. Following [14], let us discuss these results separately for Na and Li. First we consider

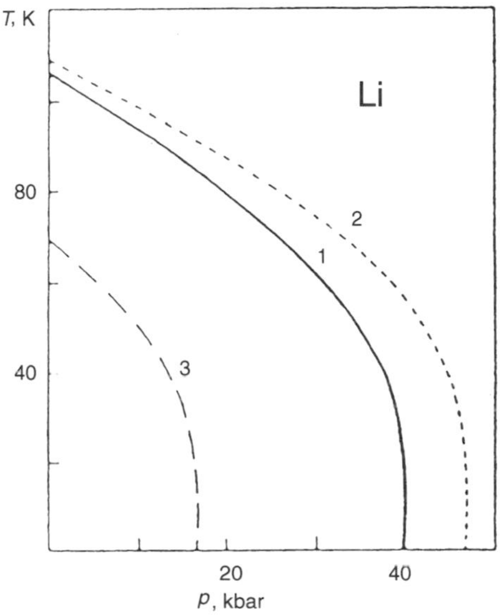

FIGURE 4.2 Phase diagrams of lithium calculated in the frame of the M2A model: 1 – $T_c(HB)$; 2 – $T_c(FB)$; 3 – $M_s(HB)$ (after [13]).

lithium. As it was already mentioned, the second-order model with the local form factor does not provide a very accurate description of some properties of Li, like the volume dependences of the shear moduli and the absolute value of the cohesive energy (see Chapter 3). These moduli themselves, as well as the volume properties, in particular, the equation of state $p(\Omega, T)$, have been described as well as, for example, Na and K (see section 2). In good agreement with experiment (Table 4.1) it was found that not only the value of the temperature $T_c \approx M_d \approx 107\,\mathrm{K}$ but also the fact discussed above that the values of the free energy for the FCC and HCP phases appare to be very close together. Together with good accuracy obtained in calculations of the lattice parameters of Li, these facts seem to be enough to suppose that the accuracy in the description of the phase transitions in Li would remain good at non-zero pressures ($p \neq 0$), too. For Li and Na, as a consequence of the small size of ion cores and a small number of valence electrons in the cores, the short-range repulsion term E_{sr} would not practically affect the results obtained without this term [13, 14]. Hence, the phase diagram of lithium calculated in Figure 4.2 seems to be well-founded. However, as will be shown in the next subsection, the behavior of Li under pressure differs from that of Na under the same conditions and cannot be described in the framework of the second-order model.

The transition temperature M_s for the phase transition from the high- to the low-temperature phase which can be easily measured experimentally, lies somewhat below the value $T_c \approx M_d$. At $p = 0$, Table 4.1 gives $\Delta T_c = M_d - M_s \approx 35\,\mathrm{K}$. To evaluate the behavior of $M_s(p)$, a simple supposition is that at small $p \leq B/3$ (B is

TABLE 4.4 Differences in the energy between different structures ΔE_{ij} (in 10^{-5} Ry/atom), the temperature T_c (in K) of the phase equilibrium at $p = 0$ and values of the pressure p_c (in kbar) for the phase equilibrium at $T = 0$

value	Li			Na			K		
	HCP–BCC (HB)	FCC–BCC (FB)	HCP–FCC (HF)	HCP–BCC (HB)	FCC–BCC (FB)	HCP–FCC (HF)	HCP–BCC (HB)	FCC–BCC (FB)	HCP–FCC (HF)
ΔE_{ij}	−4,07	−3,76	−0,31	−0,15	−0,26	0,11	3,56	3,23	0,33
ΔE_{ij}^{zp}	3,32	3,21	0,11	1,35	1,31	0,04	0,70	0,67	0,03
$\Delta E_{0ij}^{(0)}$	−9,6	−8,8	−0,8	−2,5	−2,4	−1,1	2,5	2,3	0,2
T_c	107	109		17,4	23,4				
T_{cv}	110	112		17,8	23,7				
p_c	40,0	47,0		0,49	0,95				

Comment: The difference in the energy $\Delta E = \Delta E_0 + \Delta E^{zp}$, where ΔE_0 is the difference in the energy $E(\Omega)$(1.12) between different phases, ΔE^{zp} is the difference in the energy of zero oscillations (1.13), $\Delta E_0^{(0)}$ is the value of ΔE_0 calculated neglecting the influence of zero oscillations on parameters of the pseudopotential (Tables 2.1, 2.2, case $E^{zp} = 0$), T_c is the temperature of the phase transition at $p = 0$, T_{cv} is the same value calculated neglecting the thermal expansion; notations of the phases: H–HCP, F–FCC, B–BCC. After [13, 14].

TABLE 4.5 Values of ΔE_{ij} (in 10^{-5} Ry/atom) and T_c (in Kelvin degrees) calculated in [2–5]

value	source	Li				Na			K		
		(HB)	(FB)	(HF)		(HB)	(FB)	(HF)	(HB)	(FB)	(HF)
ΔE_{ij}		−4.6	−0.9	−3.7		−6.0	−3.8	−2.4	−6.4	−2.2	−4.2
ΔE_{ij}^{zp}	[5]	2.1	2.7	−0.5	[4]	2.0	1.8	0.2	1.3	1.1	0.1
T_c		140				90			91		
ΔE_{ij}		−12.7	−4.8	−7.9		−6					
T_c	[2]	147			[3]	260					

Comment: the same notations as in Table 4.4.

the volume elastic modulus) the difference in thermodynamic potentials $\Delta\Phi_{HB}(p, T)$ at the transition point $M_s(p)$ does not change and is equal to the experimental value of $M_s = 70$ K obtained at $p = 0$, which was used in [13, 14], i.e.

$$\Delta\Phi_{HB}[p, M_s(p)] = \Delta\Phi_{HB}[0, M_s^{exper}(0)].$$

For Li, it provides $\Delta\Phi_{HB} = 2.28 \times 10^{-5}$ Ry/atom. The transition point M_s is predominantly determined by the condition that the thermodynamic preference of the new phase $\Delta\Phi_{HB}$ must be equal to the energy of "surface tension" for the nucleus of

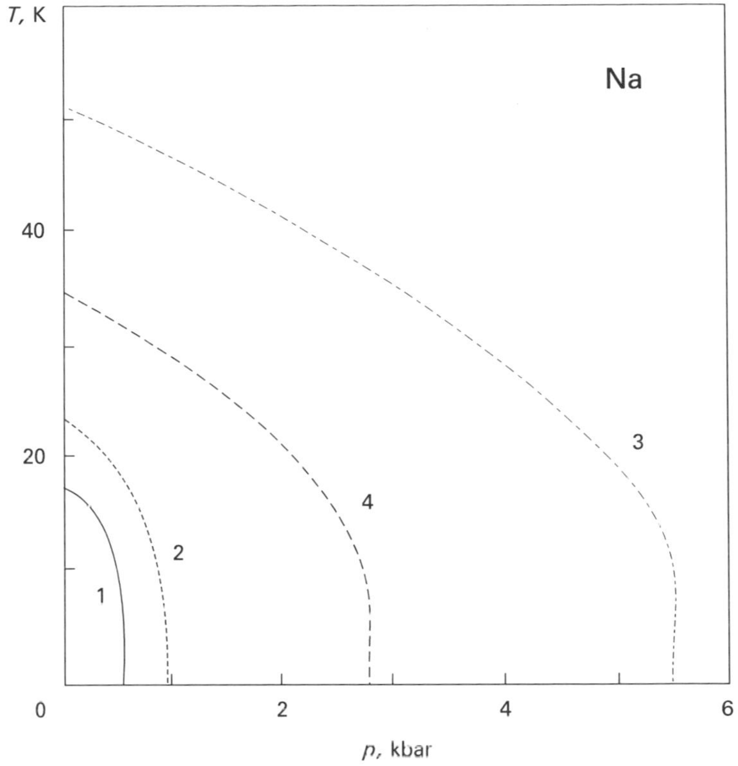

FIGURE 4.3 Phase diagram of sodium calculated in the framework of the M2A model: 1 – T_c(HB); 2 – $T_c(FB)$; 3 – $T_{s\delta}$(HB); 4 – $M_{s\delta}$(HB) (after [13]).

the new phase. However, the last values, in accordance with [13, 14], change only a little with p at small pressure. The value $M_s(p)$ evaluated using this method is shown in Figure 4.2 (and in Figure 4.3) as a dashed line. These plots show that the region of the HCP phase decreases with pressure and for $p \geq 20$ kbar this phase does not appear by cooling at all.

In the next subsection, we will show that the decrease of $M_s(p)$ with increasing p disagrees with experimental data as well as with results of more detailed (band-structure) calculations of the difference between energies of different phases. Nevertheless, we emphasize that this comparison with experiment shows that the description of the HCP–BCC phase transition at $p = 0$ is good enough.

Figure 4.4 shows the temperature behavior of the differences in thermodynamic potentials $\Delta\Phi_{ij}(T)$ for lithium at $p = 0$. With increasing T, the next energetically preferable phase (after BCC) becomes the FCC phase, and the least preferable becomes the HCP phase.

Table 4.2 shows results of calculation of the values $T\Delta S_{ij}(T)$ and $\Delta C_{ij}(T)$ within the region of the phase transition which characterize the difference in thermal

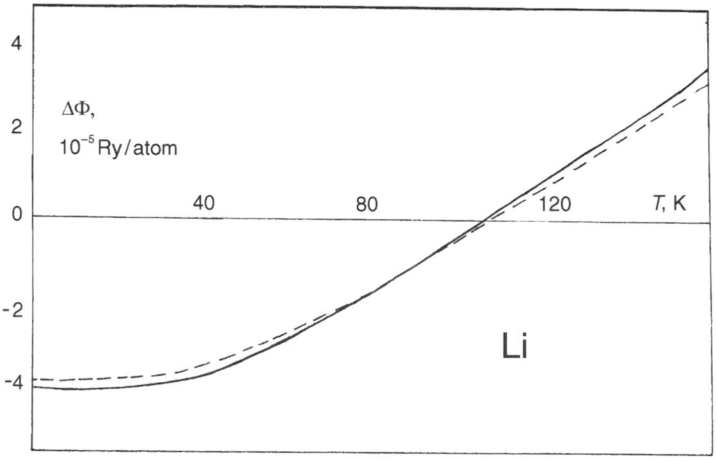

FIGURE 4.4 Temperature-dependence of the difference of thermodynamical potentials $\Delta\Phi(T)$ for lithium at $p = 0$ (the M2A model): full curve – $\Delta\Phi_{HB}$, dotted curve – $\Delta\Phi_{FB}$ (after [13]).

properties of both phases. The calculated value of $q_c \sim 7 \times 10^{-5}$ Ry/atom for Li is remarkably larger than the experimental one $q_{exper} \approx 4.5 \times 10^{-5}$ Ry/atom from [42]. However, as a consequence of the relatively large spread of the transition region in the case of Li, the accuracy of q_{exper} is not well-defined, more accurate analysis and test of experimental data would be required. Note also, that the comparison of the calculated values q_c and ΔE from Table 4.4 provides the relation $\Delta E_{ij}(T_c) = q_c \sim (1.8 \div 2)\Delta E_{ij}(0)$ mentioned above.

Let us now consider the results of calculations of the phase transitions in Na. In contrast to the case of Li, the accuracy of the second-order model in calculating the temperature T_c for Na is not good enough. Instead of the experimental value $M_d \sim 51$ K (Table 4.1), the calculated value of T_c for the HCP–BCC phase transition is equal to 17.4 K (Table 4.4). Moreover, this value of T_c lies below the calculated temperature of the FCC–BCC transition to be equal to 23.4 K, whereas, in contrast to the case of Li, the FCC phase of Na was not observed experimentally at all, even under very intensive cold working. Hence, in the same way as in the preceding calculations [3, 4], this model provides only the order of magnitude for the values ΔE and T_c for Na.

This rather poor accuracy of the description of the phase transition in Na (in comparison with Li) is unexpected because we have mentioned many times that the second-order model provides much better results for all other atomic properties of Na (as well as heavier alkali metals) than for Li. Some explanation of this fact provides a comparison of the value ΔE for the series Li–Na–K (Table 4.4). We can see, that the difference ΔE_{HB} monotonically increases in this series from negative value for Li to positive values for K (and larger positive values for Rb and Cs, see Table 3.9). For sodium, the value ΔE_{HB} is near zero, which is qualitativly confirmed by the small experimental value of T_c in Na: $T_c(Na) < 0.5T_c(Li)$. Hence, to provide

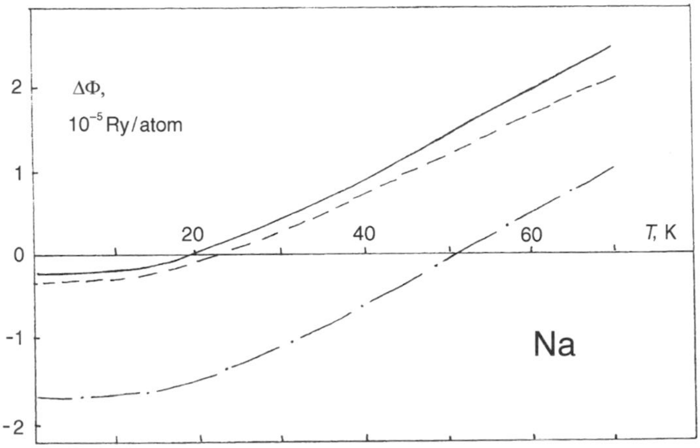

FIGURE 4.5 Temperature-dependence of the difference of thermodynamical potentials $\Delta\Phi(T)$ for sodium at $p = 0$ (the M2A model): full line – $\Delta\Phi_{HB}$, dotted line – $\Delta\Phi_{FB}$, dashed–dotted line – $\Delta\Phi^{\delta}_{HB} = \Phi_{HB} + \delta$ (after [13]).

quantitative calculations of T_c, the accuracy of calculation of ΔE_{ij} for Na must be much higher than that for other metals. For example, the decrease of ΔE_{HB} in a small value $\delta = 1.55 \times 10^{-5}$ Ry/atom wouldn't appreciably affect T_c for Li and would provide no influence on the stability conditions for the BCC phase of K, Rb, and Cs. However, as we can see from Figure 4.5, the same decrease of ΔE_{HB} results in a triple increase of $T_c(HB)$ for Na, and this new value agrees well with experiment. Hence, further improvement of the model (and the form of the pseudopotential $V(\mathbf{q})$ taking into account, for example, the effects of non-locality) would be necessary for quantitative calculations of T_c for Na.

Except $T_c(HB)$, all properties of the HCP phase of Li, which can be compared with experiment, are described by the model with good accuracy. We have already mentioned the excellent agreement of the calculated values of a, c, and $\Delta\Omega$ for the HCP phase with experiment (see Table 4.3).

The values $T\Delta S$ at $T = M_d \sim 50$ K (Table 4.2) are in good agreement with the values q_{exper} obtained in measurements [40] (probably, as a consequence of a smaller spread of the phase transition region, for Na this value has been measured with higher accuracy than for Li).

Hence, one can expect that the accuracy of the second-order model in the description of the volumic and thermodynamic properties of the HCP phase at $p = 0$ would be as good as in the case of the BCC phase. However, this accuracy would be insufficient (because of the reason mentioned above) only for the calculation of the value ΔE, i.e. the value T_c at $p = 0$. Hence, after introducing the empirical correction δ (see above) to the value ΔE to satisfy the condition $T_c(HB) = M_d$, the second-order model becomes applicable for calculations of the phase diagram of Na, too.

The HCP phase of Na exists only at small p (see Figure 4.3), much smaller than the elastic modulus $B = 75$ kbar (this modulus has approximately the same value for

TABLE 4.6 Static properties of Li and Na in the HCP and FCC phases

metal	structure	$-E$	p	B	B_{33}	B_{44}	B_2	B_{22}	B_{12}	B_{12}^{dyn}	θ_D
Li	HCP	0.55076	0.0000	134.7	39.11	47.72	0.09	142.5	0.771	−1.02	366.74
	FCC	0.55075	0.0000	134.8	17.07	107.4					340.64
Na	HCP	0.46573	0.0000	74.03	19.47	25.78	0.002	81.35	0.362	−0.362	160.95
	FCC	0.46573	0.0000	74.17	8.183	60.92					156.44

metal	structure	$\tilde{\omega}_c^2$	$\tilde{\omega}_a^2$	B'	B'_{dyn}	B'_{33}	B'_{44}	B'_{22}	B'_{12}	$(\tilde{\omega}_c^2)'$	$(\tilde{\omega}_a^2)'$	γ_0
Li	HCP	0.3163	0.0452	3.526	3.340	−1.518	1.848	2.168	0.628	2.894	1.912	0.743
	FCC			3.526	3.296	1.383	2.112					0.807
Na	HCP	0.3453	0.0537	3.786	3.200	−1.272	1.692	2.006	−0.882	3.305	1.753	0.648
	FCC			3.785	3.030	1.152	1.953					0.725

Comments: Following notations are used: $F' = -(\partial \ln F/\partial \ln \Omega)$, the calculations for Li (Na) were performed using the GV (GT) approximation for the polarization operator. Units: elastic constant B_{ik} in kbar, the Debye temperature θ_D in K, all frequencies are given in units of the plasma frequency, other values are dimensionless.

different phases, see Table 4.6). Hence, the thermodynamic potential $\Phi(p, T)$ of each phase within the region considered can be written as a power series in p. If we additionally neglect for simplicity the variation of the ratio c/a for the HCP phase with pressure (this variation is proportional to the modulus B_{12} in Table 4.6 and, as is simple to demonstrate, is really negligibly small), then the expression for the difference $\Delta\Phi(p, T)$ between two phases A and B becomes

$$\Delta\Phi_{AB}(p, T) = \Delta F(T) + p\Delta\Omega(T) - \frac{p^2}{2}\Delta\left(\frac{\Omega}{B}\right) + \frac{p^3}{6}\Delta\left[\frac{\Omega(1 + dB/dp)}{B^2}\right]. \quad (4.8)$$

Here $\Delta\Phi_{AB} = \Phi_A(T) - \Phi_B(T)$ is the difference between values corresponding to different phases (A and B) at $p = 0$, ΔF is the difference in the free energies. Evaluations show that the two last terms in (4.8) are much smaller than the two first ones so that actually the form of the phase diagram of Na is predominantly determined by the term ΔF_{AB} and $\Delta\Omega_{AB}$. But, as was mentioned above, both $\Delta\Omega_{AB}$ and the temperature dependence of $\Delta F_{AB}(T)$ can be described by the present theory with great accuracy. Hence, we can expect that, after adding the constant δ mentioned above to the value ΔF_{AB}, the phase diagram Na would be calculated well within the whole region of existence of the HCP phase.

The dashed–dotted line in Figure 4.3 shows the thermodynamic equilibrium line $T_c(p)$ obtained by replacing $\Delta\Phi_{HB}(p, T)$ by the value $\Delta\Phi_{HB} + \delta$. This line is marked as $T_\delta(HB)$. The dashed line below is the plot of the value $M_s(p) = M_{s\delta}$ obtained, as in Figure 4.2, using the supposition about the independence of $\Delta\Phi_{AB}(p, M_s(p))$ from p.

Finally, Table 4.6 shows the results of calculations of the static properties of all phases considered. The accuracy of these results is, probably, good enough. Hence, these results provide important information about the properties of these phases for which experimental studies, as was already mentioned, seem to be difficult.

4.1.3 Stability properties of crystalline phases of lithium and sodium under pressure obtained in the framework of the band structure calculations

The pseudopotential calculations [13, 19] have predicted a decrease of M_s and the disappearence of the phase transitions with increasing p, in contrast to a simple geometrical picture which predicts an increase of the stability of close-packed phases under pressure. The experimental studies [18] have confirmed this prediction for Li. However, the value of M_s for Li doesn't decrease, but rather increases with compression, at least up to $p = 17$ kbar. This increase of M_s with p agrees with the observation of the structural phase transitions in Li at $T = 296$ K, $p = 69$ kbar [20]. Measurements of the electrical resistivity $R(p, T)$ for Li have confirmed the increase of M_s with p [21] and provide an indication of an additional structural phase transition at low T at $p \approx 260$ kbar [22]. The data for $R(p, T)$ for alloy $Li_{0.9}M_{0.1}$ also demonstrate an eventual quantitative change of the dependence $M_s(p)$ (from increasing to decreasing) at rather small $p \approx 2$ kbar [23]:

The structures of the low-temperature phases of Li and Na at $p = 0$ have been identified by Barret [17] with the HCP phase, although he has mentioned that cold work of Li transforms the HCP phase to the FCC one. Further investigations [24–27] have shown, that the low-temperature phase of Li at $p = 0$ is close to the 9R structure (Sm-type, see below) with a large amount of stacking faults. On the other hand, the structure of Li at $T = 296$ K and $p > 69$ kbar, has been identified with the FCC phase [20]. Hence, we can conclude that the general features of the phase diagrams of Li and Na under pressure are not sufficiently investigated. In the first place, this situation is a consequence of technical problems in studying alkali metals under extreme conditions (high pressures, low temperatures) as well as very small changes of the macroscopic characteristics by the phase transitions and the complicated kinetics of the martensitic transformations.

The stability of different phases of Li and Na was studied theoretically using the empirical pseudopotentials [13, 19], as well as the *ab initio* ones [28–30]. But, since the differences between energies of different structures ΔE_{ij} in Li and Na are extremely small, $\Delta E \sim 0.01 - 0.1$ m Ry/atom [13, 14], the accuracy of *ab initio* calculations of these values was actually insufficient and the results for $\Delta E_{ij}(p)$ from [28–30] differ quantitatively, as well as qualitatively. On the other hand, the accuracy of pseudopotential calculations of properties of alkali metals is actually large enough (see [37] and the references therein). For this reason, an explanation of this qualitative disagreement of the results of [13, 14] for the dependence $M_s(p)$ for Li with experiment (whereas in the case of Na this agreement was good) would be very important for the theoretical studies of the stability of crystal structures in metals.

In this subsection, we will show, in particular, that different behaviors of $M_s(p)$ in Li and Na relate to qualitatively different characters of the change of band structures with pressure. In Li, the increase of p is accomplished by the approach of the Fermi level E_F to the point E_c, the peak in the density of states. However, this fact cannot be taken into account in pseudopotential calculations (see Subsection 2). The band effects [35, 46–49] result in the loss of stability of the BCC phase and stimulate the BCC–9R or BCC–FCC phase transitions. But in the case of sodium, where there exist no effects provided by the approaching of E_F to E_c, in accordance with our consideration, decreasing of the relative contribution of the band-structure energy with compression results in the increase of the stability of the BCC phase. These band-structure effects in Li must result in the "softening" of the shear moduli $C_{ss'}$ with increasing pressure, too. This fact agrees well with a series of experimental results and explains the general nature of the "pre-martensitic" softening of $C_{ss'}(p)$ observed in many systems (see, e.g., [50]).

In the present subsection, we will also discuss the results of calculations of the energy differences and the structures of different phases of Li and Na [33, 46–49]. The main account will be devoted to a qualitative consideration of the relation between the stability and the band structure. Considering these problems, we can restrict our analysis to the case of zero temperature $T = 0$ only. At $T \neq 0$, all characteristic features of phase diagrams (p, T) remain qualitatively the same, and quantitative calculations can be provided in the same manner.

The stability of the phase i relative to the phase j under the pressure p is determined by the condition $\Delta\Phi_{ij}(p) = \Phi_i - \Phi_j$, where $\Phi_i = E_i + p\Omega_i$ is the Gibbs potential (per atom) for the phase i. But, since it becomes apparent that atomic volumes of both considered phases are close, $\Delta\Omega/\Omega \sim 10^{-3}$, then $\Delta\Phi_{ij}$ is close to the energy difference at constant volume $\Delta E_{ij}(\Omega) = E_i(\Omega)_j(\Omega)$, i.e., $\Delta - t\Phi_{ij} - \Delta E_{ij} \approx \frac{1}{2}B(\Delta\Omega/\Omega^2)$, ($B$ is the volume elasticity modulus). Since this difference is actually not larger than a few per cent of ΔE_{ij}, then here (in the same manner as in [28–30]), for simplicity, only results for $\Delta E_{ij}(\Omega)$ will be presented. We also suppose that the pressure dependence of Ω corresponds to the BCC phase.

Pseudopotential calculation Here, in the same manner as above, the second-order pseudopotential model is used. But, besides BCC, FCC, and HCP phases discussed above, we consider here also 9R, 4H, and 6H structures. If we denote, as usual, the sequences of stacks of "hard spheres" in the HCP and FCC phases as ABAB... and ABCABC..., then for 4H ("double HCP"), 6H ("triple HCP"), and 9R ("close-packed rhombohedral") lattices, these stacking structures have the forms ABAC..., ABCACB... and ABCBCACAB... [25], respectively.

Remember that, in the framework of the second-order model in the pseudopotential V considered here, the difference in the free energy of both phases is

$$\Delta F_{ij} = F_i(\Omega, T) - F_j(\Omega, T), \tag{4.9}$$

$$\Delta F_{ij} = \Delta F_{ij}^{Mad}(\Omega) + \Delta E_{ij}^{bs}(\Omega) + \Delta F_{ij}^{ph}(\Omega, T). \tag{4.10}$$

Here E^{Mad}, E^{bs} and F^{ph} are the Madelung energy, the band structure energy, and the phonon contributions to F (see e.g., [1, 51] and the previous chapters):

$$E_i^{Mad} = -\alpha_i z^2 e^2/(2r_{WS}) \tag{4.11}$$

$$E^{bs} = -\sum_{\mathbf{q}} F(\mathbf{q})\,|S(\mathbf{q})|^2; \quad F(\mathbf{q}) = \frac{1}{2}\Omega V^2(\mathbf{q})\frac{\pi(q)}{1+\frac{4\pi e^2}{q^2}\pi(\mathbf{q})} \tag{4.12}$$

$$F^{ph} = E^{zp} + \frac{1}{N}T\sum_{\mathbf{k}}\ln[1-\exp(-\beta\hbar\omega_{\lambda\mathbf{k}})]; \quad E^{zp} = \frac{1}{2N}\sum_{\mathbf{k}\lambda}\hbar\omega_{\lambda\mathbf{k}}. \tag{4.13}$$

In formulae (4.11)–(4.13) $r_{ws} = (3\Omega/4\pi)^{1/3}$ is the Wigner–Seitz radius, α_i is the Madelung constant, $\mathbf{q}$ are vectors of the reciprocal lattice, $S(\mathbf{q})$ is the structural factor of the cell, $\pi(\mathbf{q})$ is the polarization operator, N is the total number of atoms, and $\omega_{\lambda\mathbf{k}}$ are the phonon frequencies. Let us now discuss the sign and value of each of these three contributions in (4.10).

Table 4.7 gives values of the Madelung constant α_i. For hexagonal phases they have been calculated for an ideal value of $\mu = c/a$ (for the 6H, 4H, and 9R phases for the first time) in [33] because the values of μ for the hexagonal phases of Li and Na are close to the ideal value $\mu_0 = 1.633$. Using the notations $i = B$, F, and H, for the

TABLE 4.7 Value of the Madelung constant α_i in equation (4.11)

phase	BCC	FCC	6H	4H	9R	HCP
$10^4(1.792-\alpha_i)$	1.41	2.53	2.77	2.88	3.00	3.24

BCC, FCC, and HCP phases, respectively, Table 4.7 shows that the values α_i satisfy the following inequalities:

$$\alpha_B > \alpha_F > \alpha_{6H} > \alpha_{4H} > \alpha_{9R} > \alpha_H. \tag{4.14}$$

The ratios of differences $\Delta\alpha_{ij}$ between neighboring α_i in the series (4.14) are approximately 9:2:1:1:2. This relative proximity of values α_i for all close-packed (CP) phases reflects a geometrical similarity of those. A remarkably larger value of α_B corresponds to the fact that for the BCC phase the Wigner–Seitz cells are closer to a sphere (for which the maximum value $\alpha = 1.8$ is reached) than for other CP phases.

The phonon contributions F^{ph} decrease with decrease of the frequencies $\omega_{\lambda\mathbf{k}}$. In particular, at $T=0$ $F^{ph} = E^{zp} = \frac{3}{2}\omega_0$ and at high T $F^{ph} = -3\,\mathrm{T}\ln(T/\omega_h)$, where ω_0 and ω_h are the mean arithmetic and mean geometric values of the frequencies $\omega_{\lambda\mathbf{k}}$, respectively. The phonon spectrum of BCC non-transition (in particular, alkali) metals contains soft phonon branches $\sum_4$ (see, e.g., Chapter 3) which cause the increase of the thermodynamic preference of the BCC phase of these metals with increasing T [52]. Values of the differences ΔE_{ij}^{zp} for BCC, FCC and HCP phases of Li and Na at the equilibrium volume $\Omega_0 = \Omega(T=0)$ (see Table 4.1) are shown in Table 4.8.

The signs of the calculated values ΔE_{ij}^{zp} are the same as for the differences ΔE_{ij}^{Mad}. The differences in the energy ΔE_{ij}^{zp} between CP phases are very small. A comparison of results from Table 4.8 with Figure 4.6 have shown that the term ΔE_{ij}^{zp} in (4.10) is important only at small compression $u = 1 - \Omega/\Omega_0$. Hence, in calculating $\Delta E_{ij}^{zp}(u)$ the next simple estimations can be used

$$\Delta E_{ij}^{zp}(u) = \Delta E_{ij}^{zp}(0)\, E_B^{zp}(u)/E_B^{zp}(0). \tag{4.15}$$

The difference ΔE_{ij}^{zp} between energies for different CP phases has been estimated as

$$\Delta E_{ij}^{zp}(u) = \Delta E_{FH}^{zp}\, \Delta\alpha_{ij}, \tag{4.16}$$

where $\Delta\alpha_{ij} = \alpha_i - \alpha_j$ is the difference between α_i from Table 4.7.

The signs of the difference ΔE_{Bi}^{bs} between values corresponding the BCC and CP phases in (4.10) are opposite to the signs of ΔE_{Bi}^{Mad} and ΔE_{Bi}^{zp}. This fact relates to the increase of the function $F(g)$ in (4.12) by the phase transition from the BCC to CP structures (see Figures 1 and 2 in [53]). For Li and Na, CP phases become energetically preferable at $T=u=0$ (see [13] and Figure 4.6). However, the ratio $R_i = E_i^{bs}/E_i^{Mad}$ calculated in the framework of the pseudopotential perturbation theory, generally speaking, decreases with increasing compression. Indeed, the

TABLE 4.8. Differences in the energy of zero oscillations ΔE_{ij}^{zp} (in mRy) at $\Omega = \Omega_0$ (see Table 4.1)

phases	FCC–BCC	HCP–FCC	Ω_0, a.u.
Li	0.0321	0.0011	143.4
Na	0.0131	0.0004	254.5

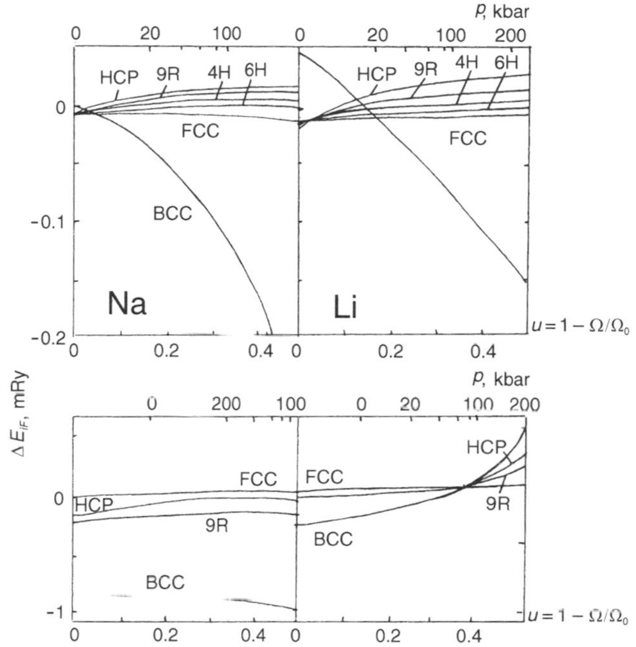

FIGURE 4.6 The difference $\Delta E_i F = E_i - E_F$ between the energies of some structures (i) and the energy of the FCC (F) phase for Na and Li against compression u: (a) calculation for the M2A model pseudopotential, (b) LMTO–ASA calculation. The scale of the plot of ΔE_{iF} in Figure (b) is 10 times smaller than in Figure (a) (after [33, 46]).

pseudopotential $V(r)$ is the Coulomb potential of the electron-ion interaction somehow "truncated" near the ion core. Hence, the value R_i is of the order of ze^2/v_F, where v_F is the Fermi velocity, i.e. this value changes with volume Ω approximately as $\Omega^{1/3}$. Hence, if there exist no specific "band structure" effects provided by E_F approaching singular points of the band-structure contribution (this was supposed in subsection 2 in considering mono-valent metals) then the term ΔE^{bs} in (4.10) must decrease with pressure and at large enough u the sequences of energetically preferences of the phases must be the same as for the sequences of α_i in inequality (4.14). These facts explain the characteristic features of the dependences $\Delta E_{ij}(u) = \Delta F_{ij}(u, T)$ (see Figure 4.6), as well as the form of phase diagrams (p, T) for Na and Li (see Figures 4.2 and 4.3).

As was mentioned above, for Na this conclusion about increasing the stability of the BCC phase under pressure agrees with experiment [18]. However, the accuracy of calculation is insufficient to provide a quantitative calculation of the equilibrium temperature M_d. As we have already mentioned is Subsection 2, this fact relates to an extremely small value of ΔE_{ij} for Na. The value $M_d \cong 50$ K observed in experiment at $u = 0$ [17] corresponds to the value $\Delta E_{BH} \cong 16\mu$ Ry in pseudopotential calculations, whereas in [13] values $\Delta E_{BH} = 1.5\mu$ Ry, $E_{HF} = 1.1\mu$ Ry have been obtained. The same model as in [13], but with better numerical accuracy, gives

$\Delta E_{BH} = 5\mu\,\text{Ry}$, $\Delta E_{HF} = 0,5\mu\,\text{Ry}$. However, if we, as was already shown, introduced some constant (independent of u) correction to the value $\Delta E_{BH}(u)$ then the phase diagram of the BCC–HCP transition in Na could be described well by this (second order) model [13, 18].

For Li at $u = 0$, calculations of ΔE_{ij} in the framework of the perturbation theory are in rather good agreement with the experimentally observed value $M_d \cong 105\,\text{K}$. In [13], values $\Delta E_{BH} = 41\mu\,\text{Ry}$ and $\Delta E_{FH} = 3\mu\,\text{Ry}$ have been obtained, whereas in calculations [3] $\Delta E_{BH} = 37\mu\,\text{Ry}$, $\Delta E_{FH} = 4\mu\,\text{Ry}$, $\Delta E_{9R-H} = 1.4\mu\,\text{Ry}$ (this last value, apparently, once again lies within the limits of the accuracy of the calculation, since in experiment $\Delta E_{9R-H} < 0$). However, the main conclusion provided by the perturbation theory, i.e. increasing the stability of the BCC phase under compression, is wrong (see Figure 2 in [18]). As will be discussed below, this result is a consequence of the fact that for Li (in contrast to Na) band-structure effects become important under compression: actual perturbation theory cannot take into account the effects arising by the approaching of the Fermi level to a singular (peak) of the density of states. These effects will be considered below using the results of full-scale band structure calculations.

Band structure calculations The electronic band structures of Li and Na have been calculated using the self-consistent linearized "muffin-tin orbital" method (LMTO) in the frame of the atomic sphere approximation (ASA) [54, 55, 32]. The exchange and correlation effects were taken into account using the local Barth–Hedin potential [56]. The "frozen core" of Li (Na) includes the 1*s* (1*s*, 2*s*, 2*p*) states, and the basic set of valent orbitals includes the 2*s*, 2*p*, 3*d* (3*s*, 3*p*, 3*d*) states. The calculations were performed for four structures: BCC, FCC, HCP, and 9R. The last structure was considered as the trigonal lattice with three atoms in the elementary cell [57], the values of lattice parameters have been taken from [27]. The procedure of numerical integration over **k** uses equidistant mash points within the irreducible part of the Brillouin zone (BZ), which includes 140, 95, 112, and 140 **k**-points for the BCC, FCC, HCP, and 9R structures, respectively.

The total energy E_i of each structure was calculated using the density functional method (see [58]) taking into account only the spherical part of the charge density within the atomic sphere. The pressure p was calculated with applying the Pettifor formula [59]. The accuracy of these self-consistent calculations was of the order $10^{-3}\%$ for E_i and of the order 1% for p. The energy and the pressure of zero oscillations were supposed to be the same as in calculations considered above.

Comparing results of different LMTO calculations for Na and Li [16, 28, 46], we conclude that in [46] the effects of "non-rigidity" of 2*s* and 2*p* shells in Na (which can be important at large compressions) was not taken into account as well as the so-called "Madelung" and "combined" corrections to the LMTO-ASA method. But, in contrast to [16, 28], the approximation of "the local forces theorem" was not used in calculation of ΔE_{ij} in [46, 33], i.e. whole self-consistent calculations of E_i were performed for each structure. The results of band-structure calculations in [46] shows that all approximate versions of the LMTO method provide a large enough accuracy for the electron density of states $\nu(\varepsilon)$. In particular, the results of calculations [33, 46] (which we consider as to be the best) of $\nu(\varepsilon)$ for BCC Na and Li at $p = 0$ (Figure 4.7) are very close to the results of "standard" calculations by Morocci et al. [60] provided using neither the atomic sphere approximation nor any linearization of energy-dependent terms. Actually, errors of band-structure calculations of the differences ΔE_{ij} were not less than 0.1–1 mRy (see, e.g., [29, 61]). This accuracy does not allow us to provide quantitative calculations of ΔE_{ij} in Na and Li. The large spread of results of calculations [28–30] of ΔE_{ij} confirms this conclusion. However, for qualitative explanations of the change of the dependences $\nu(\varepsilon)$ and ΔE_{ij} with deformation, the accuracy of the version of the LMTO–ASA method used here is, apparently, good enough. This conclusion, in particular, is confirmed by successful applications of this method in describing structural phase transitions in other systems [61], as well as by comparison of the results obtained in [3, 46] with results of a more accurate LMTO-calculation [16] (see below).

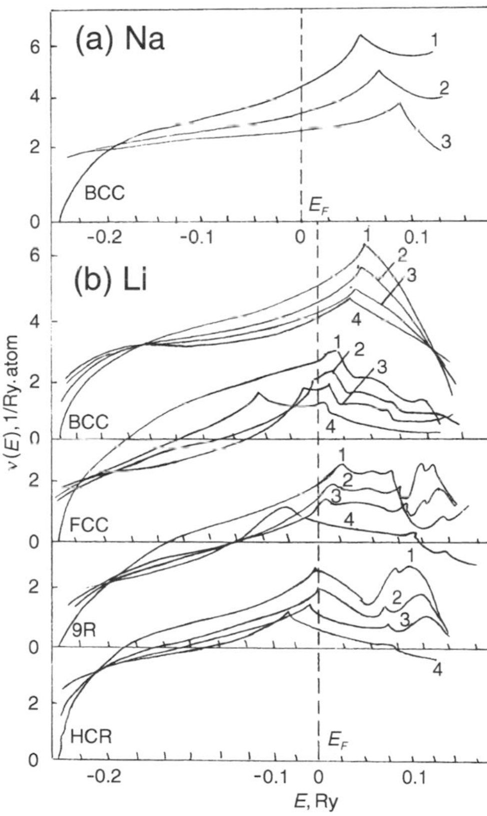

FIGURE 4.7 Electronic density of states $\nu(E)$ for the BCC phases of (a) Na and (b) Li calculated using the LMTO-ASA method. The values E are shown relating to the Fermi level. Curves 1, 2, 3 for Na correspond to the values of the compression $u = 0$; 0.2; 0.5, respectively; curve 1, 2, 3, 4 for Li – to $u = 0$; 0.2; 0.4; 0.6 (after [33, 46]).

The calculated density of states $\nu(\varepsilon)$ under compression u (Figure 4.7) has qualitatively different forms for Na and Li. For BCC Na, the difference $\Delta_{CF} = E_C - E_F$ between E_F and the singularity (peak) of the function $\nu(\varepsilon)$, E_C, is remarkably larger than for Li (although the free electron approximation provides $\Delta_{CF}(\text{Li}) \cong 1.5\Delta_{CF}(\text{Na})$), and the increase of u provides increasing Δ_{CF}, i.e., E_C moves to the right of E_F. The same effect takes place for FCC, HCP, and 9R phases of Na. However, for Li the value E_C moves under compression to the left of E_F. Hence, in the BCC phase an "overlapping" of E_F and the peak of $\nu(\varepsilon)$ takes place. This effect is accomplished by non-analytical "band structure" contributions to the energy which, at small Δ_{CF}, result in increasing E^{bs} ([33]). On the other hand, increasing u in CP phases of Li moves the peaks to the left of E_F. That corresponds to a touch of the Fermi surface by the face of the BZ and a decrease of the energy. These band-structure effects (which were not taken into account in calculations using the perturbation theory, see subsection 2) explain the increase of the energy of the BCC phase of Li (in comparison with other CP phases) with compression (see Figure 4.6.) On the other hand, for Na, where there are no such band-structure effects, the characteristic features of the dependence of ΔE_{Bi} on u are qualitatively the same as for the LMTO as for the perturbative calculations.

The difference in behavior of $\Delta_{CF} = E_C - E_F$ as a function of compression between Li and Na relates to a different role of the electron–ion interaction. For Na, this interaction is much weaker than for Li (see Chapter 3) and, in the first approximation, the dependence of Δ_{CF} on u has to be the same as for free electrons: $\Delta_{CF} \sim (\Omega_0/\Omega)^{2/3}$. But for Li, the interaction of p-electrons with the nucleus is much stronger and has a resonance-like character. The states corresponding to E_C (e.g. for the BCC phase, E_C is located near the point N in the BZ) are bonded p-type orbitals [1, 31]. For this reason, E_C increases with compression much slower than E_F and the difference Δ_{CF} decreases.

Let us compare the results of [33, 46] with other band structure calculations of ΔE_{ij} for Li and Na. The dependences $\Delta E_{BF}(u)$ and $\Delta E_{HF}(u)$ (Figure 4.6) are qualitatively similar to the results of the most accurate LMTO calculations [16] (for Na at $u \ll 0.3$). However, the band-structure calculation with *ab initio* pseudopotential [29] yield a decrease of ΔE_{BF} and ΔE_{BH} for Li with increasing u, at least up to $u \simeq 0.2$. But, if the energy of the 9R-phase observed in Li, as well as those of other CP phases, were close together (this follows from all calculations shown in Figure 4.6, as well as from simple geometrical reasons), then the decrease obtained in [29] would contradict the experimental phase diagram of Li (see Figure 2 in [18]). Moreover, the values of ΔE_{BH} calculated in [29] become 10–15 times larger than the values obtained from experimental data for M_d (see above). Hence, the accuracy of the method used in [29] were insufficient for calculations of $\Delta E(u)$ for both Li and Na. On the other hand, the results of cluster band-structure calculations of $\Delta E(u)$ for Li [30] disagree with results [33, 46] as well as with [16] and [29].

The results discussed in this subsection give the conclusion that the band structure effects clearly influence the martensitic phase transitions in Li under pressure. These effects result in decreasing structural stability as E_F approaches peaks of the density of states. Earlier, these effects were discussed for Cs under pressure in [16] and for

NiTi-based alloys in [62]. Below we, following [47], will discuss another demonstration of the band-structure effects, in particular for Li.

4.1.4 Pre-martensitic softening of shear moduli for alkali metals under pressure

The effects of peculiarities of the band structure, for example, the proximity of the Fermi level to singularity points of the density of states, upon the elastic moduli were discussed in detail [33, 47, 48]. In particular, there exist experimental indications on the influence of the band-structure effects on the shear moduli $C_{ss'}$. A well known example is the anomalies in the concentration and temperature depend-ence of $C_{ss'}$ in BCC transition metals, which were considered by many authors [62 66] to be the effect of band structure. There were many discussions of the "pre-martensitic anomalies" (i.e. appreciable softening of the moduli $C_{ss'}$ in many metals and alloys with the approach towards the points of structural phase transitions by varying either pressure or concentration [66–69]) which also relate to the band-structure effects [47, 48].

The progress reached for *ab initio* methods of the band structure and the total energy calculations [54, 60, 67] allows us to provide effective and consistent calculations of some volume characteristics of metals like the pressure p and the elastic modulus B. On the other hand, even today there are no simple enough methods for calculating the shear moduli $C_{ss'}$. A few moduli $C_{ss'}$ were calculated using the *ab initio* pseudopotentials [68, 69] or "not-muffin-tin correction" in the framework of the LMTO approach [70]. However, all methods used are very cumbersome and less suitable for applications and generalizations.

In qualitative discussions of the influence of the band-structure effects on $C_{ss'}$ in metals (see [33, 47–48, 63–65]), some model approximations were used. Actually, the starting point was the definitions (1.74) which at $T = 0F = E$ provide the next formulae for the energy and the elastic modulus

$$E = E_{bs} + E_{es}, \quad C_{ss'} = C_{ss'}^{bs} + C_{ss'}^{es}, \tag{4.17}$$

where E_{bs} is the "band-structure" energy which is the sum of the one-electron energy $E_{\mathbf{k}}$ over occupied states, E_{es} is the sum of all other contributions to the energy called, briefly, the electrostatic term. The next supposition was that these band structure anomalies would be determined by the "band structure" contribution $C_{ss'}^{bs}$ only, whereas for $C_{ss'}^{es}$ these anomalies would be negligible. In [63, 65], the terms $C_{ss'}^{es}$ were not considered at all, only the contributions $C_{ss'}^{bs}$ were calculated (in the frame of the tight-binding method within the "rigid–band" approximation [31]). Bujar *et al.* [64] used an approximation of $C_{ss'}^{es}$ by some smooth function of the volume, $A_{ss'}f(\Omega)$, with adjustable coefficients $A_{ss'}$. The supposition about negligibility of the band-structure effects for the moduli $C_{ss'}^{es}$ seems to be plausible and, in some cases, it even can be formally justified. For example, if these effects relate only to the proximity of the Fermi level E_F to E_C (the van Hove singularity point for the density of states $\nu(E)$), i.e. if the parameter $\eta = (E_F - E_C)/E_C$ is small, then the leading singularity in η term provides the contribution $C_{ss'}^{sing} = \text{const}\,|\eta|^{1/2}$ [71] to $C_{ss'}^{bs}$, whereas singularities of $C^{es}_{ss'}$ are not stronger than $|\eta|^{3/2}$. In [72], supposing the cell potential to be "frozen" under shear deformations [73], an attempt was made to achieve a more formal foundation of the approximation $C^{es}_{ss'} = A_{ss'}f(\Omega)$ used in the calculation of the band contribution to $C_{ss'}$ in the framework of the LMTO method. However, more detailed analysis of these calculations provided in [70] has shown that this approximation for $C^{es}_{ss'}$ is not well-founded.

In the present subsection we, following [47, 48], will consider the dependence of $C_{ss'}$ on the compression $u = (\Omega_0 - \Omega)/\Omega_0$ (where Ω_0 is the value of Ω at $p = 0$) for the BCC and HCP phases of Li, Na, Cs — in full detail, as well as for Ca, Sr, and Ba — only the main interesting features (detailed consideration see in [33, 46, 47]). Analyzing these examples, we want to consider the band structure contributions to $C_{ss'}(u)$ for simple metals for which these effects play an apparently greater role than for BCC *d*-metals discussed in [33].

We consider also some peculiarities of the anomalous band-structure contribution to $C_{ss'}$ for the case when the variation of some external parameter (we will consider the pressure, but for alloys it can be a composition, too) does not correspond to the rigid-band model, but to essential change of the form of $\nu(E)$, in particular, near E_F.

It is interesting also to establish the relationship between anomalies of the band-structure contribution to the shear moduli and the structural phase transitions, as well as a microscopic mechanism of the "pre-martensitic softening" in these modules.

Here we predominantly describe qualitative effects. In calculations of [33, 47–49], which we follow, approximations similar to those described above have been used. The band structure contributions $C^{bs}_{ss'}$ in (4.17) were calculated using the LMTO–ASA method and the "frozen potential" approximation. The values $C^{es}_{ss'}$ were taken to be proportional to the contribution $C^{Mad}_{ss'}$ to the shear moduli provided by the Madelung energy of point-like ions immersed in a uniform compensating background charge:

$$C^{es}_{ss'} = AC^{Mad}_{ss'}.$$

As was shown in [33] (see below), taking the same value $A = 5/3$ for all metals considered in that work actually results in good agreement between calculated and experimental values of $C_{ss'}$ at $u = 0$ as well as available data about the dependence of the averaged shear modulus G on pressure. However, there is not a foundation for this universal interpolation for $C^{es}_{ss'}$. Nevertheless, relatively simple calculations [47–49] of $C^{bs}_{ss'}$ based on the LMTO–ASA method in combination with the interpolation mentioned for $C^{es}_{ss'}$ can be useful not only for qualitative investigations of the dependence of $C_{ss'}$ on external parameters but also for semi-quantitative evaluations of absolute values of $C_{ss'}$, apparently, for some classes of metals and alloys only. We take this assumption to be a working hypothesis and will use the results of works [33, 47–49].

Method and approximations The electronic band structure was calculated for each value of the atomic volume Ω using the self-consistent LMTO-method within the atomic sphere approximation (ASA) [54]. In the basis of orthogonal MT-orbitals, the effective two-center Hamiltonian determining the energy spectrum of the crystal [74] is:

$$H_{lm,l'm'}(\mathbf{k}) = C_l + \Delta_l^{1/2}\tilde{\tilde{S}}_{lm,l'm'}(\mathbf{k})\Delta_{l'}^{1/2}(\mathbf{k}),$$
$$\tilde{\tilde{S}}(\mathbf{k}) = \tilde{S}(\mathbf{k})\left[1 - \gamma_l\tilde{S}(\mathbf{k})\right]^{-1}. \tag{4.18}$$

Here C_l and Δ_l are the parameters characterizing the center and the width of the non-hybridized l-band, $S(\mathbf{k})$ is the matrix of structural constants of the LMTO method and γ_l are the deformation parameters of the energy bands. The Hedin–Lundquist expression [74, 75] for the exchange-correlation potential was used. The pressure $p(\Omega)$ (shown in the figures below in upper scale) was calculated using the Pettifor formula [59]. The equilibrium volume $\Omega_0 = \Omega\ (u = 0)$ was taken to be 143.4, 254.5, 747.7, 293.2, 379.1, and 426.9 a.u., for Li, Na, Cs, Ca, Sr, and Ba respectively. For alkali metals, these values correspond to $T = 0$, for alkaline-earth metals to the room temperature.

In calculations of the shear modules $C_{ss'}$ authors of [33.47], as well as [70], have started from the forces theorem [76, 77]. In accordance with this theorem, the generalized forces F_s, corresponding to the deformation u_s, can be written (in the framework of the local density functional approximation) as

$$F_s = \frac{\partial E}{\partial u_s} = \sum_i \theta(-\xi_i)\left[\frac{\partial \varepsilon_i'}{\partial u_s}\right]_V + \int d^3r\,d^3r'\rho(\mathbf{r})\rho(\mathbf{r}')\frac{\partial}{\partial u_s}\frac{1}{|\mathbf{r}-\mathbf{r}'|}$$
$$= F_s^{bs} + F_s^{es}. \tag{4.19}$$

Here $\xi_i = \varepsilon_i - E_F > \varepsilon_i$ are one-electron energy levels for the self-consistent crystal potential $V(\mathbf{r})$, $\theta(x)$ is the step function (is equal 1 for $x > 0$ and 0 for $x < 0$), and $(\partial\varepsilon_i/\partial u_s)_V$ means the derivative of ε_i calculated for the "frozen" (not changing with deformation) potential $V(\mathbf{r})$. In accordance with the Hellman–Feynmann theorem (see, e.g., [58]) this derivative can be also expressed through the derivative of the kinetic energy operator $\hat{T}$:

$$\left[\frac{\partial \varepsilon_i'}{\partial u_s}\right]_V = \left\langle i\left|\frac{\partial \hat{T}}{\partial u_s}\right|i\right\rangle. \tag{4.20}$$

The value $\rho(\mathbf{r})$ in the second term in (4.19) is the total charge density of nuclei and electrons calculated for the ground state of the metal.

Hence, in accordance with (4.19) and (4.20), the force F_s is the sum of two contributions with different physical natures: the "band-structure" term F_s^{bs} relating to the variation in the kinetic energy of electrons in the Fermi-region under the deformation u_s and the "electrostatic" term F_s^{es} to be equal to the derivative of the classical electrostatic energy with respect to u_s at constant density $\rho(\mathbf{r})$.

The elastic modulus $C_{ss'}$ at $T = 0$ is determined by the second derivative $\partial^2E/\partial u_s\partial u_{s'}$ (see Chapter 1). For this value there is no simple separation of contributions similar to (4.19) as a consequence of a self-consistent nature of the Kohn-Sham levels (see [58]), as well as the fact that, in addition to the terms C_s^{bs} with derivatives at constant V and to the "electrostatic" contribution $C^{es}_{ss'}$, there appear also the terms $C^{m}_{ss'}$ provided by mixed derivatives of the kinetic and potential energies of electrons:

$$C_{ss'} = C^{bs}_{ss'} + C^{m}_{ss'} + C^{es}_{ss'}; \tag{4.21}$$

$$C^{bs}_{ss'} = \frac{1}{N\Omega}\sum_i \left[\theta(-\xi_i)\left(\frac{\partial^2 \varepsilon_i}{\partial u_s\, \partial u_{s'}}\right)_V - \delta(\xi_i)\left(\frac{\partial \xi_i}{\partial u_s}\right)_V\left(\frac{\partial \xi_i}{\partial u_{s'}}\right)_V\right]; \tag{4.22}$$

$$C^{m}_{ss'} = \frac{1}{N\Omega}\sum_i \left[\theta(-\xi_i)\int d^3r\, d^3r' \varphi_i(\mathbf{r})\frac{\partial \hat{T}}{\partial u_s}\left(\frac{\delta\varphi_i(\mathbf{r})}{\delta V(\mathbf{r}')}\right)\frac{\partial V(\mathbf{r}')}{\partial u_{s'}} - \delta(\xi_i)\left(\frac{\partial \hat{T}}{\partial u_s}\right)_{ii}\left(\frac{\partial V}{\partial u_{s'}}\right)_{ii}\right]; \tag{4.23}$$

$$C^{es}_{ss'} = \frac{1}{N\Omega}\frac{\partial F^{es}_s}{\partial u_{s'}}. \tag{4.24}$$

In the frame of the LMTO–ASA method, the derivatives $(\partial\varepsilon_i/\partial u_s)_V$ and $(\partial^2\varepsilon_i/\partial u_s \partial u_{s'})_V$ in (4.21) are determined only by variations of the structural constants $\tilde{\hat{S}}(\mathbf{k})$ in the Hamiltonian (4.18) with the deformations u_s. Then, the parameters C_l, Δ_l, γ_l of the potential do not change; their values correspond to a self-consistent calculation for non-deformed crystal. This fact allows relatively simple calculations of these derivatives using numerical differentiation of the values ε_i obtained by the diagonalization of the Hamiltonian $\hat{H}$ (4.18) in which $\hat{S}(\mathbf{k})$ corresponds to a deformed lattice. On the other hand, evaluations of mixed derivatives in (4.22) need, generally speaking, self-consistent calculations for a deformed lattice [78], i.e., with no actual approximations of the LMTO–ASA method. In particular, such calculations must take into account the non-spherically-symmetric components of the potential. This fact results in remarkably complicated calculations. However, such self-consistent calculations for Pd and Au [70] have shown that for the modulus C' considered by Chistiancen the contributions $C^{m}_{ss'}$ are negligibly small in comparison with $C^{bs}_{ss'}$. This fact reflects relatively small values of contribution to $\partial\varepsilon_i/\partial u_s$ provided by non-spherical-symmetric components of the potential $\partial V/\partial u_s$ in comparison with contributions $\partial\hat{T}/\partial u_s \sim \partial S(\mathbf{k})/\partial u_s$. Moreover, since the main purpose of works [33, 47] was to consider qualitative effects then, in calculations discussed below, the contribution $C^{m}_{ss'}$ is omitted.

Quantitative calculations of the electrostatic contribution $C^{es}_{ss'}$ in (4.22)–(4.24) also need appreciable efforts [70]. But, as was already mentioned, general physical considerations show that it is possible to suppose (according to [33, 47]) that the band-structure anomalies for $C_{ss'}(u)$ are predominantly determined by the contribution $C^{bs}_{ss'}$, whereas the contribution $C^{es}_{ss'}$ can be considered as a smooth function of the volume Ω. In the same manner as in [72], we suppose that

$$C^{es}_{ss'} = AC^{Mad}_{ss'}(Z,\Omega), \quad C^{Mad}_{ss'} = b_{ss'}Z^2e^2\Omega^{-4/3}, \tag{4.25}$$

TABLE 4.9 Values for the coefficients $b_{ss'}$ in (4.25)

structure	BCC	FCC
$b' = \frac{1}{2}(b_{11} - b_{12})$	0.03958	0.03333
b_{44}	0.29462	0.29870

where $C_{ss'}^{Mad}$ is the "Madelung" contribution, A is some empirical constant, Z is the valence, $b_{ss'}$ are some numerical coefficients depending on the lattice geometry only.

For the moduli $C' = \frac{1}{2}(C_{11} - C_{12})$ and C_{44}, consided below, the values of these coefficients for the BCC and HCP structures are listed in Table 4.9 (taken from [33]).

In calculations of the band structure contributions $C_{ss'}^{bs}$, a technique similar to that described in [65] was used. The modulus C' corresponds to the tetragonal deformation η_1, and C_{44} to the trigonal deformation η_2. The corresponding deformation matrices $\hat{a}_1$ and $\hat{a}_2$ have the form

$$\alpha_1 = \begin{pmatrix} \zeta_1^{-1/3} & 0 & 0 \\ 0 & \zeta_1^{-1/3} & 0 \\ 0 & 0 & \zeta_1^{2/3} \end{pmatrix}; \quad \hat{\alpha}_2 = \frac{1}{3}\zeta_2^{-1/3} \begin{pmatrix} 2+\zeta_2 & 1+\zeta_2 & 1+\zeta_2 \\ 1+\zeta_2 & 2+\zeta_2 & 1+\zeta_2 \\ 1+\zeta_2 & 1+\zeta_2 & 2+\zeta_2 \end{pmatrix}, \tag{4.26}$$

where $\zeta_i = (1 + \eta_i)^{-1}$. Explicit expressions for the band-structure contributions (4.22) to the moduli C_α ($\alpha = 1, 2,\ C_1 = C',\ C_2 = C_{44}$) have the form

$$C_\alpha^{bs}(E_F) = C_\alpha^{(1)} + C_\alpha^{(2)} = \frac{3}{4}\sum_n d^3k \left[\theta(E_F - \varepsilon_{n\mathbf{k}}) \frac{\partial^2 \varepsilon_{n\mathbf{k}}}{\partial \eta_\alpha^2} - \delta(\varepsilon_{n\mathbf{k}} - E_F)\left(\frac{\partial \varepsilon_{n\mathbf{k}}}{\partial \eta_\alpha}\right)^2\right]. \tag{4.27}$$

Here $\varepsilon_{n\mathbf{k}}$ is the energy corresponding to the state with the quasi-momentum $\mathbf{k}$ in the nth band, and $C_\alpha^{(1)}(E_F)$ and $C_\alpha^{(2)}(E_F)$ correspond to the first and the second terms in the square brackets in (4.27), i.e., the contributions from states inside the Fermi-surface and on the Fermi surface, respectively. If we replace the value E_F in (4.27) by some variable E then the function $C_\alpha^{(1)}(E)$ becomes similar to the density of states $\nu(E)$ (the only difference is the factor $-\frac{3}{4}(\partial\varepsilon_{n\mathbf{k}}/\partial\eta_\alpha)^2$ in the integrand). Now the value $C_\alpha^{(2)}(E)$ is similar to the number of states $N(E)$:

$$N(E) = \int_0^E d\varepsilon' \nu(\varepsilon'). \tag{4.28}$$

In this event, all characteristic features of singularities of the functions $C_\alpha^{(1)}(E)$ would be the same as, for example, the van Hove singularities of $\nu(E)$. The same similarity takes place for the functions $C_\alpha^{(2)}(E)$ and $N(E)$ (see Figure 4.8).

In calculation of C' in [33, 47], the body-centered-tetragonal (BCT) lattice was used for which the values of the tetragonal parameter $c/a = 1$ and $c/a = \sqrt{2}$

correspond to the BCC and FCC lattices, respectively. The modulus C_{44} was calculated for the trigonal lattice, for which $c/a = (1/4)\sqrt{6}$ and $c/a = \sqrt{6}$ correspond to the BCC and FCC structures, respectively. The self-consistent potential $V(\mathbf{r})$ was calculated for each value of the volume Ω of the unit cell of the non-deformed BCC or FCC lattice. The first and the second derivatives with respect to η_α in (4.27) were calculated applying the five-point numerical differentiation formula to the band-structure energy $\varepsilon_{n\mathbf{k}}$ with the increment $\Delta(c/a) = 0.0701$ and 0.02. The changes of $\varepsilon_{n\mathbf{k}}$ with lattice deformations were calculated taking into account the "symmetry" of the splitting of the one-electron spectrum and the requirement of the continuous variation of the wave functions by small deformations of the crystal. The number of mesh-points within the irreducible part of the Brillouin zone (BZ) by numerical integrations via the method of tetrahedra was chosen to be 104 for the BCC and FCC, 244 for the BCT, and 288 for the trigonal structures (in all cases this number corresponds to 3375 mesh-points within the whole BZ).

For alkali and alkaline-earth metals considered in [33, 47], taking the same value $A = 5/3$ actually provides good agreement with available experimental data. Let us consider some foundations for the possibility of satisfying approximate relations like (4.25) with $A = 5/3$. The value E_{es} in (4.17) is the difference between the Madelung energy of ions E^{Mad} and the energy of the electron–electron interaction (this last term was taken doubly into account in E_{bs}):

$$E_{es} = E^{Mad} - U_{ee}$$

If the term $C^{bs}_{ss'}$ were the full derivative $\partial^2 E_{bs}/\partial u_s\,\partial u_{s'}$ of the band structure energy E_{bs} in equation (4.17) (but not the derivative at constant potential, as has factually been made in (4.22)) and if calculations of $C^{es}_{ss'} = \partial^2 E_{es}/\partial u_s\,\partial u_{s'}$ in (4.17) were performed only for the electrostatic contribution U_{ee} to U^{es}_{ee} (neglecting the exchange-correlation contributions U^{xc}_{ee}), then the value $A = 5/3$ in (4.25) would correspond to the following relation

$$\frac{\partial^2 E_{es}}{\partial u_s\,\partial u_{s'}} = \frac{\partial^2}{\partial u_s\,\partial u_{s'}}\left(E^{Mad} - U^{es}_{ee}\right) = \frac{5}{3}\frac{\partial^2}{\partial u_s\,\partial u_{s'}}E^{Mad}. \tag{4.29}$$

To evaluate U^{es}_{ee} and E^{Mad} in (4.29), we can use a well-known formula for the electrostatic energy of a neutral Wigner–Seitz sphere with a constant electron density Z/Ω and a point-like charge Z in its center [79]. It gives

$$E^{Mad} - U^{es}_{ee} = \frac{5}{3}E^{Mad}. \tag{4.30}$$

For this reason, if there were not any differentiation in (4.29), an approximate validity of this equality would be rather natural. However, it is impossible to prove similar relations for derivatives participating in (4.29) as well as to motivate the supposition about negligible values of terms with $\partial V/\partial u_s$ and U^{xc}_{ee} made in [47]. Nevertheless, a supposition about the approximate validity of relations like (4.25) with $A = 5/3$ is possible and works well.

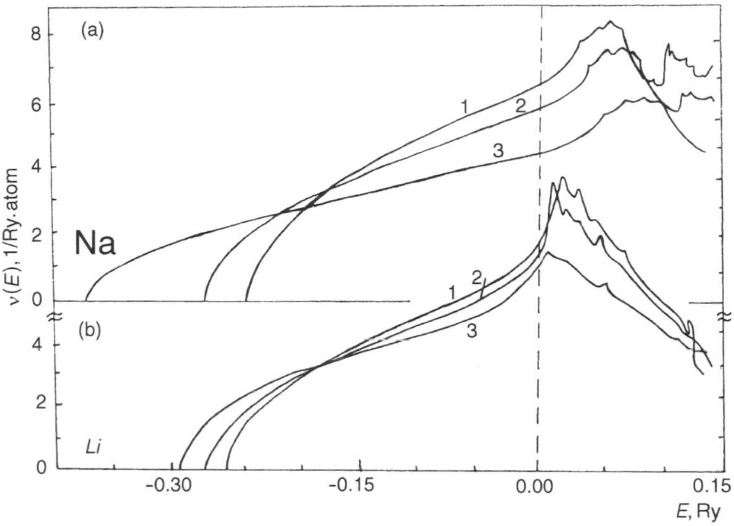

FIGURE 4.8 Electronic density of states $\nu(E)$ for BCC phases of Na (a) and Li (b). The energy is shown relating to the Fermi level. Curves (a), (b), (c) correspond to the values of the compression $u = 0, 0.2, 0.5$, respectively (after [33]).

Variations of the Electronic Structure under Compression The results of calculations of the total electron density of states $\nu(E)$ and the partial densities $\nu_l(E)$ for s-, p- and d-electrons for different values of the compressions u for alkali and alkaline-earth metals have been presented in [33, 47]. We will not discuss here these results in detail and refer the reader to these original papers. Here, we consider briefly the results for lithium and sodium only. The dependences $\nu(E, u)$ for BCC Li and Na (Figure 4.8) are shown here for comparison. These metals have simple band structures [80, 81] constructed from s- and p-states, with a peak of $\nu(E)$ at point E_C, corresponding to the van Hove singularity relating to the point N within the BZ. Increasing the compression increases the difference $\Delta_{CF} = E_C - E_F$ for Na in accordance with a nearly-free-electron (NFE) character of the band structure: $\Delta_{CF} \sim \Omega^{-2/3}$. Hence, the states with $E = E_C$ remain non-occupied and the respective singularity in $\nu(E)$ plays no role for sodium. On the other hand, an intensive resonance-like attraction of p-electrons in Li, which corresponds to point N, results in a relatively weak dependence of the values E_C on the compression. But, simultaneously, the energies of NFE s-states, which play a dominant role in E_F, increase with compression approximately as $(\Omega_0/\Omega)^{2/3} = (1 - u)^{2/3}$. As a result, the value Δ_{CF} for the BCC Li decreases with increasing u and the Fermi-level overlaps the peak of $\nu(E)$. This behavior results in decreasing band-structure energy of the BCC phase relating close-packed phases and provides structural phase transitions to one of these close-packed phases (see subsection 2). This "band-structure" contribution to the structural instability can be observed also as an appreciable softening of shear modules by approaching E_F to E_C.

The influence of the compression on the band structure of cesium was discussed by many authors (see e.g. [16] and references therein). The results of [33, 46–48] agree in general with those of previous calculations. They demonstrate the general feature of the change of the band structure under compression (which is also a characteristic feature of alkaline-earth metals): decreasing the energy of resonant *d*-states with respect to that of *s*- and *p*-states. This behavior is similar to decreasing the energy of *p*-states relative to *s*-states in lithium, discussed above. For cesium, the Fermi level moves with increasing u into the *d*-band and, as a consequence, structural phase transitions take place (see section 2). The value of the Fermi energy E_F at $u \geq 0.3$ for FCC Cs lies within the pseudo-gap which corresponds to the point X in the BZ. Further increase of u, considered in [16], results in the actual "band-structure" mechanism (like that suggested by Jones [82]) of the transition from the BCC to FCC phase, which have been observed experimentally at $u = u_C^{exper} \cong 0.4$ [83].

Before we start the discussion of the phase transition from BCC to FCC phase of Cs, let us show that the change of $\nu(E)$ results in drastic anomalies in behavior of the shear moduli $C_{ss'}$ under compression, especially near the points of structural phase transition.

Changes of shear moduli with pressure Figures 4.9–4.10 show the results of calculations [33, 47] of the dependences $C_{ss'}(u)$ for the BCC or FCC phases of Li, Na, Cs, as well as available experimental data. Experimental values for $C_{ss'}$ at $u = 0$ for alkali metals correspond to $T = 0$ (see Chapter 3).

Let us discuss the results obtained in [33, 47]. Figures 4.9–4.10 show that at $u = 0$ the calculated values of $C_{ss'}$ are in good agreement with experimental data. However, the results for BCC cesium obtained for $A = 5/3$, $Z = 1$ in (4.25) disagree with

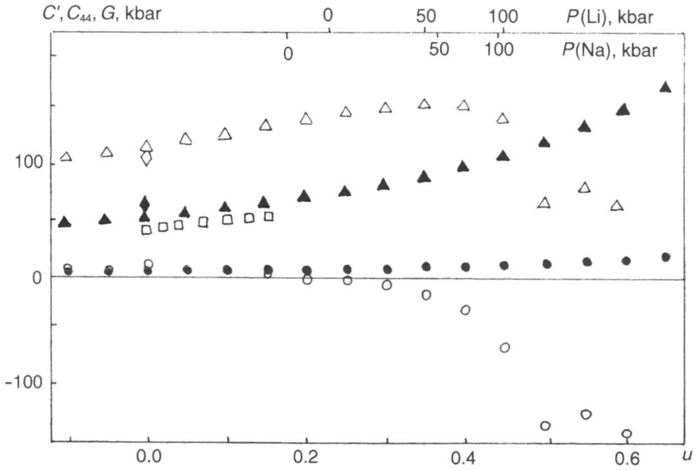

FIGURE 4.9 Shear moduli for the BCC phases of Li (open symbols) and Na (heavy symbols) against compression u. Results of calculations: triangles – C_{44}, circles – C'. Experimental: rhombs – C_{44} at $u = 0$; squares – $G(u)$ for polycrystalline samples [84] (after [33, 47]).

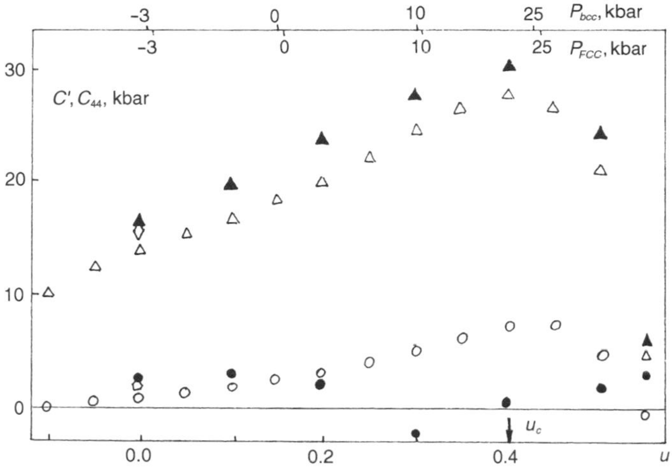

FIGURE 4.10 Shear moduli C' and C_{44} for Cs against compression u: open symbols – the BCC phase, heavy symbols – FCC. Other notations are the same as in Figure 4.9 (after [33.47]).

experiment: instead of experimental values $C_{44} = 16$ kbar and $C' = 2.2$ kbar, calculated values are 8 and 0.3, respectively. However, taking into account a non-point-like structure of the electron core of cesium and using in (4.25) instead the nominal valence $Z = 1$, for example, the value $Z_0 = 1.123$ (estimated in [72] using the relation for the electron density in the interstitial space) provides $C_{44} = 14$ and $C = 1.1$ kbar. If we take into account the large degree of cancellation of contributions C'_{es} and C'_{bs} in cesium ($C'_{bs} = -2.6$ kbar), then this last value of C' can be considered to be close enough to experimental data. For this reason, to provide a more realistic description of $C_{ss'}$ for Cs in calculations shown in Figure 4.10, the values $C^{es}_{ss'}$ in (4.25) were calculated for $A = 5/3$ and $Z = Z_0 = 1.123$.

A comparison of the calculated averaged shear modulus $G(u)$ at $u \neq 0$ with experimental data for polycrystalline samples of Li (see Figure 4.9) is difficult because of the well-known problem of a vaguely-determined relation between G and the values C' and C_{44} calculated for an ideal crystal. However, note that calculated values of the compression u_c which corresponds to rapidly decreasing or vanishing of one or both shear moduli C_{ss} (Figure 4.9) takes place, as a rule, in the vicinity of the values of u_c corresponding to experimentally observed phase transitions. The value $u_c \cong 0.25$ obtained for lithium (with re-calculation to the value $\Omega_0 = \Omega(T = 295\,\mathrm{K})$) becomes somewhat smaller than the observed value $u_c \cong 0.3$. This fact reflects, apparently, a general under-estimation of the value C' in the case of calculations for lithium. This effect has a remarkable value at $u = 0$, too. But, since the stability of phases in lithium depends on thermal effects, which are not considered in our analysis, this metal is not a very suitable object for such evaluations of the structural stability.

To illustrate the relative importance and characteristic features of both contributions $C^{(1)}_{ss'}$ and $C^{(2)}_{ss'}$ to the total band structure contribution $C^{bs}_{ss'}$ (4.27), note that often the "Fermi-surface" contributions $C^{(2)}_{ss'}$ are the main ones. These terms vary with E similar (but with the opposite sign) to the density of states $\nu(T)$ shown in Figure 4.8 as follows from the analogy in behavior of these values mentioned above. However, the variations of the functions $C^{(2)}_{ss'}$ (for example, near the singularity point E_c of $\nu(E)$) are actually more rapid than for $\nu(E)$. This fact determines the possibility to observe these "band-structure" anomalies in the behavior of the shear moduli.

The dependences $C_{ss'}(u)$ for alkali metals For sodium, these dependences (Figure 4.9) are close to results of earlier calculations performed in the frame of the pseudopotential perturbation theory [85–89]. For $u \leq 0.1$, they are also close to the experimental data for $C_{ss'}(u)$ from [85, 90]. All these moduli smoothly and monotonically increase with compression reflecting an absence of singularities in $\nu(E)$ at $E < E_F$ (Figure 4.8). On the contrary, in the BCC lithium, E_F approaches to the value E_c corresponding to a peak in $\nu(E)$ with increasing compression. This effect provides a decrease of the structural stability which can be observed as a rapid decrease of the modulus C' corresponding to a phase transition to close-packed phases. This decrease of C' in lithium is totally determined by the "Fermi-surface" band-structure contribution $C'^{(2)}$ (Table 4.10). In the case of a simple van Hove singularity considered here this contribution obeys the most rapid change with decreasing $E_F - R_c$. For the modulus C_{44} of lithium, these "band structure" contributions to the softening at small $u \leq 0.25$ is much weaker although they can be observed (Figure 4.9) as a negative curvature of the function $C_{44}(u)$ (in contrast to sodium where this curvature is positive). These band-structure effects also explain a well-known over-estimation of the values of the derivatives $\partial C_{ss'}/\partial p$ for Li obtained in calculations using the pseudopotential perturbation theory (in factor three for $\partial C'/\partial p$ and 1.5 for $\partial C_{44}/\partial p$, see Chapter 3). For all other alkali metals for which these band-structure effects at $p = 0$ are small, such calculations actually provide good agreement with experiment (see Chapter 3).

The dependences $C_{ss'}(u)$ for Cs, shown in Figure 4.10, differ in some details from those obtained in [33, 47] for other metals. By approaching the compression u for the BCC phase to the value $u_c \cong 0.4$, which corresponds in experiment to the transitions to the FCC phase, the moduli $C_{ss'}$ increase monotonically without any indication of softening. This fact reflects a smooth variation of $\nu(E)$ and $C^{bs}_{ss'}$ for the

TABLE 4.10 Different contributions to the modulus $C'(u)$ (in kbar)

metal, structure	$\frac{\Omega_0 - \Omega}{\Omega_0}$	$C'^{(1)}$	$C'^{(2)}$	C'_{bs}	C'_{es}	C'
Li	0	51	−57	−6	12	6
BCC	0.2	72	−81	−8	16	8
Na	0	97	−115	−18	26	8
BCC	0.2	134	−167	−38	35	2
	0.3	163	−211	−48	42	−7

BCC phase at $u \leq u_c$ [47]. At the same time, for the FCC phase, the decrease of the compression from $u \sim 0.5$ to $u \leq u_c$ provides a rapid decrease of the modulus C' up to large negative values at $u \cong 0.3$. This rapid decrease is a typical illustration of the "band-structure" softening of C' which relates to passing E_F through the peak of $\nu(E)$ corresponding to the point E in the BZ. Hence, in accordance with results of calculations [47], the phase transition in the BCC phase of cesium under pressure is an example of the phase transition without a pre-martensitic anomaly (in terms of the classification submitted, for example, in [66]), whereas such anomalies must be strictly expressed by an inverse phase transition from the FCC to BCC phase. An experimental test of this asymmetry of the pre-martensitic anomaly would be very important.

Further increase of the compression for FCC cesium provides (at $u_{c2} = 0.56$) an iso-structural phase transition (the "*s–d* transition"). This transition is accomplished by a large change in the volume $\Delta u = 0.04$ (see, e.g., [83] and the next section). A non-trivial prediction of calculations [33, 47] is appreciable softening of the modulus C_{44} by approaching u to the value u_{c2}, mentioned above (see Figure 4.10). Since this softening of C_{44} is a band-structure effect relating to the overlapping of E_F by the *d*-peak in the density of states, this effect cannnot be very sensitive to approximations used in calculations.

For higher compressions, $u \geq 0.55$–0.6, corresponding to the structure after the *s*–*d* transition, it was impossible to provide a calculation of the band structure and $C^{bs}_{ss'}$ for cesium. At such compressions (as well as at $u \geq 0.4$ for Ba or at $u \geq 0.55$ for Sr and Ca) there exists a remarkable overlap of the valence electronic states considered here with "core" states as in real space, as in energy. For this reason, approximations actually used in the framework of the LMTO–ASA method, in particular, the "frozen" core approximation, become useless. On the other hand, "de-freezing" some part of core states performed in [33] in calculations, where the valence band also includes the 5*p*-states of cesium, results in large difficulties relating to the stability of self-consistent calculations. Maybe these methodical difficulties reflect a remarkable change of core states at large $u > 0.55$, in particular, due to phase transitions taking place in this region. However, for $u \leq 0.55$ these methods are apparently applicable. For this reason, softening of the modulus C_{44} at large $u \geq 0.5$ for cubic phases of alkaline earth metals and cesium predicted in [47] (the same effect, possibly, takes place for potassium and rubidium) can be considered as a common feature of the band-structure contribution for all these phases. This effect demonstrate a general trend to undergo phase transitions to some complicated structures like, for example, Cs–IV.

4.2 Polymorphic BCC–FCC Transitions in Heavy Alkali Metals

A quantitative description of polymorphic phase transitions under pressure is one of the most important but very difficult problems of the microscopic theory of metals. The complexity of the problems relates to at least three important features:

1) relatively small differences in the energy between different phases (10^{-4}–10^{-5});
2) the necessity of providing total calculations of the band structure (but not in the second order in the pseudopotential, see the preceding section);
3) the necessity of considering an influence of core states as well as non-adiabatic effects.

For this reason, even preliminary calculations of polymorphic transitions under pressure and evaluations of regions of stability of different phases for some metals or for groups of similar metals would by very important.

Properties of heavy alkali metals K, Rb, and Cs under pressure, in particular, equations of state and p–T phase-diagrams have been investigated experimentally (see reviews [33–35]). Some important results have been established: the similarity of the equations of state, and existence of polymorphic phase transitions under pressure. In particular, measurements of the change of electrical resistivity or volume of K, Rb, and Cs under pressure show some phase transitions, the structures of which have not be completely explained. For Cs there exist five such phase transitions under pressure which are used as reference points for measurements of the pressure [35].

There were many attempts to describe theoretically (predominantly phenomenologically) the properties of these metals under pressure. However, there are only a few works relating to the microscopic theory of the influence of the pressure on ideal crystals of alkali metals (see section 1). Although calculations of the zero-temperature pressure $p(\Omega, 0)$ as well as the isotherms $p(\Omega, T)$ at $T = 100$ and $T = 300\,\mathrm{K}$ [37] agree rather well with experiment, nevertheless, the pseudopotential models used in [37, 96] are rather rough and provide a bad description of other properties of alkali metals.

The pseudopotential M2A model (which does not take into account the energy E_p, see sections 2 and 3, Chapter 2) was used in calculations of equations of state of all alkali metals up to $p \sim 400\,\mathrm{kbar}$ [86]. Some deviations of theoretical results from experimental data were observed for Rb and Cs. However, the accuracy of the experimental curves $p(\Omega)$ was not good enough to compare those with results of calculations of the volume dependence of the cohesive energy. It was impossible to make any definite conclusion about the applicability of the model at $p \neq 0$.

There was also an attempt to calculate the BCC–FCC phase transition in Cs under pressure in the frame of the M2A model using different pseudopotentials (see [58] and section 2 Chapter 2). It was shown that the difference in the energy between different structures is the most sensitive characteristic for the form of the pseudopotential and for the approximation of the polarization operator $\pi(\mathbf{q})$. However, a good enough accuracy was not been obtained in [58]. A similar attempt made in [88] has also shown a total inapplicability of the M2A model to this problem and the necessity of modifying the model to provide a more suitable description of metals under compressions which are not small (see also section 1).

The idea that is very fruitful for the solution of this problem is based on the supposition that the cohesive energy of a stressed state of the metal must include the energy of core polarization. The validity of this supposition was never trivial: general estimations show that overlapping of electron densities relating to different ions in non-transition metals is actually negligible as a consequence of the small size of ion cores (see, e.g., [57]). Walles [96] has provided calculations including the energy of the short-range interaction E_{sr} to the cohesive energy and the dynamical matrix by evaluating the thermal expansion coefficient and elastic constants of Na and K at $p = 0$. However, this work has not explained the role of the energy E_p.

Nevertheless, the fact of an exponential dependence of the energies $E_p \sim E_{sr} \sim \exp(-\alpha r)$ on the inter-ionic distances r allows us to suppose that this suggestion is wrong. For this reason, in the present section, we will provide concrete microscopic calculations of the contribution E_{sr} to different properties of alkali metals. Since we, using the M2B model (Chapter 2 section 2), do not take into account details of the band structure, there is no sense of considering in detail the polarization energy E_p (see Chapter 3 section 7). Here we restrict our consideration to a simple approximation

$$E_p \cong E_{sr} = A/2 \sum_{\mathbf{R}} \exp(-R/\rho) \tag{4.31}$$

(see (3.2)) with the parameters A and p from Table 2.3.

The main aim of this section is to consider the phase transition from the BCC to FCC phase under pressure. Equations of state of alkali metals will be calculated and compared with experiment. Methodical questions relating to the absolute instability of the BCC lattice in alkali metals at large compression will be considered: the existence of a "critical" branch in the phonon spectrum and the loss of mechanical stability of the lattice due to "softening" of elastic moduli. In conclusion, we discuss the most important results of calculations and experiments performing of which would be useful for the theory.

So, in the present section, we perform an attempt to provide quantitative calculations of the differences in energy between different phases without any detailed band-structure calculation (that, as we have seen in section 1, would be very difficult). Our aim is to investigate the role of core states for the adiabatic approximation.

4.2.1 *Equations of state*

The isotherms $p(\Omega, T)$ of alkali metals were calculated using the following equation

$$p(\Omega, T) = p_{st}(\Omega) + p^*(\Omega, T), \tag{4.32}$$

where $p_{st} = -\partial E/\partial\Omega$ is the value of p at $T = 0$, which, in the framework of the model used, includes the pressure of the short-range repulsion of ions $p_{sr} = -\partial E_{sr}/\partial\Omega$; $p^* = -(\partial F^*/\partial\Omega)_T$ is the thermal phonon pressure (F^* is the phonon part of the free energy). In equation (4.32), as above, we neglect the thermal pressure of electronic excitations and the anharmonic contributions to p, which are small in all alkali metals [87, 97]. We also neglect the pressure of zero oscillations p^{zp}, which is small in heavy alkali metals [86, 98]. In the same manner as in [86], the contribution of electronic excitations was extracted from experimental data. Calculations of isotherms at $T \neq 0$ need calculations of the term $p^*(T)$ in the equation of state (4.32). To be a volume derivative of F^*, the value p^* is an integral characteristic of the microscopic Grüneisen parameters $\gamma_{\lambda\mathbf{q}}$. Hence, the calculation of p^* needs an integration over the Brillouin zone. All integral characteristics in this section have been calculated using the most rapid method of integration over the Brillouin zone: the "principal values" method [101, 102]. Figure 4.11 shows the isotherms for K, Rb, and Cs, calculated using the V_{pC} pseudopotential (see Table 2.1) in the frame of the M2A (dashed-line), as well as the M2B (full line) models, as a function of the relative change in the volume $\Delta\Omega/\Omega_0 = 1 - \Omega(p, T)/\Omega_0(T)$. Here $\Omega_0(T)$ is the equilibrium value of the cell volume Ω at $p = 0$ and $T =$ const for the BCC phase. This value was calculated using the equation of state (4.32). For the M2B model (with the contribution E_{sr}), Figure 4.11 shows the transition pressure p_c from the BCC (B) to the FCC (F) phase, above which only isotherms for the FCC phase are shown. By calculating these isotherms the value $\Omega_0(T)$ for the BCC phase at $p = 0$ was taken as an initial volume. For both the BCC and FCC phases, the curves $p(\Omega, T)$ are close (Figure 4.11); there is only a small difference in the slope. The

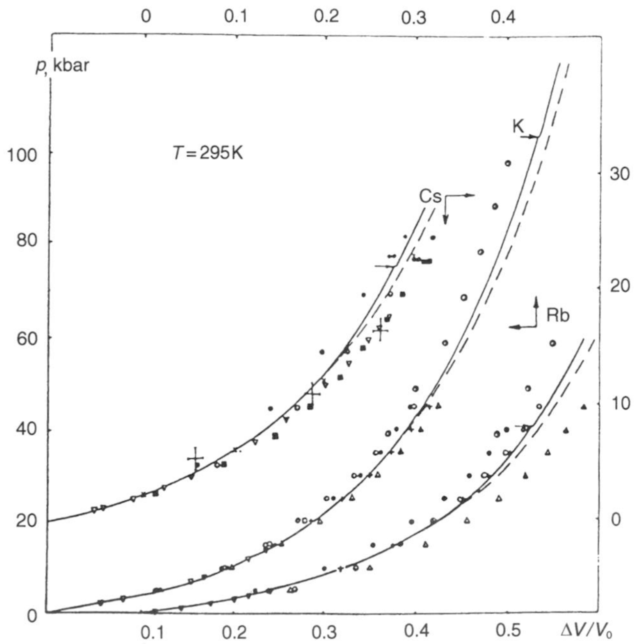

FIGURE 4.11 Room-temperature isotherms of potassium, rubidium, and cesium, calculated taking into account the core polarization energy E_{sr} (full curve) and for $E_{sr} = 0$. Sources of experimental data see in [100]. The zero-temperature isotherms have been calculated in [99, 100] using the supposition $p^* = 0$ in equation (4.32). But, since there are only a few low-temperature data for the dependence $p(\Omega, T)$ in alkali metals and the re-calculation of values measured at room temperature to $T = 0$ decreases the accuracy of experiment, we consider here only the isotherms at $T = 295\,\mathrm{K}$ (see Figure 4.11). The horizontal arrows show the BCC–FCC phase transition points.

curves calculated with taking into account the energy E_{sr} (the M2B model) lie above the curves for the M2A model and the difference between these curves increases from K to Cs (i.e., with increasing energy E_{sr}), as well as with increasing p.[1]

Under normal conditions, the contributions of the short-range repulsion (polarization) to the pressure and to the elastic modulus B for K, Rb, and Cs are of the order 10^{-3}–10^{-2} kbar (10^{-4}–10^{-3} of the Coulomb contribution). Hence, taking into account the energy E_{sr} provides no change in isotherms obtained for the M2A model at low p and, since $p^*(T_m)$ is small (≤ 2 kbar) for all alkali metals, this energy does not influence the anharmonic effects which are determined by the phonon frequencies only.

For K, the isotherm calculated using the M2B model agrees well with existing (up to $\Delta V/V_0 \sim 0.4$) experimental data, as well as with results obtained using the M2A model. In the case of Rb, taking into account the core interaction energy E_{sr} appreciably improves agreement with experiment, in particular, with the most accurate measurements [103] (crosses in Figure 4.11). For Cs, even the isotherm of the M2A model lies above a majority of experimental points. This fact can be explained as more intense, in comparison with K and Rb, influence on the d-band (see section 1). With increasing compression, the contribution of d-bands provides an increase of "softening" of the s-band, especially at the border of the Brillouin zone ([104, 105] and section 1). Closed bands become narrower and the Fermi energy decreases. The band-structure contribution E_d to the energy decreases with increasing p. It is clear, that the pressure $p_d = -\partial E_d/\partial\Omega$ becomes negative and $|\, p_d \,|$ increases with increasing compression (up to the value corresponding to the isomorphic phase transition [106, 107]). This effect must provide decreasing theoretical values of $p(\Delta V/V_0)$. Neglecting the energy E_d is one of the reasons why our consideration of isotherms of Cs is restricted here to small p. The second reason is some "non-rigidity" of ion cores of Cs (see section 7, Chapter 3). However, calculations for the model considered here show that the BCC–FCC phase transition takes place at compressions smaller than the value $\Delta V/V_0$ corresponding to "overlapping" of cores.[2]

We see that the difference between experimental curves in Figure 4.11 can be larger or of the same order as the differences between isotherms calculated for the M2A and M2B models. The accuracy of experimental measurements was discussed in [87, 100]. The less accurate seem to be results of shock–wave experiments, the most accurate are measurements performed in [103, 110]. In general, taking into account the core polarization improves the description of isotherms for K and Rb, but for Cs this description becomes worse.

---

[1] For Li and Na the repulsive term E_{sr} is small and the curves calculated for both models M2A and M2B coincide within the accuracy of the plot.

[2] These defects can be removed by taking into account the effect of d-levels on the cohesive energy, for example, using the scheme submitted in [108], and using a "non-rigid" core submitted in [109]. However, we think to be perspective an application of the density functional theory (as, for example, in [12]) to the investigations of heavy alkali metals under pressure. The band structure approach, considered in Section 1, is physically transparent, but its accuracy is not sufficient for quantitative calculations for such problems.

The similarity of the equations of state of alkali metals,

$$\frac{p}{B(p)} \approx f(\Delta V/V_0), \tag{4.33}$$

has been established empirically using results of measurements for whole group of alkali metals and, further, has been confirmed theoretically [51]. Especially good is this similarity for K, Rb, and Cs as a consequence of the similar structure of their valence and lowest conduction bands. However, the difference in the structure of ion cores for these metals must affect the repulsive energy and destroy this similarity $p/B(p)$ at large p. Really, for K, Rb, and Cs the values of $p/B(p)$ are equal to 0.09, 0.16, and 0.21 (with the accuracy 0.002) for $\Delta V/V_0$ equal to 0.1, 0.2, and 0.3, respectively. But at $\Delta V/V_0 = 0.4$, the ratio $p/B(p)$ is equal to 0.26 for K and 0.25 for Cs. Moreover, this similarity would be broken by phase transitions. Since, as will be shown below, the values p_c are predominantly determined by the short-range term E_{sr}, which depends on the structure of cores, then the ratio $p_c/B(p_c)$ can have different values: 0.30, 0.25, and 0.22 for K, Rb, and Cs, respectively.

4.2.2 Main properties of the phase transitions

The difference in the thermodynamic potential Φ (per atom) between the FCC (F) and BCC (B) phases is

$$\Delta\Phi_{FB}(p,T) = \Delta F_{FB}(\Omega(p,T),T) + p\Delta\Omega_{FB}(p,T). \tag{4.34}$$

Here F is the free energy:

$$F(\Omega,T) = E_{st}(\Omega) + F^*(\Omega,T) \tag{4.35}$$

E_{st} is the cohesive energy at $T = 0$, and F^* is the phonon contribution to the free energy (1.14).

Remember, evaluation of the value $\Delta F_{FB} = F_F - F_B$ needs calculations not only for different volumes of elementary cells: Ω_F and Ω_B, but the vectors of both direct and reciprocal lattices and, consequently, the phonon frequency $\omega_{\lambda\mathbf{q}}$ have different values.

The energy of zero oscillations has been neglected in (4.34) because calculations show that this term provides only a small influence on the behavior of K, Rb, and Cs under pressure and on the characteristics of the phase transition. The volume $\Omega(p,T)$ is calculated for given p and T using the equation of state (4.32). Since the anharmonic contribution to the pressure $p_{an} \ll p^*$, where the thermal pressure $p^* \leq 2\,\text{kbar}$ does not appreciably change within the pressure interval considered for all alkali metals, we can expect that both anharmonic contributions to the energy and pressure are small ([36], see data [111, 112]) Hence, we can restrict our consideration to the term $F^*(\Omega,T)$ in (4.35).

Main features of the polymorphic phase transition from B to F phase can be considered at $T = 0$. The function $\Phi(p,T)$ at $T = 0$ (Figure 4.12) has similar forms for both models M2A and M2B but the values of the transition pressure $p_c(0)$ are quite different. The M2B model provides for the transition pressure $p_c(0)$ to be 96, 31, and 14 kbar for K, Rb, and Cs, whereas the M2A model provides 247, 164, and

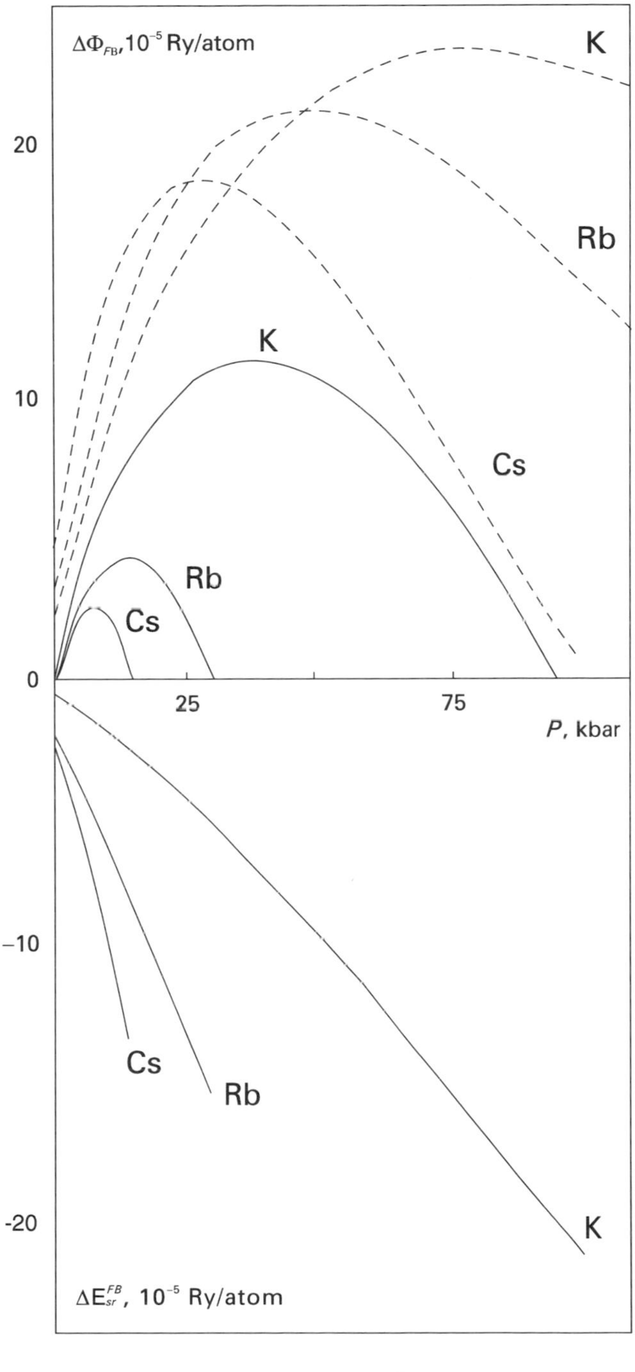

FIGURE 4.12 Difference in the thermodynamic potential $\Delta\Phi_{FB}$ and the energy $E_p^{FB} \approx E_{sr}^{FB}$ against the pressure for potassium, rubidium, and cesium (full lines). Dashed lines – $\Delta\Phi_{FB}$ for $E_p = 0$.

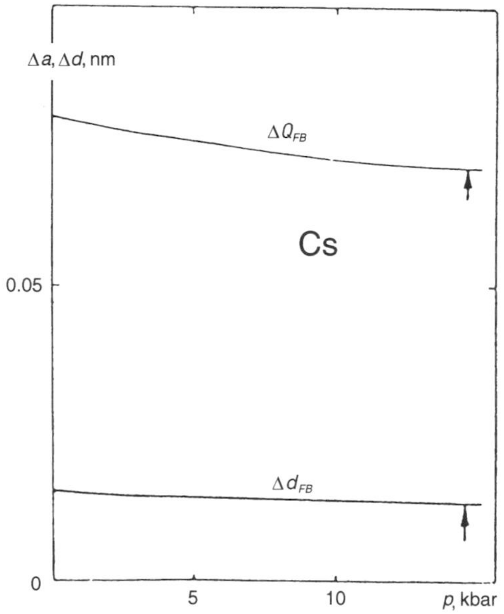

FIGURE 4.13 Pressure dependence of variations of the lattice constant Δa and the inter-ion distance Δd for the (F) FCC and (B) BCC phases of cesium.

101 kbar, respectively. This appreciable difference in $p_c(0)$ between results for two models M2A and M2B is the consequence of the fact that the difference ΔE_{sr}^{FB} is negative and that the derivative $|d(\Delta E_{sr}^{FB})/dp|$ rapidly increases from K to Cs (Figure 4.12). For Na, the value $\Delta\Phi_{FB}(p,0)$ does not vanish at least up to $p \sim 500$ kbar. Experimental measurements (up to $p \sim 350$ kbar) also show no phase transition in Na [35].

A very important characteristic feature of the BCC–FCC phase transition in heavy alkali metals is the extreme proximity of the cell volumes for both phases [99, 100], corresponding to a small value of the structure-dependent contribution to the energy in comparison with the volume-dependent part. At $p = 0$ the relative difference in inter-ion distances $\Delta d_{FB}/d_B$ is very close to the value $2^{5/6}3^{-1/2} - 1 = 0.03$.[3] The value $\Delta d_{FB}(p)$ decreases with increasing p (approximately linear, see Figure 4.13). This feature can be simply explained since $\Delta d_{FB}(0) > 0$ at $\Omega_F \cong \Omega_B$ and the crystal in the F phase is "softer" (i.e., the compressibility $\kappa_F > \kappa_B$). The value Δd_{FB} determines the sign of ΔE_{sr}^{FB} and provides a decrease of ΔE_{sr}^{FB} with increasing p. In the framework of the M2B model, it results in a rapid decrease of $p_c(0)$. Emphasize that the large coordination number of the FCC structure (larger than for the BCC one) prevents the increase of $|\Delta E_{sr}^{FB}|$ (decrease of $p_c(0)$). The large change of the lattice constant $\Delta a_{FB}/a_B \cong 2^{1/3}$–1 (26%) by the BCC–FCC phase transition (in comparison

[3] This is simply the relative change of the interatomic distance in the phase transition from the BCC to FCC lattice at constant cell volume.

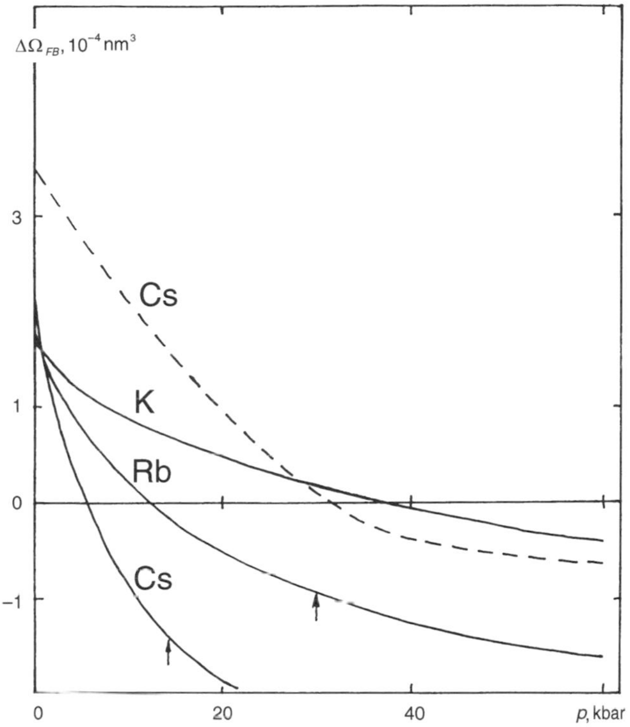

FIGURE 4.14 Pressure dependence of the variation of the cell volume for (B) BCC and (F) FCC phases of potassium, rubidium, and cesium. Arrows indicates the pressures of the BCC–FCC phase transition.

with a small change of the volume $\Delta\Omega_{FB}/\Omega_B$, 0.1% only) shows a preference of x-ray investigations (in comparison with volumetric measurements) of this transition.

The value $\Delta\Omega_{FB}(p)$ (Figure 4.14) determines the second term in (4.34) which provides for the vanishing of $\Delta\Phi_{FB}$ for both models at all T (Figures 4.12 and 4.13). Despite the fact that $\Delta\Omega_{FB} > 0$ at $p = 0$, at the transition point, $\Omega_F < \Omega_B$ (Figure 4.14). Including the pressure $p_{sr} = -\partial E_{sr}/\partial\Omega$ in the equation of state provides only a small effect on Ω_B and Ω_F and on the behavior of $\Delta\Omega_{FB}(p)$. However, inclusion of this term shifts the point $\Delta\Omega_{FB}(p) = 0$ far to the left (dashed line in Figure 4.14). Although the transition to the FCC structure is accomplished by decreasing Ω, the value d increases, i.e., the distance between cores becomes larger and at $p \sim p_c$ there is no touching of cores even for Cs.

The fact that there exists a theoretically calculated value of the pressure where both phases B and F have the same volume gives the possibility of obtaining a direct expression for the difference in the structural energy between these phases ΔE_{str}^{FB}. Figure 4.15 shows that for Cs $\Delta E_{str}^{FB} = 2.5 \times 10^{-5}$ Ry/atom which is 7×10^{-5} of the cohesive energy E. Both the main contributions to E_{str}: the band structure energy $E_e^{(2)}$ and the energy E_{sr} are 1% and 0.1% of the value of E, respectively. Despite the fact that in this point $E_{sr} \sim 0.16E_e^{(2)}$, absolute values of the differences ΔE_{sr} and

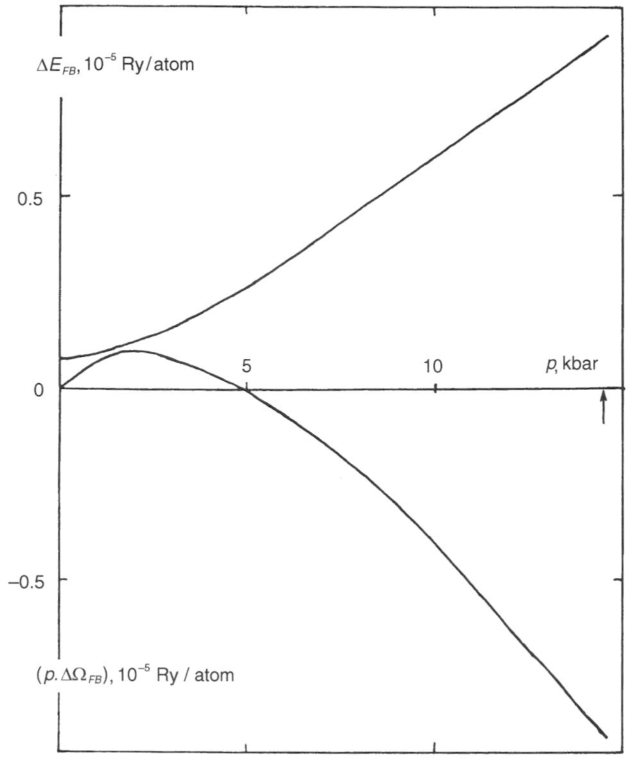

FIGURE 4.15 Pressure dependence of the contributions E_{FB} and $p\Omega_{FB}$ to the thermodynamic potential.

$\Delta E_e^{(2)}$ are of the same order in magnitude, because a structure-dependence of the energy E_{sr} is much higher than that of the electron energy (see also Table. 4.2). The difference $\Delta E_{sr}^{FB} < 0$ rapidly decreases with increasing pressure p (Figure 4.12), the difference $\Delta E_{FB}^{(2)} > 0$ smoothly increases at large p. As a result of their mutual compensation, the valueΔE_{str}^{FB} becomes much smaller at large p.

Vanishing $\Delta\Phi_{FB}(p, 0)$ can be considered from the point of view of the structure-dependence of the energy. But, since the volume-dependent contributions E_Ω to the cohesive energy E are appreciably larger than the structure–dependent terms E_{str} ($\sim 0.01E$) than even for small variations in $\Delta\Omega_{FB}$. the result is that the ΔE_{FB} in the same measure depends on the change of the structure and on the volume difference $\Delta\Omega$. writing E_{st} as

$$E_{st}(\Omega) = E_\Omega(\Omega) + E_{str}(\Omega), \tag{4.36}$$

and expanding the difference $\Delta E_{FB}(\Omega)$ (4.34) as a power series in the small parameter $\Delta\Omega_{FB}/\Omega_B$, we obtain in the first order

$$\Delta E_{str}^{FB} \approx \Delta\Phi_{FB} \tag{4.37}$$

Hence, the form of the curve $\Delta\Phi_{FB}(p,0)$ (Figure 4.12) reflects the intense competition between the negative difference ΔE_{sr}^{FB} and the positive term $\Delta E_{FB}^{(2)}$. But, factually, these differences become equal somewhat earlier than the phase transition point because the energy of direct Coulomb interaction of ions, which weakly depends on the structure but strongly on Ω, provides a negative contribution to the energy E_{str} (which factually results in a vanishing value $\Delta\Phi_{FB}$ for the M2A model at large compressions). Figure 4.12 shows that with increasing p the BCC structure initially becomes more stable than the FCC one and the difference ΔE_{str}^{FB} reaches almost its own maximum at the value of p corresponding to $\Delta E_{FB} = 0$. The difference $\Delta E_{str}^{FB} = 0$ at $p = p_c$, i.e., within the accuracy $\Delta\Omega_{FB}/\Omega_B$ the structural energies of phases have the same values at the transition point. The energy E_{str} itself remains small ($\sim 1\%E$) at any p for all alkali metals. At $p = p_c$ for Cs $E_{sr} = 0.3E_e^{(2)}$.

In calculations of $\Delta\Phi_{FB}(p, T \neq 0)$ the quasi-harmonic approximation was used. To provide it, the thermal expansion $\Omega(T)$ was taken into account in formulae of the harmonic approximation (4.34)–(4.35). It allows us to take into account almost completely anharmonic contributions to temperature variation of many thermodynamic values at large T.

The dependence $\Delta\Phi_{FB}(p, T)$ at small p changes appreciably with variation of T. For each metal considered, starting from some value of T, the difference in volumes between phases at $p = 0$ is negative and the point $\Delta\Omega_{FB} = 0$ does not exist. The value $|\Delta\Omega_{FB}|$ increases with increasing T at $p = 0$ and the slope of the curve $\Delta\Omega_{FB}(p)$ at $T = \text{const}$ rapidly increases. However, with increasing p, the behavior of $\Delta\Omega_{FB}(p, T = \text{const})$ becomes smoother and, starting with $p \sim p_c(0)$, there exist only a small change of $\Delta\Omega_{FB}(p, T)$ at all T for every metal.

The behavior of the difference in the thermodynamic potentials between FCC and BCC phases at $T \neq 0$ differs from that shown in Figure 4.11. Since the phonon part of the free energy $F^*(p, T)$ becomes non-zero and increases with increasing T, then the difference $\Delta\Phi_{FB}(p, T)$ in (4.11) must include the difference $\Delta F_{FB}^*(p, T)$ which is positive for all p and T for K, Rb, and Cs and at $p = 0$ increases with increasing T. With constant T, the curve $\Delta F^*(p, T)$ rapidly decreases at small p. With increasing p, this curve smoothly varies and reaches a minimum. This variation results in a deformation of the curve $\Delta\Phi_{FB}(p, T)$ (Figure 4.11) within the region of small p and the curve $\Delta\Phi_{FB}(p, T = \text{const})$ becomes a monotonically decreasing function of p. Hence, the stability of the BCC lattice under pressure decreases with increasing T to be always smaller than for $T = 0$. The curve $\Delta F_{FB}^*(p, T)$ at constant T crosses the curve $\Delta\Phi_{FB}$ twice. In accordance with (4.11) it means that there are two points where $\Delta E_{FB} = p\Delta\Omega_{FB}$. The first crossing point at small p relates to a larger value of the difference $\Delta\Omega_{FB}(p, T)$ near $p = 0$, in comparison with the value at $p = p_c$. The second crossing point of the curve $\Delta\Phi_{FB}$ with ΔF_{FB}^* corresponds to $p = p_c(0)$. This fact can be explained as a consequence of the close values of the differences $\Delta\Omega_{FB}(T)$ at this pressure for all T for each metal mentioned, as well as taking into account a small value of p^* itself and its weak variation with p. The increase of the pressure p_c with increasing T is very small.

4.2.3 Phase Equilibrium and the Thermodynamics of Phase Transitions

Note a similarity of phase diagrams of Cs, Rb, and K (Figure 4.16): within the accuracy of calculations they are almost straight lines except regions near $T = 0$ and near the triple point of Cs. The same form has the phase diagram of Cs obtained from experimental data [35]. At room temperature, the value of p_c for Cs for the M2B model coincides with experimental data. Near the triple point ($T = 464$ K) the deviation becomes 8% (Tables 4.11 and 4.12). A comparison of theoretical and experimental results for the region of small T provides only a little information because of a large hysteresis observed in experiment. On the other hand, at low T theoretical results turn out to be rather sensitive to details of the model. The calculations with the same parameters as for the pseudopotential model M2A, but including the energy E_{sr} [100], provide the values of p_c for Cs which differ from results of the M2B model by 40% at $T = 0$ and by 10% at $T = 295$ K. The value of dT/dp at the experimental curve of the phase equilibrium becomes larger than for the M2B model (Tables 4.11 and 4.12) and much larger than for the M2A model for which $dT/dp \approx 17$ kbar.

The first-order polymorphic phase transition destroys the continuity of thermodynamic functions (F^*, S, C_p, V etc.) and is accomplished by production or absorption of heat. Table 4.11 shows the change of the phonon contributions to the thermodynamic characteristics by the BCC–FCC transition in K, Rb, and Cs. The

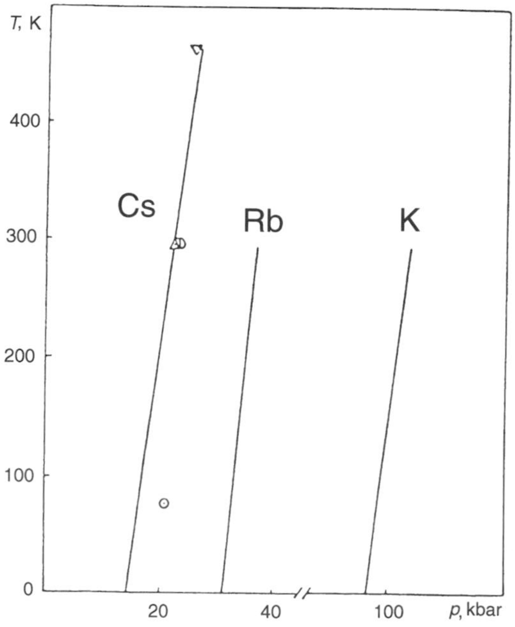

FIGURE 4.16 Phase diagrams of potassium, rubidium, and cesium. Experimental data for cesium at $T = 295$ K from [107, 113–116] are the same within the accuracy of the plot. Other data after [54], [52].

TABLE 4.11 Thermodynamic characteristics of potassium, rubidium, and cesium at the transition point ($p = p_c$) at $T = 295$ K, calculated in the framework of the M2B model (see text)

metal	p_c, kbar	$\frac{\Delta V_B}{V_{OB}}$	$-\Delta V_{FB}$, cm^3/mol	$-\Delta S_{FB}$, J/(mol/K)	$-\Delta C_p^{FB}$, J/(mol/K)	$-q$, J/mol	dT/dp, J/mol
K	104	0.53	0.04	0.11	0.033	32	35
Rb	41	0.43	0.06	0.22	0.021	64	30
Cs	22	0.38	0.10	0.26	0.019	77	38
Cs (exper.)	22.6 ± 0.6	0.40 ± 0.03	0.5	~ 2.5			~ 50

Comment: Experimental data after [98, 107, 111–115].

TABLE 4.12 Parameters of the phase diagram of cesium and the values determining dT/dp

T, K	$p_c(T)$, kbar		dT/dp, K/kbar		$-\Delta V_{FB}$, cm^3/mol		$-\Delta S_{FB}$, J/(mol·K)	
	theory	exper.	theory	exper.	theory	exper.	theory	exper.
0	14.5	21			0.083		0	
100	17.1	21.3	37.9		0.095	0.57	0.250	
200	19.5	21.8	37.6		0.097	0.51	0.254	~ 2.5
295	22.0	22.6 ± 0.6	37.9	69				
464	27.4	25.4	33.0	~ 50	0.093		0.282	

Comment: Theoretical calculations were performed in the framework of the M2B model, experimental data for dT/dp after [99, 100], other values after [113].

entropy, as well as the volume, decreases by the transition. Hence, the latent heat $q = T\Delta S$ is negative and the transition is accomplished by production of heat.

The theoretical value of the volume change by the phase transition ΔV_{FB} is smaller than the measured one (as well as the change of the entropy ΔS_{FB} which experimentally was measured as $\Delta S = \Delta V(dT/dp)$). Nevertheless, the theory confirms weak temperature dependences of ΔV and ΔS at the transition point $p = p_c$ (Table. 4.12) which provide this almost constant slope of the experiment curve of phase equilibrium between BCC and FCC phases of Cs. As a result, the values $(dT/dp)_{exper}$ agree much better with the theory than ΔV_{FB} (data for $\Delta V(T)$ at $T \leq 100$ K are rather doubtful because a large hysteresis of the curve of phase equilibrium $p_c(T)$). For the values dT/dp, it is also difficult to make any conclusion about the accuracy of the theory because experimental data for these values are rare and doubtful [35]. A small value of the theoretical volume change relates to small values of the band structure energy $E_e^{(2)}$ (to be proportional to the square of the pseudopotential form factor) and E_{sr}.

As we have already seen in section 1, the calculation of the band structure energy using the LMTO method (i.e. taking into account the scattering of electrons by ions) instead of the second order perturbation theory in the pseudopotential, appreciably changes the results for Cs as a consequence of a proximity of the *d*-band to E_F. For this reason, theoretical results considered here cannot provide any quantitative accuracy. Nevertheless, the fact, that the value of the lattice constant a_F measured for the FCC phase of Cs at $p = 25\,\text{kbar}$ $T = 295\,\text{K}$ [116] agrees with results of our calculations within the accuracy of 1.5% corresponding to the accuracy of calculations of the variation of $\Omega(T)$, seems to be an advantage of this theory. Hence, to obtain more accurate values of ΔV_{FB}, further experimental investigations are necessary.

For calculated values of p_c (Table 4.12) for K and Rb, there were no direct experimental observations of phase transitions. Nevertheless, the applicability of the theory demonstrated and the similarity of many properties of K, Rb, and Cs would be good arguments for the existence of the BCC–FCC phase transitions for these metals. For this reason, clear and intrinsic experimental investigations of these transitions would be decided arguments confirming the theory. There were some indications of the phase transitions in K at ~ 250 and $\sim 350\,\text{kbar}$ and in Rb at 70 and $100 - 140\,\text{kbar}$ [35]. It is not clear if any of these transitions really are the BCC–FCC transitions (such phase transitions until today have not been identified as a consequence of rather small, smaller than for Cs, changes in the volume, see Table 4.11). Until today, the structure of these transitions mentioned have not be explained. There were also observed some anomalies on the curves of the dependence of electrical resistivity on the pressure for K and Rb in the regions near theoretically predicted values of p_c (see review [34]), but the nature of these anomalies is also not clear.

4.2.4 The Absolute instability of the BCC lattice

For the BCC–FCC transition under pressure in heavy alkali metals, described above, the BCC lattice becomes less energetically preferable than the FCC one but it still remains stable. Here we consider the case of absolute instability of the BCC lattice, i.e. when the minimum of the potential energy disappears.

The stability condition of the crystal lattice relative to mechanical perturbations means that the energy must be a positive determined quadratic form by expansion in a power series in the deformation parameters [117]. The lattice is stable relative to the uniform deformation if

$$C_{\alpha\beta}\, u_\alpha\, u_\beta > 0. \tag{4.38}$$

The stability criterion for non-uniform deformation is

$$D_{\alpha\beta}(\mathbf{q})\, u_\alpha^{\mathbf{q}}\, u_\beta^{-\mathbf{q}} > 0, \tag{4.39}$$

Here $C_{\alpha\beta}$ are the elastic constants, $D_{\alpha\beta}(\mathbf{q})$ is the dynamical matrix, u_α are the deformation parameters.

For stability of cubic lattices with respect to the uniform deformation it is necessary to satisfy the next conditions

$$C_{11} > 0; \quad C_{11}^2 - C_{12}^2 > 0 \text{ or } C' > 0; \tag{4.40}$$

$$C_{44} > 0; \quad C_{11} + 2C_{12} > 0 \text{ or } B > 0. \tag{4.41}$$

For the general case of arbitrary deformations the condition (4.39) is equivalent to the requirement of positive values of squares of the phonon frequencies. An imaginary value of the frequency means that atomic displacements in the lattice, initiated by a small deformation, exponentially (not periodically) increasing with time. Considering the potential energy of metals, this softening of "critical" modes corresponds to a smoother potential minimum which determines the characteristics of atomic oscillations near the equilibrium point. The absolute instability corresponds to a degenerate minimum in one direction, and imaginary phonon frequencies correspond to a maximum of the energy.

An investigation of absolute instabilities is especially interesting if the first-order phase transition, which precedes the absolute instability of the structure, is close to it. In the present subsection, we investigate the characteristics of this absolute instability in BCC alkali metals and the influence of core polarization E_{sr} on this effect.

If there were no BCC–FCC phase transitions then the vanishing of the frequency of the "critical" branch Σ_4 would be observed at the point $\mathbf{q}_c = (0.4; 0.4; 0)$ at $p > p_c$ for values of the compression $\Delta V/V_0$ to be equal to 0.62, 0.56, and 0.52 for K, Rb, and Cs, respectively, (see Figure 4.17). Hence, this absolute instability demonstrates the necessity of the phase transition. For Na, the BCC lattice remains stable up to compression ~ 0.7. For all FCC alkali metals calculations do not show an absolute instability of the lattice even at much higher compressions.

In [86], vanishing $\omega(\Sigma_4)$ has been observed for the compression equal to 0.68 but at the border of the Brillouin zone. For the model M2B, the similarity of properties of K, Rb, and Cs is broken because for Cs the instability at $\mathbf{q} = 0$ appears before the instability near the border of the Brillouin zone: the vanishing of the modulus $B_{33} = 0$ takes place at a somewhat smaller compression than that which corresponds to $\omega(\Sigma_4) = 0$ at the border of Brillouin zone at the point N_4' (Figure 4.17). Moreover, for Cs, the frequency $\omega(\Sigma_4)$ vanishes near $q = 0$ before the point $\mathbf{q}_c = (0.4, 0.4, 0)$, because its long wavelength limit is determined by the elastic modulus B_{33}. Some deformations of the phonon branch Σ_4 as $\mathbf{q} \to 0$ near the absolute instability of the BCC lattice have been observed for all heavy alkali metals (Figure 4.17) for both models. However, the core interaction affects the character of instability for Cs only. Note large differences between values of p corresponding to the critical compression obtained in the frame of M2A and M2B models, although for values $(\Delta V/V_0)_c$ themselves these differences are not very large (see above). It demonstrates once again a strong influence of the core polarization on the properties of alkali metals at large compression, in particular, on the equation of state.

An analysis of the mechanical stability only cannot solve the question about real crystal structure which by phase transitions under pressure is determined by the

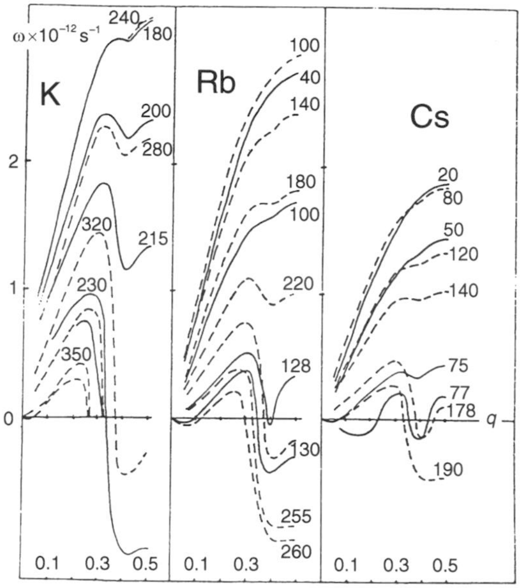

FIGURE 4.17 Phonon dispersion curves for the branch Σ_4 of potassium, rubidium, and cesium at different pressures. Notations are the same as in Figure 4.12.

thermodynamic potential (the Gibbs free energy). Our analysis in this chapter shows that the first-order BCC–FCC phase transitions precede the absolute instability. Hence, these critical values of compression show only the upper limit of this value corresponding to the phase transition, in particular, confirm once again the necessity of phase transitions in K and Rb. However, only a concrete comparison of Φ for different phases can provide complete information about possible phase transitions and their properties.

4.2.5 Behavior of elastic constants

Starting from the general theory of the crystal stability, we can suppose (see, however, section 1) that, by approaching the structural phase transition point, at this point, and after it, the elastic modului can undergo remarkable changes. For the second-order phase transitions, using the phenomenological theory, it is possible to show what would be the change in different elastic moduli under action of different deformations [118]. But there exists no such theory for the first-order structural phase transitions considered. Information about behavior of elastic moduli by a structural phase transition and near it can be obtained using results of microscopic calculations. For this reason, the models developed have been applied to study behavior of elastic moduli of heavy alkali metals under pressure. On the other hand, it was necessary to investigate the influence of the ion polarization on elastic properties of alkali metals at these conditions.

By definition, the isothermal elastic moduli are the coefficients of expansion of the Helmholtz free energy $F(\Omega, T)$ in a power series in the deformation tensor $u_{\alpha\beta}$ (see Chapter 1) normalized to the equilibrium volume of the crystal. For cubic crystals, considered here, there are only three non-zero second-order elastic constants C_{ik}: C_{11}, C_{12}, and C_{44}. Super-sonic measurements under pressure give intrinsically the Birch elastic moduli $\mathcal{B}_{ik}$ (see Chapter 1). Actually the Fuchs elastic moduli B_{ik} are also used because the deformation parameters in the expansion of the free energy have a more transparent physical sense. It provides the possibility of extracting the pure shear moduli B_{33} and B_{44} and the modulus of pure hydrostatic compression B_{11}.

In this subsection the Fuchs elastic moduli have been calculated at $T = 0$. The energy of zero oscillations was neglected. Figures 4.18–4.21 show the volume behavior of the shear elastic modules B_{33} and B_{44} for Na, K, Rb, and Cs at $T = 0$ for both models: M2B (see calculations [8, 39]) and M2A (see calculations [86]).

With no phase transition, the modulus B_{33} would vanish with increasing p for the BCC phase (see subsection 4) but increase almost linearly for the FCC phase (the curvature of the plot of $B_{33}(p)$ for the FCC phase in the figures relates to a non-linear scale for p). Qualitatively similar results have been obtained in calculations of a volume dependence of B_{33} for both models as well as for both phases B and F (see Figure 4.19 for Rb). However, for the M2A model, the BCC–FCC phase transition takes place in K, Rb, and Cs when the modulus B_{33} suffers "softening", i.e., p_c corresponds to a decrease of $B_{33}(p)$ (see Figure 4.14, plot for Rb). For the M2B model it takes place for K only, whereas for Rb and Cs the value $B_{33}(p)$ at $p = p_c$ has not yet reached its maximum. A general feature of both models is the rapid change of the slope of the curve $B_{33}(p)$ at $p = p_c$. For K the derivatives dB_{33}/dp calculated for B and F phases in the frame of the M2B model differ not only in absolute values (by a factor 5–10), but also in the sign. Hence, although p_c is very close to the value of p at which the modulus B_{33} has the same value for both phases B

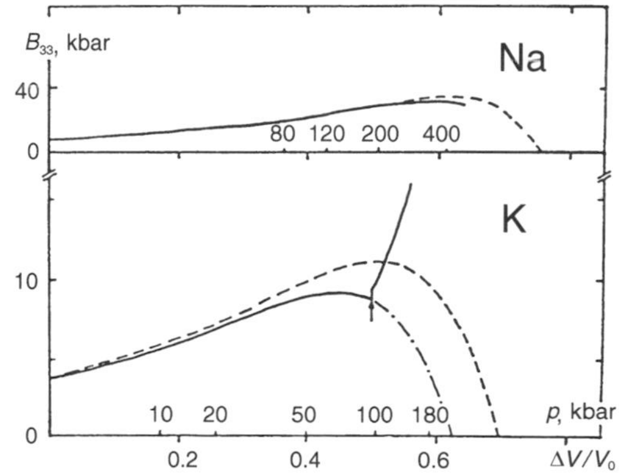

FIGURE 4.18 Shear elasticity modulus B_{33} of sodium and potassium against the pressure. The lower horizontal scale is the compression, the upper one is the pressure for the M2B model. The same notations as in Figure 4.12. Dashed–dotted lines are values for metastable phases (F or B).

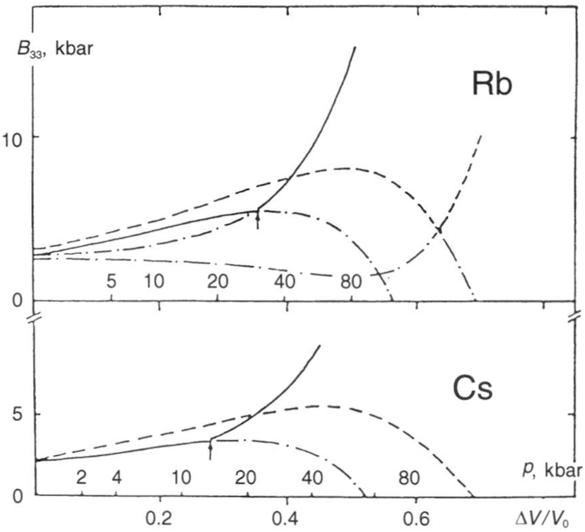

FIGURE 4.19 Pressure-dependence of the elasticity modulus B_{33} for rubidium and cesium. The same notations as in Figure 4.18.

and F, relative changes of these values by the phase transitions are 9, 12, and 6% for K, Rb, and Cs, respectively. At $T = 0$ the behavior of elastic moduli is qualitatively the same [98].

The volume dependence of the modulus B_{44} (Figure. 4.20 and 4.21), which determines the slope of the transversal phonon branch in [100] and [110] directions,

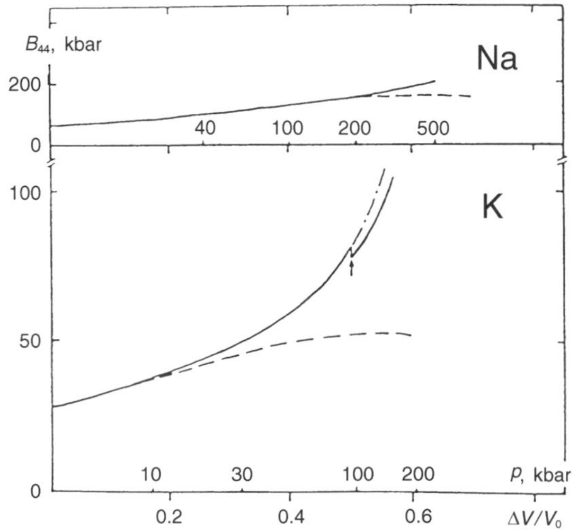

FIGURE 4.20 Volume-dependence of the shear elasticity modulus B_{44} for sodium and potassium. The same notations as in Figure 4.18.

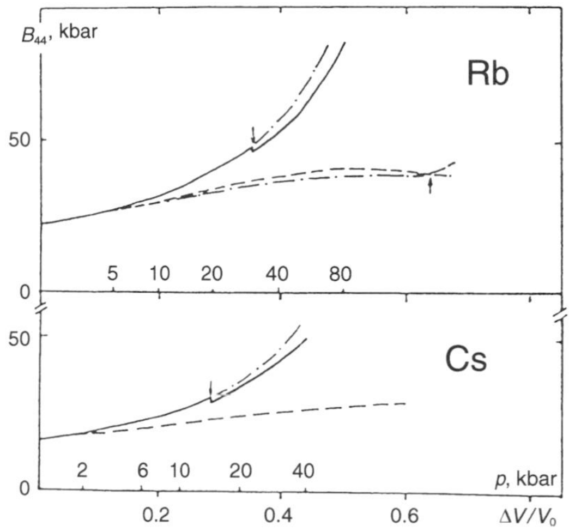

FIGURE 4.21 Volume-dependence of the elasticity modulus B_{44} for rubidium and cesium. The same notations as in Figure 4.18.

differs from the dependence $B_{33}(\Omega)$. This behavior relates to the fact that, for the M2B model used here, the frequencies of all branches (except Σ_4 for the BCC phase) increase with increasing p for both phases. This fact explains the similarity in behavior of the curve $B_{44}(p)$ for both BCC and FCC phases. The difference in results of calculations for M2A and M2B models for the frequency of the branch Σ_3 under pressure, as was already mentioned in section 2, Chapter 3, is a consequence of the core polarization E_{sr}. Taking into account this term appreciably changes results of calculations of $B_{44}(p)$ (even for Na at large p). A conclusion made in [86] that variations of $B_{33}(p)$ and $B_{44}(p)$ in the BCC phase have qualitatively similar form, seems to be questionable not only from the point of view of results for the M2B model but also for M2A model because the value B_{44} does not decrease but instead increases with increasing compression (see Figure 4.21 for Rb). This fact confirms once again the absence of the critical decrease of the frequency of the branch Σ_3 under pressure discussed in [86, 119]. A general feature of results of calculations for both these models is a small negative change of B_{44} at the transition point. The absolute value of the change B_{44}, as well as B_{33}, increases from Cs to K, but the relative change remains approximately constant ($\sim 4-5\%$). The changes of the derivative dB_{44}/dp at $p = p_c$ are equal to 22, 15, and 13% for K, Rb, and Cs, respectively.

For each metal, the pressure dependence of the modulus $B_{11}(p)$ (calculated for both models, see Figure 4.22), which determines the slope of the longitudinal phonon branches as $\mathbf{q} \to 0$, is similar to the dependence $B_{44}(p)$ calculated for the M2B model. The modulus $B_{11}(p)$ has close values for both phases. Its change by the phase transition, as well as the change of the derivative dB_{11}/dp, are less than 1% [14]. The disagreement in results for $B_{11}(p)$ between the M2A and M2B models increases with

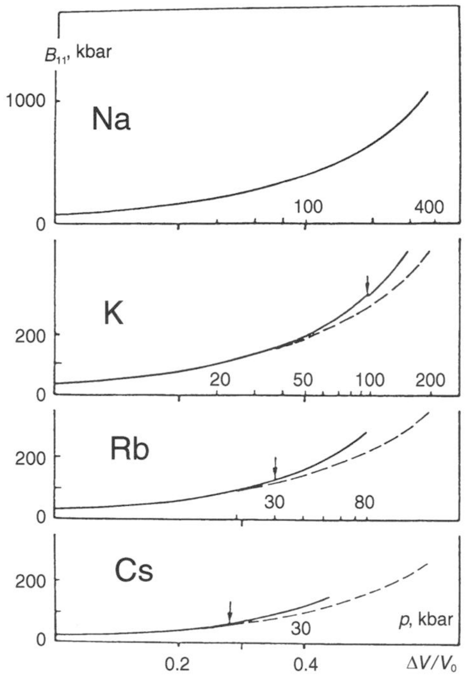

FIGURE 4.22 Volume-dependence of the compressibility B_{11} for alkali metals. The same notations as in Figure 4.18.

increasing p. A small change of the modulus B_{11} by the phase transition relates to a small change of Ω and, respectively, to a small difference between volume-dependent contributions to E, which predominantly determine $B_{11}(p)$. The strong dependence of the modulus B_{11} on p relates to a drastic decreasing of the mutual cancellation of electronic and ion contributions with compression which gives $B_{11} \sim 1/\Omega^2$ [86].

4.3 Resumé

Thc prcdiction of the form of the phase diagrams for Li and Na is the main result of the first section. Note that before these calculations there were not even qualitative calculations of the form of these diagrams. The results obtained for Na (and less for Li) also have a quantitative accuracy. Experimental tests of these results would be very interesting.

This phase diagram for Na shows that rather rough models, like the "hard–sphere" model, cannot be used here even for qualitative conclusions about the form of the phase diagram. In accordance with these models, increase of p would result in increasing the "close-packed" nature of the crystal, i.e., the temperature interval of existence of the HCP and FCC phases would increase with increasing p. This conclusion is valid for Li but not for Na, at least for the compression $u \leq 0.4$. Hence, a non-trivial result is that the influence of the pressure on the structural stability has the opposite signs for Li and Na, despite the proximity of all other properties of these metals at normal pressure. This difference can be explained as the effect the Fermi level approaching singularities of the band structure which takes place in Li under pressure but does not in Na. Both the "electron" mechanism of the phase transitions itself as well as the pre-martensitic softening of shear modules, considered here, are close to the well-known Hume-Rothery–Jones rule which relates to the phase stability [91] and can be considered as a general feature for structural phase transitions in metals and alloys. Further studies of these phase transitions for relatively simple Li and Li–Mg systems would be very useful in understanding the nature and properties of these phase transitions.

Some comments about calculations of phase diagrams at large compressions $u > 0.5$. As was already mentioned at the end of section 1 by discussion of the phase diagram of Cs, at large compressions it is necessary to take into consideration core states. At such compressions, these core states approach the region of nearly free (valence) electrons. It immediately gives a question about the influence of non-adiabaticity [91–94] which needs additional theoretical and numerical investigations. Nevertheless, for light metals like Li and Na, it would be very useful to perform a numerical analysis similar to that made in [95] for solid neon.

Here, it is useful to make a comment about the models used in calculations of the phase diagrams (the energy of the structure). Earlier (see [57]) calculations of static properties of Li (and Al) have been performed for the third (M3) and the fourth (M4) order in the pseudopotential because the accuracy of the second-order model was insufficient. These calculations have shown that usage of M3 and M4 models appreciably improves the agreement with experiment. It would be possible to provide exact calculations (from the theoretical point of view). However, a disagreement between theory and experiment would remain remarkable for such calculations, too. The complete band-structure calculations (without expansion in $V_{\mathbf{q}}$, see section 1) provide correct qualitative results but it seems to be very difficult to provide quantitative calculations for this method.

A very important result of the present chapter is the relationship between variations of shear moduli, in particular, their pre-martensitic softening near the points of the structural phase transition, and the change of the band structure. For all metals considered, except BCC Cs, the approach of the compression u to the transition point u_c is accomplished by the "band-structure softening" of one of the shear moduli. Actually, this "softening" relates to the approach of the Fermi level E_F to a peak or other maximum of the density of states $\nu(E)$. Similar relations between the pre-martensitic anomaly and the proximity of E_F to peaks of $\nu(E)$ were also discussed in publications relating to many other systems, for example, Ni–Ti alloys and systems like A-15. At the same time, an absence of these singularities in the band structure results, apparently, to the absence the anomalies considered, as was shown above for BCC Cs. For this reason, we can suppose that the proximity of E_F to peaks or other singularity points of $\nu(E)$ (or an absence of this proximity), as a rule, determine the existence (or absence) of softening the lattice near the points of structural phase transitions in metals and alloys. Theoretical and experiment tests of this hypothesis submitted by Vaks and Trefilov would be very important for microscopic understanding of the nature of pre-martensitic phenomenon since, up to the present, they were discussed predominantly phenomenologically (see, e.g., [66, 67]).

The proximity of E_F to singularity points of $\nu(E)$ explains also the large lability of elastic properties, i.e. large variations of the moduli $C_{ss'}$ following a change of external parameters. For "rigid-band"-like models, this fact was observed earlier (see, e.g., [65] and Refs. in [33]). The results of subsection 3, section 1 illustrate these effects for a more realistic case of the "non-rigid-band" model. For Na and BCC Cs at $u < u_c$, smooth variations of $C_{ss'}$ were observed with no proximity of E_F to singularities of $\nu(E)$.

Very important seem to be concrete predictions of anomalies in behavior of $C_{ss'}(u)$ in simple metals [33, 47, 48]. The most important of those are: a prediction of a rapid softening of the module C' on the both sides (in pressure) of the FCC–BCC phase transition point in Sr and Ca; the existence of similar softening of C' by decreasing the pressure p in the FCC phase of Cs and an absence of this effect by increasing p in the BCC phase; as well as the existence at large compressions $u > 0.4$, the effects of appreciable softening of the modulus $C_{44}(u)$ with monotonic increasing $C'(u)$ in the BCC Sr and Ca and in the FCC Cs. Experimental tests of these predictions would be very important for the development of an adequate description of the influence of singularities of the electron structure on properties of these metals.

Note an approachable difference between the values of the shear modulus C' for Li and Na obtained either in the frame of the M2 model or taking into account the band structure energy totally (Figure 4.23). We can see that the modulus $C'(u)$ calculated for the M2 model demonstrates under compression similar behavior (increases) for both Na and Li, but the result obtained for Li, taking into account the pseudopotential totally, shows a rather rapid decrease of C' with increasing u.

Let us now discuss the experiments which would be very important for improving the theory of the phase transition from the BCC to FCC phase as well as for studying the influence of the core polarization on the equation of state and on other metal properties under pressure.

Our analysis of the equation of state calculated in the framework of the model used shows that the description is rather good at $p \leq 50\,\text{kbar}$ for all alkali metals (except Cs). Some disagreements between the theory and experimental results for Cs, possibly, can be removed by taking into account the effect of empty d-bands on the cohesive energy. This effect results in the isomorphic electron transition in Cs at $p = 41.5\,\text{kbar}$ [35]. Taking into account the d-bands of Cs can be provided by methods of the band structure calculations described in section 1 where some conclusions about the phase diagram of Cs have been made, too. However, to obtain quantitative results it is necessary to improve experimental data for the equation of state for Cs. An absence of reliable experimental information about isotherms at large compressions does not allow us to provide undoubted conclusions about the accuracy of those description at large p and about applicability of the Born–Mayer form of E_{sr} used in describing the inter-atomic interaction in alkali metals. For this

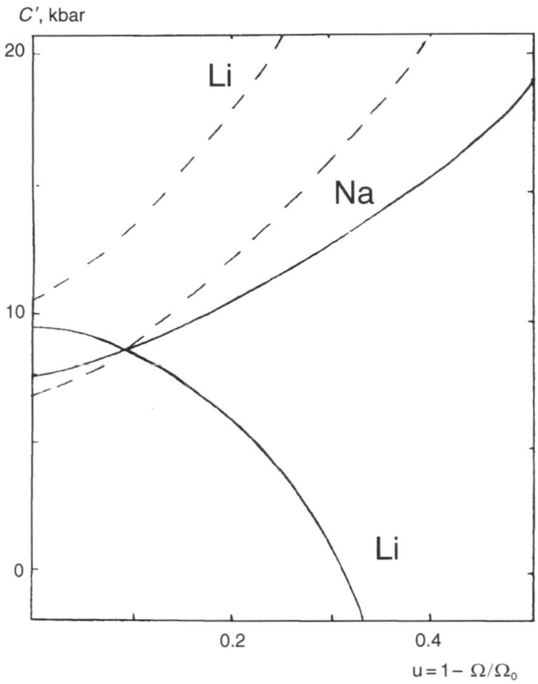

FIGURE 4.23 Shear modulus C' against the compression u for BCC lithium and sodium. Dashed lines – calculation for the M2A model (in the second order in the pseudopotential), full lines – results of band structure calculations (after [46]).

reason, increasing the accuracy of measurements of the dependence $p(V)$, although general features of that are well-known, is necessary for present-day microscopic theories of solids.

In section 2, the possibility to solve theoretically the problems relating to the structure of crystals, like the polymorphic BCC–FCC transition in Cs under pressure, have been demonstrated for the first time. We have considered in detail the role of different mechanisms of the interaction of ions and electrons in metals from the point of view of the structural phase transitions under pressure. It was shown that one of the most important factors determining the structural transformation is the core polarization of ions E_p which was taken into account here as the Born–Mayer repulsion term. This form of E_p, as was shown in section 7, Chapter 3, can be used only at relatively small compressions for which core levels are far away from the band of nearly-free electrons. If this condition were not satisfied, further development of the theory of this phenomenon taking into account non-adiabatic effects (the electron-phonon interaction) would be necessary.

Paying attention to the similarity of many properties of heavy alkali metals K, Rb, and Cs, the possibility of the BCC–FCC transitions under pressure in K and Rb has been investigated. Further experimental studies of eventual BCC–FCC transitions under pressure in K and Rb would by necessary. First, these experiments could confirm the predictability of the theory. Second, the values of $p_c(T)$ and other characteristics of these transitions would provide quantitative information about contributions of different mechanisms to the cohesive energy of these metals.

Since some disagreements between theoretical and experimental values of ΔV were established, more accurate measurements of the behavior of $p(V)$ and $p_c(T)$, values dT/dp, and, especially, the change of the volume at the BCC–FCC transition in Cs would be very important to investigate the limits of applicability of the theory.

Appendix A
Dielectric Susceptibility of Ion Cores

Here we will obtain the formula for the effective inter-electron interaction taking into account the ion polarization.

Let the point denotes the dipole vertex $\mathbf{q}^2\alpha(i\omega)/\Omega$ and the bold dashed line, as before, the screened Coulomb interaction $w_C(\mathbf{q})$. The sequence of diagrams [1–4]

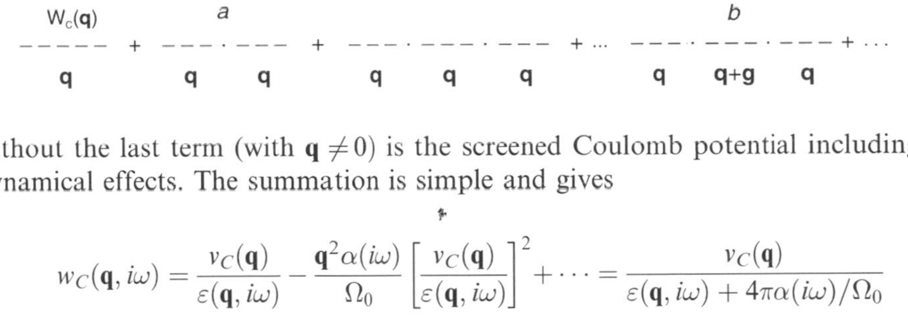

$$
\underset{\mathbf{q}}{\underset{W_c(\mathbf{q})}{-----}} + \underset{\mathbf{q}\quad\mathbf{q}}{\overset{a}{---\,\cdot\,---}} + \underset{\mathbf{q}\quad\mathbf{q}\quad\mathbf{q}}{---\,\cdot\,---\,\cdot\,---} + \cdots \quad \underset{\mathbf{q}\quad\mathbf{q+g}\quad\mathbf{q}}{\overset{b}{---\,\cdot\,---\,\cdot\,---}} + \cdots \tag{1}
$$

without the last term (with $\mathbf{q} \neq 0$) is the screened Coulomb potential including the dynamical effects. The summation is simple and gives

$$
w_C(\mathbf{q}, i\omega) = \frac{v_C(\mathbf{q})}{\varepsilon(\mathbf{q}, i\omega)} - \frac{\mathbf{q}^2\alpha(i\omega)}{\Omega_0}\left[\frac{v_C(\mathbf{q})}{\varepsilon(\mathbf{q}, i\omega)}\right]^2 + \cdots = \frac{v_C(\mathbf{q})}{\varepsilon(\mathbf{q}, i\omega) + 4\pi\alpha(i\omega)/\Omega_0} \tag{2}
$$

It is possible to inset any number of lines with non-zero momentum transferred to each dipole vertex between two dashed lines with momentum $\mathbf{q}$. The simplest diagram of this type is diagram "b" in (1). Contributions of this diagram corresponds to taking into account the local field effects. Let us neglect for simplicity the influence of the screening provided by conducting electrons on these effects. Then the expression corresponding to diagram "b" has the following form

$$
\left[\frac{4\pi\alpha(i\omega)}{\Omega_0}\right]^2\frac{q_\alpha q_\beta}{\mathbf{q}^2}\sum_{\mathbf{g}\neq 0}\frac{(\mathbf{q}+\mathbf{g})_\alpha(\mathbf{q}+\mathbf{g})_\alpha}{|\mathbf{q}+\mathbf{g}|^2}. \tag{3}
$$

Rewriting the sum in **g** as

$$S_{\alpha\beta} = \sum_{\mathbf{g}\neq 0} \frac{(\mathbf{q}+\mathbf{g})_\alpha(\mathbf{q}+\mathbf{g})_\beta}{|\mathbf{q}+\mathbf{g}|^2} = -\frac{\partial^2}{\partial\rho_\alpha\partial\rho_\beta} \sum_{\mathbf{g}\neq 0} \frac{(\mathbf{q}+\mathbf{g})\boldsymbol{\rho}}{|\mathbf{q}+\mathbf{g}|^2}\Bigg|_{\boldsymbol{\rho}=0}, \tag{4}$$

we obtain

$$S_{\alpha\beta} = -\frac{1}{4\pi}\frac{\partial^2}{\partial\rho_\alpha\partial\rho_\beta}\left[\Omega_0 \sum_{\mathbf{R}} \frac{\exp(-i\mathbf{q}\mathbf{R})}{|\boldsymbol{\rho}+\mathbf{R}|} - \int d^3r \frac{\exp(-i\mathbf{q}\mathbf{r})}{|\boldsymbol{\rho}+\mathbf{r}|}\right]_{\boldsymbol{\rho}=0}. \tag{5}$$

Now expression (3) for cubic lattices ($S_{\alpha\beta} = -\delta_{\alpha\beta}/3$) becomes

$$-\frac{4\pi}{\Omega_0}\alpha(i\omega)\frac{4\pi}{\Omega_0}\alpha(i\omega). \tag{6}$$

Similar transformations show that the summation of all diagrams with non-zero momentum transfer between two lines $w_c(\mathbf{q})$ in the dipole vertex is equivalent to replacing $\alpha(\omega)$ in (2) by $\tilde{\alpha}(\omega)$, where

$$\tilde{\alpha}(i\omega) = \alpha(i\omega)\left[1 + \frac{4\pi\alpha(i\omega)}{3\Omega_0} + \left(\frac{4\pi\alpha(i\omega)}{3\Omega_0}\right)^2 + \cdots\right] = \frac{\alpha(i\omega)}{1 - \frac{4\pi\alpha(i\omega)}{3\Omega_0}}. \tag{7}$$

This result corresponds to the actual local field correction. It is possible to show that taking into account the effects of electron screening on the local field gives

$$\tilde{\alpha}(i\omega) = \frac{\alpha(i\omega)}{1 - \frac{4\pi}{3\Omega_0}\alpha(i\omega)\left[1 + \sum_{\mathbf{g}\neq 0}[1 - \varepsilon^{-1}(\mathbf{g}, i\omega)]\right]}, \tag{8}$$

The same result has been obtained by another method in [5].

Appendix B
A Method of Calculation of Van Der Waals Sums

In [6, 7] a method, similar to the Ewald method [8, 9] for calculating slowly converging sums like

$$S_{BB} = \sum_{\mathbf{R} \neq 0} R^{-6}, \tag{1}$$

has been proposed (see also [10]). Let us consider the following slowly converging sum

$$S = \sum_{\mathbf{R}}{}' \Phi(\mathbf{R}) \tag{2}$$

Rewrite $\Phi(\mathbf{R})$ as

$$\Phi(\mathbf{R}) = \Phi_1(\mathbf{R}) + \Phi_2(\mathbf{R}); \quad (|\Phi_2(0)| < \infty), \tag{3}$$

where

$$S_1 = \sum_{\mathbf{R}} \Phi_1(\mathbf{R}) \tag{4}$$

rapidly converges and $\Phi_2(\mathbf{R})$ has the Fourier transformation $\varphi_2(\mathbf{q})$ rapidly decreasing as $|\mathbf{q}| \to \infty$. Then

$$S = \sum_{\mathbf{R}}{}' \Phi_1(\mathbf{R}) + \sum_{\mathbf{G}} \varphi_2(\mathbf{G}) - \Phi_2(0) \tag{5}$$

Both sums in (5) are supposed to be rapidly converging. In our case $\Phi(\mathbf{R}) = \mathbf{R}^{-6}$, and it is convenient to take Φ_1 and Φ_2 as

$$\Phi_1(\mathbf{R}) = \mathbf{R}^{-6} \exp(-a\mathbf{R}^2)(1 + a\mathbf{R}^2 + a^2\mathbf{R}^4/2),$$

$$\Phi_2(\mathbf{R}) = \mathbf{R}^{-6}\left[1 - \exp(-a\mathbf{R}^2)(1 + a\mathbf{R}^2 + a^2\mathbf{R}^4/2)\right] = \int_0^a dx \frac{x^2}{2} e^{-xR^2}. \tag{6}$$

where

$$\varphi_2(q) = \frac{\pi^{3/2}}{2\Omega_0} \int_0^a dx \sqrt{x} \exp\left(-\frac{q^2}{4x}\right) = \frac{\pi^{3/2} q^3}{16\Omega_0} \int_{q^2/4a}^{\infty} dx e^{-x} x^{-5/2}. \tag{7}$$

Now we have the following expression for S:

$$S = \sum_{\mathbf{R}\neq 0} \mathbf{R}^{-6} \exp(-a\mathbf{R}^2)(1 + a\mathbf{R}^2 + a^2\mathbf{R}^4/2) + \frac{\pi^{3/2} a^{3/2}}{2\Omega_0} \sum_{\mathbf{G}} f\left(\frac{\mathbf{G}}{2\sqrt{a}}\right) - \frac{a^3}{6},$$

$$f(x) = \frac{2}{3} e^{-x^2} - \frac{4}{3} x^2 e^{-x^2} + \frac{8}{3} x^3 \int_x^{\infty} dt\, e^{-t^2}. \tag{8}$$

References

Chapter 1

1. J. M. Ziman, *Principles of the Theory of Solids*, second edition Cambridge University Press, 1972.
2. J. M. Ziman, *The Calculation of Bloch Functions*, in *Solid State Physics* (eds H. Ehrenreich, F. Seitz and D. Turnbull), **26**, Academic Press, NY, 1970, 1–101.
3. V. V. Nemoshkalenko, V. N. Antonov, *Metody Vichislitelnoj Fiziki v Teorii Tverdogo Tela. Zonnaya Teoriya Metallov* (Methods of Computational Physics in the Solid State Theory), Kiev, Naukova Dumka, 1985 (in Russian).
4. L. D. Landau and E. M. Lifshitz, *Course of Theoretical Physics*, **3**, *Quantum Mechanics — Nonrelativistic Theory*, third edition Pergamon, London, 1962.
5. N. H. March, W. H. Young, S. Sampanthar, *The Many-body Problem in Quantum Mechanics*, Cambridge University Press, 1967.
6. E. V. Zarochentsev, K. B. Tolpygo and E. P. Troitskaya, *Sov. Phys. — Low Temp.* (1982), **8**, N8, 885–861.
7. V. Heine, J. H. Samson and C. M. M. Nex, *J. Phys. F.: Met. Phys.* (1981), **11**, N12, 2645–2662.
8. J. C. Phillips, L. Kleinman, *Phys. Rev.* (1959), **116**, N2, 287–294; N4, 880–884.
9. W. A. Harrison, *Pseudopotentials*, Benjamin, NY, 1966
10. V. Heine, M. L. Cohen, D. Weaire, in *Solid State Physics* (eds H. Ehrenreich, F. Seitz, D. Turnbull) **24**, Academic Press, NY, 1970.
11. A. A. Abrikosov, *An Introduction to the Theory of Normal Metals*, in *Solid State Physics* (eds H. Ehrenreich, F. Seitz, D. Turnbull), Suppl. **24**, Academic Press, NY, 1971.
12. E. V. Zarochentsev, V. P. Safronov, E. P. Troitskaya, *Sov. Phys. — Solid State* (1978), **20**, N10, 2881–2888.
13. R. A. Cowley, *Advan. Phys.* (1963), **12**, N45–48, 421–480.

14. V. G. Vaks, S. P. Kravchuk, A. V. Trefilov, *J. Phys. F.: Met. Phys.* (1980), **10**, N11, 2325–2344.
15. V. G. Vaks, S. P. Kravchuk, A. V. Trefilov, *J. Phys. F.: Met. Phys.* (1980), **10**, N10, 2105–2124.
16. V. G. Vaks, S. P. Kravchuk, A. V. Trefilov, *Sov. Phys. — Solid State* (1979), **21**, N11, 3370–3378.
17. L. D. Landau, E. M. Lifshitz, *Course of Theoretical Physics*, **5**, *Statistical Physics*, Pergamon, London, 1958.
18. G. Leibfried, W. Ludwig, in *Solid State Physics* (eds. H. Ehrenreich, F. Seitz, D. Turnbull) **12**, Academic Press, NY, 1961.
19. E. G. Brovman, Yu. Kagan, *Sov. Phys. — Uspekhi* (1974), **112**, N3, 369–426.
20. M. Born, K. Huang, *Dynamical Theory of Crystal Lattices*, Oxford University Press, London, 1954.
21. K. Brugger, *Phys. Rev.* (1964), **113A**, N6, 1611–1612.
22. D. Wallance, in *Solid State Physics* (eds H. Ehrenreich, F. Seitz, D. Turnbull) **25**, Academic Press, NY, 1961.
23. A. A. Abrikosov, L. P. Gor'kov, I. E. Dzyaloshinskii, *Quantum Field Theoretical Methods in Statistical Physics*, Pergamon Press, NY, 1965.
24. E. G. Brovman, Yu. Kagan, A. Kholas, *Sov. Phys. — JETP* (1971), **61**, N2. 737–752.
25. V. N. Varyukhin, E. V. Zarochentsev, S. M. Orel, *K Teorii Staticheskih Svoistv Kristallov pri Konechnom Napryazhenii* (About the Theory of Statical Properties of crystals under Finite Stresses) in book: Diagrammy Sostoyaniya v Materialovedenii, Kiev, Naukova Dumka, 1979 (in Russian), **2**, 103–114.
26. E. G. Brovman, Yu. Kagan, *Sov. Phys. — JETP* (1967), **52**, N2, 557–574.
27. B. T. Geĭlikman, *Sov. Phys. — Uspekhi* (1975), **115**, N3, 403–426.
28. B. T. Geĭlikman, *J. Low Temp. Phys.* (1974), **4**, N2, 189–208.
29. H. Haken, *Quantentheorie des Festkörpers*, Teubner, Berlin, 1973.
30. V. V. Moshchalkov, N. B. Brant, *Sov. Phys. — Uspekhi* (1986), **149**, N4, 585–634.
31. I. E. Dragunov, E. V. Zarochentsev, S. M. Orel, *Energiya Metalla. Vyhod za Predely Adiabaticheskogo Priblizheniya v Teorii Metallov*, (Energy of Metals. Escape of the Adiabatic Approximation in the Metal Theory) Donetsk, 1987 (Preprint Phys.–Tech. Inst. Ac. Sc. USSR, N7 (127)) (in Russian).
32. I. E. Dragunov, E. V. Zarochentsev, S. M. Orel, *Sov. Phys. — Metal Phys.* (1989), **67**, N5, 837–844.
33. B. T. Geĭlikman, *Issledovaniya po Fizike Nizkih Temperatur* (Studies in the Low Temperature Physics), Atomizdat, Moscow, 1979 (in Russian).
34. B. T. Geĭlikman, M. Yu. Raizer, *Sov. Phys. — Solid State* (1974), **16**, N1, 152–161.
35. A. S. Davydov, *Kvantovaya Mechanika* (Quantum Mechanics) Nauka, Moscow, 1973 (in Russian).
36. A. O. E. Animalu, *Intermediate Quantum Theory of Crystalline Solids*, Prentice–Hall, NJ, 1972.
37. J. M. Ziman, *Electrons and Phonons*, Clarendron, Oxford, 1976.
38. A. B. Midgal, *Sov. Phys. — JETP* (1958), **34**, N6, 1438–1446.
39. H. Bethe, A. Sommerfeld, *Electronentheorie der Metalle*, Springer, Berlin, 1967.

40. J. M. Ziman, *Electrons and Phonons*, Clarendron, Oxford, 1976.
41. C. B. Tyablikov, V. V. Tolmachev, *Sov. Phys. — JETP* (1958), **34**, N5, 1254–1257.
42. E. V. Zarochentsev, *Sov. Phys. — Solid State* (1981), **23**, N6, 1600–1605.
43. I. B. Bersuker, V. E. Polinger, *Vibronic Interaction in Molecules and Crystals*, Springer, Berlin, 1989.
44. H. Böttger, *Principles of the Theory of Lattice Dynamics*, Akademie–Verlag, Berlin, 1983 1982.
45. D. Thouless, *The Quantum Mechanics of Many-body Systems*, Academic Press, NY, 1961.
46. E. G. Brovman, Yu. Kagan, *Sov. Phys. — JETP* (1969), **57**, N4, 1329–1341.
47. V. G. Bar'yakhtar, E. V. Zarochentsev, V. O. Safronov, *Sov. Phys. — Metal Phys.* (1978), **46**, N4, 685–698.
48. B. T. Geĭlikman, *Sov. Phys. — Uspekhi* (1975), **115**, N3, 403–426.
49. I. E. Dragunov, S. M. Orel, *Sov. Phys. — High Pressure* (1989), **31**, 15–21.
50. G. Leibfried, W. Ludwig, in *Solid State Physics* (eds H. Ehrenreich, F. Seitz, D. Turnbull) **12**, Academic Press, NY, 1961.
51. D. Wallance, in *Solid State Physics* (eds H. Ehrenreich, F. Seitz, D. Turnbull), **25**, Academic Press, NY, 1970, 301–404.
52. A. A. Maradudin, E. W. Montroll, G. H. Weiss, in *Solid State Physics* (eds H. Ehrenreich, F. Seitz, D. Turnbull) Suppl. **3**, Academic Press, NY, 1963.
53. V. G. Vaks, *Vvedenie v Mikroskopicheskuyu teoriyu Segnetoelectricov* (Introduction to the Microscopic Theory of Ferroelectrics), Nauka, Moscow, 1973 (in Russian).
54. *Theory of the Inhomogeneous Electron Gas* (eds S. Lundqvist, N. H. March), Plenum, NY, 1983.
55. E. G. Brovman, Yu. Kagan, A. Kholas, *Sov. Phys. — JETP* (1969), **57**, N11, 1635–1645.

Chapter 2

1. W. A. Harrison, *Pseudopotentials*, Benjamin, NY, 1966.
2. J. M. Ziman, *The Calculation of Bloch Functions*, in *Solid State Physics*. (eds H. Ehrenreich, F. Seitz, D. Turnbull), **26**, Academic Press, NY, 1970, 1–101.
3. *The Physics of metals*, in *Solid State Physics* (ed. J. M. Ziman), **1**, Electrons, Cambridge University Press, London, 1968. mIR, 1972).
4. V. Heine, M. L. Cohen, D. Weaire, in *Solid State Physics* (eds H. Ehrenreich, F. Seitz, D. Turnbull) **24**, Academic Press, NY, 1970.
5. E. G. Brovman, Yu. Kagan, *Sov. Phys. — Uspekhi* (1974), **112**, N3, 369–426.
6. C. M. Bertoni, V. Bortolani, C. Calandra et al., *J. Phys. F.: Met Phys.* (1974), **4**, N1, 19–38.
7. G. V. Bachelet, D. R. Hamman, M. Schluter, *Phys. Rev. B.* (1982), **26**, N8, 4199–4228.
8. F. Bonsignori, A. Magnaterra, *Sol. State Commun.* (1982), **44**, N7, 1095–1096.

9. M. Hasegavwa, M. WatabeA, *J. Phys. C: Sol. State Phys.* (1983), **16**, N2, L29–L34.
10. V. Heine, I. V. Abarenkov, *Phil. Mag.* (1964), **9**, N99, 451–465.
11. V. Heine, I. V. Abarenkov, *ibid.* (1965), **12**, N117, 529–537.
12. M. Rasolt, R. Taylor, *J. Phys. F.* (1972), **2**, N2, 270–276.
13. M. Rasolt, R. Taylor, *Phys. Rev.* (1975), **11**, N8, 2717–2725.
14. L. Dagens, M. Rasolt, R. Taylor, *ibid.* (1975), **11**, N8, 2726–2734.
15. S. S. Cohen, M. L. Klein, M. S. Duesbery, et al., *J. Phys. F.* (1976), **6**, N10, L271–273.
16. R. W. Shaw, *Phys. Rev.* (1968), **174**, N3, 679–781.
17. A. O. Animalu, V. Heine, *Phil. Mag.* (1965), **12**, N120, 1249–1270.
18. M. Rasolt, R. Taylor, *J. Phys. F.* (1973), **3**, N1, 67–74.
19. D. R. Hamman, M. Schluter, C. Chiang, *Phys. Rev. Lett.* (1979), **43**, N20, 1494–1497.
20. *Theory of the Inhomogeneous Electron Gas* (eds S. Lundqvist, N. H. March), Plenum, NY, 1983.
21. R. W. Shaw, *J. Phys. C: Solid State* (1970), **3**, N5, 1140–1158.
22. A. Zunger, M. L. Cohen, *Phys. Rev. B.* (1978), **18**, N10, 5449–5472.
23. L. Dagens, *J. Phys. C.* (1972), **5**, N17, 2333–2344.
24. P. Hohenberg, W. Kohn, *Phys. Rev. B.* (1964), **136**, N3, 864–871.
25. W. Kohn, L. J. Sham, *Phys. Rev. A* (1965), **140**, N4, 1133–1138.
26. W. Kohn, L. J. Sham, *Phys. Rev.* (1966), **145**, N2, 561–567.
27. J. P. Perdew, S. H. Vosko, *J. Phys. F: Met. Phys.* (1976), **6**, N8, 1421–1431.
28. M. Appapillai, A. R. Williams, *ibid.* (1973), **3**, N4, 759–771.
29. S. Benckert, *Phys. Stat. Sol. B.* (1975), **69**, N2, 484–490.
30. A. O. Animalu, *Proc. Roy. Soc. (London)* (1966), **294**, N1438, 376–392.
31. C. Kittel, *Introduction to Solid state Physics*, Wiley, NY, 1971.
32. V. G. Vaks, E. V. Zarochentsev, S. P. Kravchuk et al., *Phys. status solidi (b)* (1978), **85**, N1, 63–74.
33. N. W. Ashcrofts, *Phys. Lett.* (1966), **23**, N1, 48–50.
34. G. L. Krasko, Z. A. Gurskij, *Sov. Phys. — JETP* (1969), **9**, N10, 596–601.
35. G. L. Krasko, Z. A. Gurskij, *Sov. Phys. — Doklady* (1971), **197**, N4, 810–813.
36. J. M. Ziman, *Principies in the Theory of Solids*, Cambridge University Press, London, 1964.
37. M. I. Katsnelson, A. V. Trefilov, *Sov. Phys. — Solid State* (1988), **30**, N11, 3299–3310.
38. T. M. Eremenko, E. V. Zarochentsev, *Sol. State Commun.* (1979), **30**, N6, 785–789.
39. D. Geldart, R. Taylor, *Can. J. Phys.* (1970), **48**, N1, 155–192.
40. R. Taylor, *J. Phys. F: Met. Phys.* (1978), **8**, N8, 1699–1702.
41. V. G. Vaks, A. V. Trefilov, *Sov. Phys. — Solid State* (1977), **19**, N1, 244–258.
42. D. Geldart, S. H. Vosko, *Can. J. Phys.* (1966), **44**, N9, 2137–2171.
43. F. Toigo, T. O. Woodruff, *Phys. Rev.* (1970), **2**, N10, 3958–3966 (See correction: *Phys. Rev. B.* (1971), **4**, N12, 4312–4315).
44. P. Vashishta, K. S. Singwi, *Phys. Rev. B.* (1972), **6**, N3, 875–887; N12, 4883–4885.

45. E. V. Zarochentsev, S. P. Kravchuk, T. M. Tarusina, *Sov. Phys. — Solid State* (1976), **18**, N2, 413–422.
46. V. G. Vaks, E. V. Zarochentsev, S. P. Kravchuk et al., *K Teorii Atomnyh i Teplovyh Swoitv Shchelochnyh Melallov* (About Theory of Atomic and Thermal Properties of Alkali Metals), Preprint IFM SO Ac. Sc. USSR Sverdlovsk, 1976 (in Russian).
47. V. G. Vaks, E. V. Zarochentsev, S. P. Kravchuk et al., *J. Phys. F: Met. Phys.* (1978), **8**, N5, 725–742.
48. T. M. Eremenko, E. V. Zarochentsev, *Ukr. Phys. Zourn.* (1980), **25**, N2, 262–270.
49. T. M. Eremenko, E. V. Zarochentsev, *Sov. Phys. — High Pressure*, (1980), N2, 9–15.
50. V. G. Vaks, A. V. Trefilov, *Sov. Phys. — Solid State* (1978), **20**, N11, 631–632.
51. E. V. Zarochentsev, S. V. Teplov, V. O. Safronov, *Sov. Phys. — Low Temp.* (1978), **4**, N2, 382–393.
52. V. G. Vaks, S. P. Kravchuk, A. V. Trefilov, *Sov. Phys. — Solid State* (1977), **19**, N5, 1271–1278.
53. E. V. Zarochentsev, S. V. Teplov, *Sov. Phys. — High Pressure*, (1980), N2, 37–43.
54. D. L. Price, K. S. Singwi, M. P. Tosi, *Phys. Rev. B.* (1970), **2**, N8, 2983–2999.
55. A. D. Woods, B. N. Brockhouse, R. N. March et al., *Phys. Rev.* (1962), **128**, N3, 1112–1120.
56. R. A. Cowley, A. D. B. Wools, G. Dolling, *Phys. Rev.* (1966), **150**, N2, 487–494.
57. J. R. D. Copley, B. N. Brockhouse, *Can. J. Phys.* (1973), **51**, N3, 657–665.
58. J. Mizuri, C. Stassis, *Phys. Rev. B.* (1986), **34**, N8, 5890–5893.
59. R. Benedek, *ibid.* (1977), **15**, N7, 2902–2913.
60. V. G. Vaks, S. P. Kravchuk, A. V. Trefilov, *Sov. Phys. — Solid State* (1977), **44**, N6, 1151–1162.
61. D. Pines, Ph. Nozieres, *The Theory of Quantum Liquids*, Addison-Wesley, NY, 1967.
62. J. S. Dugdale, D. Gudan, *Proc. Roy. Soc.* (1960), **A254**, N1277, 184–204.
63. J. D. Filby, D. L. Martin, *ibid.* (1963), **A276**, N1365, 187–203.
64. E. V. Zarochentsev, S. V. Teplov, *K Teorii Adiabaticheskogo Potenciala Prostyh Metalov* (About Theory of the Adiabatic Potential of simple Metals), Donetsk, 1980, (Preprint. Phys. — Tech. Inst. Ac. Sc. Ukr. SSR, N10 (in Russian).
65. T. V. Zarochentsev, S. V. Teplov, *Sol. Stat. Commun.* (1980), **33**, N8, 1107–1110.
66. J. M. Luttinger, J. C. Ward, *Phys. Rev.* (1960), **118**, N5, 1417–1427.
67. E. G. Brovman, A. Kholas, *Sov. Phys. — JETP* (1974), **66**, N5, 1877–1894.
68. P. Lloyd, C. A. Sholl, *J. Phys. C.* (1968), **1**, N6, 1620–1632.
69. N. W. Ashcroft, *Phys. Rev.* (1965), **140A**, N3, 935–940.
70. E. V. Zarochentsev, O. M. Kolgushev, E. P. Troitskaya, *Sov. Phys. — High Pressure* (1992), **2**, N4, 45–53.
71. J. Mathews, R. I. Walker, *Mathematical Methods of Physics*, Benjamin, NY, 1970.
72. L. Schiff, *Quantum Mechanics*, McGraw-Hill, NY, 1955.
73. L. Kleinman, *Phys. Rev. B.* (1980), **21**, N6, 2630–2631.
74. D. M. Ceperley, B. J. Alder, *Phys. Rev. Lett.* (1980), **45**, N7, 566–569.

75. J. P. Perdew, A. Zunger, *Phys. Rew.* (1981), **23**, N10, 5048–5079.
76. G. B. Bachelet, M. Schluter, *ibid.* (1982), **25**, N4, 2103–2107.
77. V. D. Gorobchenko, E. G. Maksimov, *Sov. Phys. — Uspekhi* (1980), **130**, N1, 65–111.
78. O. V. Dolgov, E. G. Maksimov, *ibid.* (1981), **135**, N3, 441–477.
79. O. V. Dolgov, E. G. Maksimov, *ibid.* (1982), **138**, N1, 95–128.
80. M. Rasolt, S. H. Vosko, *Phys. Rev. B.* (1974), **10**, N10, 4195–4204.
81. J. Hubbard, *Proc. Roy. Soc.* (1958), **A243**, N1296, 336–357.
82. J. S. Langer, *Phys. Rev.* (1961), **124**, N4, 997–1010.
83. E. M. Lifshitz, L. P. Pitaevski, *Statistical Physics*, Pergamon Press, NY, 1980, Part 2.
84. L. Hedin, *Phys. Rev. B.* (1965), **9**, N3, A796–833.
85. M. Rasolt, R. Taylor, *J. Phys. F: Met. Phys.* (1973), **3**, N9, 1678–1682.
86. E. V. Zarochentsev, E. Ya. Fain, *Sov. Phys. — High Pressure* (1982), N10, 61–69.
87. S. V. Fomichev, M. I. Katsnelson, V. G. Koreshkov, A. V. Trefilov, *Phys. status solidi (b)* (1990), **161**, N1, 153–164.
88. B. A. Greenberg, M. I. Katsnelson, V. G. Koreshkov, Yu. N. Osetskii, G. V. Prschanskikh, A. V. Trefilov, Yu. F. Shamanaev, L. I. Yakovenkova, *Phys. status solidi (b)* (1990), **158**, N2, 441–455.
89. E. V. Zarochentsev, *Sov. Phys. — Solid State* (1981), **23**, N6, 1600–1605.
90. M. I. Katsnelson, V. G. Koreshkov, A. V. Trefilov, IFM Preprint, 86/1, Sverdlovsk 1986.
91. A. V. Trefilov, Dr. Sc. Thesis, Moscow, 1990 (in Russian).
92. J. C. Slater, *The Self-consistent Field for Moleculas and Solids*, McGraw-Hill, NY, 1974.
93. I. V. Abarenkov, I. M. Antonova, V. G. Bar'yakhtar, V. L. Bulatov, E. V. Zarochentsev, *Metody Vichislitelnoj Fiziki v Teoriii Tverdogo Tela. Zonnaya Teoriya Metallov. Elektonnaya Struktura Idealnyh i defectnyh Kristallov* (Methods of Computational Physics in Solid State Theory. Electron Structure of Ideal and Defect Crystals), Kiev, Naukova Dumka, 1991 (in Russian).
94. J. P. Perdew, A. Zunger, *Phys. Rew.* (1981), **23**, N10, 5048–5079.
95. V. G. Bar'yakhtar, E. V. Zarochentsev, E. P. Troitskaya, *Metody Vichislitelnoj Fiziki v Teoriii Tverdogo Tela. Atomnye svoistva metallov.* (Methods of Computational Physics in Solid State Theory. Atomic properties of metals), Kiev, Naukova Dumka, 1990 (in Russian).

Chapter 3

1. V. G. Vaks, A. V. Trefilov, *Sov. Phys. — Solid State* (1977), **19**, N1, 244–258.
2. V. G. Vaks, S. P. Kravchuk, A. V. Trefilov, *J. Phys. F.: Met. Phys.* (1980), **10**, N10, 2105–2124.
3. V. G. Vaks, S. P. Kravchuk, A. V. Trefilov, *Sov. Phys. — Solid State* (1979), **21**, N11, 3370–3378.

4. V. G. Vaks, E. V. Zarochentsev, S. P. Kravchuk et al., *Phys. status solidi (b)* (1978), **85**, N1, 63–74.
5. S. V. Fomichev, M. I. Katsnelson, V. G. Koreshkov, A. V. Trefilov, *Phys. status solidi (b)* (1990), **161**, N1, 153–164.
6. T. M. Eremenko, E. V. Zarochentsev, V. O. Safronov, *Ukr. Phys. Zourn.* (1978), **23**, N3, 423–432.
7. V. G. Vaks, S. P. Kravchuk, A. V. Trefilov, *J. Phys. F.: Met. Phys.* (1980), **10**, N11, 2325–2344.
8. V. G. Vaks, N. E. Zein, S. P. Kravchuk et al., *Phys. status solidi (b)* (1979), **96**, N3, 856–865.
9. S. V. Teplov, Ph.D. Thesis, Donetsk, 1981 (in Russian).
10. S. V. Teplov, V. V. Nemoshkalenko et al., *Sov. Phys. — High Pressure* (1984), N15, 54–59.
11. E. V. Zarochentsev, S. V. Teplov, V. O. Safronov, *Sov. Phys. — Low Temp.* (1978), **4**, N2, 382–393.
12. E. V. Zarochentsev, V. O. Safronov, A. V. Trefilov, *ibid.* (1977), **3**, N2, 209–216.
13. V. G. Bar'yakhtar, E. V. Zarochentsev, V. O. Safronov, *Sov. Phys. — Metal Phys.* (1978), **45**, N4, 704–710.
14. T. V. Zarochentsev, S. V. Teplov, *Sol. Stat. Commun.* (1980), **33**, N8, 1107–1110.
15. E. V. Zarochentsev, S. V. Teplov, *K Teorii Adiabaticheskogo Potenciala Prostyh Metalov* (About the Theory of the Adiabatic Potential of simple Metals), Donetsk, 1980, Preprint. Phys. — Tech. Inst. Ac. Sc. Ukr. SSR, N10 (in Russian).
16. E. V. Zarochentsev, S. P. Kravchuk, T. M. Tarusina, *Sov. Phys. — Solid State* (1976), **18**, N2, 413–422.
17. V. G. Vaks, E. V. Zarochentsev, S. P. Kravchuk, V. O. Safronov, A. V. Trefilov, *Phys. status solidi (b)* (1978), **85**, N2, 749–759.
18. V. G. Vaks, E. V. Zarochentsev, S. P. Kravchuk, et al., *J. Phys. F: Met. Phys.* (1978), **8**, N5, 725–742.
19. T. M. Eremenko, E. V. Zarochentsev, V. O. Safronov, *Ukr. Phys. Zourn.* (1978), **23**, N9, 1481–1488 (in Russian).
20. V. G. Vaks, S. P. Kravchuk, A. V. Trefilov, *Sov. Phys. — Solid State* (1977), **19**, N5, 1271–1278.
21. T. M. Eremenko, E. V. Zarochentsev, *Sol. State Commun.* (1979), **30**, N6, 785–789.
22. T. M. Eremenko, E. V. Zarochentsev, *Sov. Phys. — High Pressure*, (1980), N2, 9–15.
23. E. V. Zarochentsev, S. V. Teplov, *Sov. Phys. — High Pressure*, (1980), N2, 37–43.
24. T. M. Eremenko, E. V. Zarochentsev, *Ukr. Phys. Zourn.* (1980), **25**, N2, 262–270.
25. E. V. Zarochentsev, V. O. Safronov, E. P. Troitskaya, *Sov. Phys. — Solid State* (1978), **20**, N10, 2881–2888.
26. M. L. Cohen, *Phys. Sscripta* (1982), **1**, 31–66.
27. E. Yu. Tonkov, *Fazovye Diagrammy Elementov pod Vysokim Davleniem* (Phase Diagrans of Elements at Large Pressures), Nauka, Moscow, 1979 (in Russian).
28. F. Bonsignori, A. Magnaterra, *Nuovo cimento.* (1982), **D1**, N6, 789–801.
29. W. A. Harrison, *Pseudopotentials*, Benjamin, NY, 1966.

30. E. G. Brovman, Yu. Kagan, A. Kholas, *Sov. Phys. — Solid State* (1970), **12**, N4, 1001–1023.
31. T. Soma, *Physica.* (1979), **97B**, N1, 76–86.
32. J. M. Ziman, *Principies in the Theory of Solids*, Cambridge University Press, London, 1964.
33. E. G. Brovman, Yu. Kagan, *Sov. Phys. — Uspekhi* (1974), **112**, N3, 369–426.
34. V. G. Bar'yakhtar, E. V. Zarochentsev, V. O. Safronov, *Sov. Phys. — Metal Phys.* (1978), **46**, N4, 685–698.
35. D. Wallance, *Thermodynamics of crystals*, J. Wiley, NY, 1972.
36. J. R. D. Copley, C. A. Rotter, H. C. Smith et al., *Phys. Rev. Lett.* (1974), **33**, N6, 365–367.
37. J. Mizuri, C. Stassis, *Phys. Rev. B.* (1986), **34**, N8, 5890–5893.
38. N. Nucker, U. Buchenau, *ibid.* (1985), **31**, N8, 5479–5482.
39. J. Meyer, G. Dolling, J. Kalus et al., *J. Phys. F.: Metal Phys.* (1976), **6**, N10, 1899–1914.
40. G. Dolling, J. Meyer, *ibid.* (1977), **7**, N5, 775–779.
41. M. Born, K. Huang, *Dynamical Theory of Crystall Lattices*, Oxford University Press, London, 1954.
42. V. G. Vaks, *Vvedenie v Mikroskopicheskuyu teoriyu Segnetoelectricov* (Introduction to the Microscopic Theory of Ferroelectrics), Nauka, Moscow, 1973 (in Russian).
43. R. Taylor, H. R. Glyde, *J. Phys. F.: Metal Phys.* (1976), **6**, N10, 1915–1922.
44. J. Rosen, G. Grimvall, *Phys. Rev. B.* (1983), **27**, N12, 7199–7208.
45. R. A. Cowley, *Advan. Phys.* (1963), **12**, N45–48, 421–480.
46. A. D. Woods, B. N. Brockhouse, R. N. March et al., *Phys. Rev.* (1962), **128**, N3, 1112–1120.
47. A. J. Millington, G. L. Squires, *J. Phys. F.: Metal Phys.* (1971), **3**, N1, 244–257.
48. W. J. L. Buyers, R. A. Cowley, *Phys. Rev.* (1969), **180**, N3, 755–766.
49. J. R. D. Copley, *Can. J. Phys.* (1973), **51**, N11, 2564–2586.
50. M. M. Beg, M. Nielsen, *Phys. Rev.* (1976), **B14**, N10, 4266–4273.
51. S. H. Taole, H. R. Glyde, R. Taylor, *Phys. Rev. B.* (1978), **18**, N6, 2643–2655.
52. H. R. Glyde, J. P. Hansen, M. L. Klein, *ibid.* (1977), **16**, N8, 3476–3483.
53. H. R. Glyde, R. Taylor, *ibid.* (1972), **5**, N4, 1206–1213.
54. M. S. Duesbery, H. R. Glyde, R. Taylor, *ibid.* (1973), **8**, N4, 1372–1378.
55. L. D. Landau, Yu. Rummer, *Phys. Z. Sowjet.* (1937), **11**, 18.
56. A. I. Akhiezer, *Sov. Phys. — JETP* (1938), **8**, N12, 1318–1329 (in Russian).
57. B. Ya. Balagurov, V. G. Vaks, *Sov. Phys. — JETP* (1970), **57**, N11, 1646–1659.
58. Y. P. Varshni, *Phys. Rev. B.* (1970), **2**, N10, 3952–3958.
59. Ya. I. Frenkel, *Vvedenie v Teoriyu Metallov* (Introduction to the Metal Theory), GITTL, Moscow, 1948 (in Russian).
60. G. R. Barsch, *Phys. status solidi (b)* (1967), **19**, N1, 129–138.
61. E. V. Zarochentsev, S. M. Orel, V. N. Varyukhin, *Phys. status solidi (a)* (1979), **52**, N2, 455–462.
62. E. V. Zarochentsev, S. M. Orel, V. N. Varyukhin, *Phys. status solidi (a)* (1979), **53**, N1, 75–85.

63. D. R. Schouten, C. A. Swenson, *Phys. Rev. B.* (1974), **10**, N6, 2175–2185.
64. D. Geldart, R. Taylor, *Can. J. Phys.* (1970), **48**, N1, 155–192.
65. G. K. White, J. G. Collins, *J. Low Temp. Phys.* (1972), **7**, N1, 43–75.
66. B. Yates, C. H. Panter, *Proc. Phys. Soc.* (1962), **80**, N514, 373–382.
67. D. J. Rasky, F. Milstein, *Phys. Rev. B.* (1986), **33**, N4, 2765–2780.
68. P. W. Bridgman, *Proc. Am. Acad. Sci.* (1948), **76**, N3, 71–87.
69. M. S. Anderson, E. J. Gutman, J. R. Packard et al., *J. Phys. Chem. Solids* (1969), **30**, N6, 1587–1601.
70. H. T. Hall, L. Merrill, J. D. Barnett, *Science.* (1964), **146**, N3636, 61–63.
71. S. M. Vajdya, I. C. Getting, G. C. Kennedy, *J. Phys. Chem. Solids* (1971), **32**, N11, 2545–2556.
72. B. T. Geĭlikman, *Sov. Phys. — Uspekhi* (1975), **115**, N3, 403–426.
73. R. Srinivasan, K. S. Girirajan, *J. Phys. F.: Metal Phys.* (1974), **4**, N7, 951–959.
74. A. Kholas, Ph. D. Thesis, JINR, Dubna, 1970 (in Russian).
75. S. K. Schiferl, D. C. Wallace, *Phys. Rev. B.* (1985), **31**, N12, 7662–7667.
76. Z. S. Basinski, M. S. Duesbery, A. P. Pogany, *Can. J. Phys.* (1970), **48**, N12, 1480–1488.
77. S. S. Cohen, M. L. Klein, *Phys. Rev. B.* (1975), **12**, N8, 2984–2987.
78. S. S. Cohen, M. L. Klein, M. S. Duesbery et al., *J. Phys. F.* (1976), **6**, N3, 337–347.
79. L. Dagens, M. Rasolt, R. Taylor, *Phys. Rev. B.* (1975), **11**, N8, 2726–2734.
80. G. Leibfried, W. Ludwig, in *Solid State Physics* (eds H. Ehrenreich, F. Seitz, D. Turnbull) **12**, Academic Press, NY, 1961.
81. D. Geldart, S. H. Vosko, *Can. J. Phys.* (1966), **44**, N9, 2137–2171.
82. F. Toigo, T. O. Woodruff, *Phys. Rev. B.* (1970), **2**, N10, 3958–3966.
83. F. Toigo, T. O. Woodruff, *Phys. Rev. B.* (1971), **4**, N12, 4312–4315.
84. P. Vashishta, K. S. Singwi, *Phys. Rev. B.* (1972), **6**, N3, 875–887; N12, 4883–4885.
85. D. L. Martin, *Can. J. Phys.* (1970), **48**, N11, 1327–1339.
86. D. Wallance, in *Solid State Physics* (eds H. Ehrenreich, F. Seitz, D. Turnbull), **25**, Academic Press, NY, 1970, 301–404.
87. A. L. Jain, *Phys. Rev.* (1961), **123**, N4, 1234–1238.
88. T. Slotvinski, J. Trivinsonno, *J. Phys. Chem. Sol.* (1969), **30**, N6, 1276–1278.
89. C. A. Swenson, *ibid.* (1966), **27**, N1, 33–38.
90. H. C. Nash, C. S. Smith, *ibid.* (1959), **9**, N2, 113–118.
91. J. P. Day, A. L. Ruoff, *Phys. status solidi (a)* (1974), **25**, N1, 205–213.
92. M. E. Diederich, J. Trivisonno, *J. Phys. Chem. Solids* (1966), **27**, N4, 637–642.
93. R. I. Beercroft, C. A. Swenson, *J. Phys. Chem. Solids* (1961), **18**, N2, 329–344.
94. R. H. Martinson, *Phys. Rev.* (1969), **178**, N3, 902–913.
95. G. Fritsch, F. Geipel, A. Prasetyo, *J. Phys. Chem. Solids* (1973), **34**, N9, 1961–1969.
96. P. S. Ho, A. L. Ruoff, *ibid.* (1968), **29**, N12, 2101–2111.
97. W. R. Marquardt, J. Trivisonno, *J. Phys. Chem. Solids* (1965), **26**, N2, 273–278.
98. E. J. Gutman, J. Trivisonno, *J. Phys. Chem. Solids* (1967), **28**, N5, 805–809.
99. C. R. Monford, C. A. Swenson, *J. Phys. Chem. Solids* (1965), **26**, N2, 291–301.

100. P. A. Smith, C. S. Smith, *J. Phys. Chem. Solids* (1965), **26**, N2, 279–289.
101. G. Fritsch, H. Bube, *Phys. status solidi (a)* (1975), **30**, N2, 571–576.
102. C. A. Swenson, *Phys. Rev.* (1955), **99**, N2, 423–429.
103. F. I. Kollarits, J. Trivisonno, *J. Phys. Chem. Solids* (1968), **29**, N11, 2133–2139.
104. I. N. Makarenko, A. M. Nikolaenko, V. A. Ivanov, S. M. Stishov, *Sov. Phys. — JETP* (1975), **69**, N5, 1723–1733.
105. N. Metropolis, A. W. Rosenbluth, M. N. Rosenbluth et al., *J. Chem. Phys.* (1953), **21**, N6, 1087–1092.
106. E. R. Cowley, R. A. Cowley, *Proc. Roy. Soc. (London)* (1965), **A287**, N1409, 259–280.
107. R. N. Glaytor, B. J. Marchall, *Phys. Rev.* (1960), **120**, N2, 332–334.
108. G. N. Kamm, G. A. Alers, *J. Appl Phys.* (1964), **35**, N2, 327–330.
109. I. E. Dragunov, E. V. Zarochentsev, S. M. Orel, *Sov. Phys. — Metal Phys.* (1989), **67**, N5, 837–844.
110. K. J. Dunn, A. L. Ruoff, *Phys. Rev.* (1974), **10**, N6, 2271–2274
111. G. Leibfried, in *Handbuch der Physik* (ed. S. Flügge), **7**, Part 1, Springer, Berlin, 1958.
112. W. E. Boule , R. J. Sladek, *Phys. Rev. B.* (1975), **11**, N8, 2933–2940.
113. K. B. Tolpygo, *Sov. Phys. — Solid State* (1959), **1**, 211–227.
114. E. V. Zarochentsev, V. I. Orekhov, E. P. Troitskaya, *Sov. Phys. — Solid State* (1974), **16**, N8, 2249–2255.
115. C. M. Bertoni, V. Bortolani, C. Calandra et al., *J. Phys. F.: Met Phys.* (1974), **4**, N1, 19–38.
116. T. Suzuki, *Phys. Rev. B.* (1971), **3**, N12, 4007–4014.
117. K. Shimada, *Phys. status solidi (b)* (1974), **61**, N2, 325–335.
118. M. A. Coulthard, *J. Phys. C: Solid State* (1970), **3**, N4, 820–834.
119. J. Hafner, P. Schmunk, *Phys. Rev* (1974), **9**, N10, 4138–4150.
120. J. Hafner, *Z. Physik. B.* (1975), **22**, N2, 351–357.
121. S. Benckert, *Phys. status solidi (b)* (1975), **69**, N2, 484–490.
122. V. M. Gerasimov, *Sov. Phys. — Solid State* (1978), **20**, N9, 2570–2574.
123. T. Soma, Y. Konno, H.–M. Kagaya, *Phys. status solidi (b)* (1983), **117**, N2, 743–748.
124. T. Soma, T. Iton, H.-M. Kagaya, *Phys. status solidi (b)* (1984), **125**, N1, 107–112.
125. J. Hammerberg, N. W. Ashcroft, *Phys. Rev. B.* (1974), **9**, N2, 409–424.
126. *Theory of the Inhomogeneous Electron Gas* (eds S. Lundqvist, N. H. March), Plenum, NY, 1983.
127. D. Pines, Ph. Nozieres, *The Theory of Quantum Liquids*, Addison–Wesley, NY, 1967.
128. R. Taylor, *J. Phys. F: Met. Phys.* (1978), **8**, N8, 1699–1702.
129. M. Rasolt, R. Taylor, *Phys. Rev.* (1975), **11**, N8, 2717–2725.
130. M. Rasolt, R. Taylor, *J. Phys. F.* (1972), **2**, N2, 270–276.
131. C. Kittel, *Introduction to Solid state Physics*, Wiley, NY, 1971.
132. P. S. Ho, A. L. Ruoff, *J. Appl. Phy.* (1969), **40**, N8, 3251–3256.
133. C. Moore, *Atomic Energy Levels*, US. Nat. Bur. Stand. (1975), 1949–1958.

134. V. Sarma, J. P. Reddy, *Phys. status solidi (a)* (1972), **10**, N2, 563–567.
135. J. R. Tomas, *Phys. Rev.* (1968), **175**, N3, 955–962.
136. D. L. Waldorf, G. A. Alers, *J. Appl. Phys.* (1962), **33**, N11, 3266–3269.
137. R. A. Miller, D. E. Shuele, *J. Phys. Chem. Solids* (1968), **30**, N3, 589–600.
138. L. A. Girifalko, V. G. Weizer, *Phys. Rev.* (1959), **114**, N3, 687–690.
139. S. S. Mathur, P. N. Gupta, *Acuastica* (1974), **34**, N1, 114–118.
140. B. Prasad, R. S. Srivastava, *Phys. status solidi (b)* (1973), **57**, N2, 743–747.
141. C. Stassis, J. Zaretsky, D. K. Misemer et al., *Phys. Rev. B.* (1983), **27**, N6, 3303–3307.
142. F. Bonsignori, A. Magnaterra, *Sol. State Commun.* (1982), **44**, N7, 1095–1096.
143. M. M. Dacorogna, M. L. Cohen, *Phys. Rev. B.* (1986), **34**, N8, 4996–5002.
144. D. Sen, S. K. Sarkar, S. Sengupta et al., *Phys. status solidi* (1983), **115**, N2, 593–602.
145. N. W. Ashcroft, *Phys. Lett.* (1966), **23**, N1, 48–50.
146. E. V. Zarochentsev, O. M. Kolgushev, E. P. Troitskaya, *Sov. Phys. — High Pressure* (1992), **2**, N4, 31–39.
147. J. P. Perdew, A. Zunger, *Phys. Rev. B.* (1981), **23**, N10, 5048–5079.
148. J. M. Ziman, *Electrons and Phonons*, Clarendron, Oxford, 1976.
149. V. Heine, M. L. Cohen, D. Weaire, in *Solid State Physics* (eds H. Ehrenreich, F. Seitz, D. Turnbull) **24**, Academic Press, NY, 1970.
150. M. I. Katsnelson., A. V. Trefilov, *Sov. Phys. — Solid State* (1988), **30**, N 11, 3299–3310.
151. A. V. Trefilov, Dr. Sc. Thesis, Donetsk, 1990 (in Russian).
152. G. V. Bachelet, D. R. Hamman, M. Schluter, *Phys. Rev. B.* (1982), **26**, N8, 4199–4228.
153. N. E. Zein, V. V. Kamyshenko, G. D. Samolyuk, *Sov. Phys. — Solid State* (1990), **32**, N6, 1846–1853.
154. N. Takeuchi, C. T. Chan, K. M. Ho, *Phys. Rev. B.* (1989), **40**, N3, 1566–1570.
155. A. Zunger, M. L. Cohen, *Phys. Rev. B.* (1979), **19**, 568–582.
156. C.-L. Fu, K.-M. Ho, *Phys. Rev. B.* (1983), **28**, 5480–5486.
157. K.-M. Ho, C.-L. Fu, B. N. Harmon, *Phys. Rev. B.* (1984), **29**, N4, 1575–1587.
158. G. E. Grechnev, *Sov. Phys. — Low Temp.* (1983), **9**, 1060–1065.
159. J. J. Rehr, E. Zaremba, W. Kohn, *Phys. Rev. B.* (1975), **12**, N6, 2062–2066.
160. J. Mahantu, R. Taylor, *Phys. Rev. B.* (1978), **17**, N2, 556–559.
161. D. D. Richardson, J. Mhanty, *J. Phys. C.* (1977), **10**, N 20, 3971–3976.
162. J. C. Upadhaya, S. Wang, *Phys. Lett.* (1979), **73A**, N3, 238–240; *J. Phys. F.* (1980), **10**, 441–444.
163. J. C. Upadhaya, S. Wang, R. A. Moore, *Can. J. Phys.* (1980), **58**, N3, 905–908.
164. J. Cheung, N. W. Aschcroft, *Phys. Rev. B.* (1981), **24**, N4, 1638–1642.
165. K. K. Mon, N. W. Aschcroft, G. V. Chester, *Phys. Rev. B.* (1979), **19**, N10, 5103–5122.
166. J. C. Slater, *Quantum Structure of Molecules and Solids*, McGraw-Hill, NY, 1963.
167. R. Benedek, *Phys. Rev. B.* (1977), **15**, N6. 2902–2913.
168. R. Niemien, M. J. Puska, *Phys. Scr.* (1982), **25**, N6, 952–956.

169. V. G. Vaks, A. V. Trefilov, *Sov. Phys. — Solid State* (1978), **20**, N11, 631–632.
170. V. G. Bar'yakhtar, E. V. Zarochentsev, E. P. Troitskaya, *Metody Vichislitelnoj Fiziki v Teoriii Tverdogo Tela. Atomnye svoistva metallov.* (Methods of Computational Physics in Solid State Theory. Atomic properties of metals), Kiev, Naukova Dumka, 1990 (in Russian).
171. C. Moore, *Atomic Energy Levels*, US. Nat. Bur. Stand. (1975), 1949–1958.
172. V. G. Vaks, E. V. Zarochentsev, S. P. Kravchuk, V. O. Safronov, A. V. Trefilov, *Phys. status solidi (b)* (1978), **85**, N6, 749–759.
173. T. M. Eremenko, E. V. Zarochentsev, *Solid State Commun.* (1979), **30**, N6, 785–789.
174. T. M. Eremenko, E. V. Zarochentsev, *Ukr. Phys. Zourn.* (1980), **25**, N2, 262–270.
175. M. I. Katsnelson, A. V. Trefilov, *Sov. Phys. — Uspekhi,* (1988), **154**, N3, 523–525.
176. S. V. Vonsovsky, M. I. Katsnelson, A. V. Trefilov, *J. Magn. and Magn. Matter.* (1986), **61**, N1, 83– 87.
177. M. I. Katsnelson, A. V. Trefilov, *Physica B.* (1990), **163**, N2, 182–184.
178. M. I. Katsnelson, A. V. Trefilov, *Phys. Lett. A.* (1985), **109**, N3, 109–112.
179. A. B. Migdal, *Teoriya Konechnych Fermi-sistem*, Theory of finite Fermi-systems, Nauka, Moscow, 1983 (in Russian).
180. P. Nozieres, *Theory of Interacting Fermi Systems*, Benjamin, NY, 1963.
181. V. D. Gorobchenko, E. G. Maksimov, *Sov. Phys. — Uspekhi*, (1980), **130**, N1, 65–111.
182. E. M. Lifshitz, L. P. Pitaevski, *Statistical Phisics*, Pergamon Press, NY, 1980.
183. P. Fulde, M. Loewenhaup, *Adv. Phys.* (1985), **34**, 589–637, Part 2.
184. M. I. Katsnelson, A. A. Soldatov, A. V. Trefilov, *Sov. Phys. — Metal Phys.* (1985), **59**, N5, 883–888.

Chapter 4

1. V. Heine, M. L. Cohen, D. Weaire, in *Solid State Physics* (eds H. Ehrenreich, F. Seitz, D. Turnbull) **24**, Academic Press, NY, 1970.
2. R. P. Bajpai, M. Ono, Y. Ohno et al., *Phys. Rev.* (1975), **12**, N6, 2194–2211.
3. G. K. Straub, D. C. Wallace, *Phys. Rev.* (1971), **3**, N4, 1234–1239.
4. R. Pynn, I. Ebbsjo, *J. Phys. F* (1971), **1**, N5, 744–752.
5. T. Shneider, E. Stoll. K teorii metallicheskogo litiya, in *Vychislitelnye metody v teorii tverdogo tela*, (Computational methods of solid state physics), Mir, Moscow, 1975, 196–208 (in Russian).
6. A. O. E. Animalu, *Phys. Rev.* (1967), **161**, N2, 445–455.
7. J. Hafner, H. Nowotny, *Phys. status solidi (b)* (1972), **51**, N1, 107–114.
8. M. Appapillai, A. R. Williams, *ibid.* (1973), **3**, N4, 759–771.
9. *Theory of the Inhomogeneous Electron Gas* (eds S. Lundqvist, N. H. March), Plenum, NY, 1983.
10. G. V. Bachelet, D. R. Hamman, M. Schluter, *Phys. Rev. B.* (1982), **26**, N8, 4199–4228.

11. J. Hafner, V. Heine, *J. Phys. F.* (1983), **13**, N11, 2479–2501.
12. M. M. Dacorogna, M. L. Cohen, *Phys. Rev. B.* (1986), **34**, N8, 4996–5002.
13. V. G. Vaks, S. P. Kravchuk, A. V. Trefilov, *Sov. Phys. — Solid State* (1977), **19**, N11, 3396–3399.
14. A. V. Trefilov, Ph. D. Thesis, Moscow, 1977 (in Russian).
15. T. M. Eremenko, E. V. Zarochentsev, *Sol. State Commun.* (1979), **30**, N6, 785–789.
16. H. L. Skriver, *Phys. Rev. B.* (1985), **31**, N4, 1909–1923.
17. C. S. Barret, *Astra. Cryst.* (1956), **29**, N8, 671–677.
18. A. A. Chernyshov, V. A. Sukhoparov, A. Sadykov, *Sov. Phys. — JETP* (1983), **37**, N8, 345–348.
19. D. A. Young, M. Ross, *Phys. Rev. B.* (1984), **29**, N2, 682–691.
20. B. Olinger, J. M. Shaner, *Science* (1983), **219**, N4588, 1071–1072.
21. G. Oomi, S. V. Woods, *Phys. status solidi (b)* (1985), **130**, N1, K77–K80.
22. T. H. Lin, K. J. Dunn, *Phys. Rev. B.* (1986), **33**, N2, 807–811.
23. G. Oomi, M. A. K. Mohammed, S. V. Woods, *Sol. State Commun.* (1987), **62**, N3, 141–143.
24. C. M. McCarthy, C. W. Tomrson, S. A. Werner, *Phys. Rev. B.* (1980), **22**, N2, 574–580.
25. A. W. Overhouser, *Phys. Lett.* (1984), **53**, N1, 64–65.
26. R. Berliner, S. A. Werner, *Physica B.* (1986), **136**, N2, 481–484.
27. H. G. Smith, *Phys. Rev. Lett.* (1987) **58**, N2, 1228–1231.
28. A. K. McMahan, J. A. Moriarty, *Phys. Rev. B.* (1983), **27**, N6, 3235–3251.
29. M. M. Dacorogna, M. L. Cohen, *Phys. Rev. B.* (1986), **34**, N8, 4996–5002.
30. A. D. Zdetsis, *Phys. Rev. B.* (1986), **34**, N11, 7666–7669.
31. J. M. Ziman, *Principles in the Theory of Solids*, Cambridge University Press, London, 1964.
32. H. I. Skriver, *The LMTO method*, Springer, NY, 1984.
33. A. V. Trefilov, Dr. Sc. Thesis, Moscow, 1990 (in Russian).
34. V. V. Evdokoimova, *Sov. Phys. — Uspekhi* (1966), **88**, N1, 93–125.
35. E. Yu. Tonkov, *Fazovye Diagrammy Elementov pod Vysokim Davleniem* (Phase Diagrans of Elements at Large Pressures), Nauka, Moscow, 1979 (in Russian).
36. R. Srinivasan, K. S. Girirajan, *J. Phys. F.: Metal Phys.* (1974), **4**, N7, 951–959.
37. A. M. Bratkovskij, V. G. Vaks, A. V. Trefilov, *Sov. Phys. — JETP* (1984), **86**, N6, 2114–2131.
38. T. M. Eremenko, E. V. Zarochentsev, *Ukr. Phys. Zourn.* (1980), **25**, N2, 262–270.
39. T. M. Eremenko, E. V. Zarochentsev, *Sov. Phys. — High Pressure* (1980), N2, 9–15.
40. D. L. Martin, *Proc. Roy. Soc.* (1960), **A254**, N1279, 433–454.
41. J. S. Dugdale, D. L. Gudan, *ibid.*, (1960), **A254**, N1277, 184–204.
42. J. D. Filby, D. L. Martin, *Ibid.* (1963), **A276**, N1365, 187–203.
43. V. G. Vaks, *Vvedenie v Mikroskopicheskuyu teoriyu Segnetoelectricov* (Introduction to the Microscopic Theory of Ferroelectrics), Nauka, Moscow, 1973 (in Russian).
44. S. V. Popova, N. A. Bendeliani, *Vysokie Davleniya* (High Pressures), Nauka, Moscow, 1974 (in Russian).

45. C. A. Swenson, *J. Phys. Chem. Solids.* (1966), **27**, N1, 33–38.
46. V. G. Vaks, M. I. Katsnelson, V. G. Koreshkov, A. I. Likhtenshtein, O. E. Parfenov, V. F. Skok, V. A. Sukhoparov, A. V. Trefilov, *J. Phys.: Condens Matter* (1989), **1**, N32, 5319–5335.
47. V. G. Vaks, M. I. Katsnelson, A. I. Likhtenshtein, G. V. Peschanskikh, A. V. Trefilov, *J. Phys.: Condens Matter* (1990), **2**, N49, 9875–9899.
48. V. P. Antropov, V. G. Vaks, M. I. Katsnelson., V. G. Koreshkov, A. I. Likhtenstain, *Sov. Phys. — Uspekhi* (1988), **154**, N3, 525–528.
49. M. I. Katsnelson, A. V. Trefilov, *Sov. Phys. — Uspekhi,* (1988), **154**, N3, 523–525.
50. V. A. Goncharova, G. G. Il'in, F. F. Voronov, *Sov. Phys. — Solid State* (1982), **24**, N6, 1849–1851.
51. E. G. Brovman, Yu. Kagan, *Sov. Phys. — Uspekhi* (1974), **112**, N3, 369–426.
52. Zender C. Contributiobs to the theory of beta-phase alloys, *Phys. Rev.* (1947), **71**, N12, 846–851.
53. V. G. Vaks, S. P. Kravchuk, A. V. Trefilov, *MMT* (1977), **44**, N6, 1151–1164.
54. O. K. Andersen, *Phys. Rev. B.* (1975) **12**, N8, 3060–3083.
55. O. Cunnarson, O. Jepsen, O. K. Andersen, *Phys. Rev.* (1983), **27**, N12, 7144–7168.
56. U. Barth, L. Hedin A, *J. Phys. C.* (1972), **5**, N13, 1629–1642.
57. *Theory of the Inhomogeneous Electron Gas* (eds S. Lundqvist, N. H. March), Plenum, NY, 1983.
58. I. V. Abarenkov, I. M. Antonova, V. G. Bar'yakhtar, V. L. Bulatov, E. V. Zarochentsev, *Metody Vichislitelnoj Fiziki v Teoriii Tverdogo Tela. Zonnaya Teoriya Metallov. Elektonnaya Struktura Idealnyh i defectnyh Kristallov* (Methods of Computational Physics in Solid State Theory. Electron Structure of Ideal and Defect Crystals), Kiev, Naukova Dumka, 1991 (in Russian).
59. D. G. Pettifor, *J. Phys. F.* (1977), **74**, N2, K129–K133.
60. V. L. Moruzzi, J. L. Janak, A. R. Williams, *Calculated electronic properties of metals*, Pergamon Press, Oxford, 1978.
61. J. H. Xu, T. Oguchi, A. G. Freeman, *Phys. Rev. B.* (1987), **35**, N13, 6940–6943.
62. V. I. Anisimov, M. I. Katsnelson, A. I. Likhtenstein, A. V. Trefilov, *Sov. Phys. — JETP* (1987), **45**, N6, 285–288.
63. J. Ashkenazi, M. Dacogorna, M. Peter, Y. Talmor, E. Wolker, S. Steineman, *Phys. Rev. B.* (1978), **18**, N8, 4120–4131.
64. P. Bujard, R. Sanjines, E. Wolker, J. Ashkenasi, M. Peter, *J. Rhys. F: Metal Phys.* (1981), **11**, N3, 775–787.
65. Y. Ohta, M. Shimizi, *J. Phys. F.* (1983), **13**, N4, 761–777.
66. P. G. De Carmado, F. R. Prozen, S. Steinman, *J. Phys. F.* (1987), **17**, N5, 1065–1079.
67. M. L. Cohen, *Phys. Scripta* (1982), **T1**, 5–11.
68. P. K. Lam, M. Y. Chon, M. L. Cohen, *J. Phys. C.* (1984), **17**, 2065–2077.
69. W. Maysenholder, S. G. Louie, M. L. Cohen, *Phys. Rev.* (1983), **31**, 1817–1818.
70. N. E. Christensen, *Sol. State Commun.* (1984), **49**, N7, 701–705.
71. I. M. Lifshitz, M. Ya. Azbel, M. Kaganov, *Elektronaaya Teoriya Metallov* (Electon Theory of Metals), Nauka, Moscow, 1971 (in Russian).
72. M. Dacorogna, J. Ashkenazi, M. Peter, *Phys. Rev. B.* (1982), **26**, N8, 2860–2867.

73. V. G. Bar'yakhtar, E. V. Zarochentsev, V. V. Kolesnikov, *Sov. Phys. — Solid State* (1990), **32**, N8, 106–112.
74. O. K. Andersen, In: *The electronic structure of complex systems* (eds W. Temmerman, P. Pharisean), Plenum, NY, 1984.
75. L. Hedin, S. Lundqvist, *Solid State Physics* (1969), **23**, 1–80.
76. A. R. Mackintosh, O. K. Andersen, in: *Electrons at the Fermi surfase* (ed. M. Springford), Cambrige University Press, Cambrige, 1980.
77. E. V. Zarochentsev, E. P. Troitskaya, *Sov. Phys. — High Pressure* (1993), **3**, N3, 42–60.
78. N. E. Zejn, *Sov. Phys. — Solid State* (1984), **26**, N10, 3029–3089.
79. G. Mahan, *Many particl physics*, Plenum Press, New York, 1981.
80. E. V. Zarochentsev, E. Ya. Fain, *Sov. Phys. — High Pressure* (1982), **10**, N10, 61–69.
81. E. Ya. Fain, Ph. D. Thesis, Rostov–Don, 1979 (in Russian).
82. H. Jones, *Proc. Phys. Soc.* (1937), **49**, N1, 250–263.
83. R. Boehler, M. Ross, *Phys. Rev. B.* (1984), **29**, N9, 3673–3680.
84. F. F. Voronov, E. I. Gromnitskaya, O. V. Stal'gorova, *Sov. Phys. — Metal Phys.* (1987), **64**, N6, 1084–1088.
85. V. G. Vaks, A. V. Trefilov, *Sov. Phys. — Solid State* (1977), **19**, N1, 244–258.
86. V. G. Vaks, S. P. Kravchuk, A. V. Trefilov *Sov. Phys. — Solid State* (1977), **19**, N5, 1271–1278.
87. E. V. Zarochentsev, Dr. Sc. Thesis, Donetsk, 1980 (in Russian).
88. V. G. Bar'yakhtar, T. M. Eremenko, E. V. Zarochentsev, V. O. Safronov, in *Vysokie Davleniya i Svojstva Materialov* (High Pressures an Material Properties), Kiev, Naukova Dumka, 1980 (in Russian), 7–12.
89. V. O. Safronov, Ph.D. Thesis, Donetsk, 1977 (in Russian).
90. V. G. Vaks, E. V. Zarochentsev, S. P. Kravchuk et al., *J. Phys. F: Met. Phys.* (1978), **8**, N5, 725–742.
91. V. Heine, M. L. Cohen, D. Weaire, in *Solid State Physics* (eds H. Ehrenreich, F. Scitz, D. Turnbull) **24**, Academic Press, NY, 1970.
92. I. E. Dragunov, S. M. Orel, *Sov. Phys. — High Pressure* (1989), **31**, 15–21.
93. I. E. Dragunov, E. V. Zarochentsev, S. M. Orel, *Sov. Phys. — High Pressure* (1192), **2**, N1, 31–50.
94. I. E. Dragunov, Ph.D. Thesis, Donetsk, 1992 (in Russian).
95. E. V. Zarochentsev, E. P. Troitskaya, *Sov. Phys. — Solid State* (1985), **27**, 2474–2478.
96. D. Wallance, *Thermodynamics of crystals*, J. Wiley and Sons, NY, 1972.
97. V. G. Vaks, S. P. Kravchuk, A. V. Trefilov, *J. Phys. F.: Met. Phys.* (1980), **10**, N10, 2105–2124.
98. V. G. Vaks, E. V. Zarochentsev, S. P. Kravchuk, et al., *J. Phys. F: Met. Phys.* (1978), **8**, N5, 725–742.
99. T. M. Eremenko, E. V. Zarochentsev, *Ukr. Phys. Zourn.* (1980), **25**, N2, 262–270.
100. T. M. Eremenko, E. V. Zarochentsev, *Solid State Commun.* (1979), **30**, N6, 785–789.

101. A. Baldereschi, *Phys. Rev.* (1973), **7**, N12, 5212–5215.
102. D. G. Chadi, M. L. Cohen, *Phys. Rev.* (1973), **8**, N12, 5447–5733.
103. S. M. Vajdya, I. C. Getting, G. C. Kennedy, *J. Phys. Chem. Solids* (1971), **32**, N11, 2545–2556.
104. S. G. Louie, M. L. Cohen, *Phys. Rev.* (1974), **B10**, N8, 3237–3245.
105. A. K. McMahan, *Phys. Rev.* (1978), **17**, N4, 1521–1527.
106. D. B. McWhan, G. Parisot, D. Bloch, *J. Phys. F.* (1974), **4**, N4, L69–75.
107. H. T. Hall, L. Merrill, J. D. Barnett, *Science.* (1964), **146**, N3636, 61–63.
108. F. Bonsignori, A. Magnaterra, *Nuovo cimento.* (1982), **D1**, N6, 789–801.
109. T. Soma, Y. Konno, H.-M. Kagaya, *Phys. status solidi (b)* (1983), **117**, N2, 743–748.
110. I. N. Makarenko, A. M. Nikolaenko, S. M. Stischov, V. A. Ivanov, in *Proc. of the 3rd Int. Conf. of Liquid Metals*, Bristol, 1976, 79–89.
111. V. G. Vaks, S. P. Kravchuk, A. V. Trefilov, *J. Phys. F.: Met. Phys.* (1980), **10**, N10, 2105–2124.
112. V. G. Vaks, S. P. Kravchuk, A. V. Trefilov, *Sov. Phys. — Solid State* (1977), **19**, N5, 1271–1278.
113. P. W. Bridgman, *Proc. Amer. Acad. Sci.* (1938), **72**, N10, 207–225.
114. P. W. Bridgman, *Proc. Amer. Acad. Sci.* (1948), **76**, N3, 71–87.
115. M. S. Anderson, E. J. Gutman, J. R. Packard et al., *J. Phys. Chem. Solids* (1969), **30**, N6, 1587–1601.
116. G. C. Kennedy, P. N. La Mory, *J. Geophys. Res.* (1962), **67**, N2, 851–856.
117. G. C. Kennedy, A. Jayaraman, R. C. Newton, *Phys. Rev.* (1962), **126**, N4, 1363–1366.
118. C. F. Weir, G. J. Piermarini, S. Block, *J. Chem. Phys.* (1971), **54**, N6, 2768–2770.
119. M. Born, K. Huang, *Dynamical Theory of Crystall Lattices*, Oxford University Press, London, 1954.
120. L. D. Landau, E. M. Lifshitz, *Course of Theoretical Physics*, 5, *Statistical Physics*, Pergamon, London, 1958.
121. V. G. Vaks, S. P. Kravchuk, A. V. Trefilov, *Sov. Phys. — Solid State* (1979), **21**, N11, 3370–3378.

Appendices

1. A. A. Abrikosov, L. P. Gor'kov, I. E. Dzyaloshinskii, *Quantum Field Theoretical Methods in Statistical Physics*, Pergamon Press, NY, 1965.
2. E. M. Lifshitz, L. P. Pitaevski, *Statistical Physics*, Pergamon Press, NY, 1980.
3. N. H. March, W. H. Young, S. Sampanthar, *The Many-body Problem in Quantum Mechanics*, University Press, NY, 1967.
4. E. G. Brovman, Yu. Kagan, *Sov. Phys. — Uspekhi* (1974), **112**, N3, 369–426.
5. K. Starm, *Sol. State Commun.* (1983), **48**, N1, 29–32.
6. V. G. Vaks, S. P. Kravchuk, A. V. Trefilov, *J. Phys. F.: Met. Phys.* (1980), **10**, N10, 2105–2124.

7. V. G. Vaks, S. P. Kravchuk, A. V. Trefilov, *Sov. Phys. — Solid State* (1979), **10**, 1205–1234.
8. J. M. Ziman, *Principles in the Theory of Solids*, Cambridge University Press, London, 1964.
9. V. Heine, M. L. Cohen, D. Weaire, in *Solid State Physics* (eds H. Ehrenreich, F. Seitz, D. Turnbull) **24**, Academic Press, NY, 1970.

Index